ROUTLEDGE HANDBOOK OF INDIGENOUS PEOPLES IN THE ARCTIC

This handbook brings together the expertise of Indigenous and non-Indigenous scholars to offer a comprehensive overview of issues surrounding the well-being, self-determination and sustainability of Indigenous peoples in the Arctic.

Offering multidisciplinary insights from leading figures, this handbook highlights Indigenous challenges, approaches and solutions to pressing issues in Arctic regions, such as a warming climate and the loss of biodiversity. It furthers our understanding of the Arctic experience by analyzing how people not only survive but thrive in the planet's harshest climate through their innovation, ingenuity and agency to tackle rapidly changing environments and evolving political, social, economic and cultural conditions. The book is structured into three distinct parts that cover key topics in recent and future research with Indigenous Peoples in the Arctic. The first part examines the diversity of Indigenous peoples and their cultural expressions in the different Arctic states. It also focuses on the well-being of Indigenous peoples in the Arctic regions. The second part relates to the identities and livelihoods that Indigenous peoples in Arctic regions derive from the resources in their environments. This interconnection between resources and people's identities underscores their entitlements to use their lands and resources. The third and final part provides insights into the political involvement of Indigenous peoples from local all the way to the international level and their right to self-determination and some of the recent related topics in this field.

This book offers a novel contribution to Arctic studies, empowering Indigenous research for the future and rebuilding the image of Indigenous peoples as proactive participants, signaling their pivotal role in the co-production of knowledge. It will appeal to scholars and students of law, political sciences, geography, anthropology, Arctic studies and environmental studies, as well as policy-makers and professionals.

Timo Koivurova is Research Professor and Director at the Arctic Centre, University of Lapland. He has a multidisciplinary specialisation in Arctic law and government but has also conducted broader research on global law.

Else Grete Broderstad is Professor in Indigenous Studies and coordinates the Indigenous Master Programme, UiT, The Arctic University of Norway. Her research areas include Indigenous rights, political participation, governance of the Circumpolar North, resource management and conflicting interests between Indigenous traditional livelihoods and large-scale industries.

Dorothée Cambou is Assistant Professor in sustainability science at the faculty of law, HELSUS, University of Helsinki. Her research focuses on the rights of Indigenous peoples and the governance of land and natural resources in the Arctic and the Global South.

Dalee Dorough, an Inuk from Alaska, is Senior Scholar and Special Advisor on Indigenous Peoples in the Arctic, UAA. Specializing in international Indigenous human rights, she holds a PhD in law bestowed by University of British Columbia in 2002 and Master of Arts in law and diplomacy from The Fletcher School in 1991.

Florian Stammler is Coordinator of Anthropology and Research Professor at the Arctic Centre, University of Lapland. His research is on nomadism, human–environment relations, Indigenous peoples and extractive industries. His publications include *Reindeer Nomads Meet the Market* and *Good to Eat, Good to Live With*.

ROUTLEDGE HANDBOOK OF INDIGENOUS PEOPLES IN THE ARCTIC

Edited by Timo Koivurova,
Else Grete Broderstad, Dorothée Cambou,
Dalee Dorough and Florian Stammler

LONDON AND NEW YORK

First published 2021
by Routledge
2 Park Square, Milton Park, Abingdon, Oxon OX14 4RN

and by Routledge
52 Vanderbilt Avenue, New York, NY 10017

Routledge is an imprint of the Taylor & Francis Group, an informa business

British Library Cataloguing-in-Publication Data
A catalogue record for this book is available from the British Library

Library of Congress Cataloging-in-Publication Data
A catalog record for this book has been requested

ISBN: 978-0-367-22039-6 (hbk)
ISBN: 978-0-429-27045-1 (ebk)

Typeset in Bembo
by Apex CoVantage, LLC

CONTENTS

Contents

FIGURES

TABLES

CONTRIBUTORS

Anatoly Alekseyev was born into a nomadic Even reindeer-herding family in the Verkhoyansk Mountains, where he became in turn a local administrator, a political activist for his people and an anthropologist. He is now a research professor at the North-Eastern Federal University in Yakutsk. He has appeared in many documentary films, and his books include *Zabyty Mir Predkov* (The Forgotten World of the Ancestors).

Lukas Allemann is an anthropologist and historian at the Arctic Centre, University of Lapland, Finland. In recent years, he has focused on the history and consequences of social engineering among Northern Indigenous minorities in the Soviet Union, with recurring field stays in Northwest Russia and using mainly oral history methods.

Claudio Aporta is associate professor in the Marine Affairs Program at Dalhousie University. An anthropologist by training, he has conducted research on Inuit environmental and geographic knowledge across the Canadian Arctic since 2000. His interests also involve exploring visualizations of oral geographic knowledge through participatory mapping and community participation in marine spatial planning.

Tim Aqukkasuk Argetsinger is Iñupiaq from Alaska. He has worked in the area of Inuit policy for more than a decade and recently served as executive political advisor to Inuit Tapiriit Kanatami, the national Inuit representative organization in Canada. Argetsinger currently serves as a commissioner on the Lancet Commission on Arctic Health. He lives with his family in Nuuk, Greenland.

Malin Brännström is the director of the Silvermuseum and the Institute for Arctic Landscape Research, INSARC, in Arjeplog, Sweden. She is also affiliated as a lecturer and researcher to the Department of Law, the Vardduoe Sámi Research Center and the Arctic Research Center at the University of Umeå. Her research is focused on real estate law, the use of natural resources, property rights, Sámi land rights and Indigenous law.

Else Grete Broderstad is Professor in Indigenous studies and coordinates the Indigenous Master Programme at the Center for Sámi Studies, UiT The Arctic University of Norway.

Her research areas include Indigenous rights and political participation, governance differences and similarities in the Circumpolar North, resource management and conflicting interests between Indigenous traditional livelihoods and large-scale industries.

Dorothée Cambou is Assistant Professor in sustainability science at the faculty of law, HELSUS, University of Helsinki. Her research – focused on the rights of Indigenous peoples and the governance of land and natural resources – is published widely in peer-reviewed journals and books. She is the co-editor of *Society, Environment and Human Security in the Arctic Barents Region*. She is also the current chair of the *Nordic Network for Sámi and Indigenous Peoples Law* (NORSIL) and teaches Indigenous peoples' rights in international law at the University of Helsinki.

Parnuna Egede Dahl is a joint PhD Fellow at Ilisimatusarfik/University of Greenland and Aalborg University. In collaboration with the Inuit Circumpolar Council – Greenland, her research focuses on the utilization of Indigenous knowledge in environmental impact assessments of extractive industry projects in Greenland and the Canadian Arctic. She is also an independent consultant through Sammivik Consult, working with Greenland and Arctic policy interfaces between science, policy and Indigenous Peoples' rights.

Stefan Dahlberg is Professor in Political Science at the Department of Humanities and Social Science at the Mid Sweden University and at the Department of Comparative Politics, University of Bergen, where he is Research Director at the Digital Social Science Core Facility (DIGSSCORE). His research interests include democratic representation, political legitimacy and survey methodology. Recent publications have appeared in *Journal of Politics*, *Public Opinion Quarterly*, *Electoral Studies* and *West European Politics*.

Dalee Dorough, an Inuk from Alaska, is Senior Scholar and Special Advisor on Arctic Indigenous Peoples, UAA. Specializing in international Indigenous human rights, she holds a PhD in law bestowed by University of British Columbia in 2002 and Master of Arts in law and diplomacy from The Fletcher School in 1991.

Viktoriya Filippova is Senior Researcher at the Institute for Humanities Research and Indigenous Studies of the North, Siberian branch, Russian Academy of Sciences. Her research interests include historical geography, demography and settlement of the Indigenous peoples, traditional use of natural resources, GIS technology and climate change.

Gail Fondahl is Professor of Geography at the University of Northern British Columbia, Canada. Her research focuses on legal geographies of Indigenous territorial rights in Russia and cultural and governance dimensions of arctic sustainability. Fondahl has carried out fieldwork in Siberia since 1992. She was President of the International Arctic Social Sciences Association (2011–2014) and Canada's representative to the International Arctic Science Committee's Social & Human Sciences Working Group (2011–2018).

Leena Heinämäki works as a senior researcher in the Northern Institute for Environmental and Minority Law, Arctic Centre. She teaches and holds a docentship on Indigenous peoples' rights at the faculty of law, University of Lapland. She has published widely on Indigenous peoples' rights and led several research projects particularly related to Sámi people.

Timothy Heleniak is a senior research fellow at Nordregio, the Nordic Centre for Spatial Development. He has written extensively on migration, population change and regional development in the Arctic. He has a grant from the U.S. National Science Foundation titled Polar Peoples: Past, Present, and Future, which examines population change across the Arctic. He was formerly the Editor of the journal *Polar Geography* and is currently the Series Editor of Routledge Research in Polar Regions.

Aytalina Ivanova is Research Docent at the Faculty of Law, North-Eastern Federal University, Yakutsk, Russia. With a background in history of statehood in Siberia, she has specialized for the last ten years in legal anthropology, Indigenous rights and the relation of Indigenous peoples and extractive industries, and Russian Arctic legislation. She is the Russian PI of the international research project WOLLIE Live, Work or Leave on well-being in Arctic extractive industry towns. Her previous publications include two chapters in the recent *Routledge Handbook of Arctic Security* (2020).

Eva Josefsen is Associate Professor at the Department of Child Welfare and Social Work, Campus Alta, UiT The Arctic University of Norway. Her research areas include Sámi self-determination within a unitary state framework; the Sámi Parliament and elections and Sámi politics on local, regional and state levels. She also works on land resource management, gender equality, Sámi language use and Sámi identity in Norwegian local communities.

Galina Keptuke (1951–2019) was a senior professor in Evenki oral literature and language at the Institute for Humanities Research and Indigenous Studies of the North (Siberian Branch of the Russian Academy of Sciences) and a recognized writer of Evenki prose. From 1978, she conducted a tremendous amount of fieldwork in various Russian regions. Her research – focused on rituals, gender and so on – was published in many peer-reviewed Russian journals and books.

Timo Koivurova is Research Professor and Director at the Arctic Centre, University of Lapland. He has a multidisciplinary specialisation in Arctic law and government but has also conducted broader research on global law. Some of his research areas include the legal status of the Indigenous peoples of the Arctic, the laws and politics surrounding the opening of the Arctic Ocean and the continued development of the Arctic Council as an intergovernmental forum.

Rauna Kuokkanen (Sámi) is Research Professor of Arctic Indigenous Studies at the University of Lapland, Finland. She is also an adjunct professor of Indigenous studies at the University of Toronto. Her research focuses on comparative Indigenous politics, Indigenous feminist theory, Arctic Indigenous governance and Nordic settler colonialism. Her most recent book is *Restructuring Relations: Indigenous Self-Determination, Governance and Gender* (2019).

Steve J. Langdon is Professor Emeritus of Anthropology at the University of Alaska Anchorage. Over his 45-year career, he has conducted research projects on many public policy issues impacting Alaska Natives. He has advocated for policies that enhance and promote rural Alaska Native communities and their cultures in such areas as fisheries, lands, customary and traditional resource use, tribal government, cultural heritage, customary trade and co-management.

Alexandra Lavrillier is co-director of the CEARC (University of Paris-Saclay, UVSQ, OVSQ) and an associate professor in social anthropology. Her research – based on many years of fieldwork from 1994 and focused on nomadism, traditional economics, social organization, landscape management, perception of Nature, ritual practices, ethnolinguistics, childhood, ecological knowledge, and the adaptations to post-socialism, the market economy and climate change among the Evenki, Even and Sakha – is published in peer-reviewed journals and books.

Ulf Mörkenstam is Associate Professor at the Department of Political Science, Stockholm University. His main research interests are political theory and policy analysis, with a specific focus on Sámi politics and the rights of Indigenous peoples. He has written extensively on Swedish state's policy towards the Sámi in Sweden and was principal investigator for the first two ever election studies conducted in conjunction to the elections to the Swedish Sámediggi (Sámi Parliament) in 2013 and 2017.

Olivia Napper is a cartographer in the Humanitarian Information Unit at the U.S. Department of State's Office of the Geographer and Global Issues. She's a recent graduate of the Geography M.S. program at the George Washington University, where she worked as a graduate research assistant for the Polar Peoples: Past, Present, and Future project.

Ragnhild Nilsson is a PhD student at the Department of Political Science, Stockholm University. Her research focuses on Indigenous and Sámi self-determination. She has also written on Sámi self-constitution and on social and institutional trust among the Sámi electorate in Sweden.

Matrena Okorokova (PhD in political science) is Associate Professor of the Department of History, Social Studies and Political Science of the North-Eastern Federal University (Yakutsk, Russia). Her scientific works on the implementation of state youth policy, political socialization and youth identity are published in peer-reviewed journals. In addition to these areas of research, her areas of scientific interest are political regionalism, the electoral system and the electoral process.

Torjer A. Olsen is the academic director of the Center for Sámi Studies, UiT The Arctic University of Norway. As a professor in Indigenous studies, his research areas include Indigenous issues in education; gender, power and methodologies in Indigenous research; and Christianity in Indigenous contexts.

Paul C. Ongtooguk is a son of Tommy Ongtooguk Iñupiat and Director of Alaska Native Studies at the University of Alaska Anchorage. He graduated from high school Nome in Alaska. He was also a Des Hart Scholar at the University of Washington, a Holmes Scholar at Michigan State University and a Gordon Russell Fellow at Dartmouth College. His publications and contributions include *Living Our Cultures, Sharing Our Heritage* and (in press) *Native American Warriors*, both published with Smithsonian Press. He is also a former delegate to the Alaska Federation of Natives and a Former Council member IRA of Kotzebue.

Eleanor Peers is the Arctic Information Specialist at the library of the Scott Polar Research Institute, University of Cambridge. She holds a doctorate in sociology from the University of

Cambridge (2010) and has held fellowships at the Max Planck Institute for Social Anthropology (2010–2013) and the University of Aberdeen (2015–2017). She has conducted fieldwork in Buryatia and the Republic of Sakha (Yakutia) and has published on post-Soviet popular culture, ethnic revival in Siberia and post-Soviet shamanism.

Øyvind Ravna is Professor of Law at UiT The Arctic University of Norway. His research fields include property law and Indigenous people's law, especially with emphasis on rights related to land and natural resources in the Sámi areas. Ravna has written the book *Finnmarksloven* (The Finnmark Act), 2013, and *Same-og reindriftsrett* (Sámi and reindeer husbandry law), 2019. Human rights, property law and Sámi and Indigenous law are among his teaching subjects.

Antonina Savvinova is Head of the Laboratory of Electronic Cartography Systems and Associate Professor of the Ecology and Geography Department, Institute of Natural Sciences at North-Eastern Federal University, Yakutsk, Russia. Her research focuses on traditional use and management of natural resources by Indigenous peoples of the North, sustainable development of northern territories, Indigenous knowledge, GIS and remote sensing, climate change and natural hazards in the Arctic.

Vyacheslav Shadrin is Research Fellow at the Institute for Humanities Research and Indigenous Studies of the North, Siberian Branch of Russian Academy of Sciences. He is also Head Chief and Chair of the Council of Yukaghir Elders. His research, on the history and culture of arctic Indigenous peoples, Indigenous rights, ethnological expertise (social impact assessment), climate change, traditional land-use and traditional knowledge, has been published widely.

Lena Sidorova is Associate Professor at the Institute of Cultures and Languages of the Peoples of North-Eastern Russia at the North-Eastern Federal University in Yakutsk. She also works at the journal *Ilin* (east, in Sakha) of history and cultural studies, published in Yakutsk. Her research focuses on varying aspects of life in contemporary Russian and Siberian towns, religious life and nomad cultures within the Arctic's Indigenous communities.

Florian Stammler is the Coordinator of the anthropology research team and Research Professor at the Arctic Centre, University of Lapland, in Finland. His research and supervision work is on nomadism, Arctic pastoralism, human–environment relations, Indigenous peoples and extractive industries and Arctic urban anthropology, with a regional focus on the Russian North. Research projects under his coordination included the ORHELIA Arctic oral history, Arctic Ark and the WOLLIE Live, Work or Leave projects. He has contributed to Indigenous capacity building by training Indigenous scholars, and numerous of his publications are co-produced with Indigenous authors.

Liubov Sulyandziga, prior to completing her PhD at Kyushu University, Japan, graduated from Moscow State Linguistic University; Leuven University, Belgium and Kyushu University, Japan. Liubov is currently a member of the Young Arctic Leaders in Research and Policy program supported by the Arctic-COAST NSF research coordination network, USA. Her current scientific interests include Indigenous peoples' empowerment and rights and impacts of the extractive industries. Liubov has published articles in *Polar Science Journal* and several edited collections dedicated to Russia's Arctic.

Rodion Sulyandziga is Director of the Center for Support of Indigenous Peoples of the North, an Indigenous-led organization aiming to protect the rights of Northern peoples of the Russian Federation. He is an acting member of the United Nations Expert Mechanism on the rights of Indigenous peoples. His main research and study are focusing on self-determination and self-governance issues, climate change impacts and sustainable development of Northern communities.

Pelle Tejsner is currently Adjunct Professor of Anthropology with the Department of Anthropology and Arctic Research Centre (ARC) at the Institute of Bioscience, Aarhus University. His PhD thesis was based on long-term fieldwork in the Disco Bay area in Northwest Greenland and focused on local perceptions of climate change and renewable marine resource harvesting as part of his interests in subsistence strategies, small-scale fisheries and cultural transitions among Kalaallit Inuit. His postdoctoral research grant with the Danish Council for Independent Research (DFF No. 4001-00204B/2015) examines community impacts, whaling as subsistence and perceptions of potential future sources of food contaminants from oil exploration, mining and related industrialization of the Arctic.

Aleksandr Varlamov (PhD in philology, DSc) is Senior Researcher in oral literature and professor of philology at the Institute for Humanities Research and Indigenous Studies of the North (Siberian Branch of the Russian Academy of Sciences). His research – focused from 2004 on Evenki oral literature, linguistics, ethno-history, ethnography, rituals and games – is published in peer-reviewed Russian journals and books. He also writes poems in Russian and Evenki.

Aimar Ventsel is Senior Research Fellow at the Department of Ethnology, University of Tartu. He has published on the Russian Far East and Eastern Germany. Ventsel has authored *Punks and Skins United: Identity, Class and Economy of East German Subculture*, published by Berghahn, and *Reindeer, Rodina and Reciprocity: Kinship and Property Relations in a Siberian Village. Halle Studies in the Anthropology of Eurasia*, LIT Verlag.

Piers Vitebsky recently retired as Head of Anthropology at the Scott Polar Research Institute in the University of Cambridge and is Honorary Professor at the North-Eastern Federal University in Yakutsk. His books include *Reindeer People: Living with Animals and Spirits in Siberia*, *Dialogues with the Dead: The Discussion of Mortality among the Sora of Eastern India* and *Living without the Dead: Loss and Redemption in a Jungle Cosmos*.

Charlie Watt, President of Makivik Corporation, led the Inuit of Nunavik in the 1970s negotiations that created the James Bay and Northern Quebec Agreement. He was appointed to the Senate of Canada in 1984, where he served for 34 years. As a senator, he spearheaded the push to entrench Aboriginal and treaty rights in the Constitution. An officer of the National Order of Quebec, Watt is also a recipient of the Queen's Golden and Diamond Jubilee medals.

Emma Wilson (PhD) is an independent researcher and consultant, Director of ECW Energy Ltd. and Associate of the Consultation Institute (UK). Her work focuses on sustainability, public participation and social impact assessment in the context of energy and extractive industry development, predominantly in Russia, Central Asia and the European Arctic. She has published research on international standards, Indigenous rights and local community engagement in numerous academic journals and books, including two chapters in Johnstone, R.L. and Hansen, A.M. (eds.) (2020) *Regulation of Extractive Industries: Community Engagement in the Arctic*, Routledge.

PREFACE

Indigenous peoples have lived in their Arctic homelands for millennia and have developed numerous forms of livelihood over centuries, based upon their respective adaptation to this unique environment. Today they inhabit three different continents, live in the territories of seven out of the eight Arctic states and represent about 10% of the Arctic population of four million inhabitants. The lands and waters of the Arctic anchor Indigenous societies and provide resources upon which their cultures and specific knowledge continue to flourish.

Yet the challenges faced by these distinct and original Arctic inhabitants are vast and deep. The region is warming at twice the rate of the global average, ice and snow are melting, ecosystems are transforming – and Indigenous peoples and their cultures are struggling to adapt to this rapid change. As a consequence, Arctic sustainability is an issue of increasing concern, as is the adaptation of Arctic Indigenous peoples to these changing conditions. However, Indigenous peoples are also at the forefront of this transformation and play a vital role in promoting and fostering Arctic sustainability. The diversity of Indigenous cultures; their close relationship with the lands, waters and resources and their deeply spiritual, cultural, social and economic connections with the Arctic environment make Indigenous peoples uniquely positioned to ensure Arctic sustainability that is also crucial for their self-determination.

This research handbook is both necessary and timely, as it addresses issues pertaining to Indigenous peoples throughout the circumpolar Arctic with the goal of examining questions regarding their well-being, sustainability and self-determination in this distinct region of the world. In this regard, the general focus of this volume emphasizes interrelated questions that seek to answer, for instance: What is the viability and sustainability of Arctic Indigenous communities? What are the threats to the lands, territories and resources of such communities, and what innovative actions and initiatives are being undertaken by Arctic Indigenous peoples regarding their cultural, political, economic, social and spiritual well-being? Are these actions and initiatives sustainable and viable to ensure Indigenous political, economic, social and cultural self-determination? Reflecting on these core questions provides an opportunity to advance the diverse and direct voices of Arctic Indigenous peoples and their allies to address issues of importance and concern to them, many of those known but not often adequately heard by others.

For this purpose, the handbook is written both by Indigenous and non-Indigenous scholars and reflects a multidisciplinary approach, with contributions coming from the fields of anthropology, law, political science, education and cultural studies. By bringing together leading authors from across the Arctic to share their expertise and visions, the goal is also to contribute to defining the research agenda of international multidisciplinary research within the field of Arctic Indigenous peoples and to empower Indigenous research for future decades. In this regard, this handbook also challenges discourses that depict Indigenous peoples as passive victims of civilization and of rapid environmental changes or mere providers of Indigenous knowledge that helps scientists to understand natural and social processes. Instead, it seeks to replace these discourses with content that displays Indigenous agency, innovation, diversity and alternative views. As such, this volume celebrates, displays and supports Indigenous peoples' continued diversity in tackling the challenges of their rapidly changing conditions. More broadly, this handbook also provides a venue for demonstrating Indigenous challenges, approaches and solutions to pressing problems facing our planet but which are amplified in the Arctic and ultimately opens the path for transformative perspectives on the well-being, sustainability and self-determination of Arctic Indigenous peoples.

Today, the Arctic is mostly inhabited by humans leading a lifestyle that relies almost exclusively on imports from more temperate regions. This leads to the assumption that sustaining life there is economically unviable. In contrast, Indigenous livelihoods have a long track record of self-sufficiency and reliance on their own resources. Therefore, a deep understanding of Indigenous experience, practice and thinking can be informative also beyond the Arctic: analyzing how people not only survive but thrive in the planet's harshest climate may give us lessons for re-shaping the relations of humans to their changing environment in general.

The chapters in this handbook are organized into three topical sections, each of which covers a range of topics that the editors considered particularly important in recent and future research with Arctic Indigenous peoples. The topical orientation of the sections should be seen as emphasis on issues that are of crucial importance from the point of view of Indigenous residents themselves, for whom the Arctic is the centre of the world, while governments and administrators make decisions for them in faraway capitals that hardly ever feel the practical implications of the decisions they make.

The first section is about diversity and well-being. Here, the emphasis is on the diverse expressions that the relations of people and their natural and social surroundings have found in the contemporary Arctic. The chapters in the section show the diversity of different state approaches to classifying and governing Indigenous peoples on the one hand and the diversity of cultural expressions in different Arctic cultures on the other. Taken together, the chapters in the section can be read in the direction of showing how research should contribute to the cultural, social and spiritual well-being of Arctic Indigenous peoples themselves.

The second section is about the ways in which inhabitants of the Arctic relate to the resources in their environment, which constitute the source of life and the inspiration for Arctic Indigenous identities. Relations to resources are – obviously – densely connected to entitlements to use the land. Three of the chapters in this section go deeper into this connection, with examples from Russia and Norway. In comparison, in the North American Arctic, Indigenous livelihoods are more closely connected to water resources, which is why the chapters on Canadian and Alaskan Indigenous–resource relations focus on relations with the marine environment.

The third section is of particular importance for scholars and activists alike, as well as for the international political arena: both nationally and internationally, Indigenous self-determination has developed rapidly in the last decades, and chapters in this section give insights into some of the most recent topics in this field.

Taken together, although this handbook presents the latest research results from Indigenous peoples in the Arctic, it should by no means be taken as an exhaustive overview of the available knowledge. Rather, the editors would like to see this volume as an important contribution on the previously mentioned selected topics, introducing the state of the art in these fields as well as indicating important avenues for future research. The diversity of topics and gaps in our understanding are still significant. Therefore, we hope that the reader can become inspired not only by what is written in this handbook but also by what is *not* written here: the gaps are an invitation for future research to be conducted and published.

Timo Koivurova, Else Grete Broderstad, Dorothée Cambou,
Dalee Dorough, Florian Stammler

ACKNOWLEDGEMENTS

This handbook is the result of high-level collaborations among editors and authors and was made possible thanks to the financial and other support from various individuals, funding agencies and institutions. The editors are also grateful for the assistance and support of Mana Tugend and their publishers' team, specifically Nonita Saha and Faye Leerink in editing this handbook.

SECTION 1

Arctic Indigenous diversity and the foundations of cultural, social and spiritual well-being

Florian Stammler

This first section hosts papers of very different kinds and orientations under the heading of diversity. Celebrating diversity is part of the mission of this book, as too long Indigenous peoples have been looked down upon by colonisers as people without history and diversity. Correspondingly, as Ongtooguk argues in his epilogue to this volume, very coarse and superficial descriptions of Arctic Indigenous peoples have become very influential in shaping their image in the western-dominated public discourses. This section cannot claim to introduce the reader to the diversity of all Arctic cultures. Rather, the chapters present in-depth insights from particular examples that illustrate how diversely different Arctic Indigenous cultures express their relations to the natural and social surroundings. One important goal of research is to generate new insights that can be used to further human well-being on this planet. This is not only true of Indigenous research but should be a guiding goal of any research. Among Indigenous peoples everywhere, not only in the Arctic, well-being is mostly the result of the interplay between the emergence and influence of the dominant societies in states that Indigenous peoples do not control and of their desire to manifest their own livelihoods within the existing narrowly defined frameworks of such states.

This starts with the very definition of who is considered Indigenous and how states do or do not record or reflect Indigenous status in their own statistics – the focus of Chapter 1. Officially, this is decided by governments which are not themselves Indigenous and statistical offices that record numbers according to the respective state's ideologies. Thus, Indigenous registry is part of a state's effort to socially engineer their population to its liking and favour. As Heleniak and Napper show, a statistically separate Indigenous status can be perceived as respect for the particularity of Indigenous peoples as well as racial discrimination. The authors show how differently the Arctic states approach this issue: in Alaska, something called race is the determining factor for Indigenous status, which sounds strange in a post-fascist European context that has largely banned 'race' as category from anywhere. In his chapter in the next section on Alaska marine mammal harvesting, Langdon further problematises this categorisation. In Canada, ethnicity is the determining identifier for Indigenous status, with the three categories of Inuit, Metis and First Nations. In Fennoscandia, neither race nor ethnicity is recorded by statistics, as both are considered to bear too much historically problematic meaning. However, beyond statistic entries Indigenous self-identification has a higher importance there. Self-identification is also a crucial component in Russia's statistical category of *natsional'nost*, which is close to ethnic group and recorded in census data. However, since 7 May 2020 in Russia, the largest and ethnically most diverse Arctic country, a new registry for Indigenous peoples is in force,[1] with the goal of clearly classifying Indigenous individuals as basis for the allocation of privileges and subsidies by the state.

Another form of states socially engineering Indigenous populations is by relocating them to specific territories, as has been done in most Arctic countries in the past. As a result, few Indigenous peoples remain fully nomadic in their lifestyle in the 21st century. The chapter by Allemann provides an example of the Russian Saami and the traumatic experience of displacement (his preferred term over relocation), resulting in socially engineered villages where nobody waited for them.

As a result of such social engineering and state interference with Indigenous peoples, livelihoods, cultural expressions, languages and religion transformed radically in the 20th century throughout the Arctic. Indigenous peoples are now brought up by their parents in these dramatically changed settings as part of the educational and political systems of their respective states and at the same time their own traditions. The result of these processes is demonstrated in diverse ways by the other chapters of this section: Olsen addresses ways in which Saami education in Norway can transform more towards implementing Indigenous approaches and principles. However, the author also acknowledges that not always are the Saami and Norwegian "distinct and separate in all manners and matters". Instead, the chapter emphasises the need for fine-grained approaches, instead of "one size fits it all" generalisations even within the Saami. The author underscores the relevance of Indigenous approaches for education in general, drawing on broader theories from around the planet, which provide useful guidance for pedagogy inspired by Indigenous peoples' values without falling into the trap of simplistic Indigenous/dominant society dichotomies.

The impossibility of the task of separating Indigenous and country-wide tendencies in education is also evident in the Ivanova et al. chapter on political leadership. The authors indicate that classical categorisations such as those by Weber (1922), Adorno (1973) and Hermann (2010) apply just as well to Indigenous as to any other political leadership. On the example of Russia's most Indigenous northern region, they analysed what the most desired traits of character are for a political leader among young people in an urban and rural setting. They show how Indigenous political culture follows largely the country-wide mainstream, because the political system is just as engineered by the state as education and economy are. Even where politicians are Indigenous by origin, they are neither selected nor act according to Indigenous traditional institutions of leadership.

Under such conditions, education in Indigenous languages is a particularly pressing issue for many groups, since languages (Harrison 2007), alongside species and other diversities, die so rapidly on this planet (Tilman et al. 2017). Varlamov et al. provide the example of the Evenki and how Indigenous language education benefits from new-generation computer technology and internet-based resources. Lavrillier argues that there is a need to carefully balance the needs of scientific language documentation as a field of science with needs of Indigenous practitioners themselves, who develop e-learning tools in their languages with more applied goals.

The chapter by Peers et al. tells the history of popular music among the East Siberian Sakha people, much of which is still expressed in their own language. They show how the intimate relations between people and their environment are not only obvious in the Indigenous livelihood but also are at the core of popular music. Even today, when a large portion of the Sakha changed from a semi-nomadic agro-pastoralist livelihood to a village-based and later urban lifestyle, the connection to the land continues to inspire popular music even in urban disco and rock environments.

Along similar lines but differently topically oriented, Vitebsky and Alekseyev's chapter about Indigenous Arctic religions or spirituality identifies relation to the environment as the core foundation common to all Indigenous belief systems. In the first overview of its kind, Vitebsky and Alekseyev give an impressively dense yet broad survey of such religions, with illustrative examples of how Indigenous Arctic inhabitants shape their intimate relations to their surroundings (animals, stones, water, landscape), often with the help of shamans as spiritual experts, As different and far from each other as the Inuit, Saami and Siberian Indigenous peoples' religions are, none of them has written texts and codes like bigger "world religions". Rather, they are united by the idea of "entities that fill the environment with force and agency" as foundation for their sophisticated theologies. In recent centuries, Indigenous religions have all been impacted by more dominant colonial ideologies, whether in the form of Christianity or Marxism, but Vitebsky and Alekseyev go on to show that the principles of the earlier religions remain powerful in spite of all past attempts at domination and are being transformed today in new ethnic, urban and ecological movements. Like most other contributors in this section, Vitebsky and Alekseyev illustrate how Indigenous research should go beyond being "fascinated by archaic survivals", as it is common for research to search for the 'last remaining shaman', the 'last remaining speaker', the 'last remaining chief' among Indigenous peoples.

No one denies the global reduction of diversity of which these 'last remaining' orientations are a result: species of plants, animals, languages, livelihoods and religions, as well as entire cultural and ethnic groups, are dying out. This is the result of the global tendency of economic and social engineering that states implement in their territories, including throughout the Arctic. However, the chapters in this section confirm that Indigenous peoples are far from being passive victims and just 'losers' in the devastation of diversity. Research should therefore avoid what could be called the 'traditional trap', meaning the all-too-easy tendency of museifying Indigenous peoples in all aspects of their culture, even if this sometimes seems to 'work' as a strategy by activists for protection of Indigenous rights. Instead, the authors argue that future Indigenous research in the diverse fields introduced by this section should focus more on new developments, such as on present-day religious energy through urban rituals, pamphlets and social media (Vitebsky and Alekseyev); expressing Indigenous heritage in global hip hop and other music styles (Peers et al.); marrying scholarly and applied needs in the development of e-learning tools for Indigenous languages (Varlamov et al.) and displaying Indigenousness in political systems dominated by country-wide mainstreams (Ivanova et al.), just to provide some concrete examples.

Social, cultural and spiritual well-being of Indigenous peoples all over the Arctic can benefit from research that documents the continued diversity of all their forms of life and cultural expressions, not only in the traditional past but under the conditions of rapid climatic, economic and social change of the present and the closer future.

Note

1 https://sever-press.ru/2020/05/07/dolgozhdannyj-reestr-korennyh-narodov-vstupil-v-silu/,https://nazaccent.ru/content/32054-zakon-o-sisteme-ucheta-korennyh-malochislennyh.html, accessed 15 May 2020.

References

Adorno, T. W. 1973. *Studien zum autoritären Charakter*. Frankfurt and Main: Suhrkamp.

Harrison, K. D. 2007. *When Languages Die: The Extinction of the World's Languages and the Erosion of Human Knowledge*. Oxford and New York: Oxford University Press.

Hermann, M. G. 2010. "Assessing Leadership Style: A Trait Analysis." In *The Psychological Assessment of Political Leaders: With Profiles of Saddam Hussein and Bill Clinton*, edited by J. M. Post, 178–214. Ann Arbor: University of Michigan Press.

Tilman, D., et al. 2017. "Future Threats to Biodiversity and Pathways to Their Prevention." *Nature* 546 (7656): 73–81.

Weber, M. 1922. "Die Drei Reinen Typen Der Legitimen Herrschaft." *Preussische Jahrbücher* 187 (1): 1–12.

1

THE ROLE OF STATISTICS IN RELATION TO ARCTIC INDIGENOUS REALITIES

Timothy Heleniak and Olivia Napper

Abstract

All Arctic states categorize their populations based on some aspect of identity in government data-collection efforts. These include identity according to race, ethnicity, ethnic origin, tribe, language, religion, nationality, citizenship, place of birth, or national origin. The approaches used by Arctic states to classify the identities of peoples vary considerably, and there have been significant changes in classification over time. The United States classifies people based on race, a trait based mostly on phenotypes or observable characteristics. Canada classifies people based on ethnic origin; this includes three groups of aboriginal peoples – Inuit, Métis, and First Nations. Greenland categorizes people based on place of birth. Iceland and the Faroe Islands have never classified people by identity. Norway, Sweden, and Finland all ceased recording ethnicity, including that of the Sámi, in the censuses after World War II. The Soviet Union settled on the concept of *natsionalnost'* (ethnicity) to divide people into different groups, and it is still used in post-Soviet Russia. This chapter examines how the national statistical offices of the Arctic states categorize Arctic peoples, both currently and historically, and the uses of these classifications.

Introduction

Concepts of identity, indigeneity, race, and ethnicity are social constructions and are thus fungible over time, space, and circumstance (Schweitzer et al. 2014). People can have multiple identities or partial identities, and these identities can be self-ascribed or defined by others. Despite these difficulties, the Arctic states have long sought to categorize the Arctic peoples they encountered as they expanded state control northwards according to western norms. These modern states sought to simplify complex cultures by forcing people into narrow statistical categories. The Arctic states and regions share in common high latitude, cold climates, and remoteness and more recently a shared identity, being part of international organizations such as the Arctic Council. They differ in their histories, current and past political statuses, economic structures, and mix of peoples. The Arctic states have adopted quite different approaches to the categorization of peoples, including those found primarily in the Arctic. This chapter examines how the statistical offices of the Arctic states have constructed and used concepts of identity and the implications of these classifications.

Many of the Arctic states regarded the Indigenous peoples they encountered as being at lower levels of development (Axelsson 2011). How statistical authorities producing censuses, surveys, and administrative statistics classify people is often at odds with how people view themselves. Crucial elements of how Arctic Indigenous peoples are enumerated are the instructions given to census takers, whether how people identify themselves is included on a standard list of ethnicities, current identity politics, and whether there are incentives or disincentives for people to identify themselves as part of an ethnic or racial group. One problem when census takers encountered Arctic Indigenous peoples is that the categories were constructed by non-Indigenous peoples. The Arctic states variously categorize peoples' identity by race, ethnicity, or place of birth or have done away with these categories altogether as racist. Even the term 'Indigenous peoples' is highly contested because there are often competing claims who came first to a region. The chapter begins by defining the Arctic. This is followed by a general discussion of identity concepts. The major section of the chapter follows, which analyzes the history and current practices of identity classification of peoples in each Arctic state. A final section compares identity classification across the Arctic.

Defining the Arctic

At the level of states, this chapter encompasses the eight permanent members of the Arctic Council: the United States, Canada, the Kingdom of Denmark (which includes Greenland and the Faroe Islands), Iceland, Norway, Sweden, Finland, and Russia (Arctic Council 2019). These states are an important point of departure for analysis of the classification of Arctic peoples, as they all expanded their territories to include Indigenous peoples, some of whom are represented in the Arctic Council through six organizations of Indigenous peoples who are permanent participants.

Defining Arctic regions within Arctic states is problematic, as there have been numerous boundary changes of sub-national units within the Arctic states over time, including of designated homelands of Arctic peoples. This analysis focuses on the following units at the subnational level. In the United States, the state of Alaska. In Canada, Yukon, the Northwest Territories, and Nunavut. All of Greenland, Iceland, and the Faroe Islands. In Norway, Sweden, and Finland, analysis is not confined to any strict sub-national units, as the traditional areas of settlement of the Sámi, referred to as Sápmi (land of the Sámi), have shifted over time, and many Sámi now live outside of these areas. In Russia, to include all peoples historically and currently considered Arctic in Russia and their homelands or areas of traditional settlement, a broad and flexible definition is employed which includes most of the regions in the north along the Arctic coast, Siberia, and the Far East.

Concepts of identity

The Arctic states are all advanced countries and all have well-developed statistical systems. The purpose of a country's statistical system is to collect and disseminate data on the social, economic, and demographic situation in the country. They do this through conducting censuses and surveys and the use of administrative statistics. The United States, Canada, and Russia continue to conduct periodic population censuses, while the Nordic countries have ceased conducting censuses and rely on population registers. In addition to conceptual dilemmas, census taking in Arctic regions entails considerable logistical difficulties to overcome extreme weather conditions and large distances between settlements (Hamilton and Inwood 2011).

Based on a United Nations survey of national census questionnaires, 63 percent of countries include some form of ethnic enumeration in their population censuses (Morning 2015). These include classification by race, ethnicity, ethnic origin, tribe, language, religion, nationality, citizenship, place of birth, national origin, place identity, or other identities. There is obviously considerable overlap among the different categories. Here, the analysis focuses on the categories of race, ethnicity, ethnic origin, tribe, or similar classifications. The analysis includes whether the Arctic states have a concept of Indigenous when classifying people.

Four different approaches to ethnic enumeration have been identified: enumeration for political control, non-enumeration in the name of integration, discourse of national hybridity, and enumeration for antidiscrimination (Morning 2015). The first category is often associated with colonial census administration as an instrument of exclusion or discrimination. There are numerous examples throughout the Arctic where Indigenous peoples or other minorities were viewed as primitive or lower forms of humans. The second category of non-enumeration is seen in the current ethnic enumeration practices in the Nordic countries (though elections to the Sami Parliaments are based on electoral rolls). The third category of hybrid or mixed ethnicity started to be recognized in the United States with the 2000 census. The fourth category is one where ethnic data are collected to counter discrimination. The use of ethnic data in the latter category has implications for government programs designed to promote the welfare of certain minority groups such as African-Americans or Native Alaskans in the United States. A form of the use of ethnic enumeration for positive discrimination for Arctic peoples are territorial claims or the creation of homelands where there is a concentration of an ethnic group in a region. Finally, whether Arctic states define and recognize groups as Indigenous has implications, such as the special relationship that the United States government has with American Indians and Alaska Natives.

Categorizing Arctic peoples

Analysis of the categorization of Arctic peoples proceeds from west to east, beginning with Alaska and then proceeding around the Arctic to Russia.

Alaska

At the time of first contact with Europeans, estimates put the population of Alaska at about 80,000 (Sandberg 2013). Russian occupation of Alaska was driven more by commercial interests, mainly furs, than state-sponsored territorial expansion. As such, their demographic presence was not very significant, never totaling more than 1,000 persons during the period of Russian America. However, their presence did have a devastating influence on the Alaska Native population, whose numbers declined by 80 percent among some subgroups. The Russians made several incomplete enumerations of the population, one of which estimated the population at 39,813 in 1839. Alaska was sold to the United States in 1867.

The main statistical agency in the United States is the US Bureau. This agency and its predecessors have conducted decennial censuses since 1790. The United States classifies people based on race, a trait based mostly on phenotypes or observable characteristics. In the first US census in 1790, there were three categories – white, black, and mulatto (mixed black and white).

With immigration and the recognition of other minority groups, the number of racial categories has increased. A category for American Indian was added in 1870, and Alaska Natives were included in this category. In 2000, the category was expanded to read 'American Indian or Alaska Native', though they are tabulated together as a combined group. If a person checks this category, they are asked to indicate their enrolled or principal tribe. The Indigenous peoples of Alaska are generally divided into six tribal groupings: Alaska Athabaskan Aleut, Inupiat, Tlingit-Haida, Tsimshian, and Yup'ik. These groupings are based on suggestions from tribal leaders (First Alaskans Institute 2014).

The questions on race and ethnicity are asked and used by the federal government for a variety of purposes, including vital statistics, education, labor force, and health status. The collection of race data is guided by the Office of Management and Budget's *1997 Revisions to the Standards for Classification of Federal Data on Race and Ethnicity*, which dictated that a minimum of five race categories be used (Norris et al. 2012). This directive also allowed people to identify as more than one race. Race data are also collected by state and local government, non-governmental organizations, and the private sector. Alaska, and thus Alaska Natives, were first included in the 1880 census, just over a decade after the purchase of Alaska (Table 1.1) (US Bureau of the Census 1989). For the 1870 census, two new race categories had been added – Chinese and Indian. It was in the latter category where most Alaska Natives were classified. This category remained through the 1940 census. For the 1950 through 1990 censuses, the group was renamed American Indian, and starting with the 2000 census, the category was renamed American Indian and Alaska Native. For the 1960, 1980, and 1990 censuses (but not the 1970 census), separate categories for Aleut and Eskimo were included. These were dropped in the 2000 census when the category for Alaska Natives were added. In his chapter in section two of this volume, Langdon problematizes more the race-based definition of Alaska Natives after the Alaska Native Land Claims Settlement Act (ANCSA) of 1971, stipulating a minimum of one-fourth native ancestry in order to be eligible (Langdon 2020).

For Alaska, the 1880 census was much more than a simple head count. It included a survey of natural resources, geography and topography, history, and ethnology in addition to the population census enumeration (Petroff 1882). It was conducted by Ivan Petroff, a special agent to the Bureau. He, and or his deputies, traveled throughout the entire state. They enumerated a population of 33,426 persons, which is close to last reliable head count under Russian rule. The population was grouped into seven categories – White, Creole, Eskimo, Aleut, Athabaskan, Thinket, and Hyde.

American Indians, including Alaska Natives, have a unique relationship with the US government and by extension with the Census Bureau (Lujan 2014). The legal status of American Indians is termed "domestic dependent nations" within a nation. This is based on treaties between the federal government and Indian tribes. Based on early Supreme Court decisions, the relationship has been described as that of a ward to a guardian. The trust relationship has three components: land, tribal self-governance, and social services. Initially, Indians were not taxed and were thus excluded from the first six censuses, 1790 through 1850.

Starting with the 2000 census, the Census Bureau gave way to reality and started allowing people to identify as more than one race. In the 2010 census, there were 15 racial categories, including three areas where respondents could write in detailed information about their race (US Census Bureau 2012). People who identified as only one race are called the *race alone category*, which can be viewed as the minimal number of people (Norris et al. 2012). In 2010, 104,871 persons in Alaska identified as belonging only to the American Indian and Alaska Native population. Persons who choose more than one race category are referred to as the *race*

Table 1.1 Census race categories, 1880–2010

1880	1890	1900	1910	1920	1930	1940	1950	1960	1970	1980	1990	2000	2010
White	White	White	White	White	White	White	White	White	White	White	White	White	White
Black	Black	Black	Black	Black	Negro	Negro	Negro	Negro	Negro, or Black	Black, Negro	Black, or Negro	Black, African American or Negro	Black, African American or Negro
Mulatto	Mulatto	Mulatto	Mulatto	Mulatto									
			Other	other	Other	Other	Other		Other	Other	Other race	Some Other race	Some Other race
	Quadroon												
	Octoroon												
								Aleut		Aleut	Aleut		
								Eskimo		Eskimo	Eskimo		
Indian	Indian	Indian	Indian	Indian	Indian	Indian	American Indian	American Indian	Indian (American)	Indian American	Indian American	American Indian or Alaskan Native	American Indian or Alaskan Native
Chinese	Chinese	Chinese	Chinese	Chinese	Chinese	Chinese	Chinese	Chinese	Chinese	Chinese	Chinese	Chinese	Chinese
	Japanese	Japanese	Japanese	Japanese	Japanese	Japanese	Japanese	Japanese	Japanese	Japanese	Japanese	Japanese	Japanese
				Filipino	Filipino		Filipino	Filipino	Filipino	Filipino	Filipino	Filipino	Filipino
				Korean	Korean	Korean			Korean	Korean	Korean	Korean	Korean
				Hindu	Hindu	Hindu				Asian Indian	Asian Indian	Asian Indian	Asian Indian
										Vietnamese	Vietnamese	Vietnamese	Vietnamese
												Other Asia	Other Asia
								Hawaiian	Hawaiian	Hawaiian	Hawaiian	Native Hawaiian	Native Hawaiian
								Part Hawaiian					
									Central or South American				
					Mexican				Mexican				
									Puerto Rican				
									Cuban				
										Samoan		Samoan	Samoan

(Continued)

Table 1.1 (Continued)

1880	1890	1900	1910	1920	1930	1940	1950	1960	1970	1980	1990	2000	2010
									Other Spanish	Guamanian	Samoan	Guamanian or Chamorro	Guamanian or Chamorro
											Guamanian	Other Pacific Islander	Other Pacific Islander
										Mexican, Mexican American, Chicano	Puerto Rican, Other Asian or Pacific Islander	Mexican, Mexican American, Chicano	Mexican, Mexican American, Chicano
										Puerto Rican	Mexican, Mexican American, Chicano	Puerto Rican	Puerto Rican
										Cuban	Puerto Rican	Cuban	Cuban
										Other Spanish/ Hispanic	Cuban	Other Spanish/ Hispanic	Other Spanish/ Hispanic
											Other Spanish/ Hispanic		

Source: 200 Years of US Census Taking: Population and Housing Questions 1790–1990. US Department of Commerce. US Bureau of the Census.

in combination population. In 2010, 33,441 persons in Alaska were in this category. The maximum number of people in a racial group is the sum of the previous two and is termed *alone-or-in-combination*, of which there were 138,312 in Alaska in 2010.

Following the adverse demographic impact on Alaska Natives when they came in contact with Europeans, their total population size was just under 10,000 in the early 1800s. Regardless, given the minimal demographic presence of others, Alaska Natives, including Creoles, made up 99 percent of Alaska's population through the first census conducted under US control in 1880 (Figure 1.1). At that time, the Alaska Native population was 32,996. The Alaska Native population continued to decline, reaching a low of 23,531 in the 1890 census before beginning a period of significant growth. However, with the influx of outsiders from the discovery of natural resources and military buildup, their share of the total population declined. The Alaska Native share of the state's total population has been about 17 percent for the last 50 years, and their total population has grown their largest size ever.

Canada

Long before the arrival of Europeans in Canada and the first colonial enumeration in 1666, the Aboriginal populations in present-day Canada used oral tradition to maintain approximate counts of their populations. With the arrival of missionaries in the early 1600s, numerous attempts at enumeration used the terms "warriors" and "souls" to count Aboriginal peoples, categories which were indicative of the authorities' priorities. A 1665 enumeration by Jesuit missionaries produced a count of 2,040 warriors and 11,700 souls (Goldmann and Delic 2014). The first national censuses in Canada were the 1851/2 and 1861 enumerations (Hamilton and Inwood 2011). In these enumerations, the concept of 'origin' was developed. Given that Canada was largely an immigrant country, people could not state that they were Canadian, with the exception of the Aboriginal population, who were classified as 'Indian'. The first Canadian census to include the entire territory and all northern aboriginal peoples was in 1891.

The following decennial census in 1901 saw the inclusion of instructions to enumerators that referenced the Aboriginal population. Census takers were to record the names of tribes and denote various combinations of persons of mixed red and white blood – commonly known as "half breeds" for all Indians they enumerated (Goldmann and Delic 2014). In the years following, enumerators were given instructions on how to derive ancestry along the matrilineal side for Indians and the patrilineal side for all others. The use of matrilineal descent rules for the Aboriginal populations in Canada prior to 1941 was influenced by evolutionary theory, which perceived societies as existing on a scale between savagery and civilization, with more civilized societies using patrilineal lines to determine ancestry and family names.

While self-enumeration was introduced in the 1971 Census, it took several more years to reach the remote regions of Northern Canada. The 1980s through the 2000s saw a series of changes to the census, such as the inclusion of a separate question on Aboriginal identity and the accommodation of multiple origin responses (Goldmann and Delic 2014).

Today, data on the ethnic origin of people in Canada are collected during censuses from the long form, which is given to 20 percent of the population (Statistics Canada 2005).These are collected from the question "What were the ethnic or cultural origins of this person's ancestors?" There is a subsequent question which asks "Is this person an Aboriginal person, that is, North American Indian, Métis or Inuit (Eskimo)?" If the answer is yes, there is a question which asks "Is this person a member of an Indian Band/First Nation?" and if so, to specify. Respondents self-identified as 'North American Indian'; however, the term 'First Nations people' is used in data tabulations and analytical reports. Canada's classificatory scheme for Aboriginal ethnic

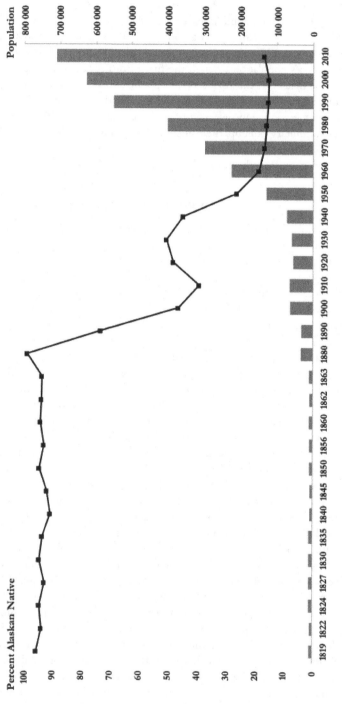

Figure 1.1 Total population and percent Alaska Native in Alaska, 1919–2010

Table 1.2 Ethnic origin of Arctic Canada, 2016 (percent of the population)

	Yukon	*NWT*	*Nunavut*
Total	100.0	100.0	100.0
North American Aboriginal origins	23.7	50.4	85.5
First Nations (North American Indian)	20.7	36.8	1.9
Inuit	1.1	11.0	84.2
Métis	2.9	7.1	0.5
Other North American origins	26.6	20.0	5.1
European origins	71.2	44.0	23.4

Source: Statistics Canada 2017. Census Profile. 2016 Census (https://www12.statcan.gc.ca/census-recensement/2016/dp-pd/prof/index.cfm?Lang=E)

origin has faced linguistic and logistical problems, but it has also had implications on Aboriginal land claims and status

Similar to US data, Canadian data on ethnic origin are difficult to compare over time becase of changes in classification and allowing persons to report more than one ethnic origin. Data from the 2016 census show the differences in ethnic origin for the three northern territories of Canada (Table 1.2) (Statistics Canada 2017). In Yukon, one-quarter of the population identified as having North American Aboriginal origins, one-half in the Northwest Territories, and 86 percent in Nunavut.

Greenland

In about 1200, the ancestors of the present-day Inuit crossed to Greenland from Canada. A Norse settlement was established in 985 but seems to have died out by about 1500. In 1721, the Norwegian missionary Hans Egede established a Danish-Norwegian presence which continues to the present. The Danish colonial presence controlled all traffic between the outside world and Greenland until World War II. During World War II, the United States built several airbases in Greenland, and following the war, Greenland became more open to the world. Greenlanders became Danish citizens, and the use of the Danish language became more common. Home rule was instituted in 1979, which gave Greenlanders more control over their affairs, and in 2009, they achieved self-rule.

The conducting of censuses and collection of other data on the population and economy of Greenland was initially carried out as part of the statistical system of Denmark (US Department of Commerce, Bureau of the Census 1943). The first census was conducted in 1789, though the results are considered unreliable (Gad 1982). Similar to other places, much of the early data collection was the responsibility of the clergy. The census excluded stationed Europeans. Counts were made of the number of 'half-breeds', typically marriages between Greenlandic women and stationed Europeans. Starting in 1830, censuses became more regular, taken every 5 or 10 years. A distinction is made in those censuses between natives and Europeans. Greenland currently categorizes people based on place of birth, the main distinction being in Greenland or outside Greenland. This distinction can be roughly thought to be native Greenlanders or non-Greenlanders or Inuit or non-Inuit. This distinction is important because there are considerable differences in age structure, health patterns, and socioeconomic status between those born in Greenland and those born outside. This applies to both Greenland and the Greenlandic diaspora in Denmark.

At the beginning of the twentieth century, the population of Greenland was less than 12,000 and the percent born outside the country was only 2 percent of the total (Figure 1.2) (Statistics Greenland 2019). The population would grow quite rapidly during the twentieth century, reaching 50,000 in 1980. This growth was mostly driven by natural increase of the native Greenland population. Net migration into the country played a smaller but still significant role, as the number of persons born outside the country increased from just 272 in 1901 to a peak of nearly 10,000 in 1975. The percent of the population born outside the country increased to a peak of 19.2 percent when Home Rule was instituted and more native Greenlanders got jobs in government and the private sector. Since then, the number of persons born outside the country has declined to less than 6,000, and the share is only 10 percent of the total population. Given the unique relationship between Greenland and Denmark, it's not surprising that a rather significant portion of the population born in Greenland resides in Denmark, going there for study, work, or marriage. The number of people born in Greenland and living in Denmark has increased from 13,865 in 2008 to 16,566 in 2019 (Statistics Denmark 2019).

Iceland and the Faroe Islands

Iceland was uninhabited until about the year 870, when it was discovered and settled by Norsemen. Thus, there was no population considered Indigenous. Following the first small group, others followed during what is termed the Age of Settlement, which lasted from 870 to 930, at which time the population was estimated to be about 60,000. At the end of the settlement period, a distinct national identity had started to take shape. Due to its geographic location, the population remained isolated and quite homogeneous until quite recently (Jóhannesson et al. 2013). Because of this almost-extreme cultural similarity, the Icelandic governments never felt a need to classify peoples according to race, ethnicity, or similar concepts. It is only in recent decades, with economic development, that there has been increased immigration into Iceland, resulting in a more diverse population.

Iceland has an impressive literary and historical record-keeping tradition, beginning with the Saga Age in the 900s. It has also long had a well-developed statistical system. The world's first complete population census of an entire country, including the recording of all names, was conducted in Iceland in 1703. Statistics Iceland was founded in 1914, and in 1952, a national Register of Persons was established, which provided a uniform, centralized registration of the entire population. However, because of the homogeneity of the population, there was no attempt and seemingly no impetus to disaggregate the population into any type of racial or ethnic categories.

Currently, the only categorizations of the population by Statistics Iceland are by place of birth, origin, and citizenship (Statistics Iceland 2019). These data do reveal an increasingly diverse population and one with a larger share of the population not having an Icelandic background (Heleniak and Sigurjonsdottir 2018).

The number and share of foreign citizens in Iceland has grown significantly with increased immigration (Figure 1.3). In 1940, there were barely 1,000 foreign citizens in Iceland, just 1 percent of the population. The share of foreign citizens would remain quite low, less than 2 percent of the population, until the mid-1990s. As the economy expanded, immigration to the country increased again, and the number and share of foreign citizens reached historically high levels in 2018, with the 37,830 foreign citizens now making up 10.9 percent of the population (Statistics Iceland 1997).

Iceland collects quite detailed data on the population of foreign origin. Just two decades ago, in 1998, 95.4 percent of the population had been born in Iceland and less than 5 percent were foreign born. In 2017, the sum of those with no foreign background and those born abroad

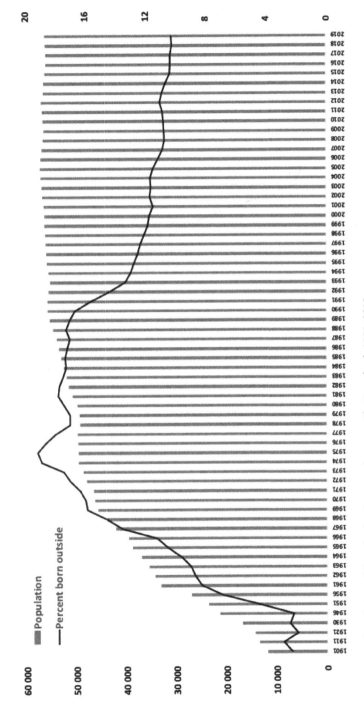

Figure 1.2 Population of Greenland and percent born outside Greenland, 1901–2019

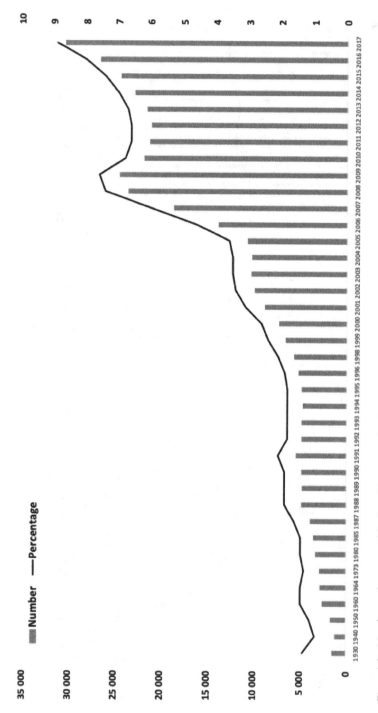

Figure 1.3 Number and percent of foreign citizens in Iceland, 1930–2018

with an Icelandic background had declined to 83 percent of the total population. Thus, the total share of the population with some foreign background is now 17 percent of the Icelandic population, a significant increase from 20 years previous, when it was just 5 percent.

The situation regarding identity in the Faroe Islands is like that in Iceland. The Faroes were uninhabited until the 800s. Since the 1300s, they have been part of the Danish Kingdom. Population growth was small, and for many centuries, there was little interaction with the outside world in terms of the movement of peoples. Like Iceland, the population of the Faroes were quite homogenous until recently, and thus there never appeared to be a need to categorize peoples according to any identity characteristics. Like Iceland, there has been a recent increase in the foreign-born population and an increasing share coming from outside Denmark. In 1985, 92 percent of the population had been born in the Faroe Islands, another 6 percent had been born in Denmark, and less than 2 percent had been born elsewhere (Statistics Faroe Islands 2019). By 2018, the share born in Denmark had increased to 8 percent, and the share born outside of either the Faroe Islands or Denmark had increased to 5 percent.

Norway, Sweden, and Finland

Norway, Sweden, and Finland are considered together because the Indigenous peoples in the northern regions are the same, the Sámi, and the treatment of them in censuses and statistical registers is similar. The Sámi people are spread across those three countries plus Russia in the Sámi homeland called Sápmi (see Figure 1.4). People began living in northern Fennoscandia about 11,000 to 10,000 BC following the retreat of the glaciers. There was a migration of Finno-Ugrian peoples, who later split into distinct Finnish and Sámi peoples.

Starting in the 1600s, there were increasing numbers of researchers, administrators, and travellers visiting Lapland. At first, this was spurred by a search for the exotic and later for the natural resources of the region. This produced a body of literature about the Sámi, including the classic work *Lapponia* in 1673 (Lehtola 2002). As the various kingdoms which preceded the current states of Norway, Sweden, and Finland expanded control northward to encompass the Sámi, they began to levy tribute taxes, usually in the form of furs. This led to an increased need for accurate counts of the number of Sámi eligible for taxation. In the early 1700s, following the Great Nordic War of 1700 to 1720, there was interest in demarcating the northern borders between Denmark–Norway and Sweden, which required knowledge of Sámi settlement areas (Hansen and Olsen 2014).

Sweden established the world's first statistical record of the entire population, the *Tabellverket*, in the 1600s (Axelsson 2011). Responsibility for keeping the records was shared between the clergy and the state. While there were no explicit instructions regarding the Sámi population, identity markers noting the Sámi were included. The Swedish parish registers began to include ethnicity in the eighteenth century. Statistics Sweden was established in 1858, and the first national census was conducted soon thereafter in 1860, which contained a category for Sámi (Axelsson 2011). Data on 'others', including Sámi, Finns, Norwegians, and Russians, were included in the name category. A law was passed in 1886 which delimited Sámi taxable lands (Kent 2014). Instructions to Swedish censuses were to use language spoken at home as the decisive factor. In the early 1900s, with increasing numbers of Sámi using Swedish or Finnish, the criteria was switched to race or a nomadic way of life rather than language. Without citing the classification method, one source cites there being 30,000 Sámi in the traditional areas in 1886.

Likewise, Norway began to include a question on ethnicity in 1845 during a period when there was increased interest in defining the non-Norwegian population (Evjen and Hansen 2009). Between the start of the nineteenth century and the early twentieth century, there was a policy of assimilation of minorities or forced 'Norwegianization'.

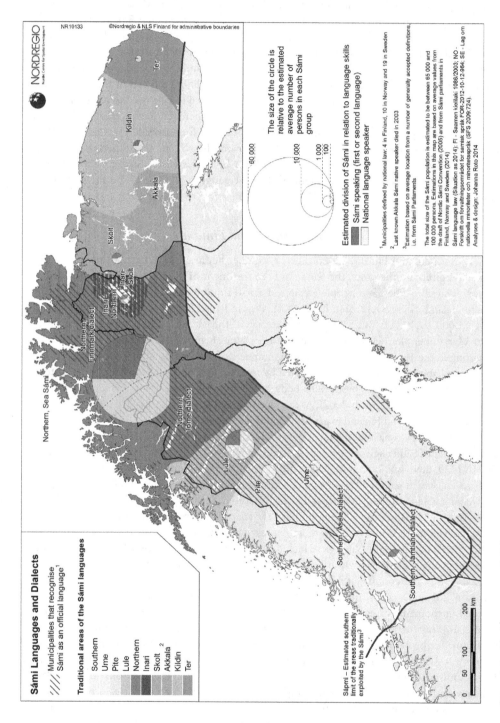

Figure 1.4 Sámi languages and dialects

There is not universal agreement as to the origins of the term 'Lapp', with some believing that the term was derived from the Finnish term 'lappi', meaning end or limit, referring to the place where the Sámi resided at the edge of the known world (Kent 2014). Regardless, the term 'Lapp' was one ascribed to these people by outsiders. It has come to be considered derogatory and has been supplanted by the self-ascribed term Sámi. Like other Arctic Indigenous peoples, Sámi were considered a lower human race, which led to discrimination (Axelsson 2011).

The situation of the Sámi becoming a minority population was similar on the Kola Peninsula in Russia. In 1897, Sámi made up 20 percent of the population, but with industrialization of the area following the Bolshevik Revolution and the large-scale migration of Russians to the area, the Sámi share of the population declined to 1.3 percent in 1933. According to the last Soviet census conducted in 1989, the 1,615 Sámi in Murmansk made up just 0.2 percent of the population.

While outsiders often speak of one Sámi people, one Sámi language, or one Sámi culture, there was and remains considerable diversity in livelihoods, language, dress, and other markers. Some of the previous classifications of Sámi as Forest Lapp, Sea Sámi, Reindeer Sámi, River Lapp, Mountain Sámi, and Eastern Sámi were misleading at the time and are certainly no longer relevant, as many Sámi no longer engage in these traditional pursuits nor reside in these former areas.

An important classification of the Sámi is on the basis of language, of which today there are nine (Lehtola 2002). These are divided into two main groups, Western and Eastern Sámi. The Western Group consists of North, Lule, Pite, Ume, and South Sámi. Eastern Sámi consists of Anar, Skolt, Akkala, Kildin, and Ter Sámi. These divisions are created by linguists, while the speakers all refer to their own language as simply Sámi. There are an estimated 50,000 Sámi speakers. North Sámi is the largest, spoken by 17,000 people (Kent 2014).

The last years of World War II had a devastating impact on the populations of Lapland, including the Sámi, which included border changes, transfer of territories, and forced migration. Sweden and Norway both began to question the difficulties of assigning ethnicity to the increasing number of those of mixed ethnicity. Sweden last included information on ethnicity in the 1945 census (Axelsson and Sköld 2011). There seemed to be a declining interest in classifying people according to race as result of the Holocaust experience (Axelsson 2010). A similar reckoning occurred regarding racial classification in Norway (Søbye 2006). This led to a re-evaluation of distinguishing various groups, including Finns, Lapps, and returning Norwegian-Americans, as well as persons with mental or physical deficiencies.

The Sámi Act in Norway in 1987 led to the establishment of the Sámi Parliament in 1989. In 1977, the Sámi were designated as Indigenous by the Swedish government (Axelsson 2010). A Sámi Parliament was established in Sweden in 1993 and in Finland in 1996. The current criteria for being Sámi is "a person who feels themselves to be Sámi and who either has Sámi as their first language or has at least one parent or grandparent who has Sámi as their first language" (Lehtola 2002). According to the Sámi Parliament, there are 45,000 Sámi in Norway, 20,000 in Sweden, 8,000 in Finland, and 2,000 in Russia. Given the imprecise classification, the total number is estimated to be between 75,000 and 100,000 (Kent 2014).

The lack of ethnic identity markers is now seen as a problem (Axelsson and Sköld 2011). The only data collected by Statistics Norway are based on geography for the Sámi areas north of Saltfjellet, not on ethnicity (Statistics Norway 2018). Statistics Sweden has some data on enrollment in Sámi schools (Statistics Sweden 2019). Statistics Finland collects data on language and elections to Sámi parliament (Statistics Finland 2019). The Nordic countries have shifted from the states attempting to place people into identity categories to where the Sámi themselves, in this case the Sámi parliaments, allow people to self-identify and opt in as members. The Norwegian Sami Parliament and the ministry responsible for Sami affairs in 2007 established a professional analysis group on Sami statistics due to the need to strengthen the foundation for policy assessments and decisions though consultations between Norwegian authorities and the Sami Parliament (Government of Norway 2018).

Russia

The Soviet Union was, and Russia remains, an extremely ethnically and linguistically complex country. This especially applies to the Arctic and northern regions where Indigenous Arctic peoples predominantly reside, many of whom continue to engage in traditional economic activities. As the Russian Empire expanded outwards after overthrowing the Tatars, an increasing number of non-Russian and non-Slavic peoples came under its rule. The rapid industrialization of the Soviet Union began with the first five-year plan in 1928, in which exploitation of the natural resources of Siberia and the Arctic played a large role. There was a demand for information to implement the centrally planned economy of the new Communist state. There was also a need for information about the diverse peoples of the newly formed country to implement 'nationalities' policy. These were promises of national self-determination for ethnic groups in reward for support of the Bolshevik regime. The methods for categorizing Arctic and other peoples and demarcating homelands which were developed in the early Soviet period remain in place in the post-Soviet period in independent Russia.

Russia and/or the Soviet Union has conducted population censuses at irregular intervals · due to various circumstances. The last census in Tsarist Russia was conducted in 1897. The Soviet Union conducted censuses in 1926, 1939, 1959, 1970, 1979, and 1989 (Clem 1986). Independent Russia conducted censuses in 2002 and 2010. At times during the Soviet period, the publications of census results, as well as many other social, demographic, and economic statistics, were severely restricted.

When the Bolsheviks came to power after the 1917 revolution, they needed to make sense of the multinational empire they were presiding over. Through a rather complicated process, they settled on the concept of *natsionalnost'* as the term used to divide people into different groups, and it is still used in post-Soviet Russia (Hirsch 1997). The term is better translated as ethnicity or ethnic-group affiliation so as not to be confused with the western concept nationality on a passport. They used the results of the first all-union census in 1926 after the creation of the Soviet Union to demarcate ethnic homelands, including those of the Arctic and Siberian peoples. The different ethnic units or communities were classified according to their stage of historical development, and the various Arctic and Siberian ethnic groups were categorized as being at the lowest level of development. Hence, the homelands of Arctic and Siberian groups were designated as autonomous okrugs, which are below autonomous oblasts and autonomous republics, the latter of which became the independent successor states of the Soviet Union. Especially baffling to the Bolsheviks were the various Arctic and Siberian groups they were now presiding over. As a result, simultaneous to the first Soviet census in 1926, a Polar census was carried out which went far beyond a usual head count and encompassed an in-depth ethnographic accounting of the lifestyles of Arctic populations (Anderson 2011). In the post-Soviet period, the various Arctic and Siberian groups are classified belonging to a legal category titled *korennye malo-chislennye narod* (small-numbered native peoples). Of these, 26 groups with populations less than 50,000 were designated as *malo-chislenny narod severa* (small-numbered peoples of the North), a number which has since grown to 37 (Table 1.3).

How the Soviet Union classified people with the census category ethnicity was a critical break from the Tsarist regime, which classified people based on religion and native language. The language question would be retained in subsequent censuses but not used as a proxy for ethnicity. According to Francine Hirschs's classic work, ethnic enumeration was an integral part of state formation that went through three stages (Hirsch 2005). The first was the period of physical conquest from 1917 to 1924. This was the period of the civil war, conquest of

Table 1.3 Small-numbered peoples of the North in Russia, 1989, 2002, and 2010

Name (English)	Number			Percent change	
	1989	*2002*	*2010*	*1989–2002*	*2002–2010*
Aleut	644	540	482	−16,1	−10,7
		−12	0		
Veps	12142	8240	5936	−32,1	−28
Dolgan	6571	7261	7885	10,5	8,6
Itel'man	2429	3180	3193	30,9	0,4
Kamchadal		2293	1927		−16
Kerek		8	4		−50
Ket	1084	1494	1219	37,8	−18,4
Koryak	8942	8743	7953	−2,2	−9
Kumanda		3114	2892		−7,1
Mansi	8266	11432	12269	38,3	7,3
Nanai	11883	12160	12003	2,3	−1,3
Nganasan	1262	834	862	−33,9	3,4
Negidal	587	567	513	−3,4	−9,5
Nenets	34190	41302	44640	20,8	8,1
Nivkh	4631	5162	4652	11,5	−9,9
Uil'ta	179	346	295	93,3	−14,7
Oroch	883	686	596	−22,3	−13,1
Saami	1835	1991	1771	8,5	−11
Sel'kup	3564	4249	3649	19,2	−14,1
Soiot		2769	3608		30,3
Telengit		2399	3712		54,7
Taz		276	274		−0,7
Teleut		2650	2643		−0,3
Tofa	722	837	762	15,9	−9
Tuba		1565	1965		25,6
Tyva-Todzha		4442	1858		−58,2
Udege	1902	1657	1496	−12,9	−9,7
Ul'cha	3173	2913	2765	−8,2	−5,1
Khanty	22283	28678	30943	28,7	7,9
Chelkan		855	1181		38,1
Chuvan	1384	1087	1002	−21,5	−7,8
Chukchi	15107	15767	15908	4,4	0,9
Chulym		656	355		−45,9
Shor	15745	13975	12888	−11,2	−7,8
Evenki	29901	35527	38396	18,8	8,1
Even	17055	19071	21830	11,8	14,5
Enets	198	237	227	19,7	−4,2
Eskimo	1704	1750	1738	2,7	−0,7
Yukagir	1112	1509	1603	35,7	6,2
Total	209378	252222	257895		102,2
Total of those existing in 1989	209378	231195	237476	110,4	102,7

Source: Rosstat, Vserossiyskaya perepis naseleniya 2010 (All-Russian census of population 2010), accessed 18 November 2018 https://rosstat.gov.ru/vpn_popul

Note: Groups shown in bold are those which existed in 1989 and previous censuses. Others were added in the 2002 census

periphery regions, and the marking of international borders. The second stage was the period of conceptual conquest from 1924 to 1928. There was a debate in the national statistical office as to whether ethnicity was a biological or cultural construct. Ethnographers came up with 172 different ethnic groups after tabulating the results of the 1926 census. Many smaller nationalities, most with fewer than 50 members, were eliminated from the list and counted as being members of neighboring groups. The third stage was the period of consolidation from 1927 to 1939. The Soviet government, mainly the Soviet of Nationalities, began to use the census results for creation of ethnic homelands – national-territorial units with titular nationalities. After discussions with geographers and ethnographers, government officials determined that borders drawn along national or ethnic lines would be more durable than those drawn according to natural geographic or economic boundaries. During the 1930s, the Karelian, Komi, and Yakut Republics were created, which were homelands of larger northern ethnicities (Kaiser 1994). Also during the 1930s, homelands were created for smaller Arctic groups, including the Nenets, Yamalo-Nenets, Taymyr, Khanty-Mansiy, Evenk, Koryak, Chukchi, Komi-Permyak, Agin-Buryat, and Ust-Ordun autonomous okrugs (Figure 1.5). By 1939, the official list of nationalities was complete, though it would be tinkered with over the years.

Under the centrally planned Soviet economic system, a persons' ethnicity became an important social marker and was included in internal passports, along with other demographic information. Passports were not used to verify ethnicity during censuses, which could lead to discrepancies between counts of official nationalities and census nationalities. People had a legal ethnicity which was determined by their parents and fixed on birth certificates. People could change their ethnicity when they turned 16 but were limited to one of their parents, thus not allowing mixed nationalities.

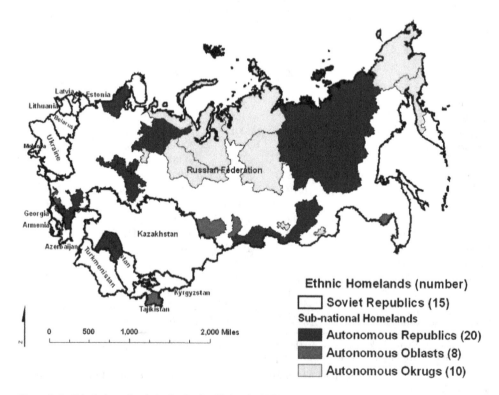

Figure 1.5 Ethnic homelands in the Soviet Union in 1989

The early Soviet period was one of *korenizatsiia* (indigenization), or promoting non-Russian nationalities. In 1924, a Committee of the North was established which promoted the welfare of the small-numbered peoples of the North. In 1959, a decree was passed on measures for improvement in economy and culture of northern peoples lead to marked increases in populations (Donahoe et al. 2008).

Reflective of the ethnic hierarchy in the Soviet Union, in the 1989 census results, the groups were not listed alphabetically but in nested hierarchy, starting with Russians, followed by titular groups of all FSU states, followed by titular peoples of autonomous groups, followed by Peoples of the North (Sokolovsky 2007).

The breakup of the Soviet Union brought new liberties and links with outside organizations for Arctic and Indigenous organizations in Russia and gave people chances to rebuild their identities and lobby for new ones. This period of national revival emerged from a period when identity categories were taken as given, and many Arctic groups were able to actively shape their own identities. The Russian Association of Indigenous Peoples of the North (RAIPON) was founded in 1990. This led to a revised list of nationalities for the 2002 census (Sokolovsky 2011). The new special list of Indigenous peoples of the Russian Federation included 45 numerically small groups. People belonging to these new legal categories of Indigenous peoples or national minorities had special rights and privileges. A somewhat arbitrary threshold of 50,000 or less was established for a group to be classified as *malochislennye narody* (small-numbered peoples). The rationale was that above this level, the ethnicity was not in danger of extinction. In 1997, the practice of recording ethnicity on passports was abolished (Donahoe et al. 2008). In the 2010 census, 46 groups were classified as small-numbered Indigenous peoples, of which 37 lived predominantly in Siberian or the Arctic. Thus, in the Arctic and Siberia, there are both small-numbered peoples of the North and larger groups such as Yakuts, Komi, and Karelians. In the most recent Russian population census, there were 1,840 groups which were consolidated into 193 groups in the final tabulation. At the end of the Soviet period, there were 10 autonomous okrugs which were ethnic homelands of different Arctic and Siberian ethnic groups. Each was administratively subordinate to a parent region. Recently, six of these regions were abolished and subsumed into the parent region, and only the Nenets, Khanty-Mansiy, Yamal-Nenets, and Chukotka okrugs remain. This has implications for political power within these regions but also for statistics, as many data series for the other okrugs are no longer published.

Comparing identity classification across the Arctic

The purpose of this chapter is to examine how the Arctic states classify people according to ethnic criteria. From this survey, the Arctic states span the range of possible ethnic categorizations. Of the four different approaches to ethnic enumeration discussed previously, none of the Arctic states use enumeration as a means of political control or exclusion, although many did in the past. Norway, Sweden, and Finland fall into the second category of non-enumeration of peoples by ethnicity. The United States and Canada have adopted practices of allowing people to declare dual or multiple identities in censuses and other data collection efforts. The United States, Canada, Greenland/Denmark, and Russia are part of the fourth category, where ethnic enumeration is used for positive discrimination, to ensure that Arctic Indigenous and other minority groups are not discriminated against. In these states, Arctic Indigenous populations also receive special benefits based on their status, though these often fail to achieve their stated goals of equal welfare with that of majority populations. All the states recognize Arctic Indigenous populations in one way or another. Figure 1.6 shows the percent of Indigenous across the Arctic according to the national definitions used by each state.

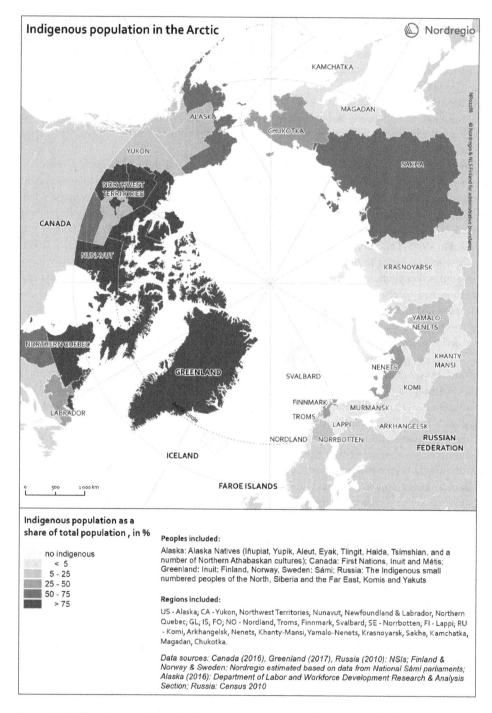

Indigenous population in the Arctic — Nordregio

Indigenous population as a share of total population , in %

- no indigenous
- < 5
- 5 - 25
- 25 - 50
- 50 - 75
- > 75

Peoples included:

Alaska: Alaska Natives (Iñupiat, Yupik, Aleut, Eyak, Tlingit, Haida, Tsimshian, and a number of Northern Athabaskan cultures); Canada: First Nations, Inuit and Métis; Greenland: Inuit; Finland, Norway, Sweden: Sámi; Russia: The Indigenous small numbered peoples of the North, Siberia and the Far East, Komis and Yakuts

Regions included:

US - Alaska; CA - Yukon, Northwest Territories, Nunavut, Newfoundland & Labrador, Northern Quebec; GL; IS; FO; NO - Nordland, Troms, Finnmark, Svalbard; SE - Norrbotten; FI - Lappi; RU - Komi, Arkhangelsk, Nenets, Khanty-Mansi, Yamalo-Nenets, Krasnoyarsk, Sakha, Kamchatka, Magadan, Chukotka.

Data sources: Canada (2016), Greenland (2017), Russia (2010): NSIs; Finland & Norway & Sweden: Nordregio estimated based on data from National Sámi parliaments; Alaska (2016): Department of Labor and Workforce Development Research & Analysis Section; Russia: Census 2010

Figure 1.6   Indigenous populations in the Arctic, 2018

Along with the eight Arctic states, there are six Indigenous organizations which are Permanent Participants of the Arctic Council: the Aleut International Association (AIA), the Arctic Athabaskan Council (AAC), Gwich'in Council International (GCI), Inuit Circumpolar Council (ICC), Russian Association of Indigenous Peoples of the North, and the Saami Council (SC) (Arctic Council 2017). These organizations are supported by the Indigenous Peoples Secretariat. Most of these organizations state that they represent the people of certain territories where Arctic Indigenous peoples reside. Many of them span several countries which differ in their classification of peoples. Most allow the groups that they represent to self-select themselves as members.

Acknowledgements

Research for this chapter is part of a project titled *Polar Peoples: Past, Present, and Future* supported by a grant from the US National Science Foundation, Arctic Social Sciences Program (award number PLR-1418272).

References

Anderson, D. G. 2011. *The 1926/27 Soviet Polar Census Expeditions*. Oxford and New York: Berghahn Books.

Arctic Council. 2017. "Permanent Participants." 22 March. Accessed 22 March 2019. https://arctic-coun cil.org/index.php/en/about-us/permanent-participants.

Arctic Council. 2019. "Arctic Council." Accessed 15 February 2019. https://arctic-council.org/index. php/en/.

Axelsson, P. 2010. "Abandoning 'the Other': Statistical Enumeration of Swedish Sami, 1700 to 1945 and Beyond." *Ber. Wissenschaftsgesch* 33: 263–279. doi:10.1002/bewi.201001469.

Axelsson, P. 2011. "'In the National Registry, All People are Equal': Sami in Swedish Statistical Sources." In *Indigenous Peoples and Demography: The Complex Relation between Identity and Statistics*, edited by P. Axelsson and P. Sköld, 117–134. New York and Oxford: Berghahn Books.

Axelsson, P., and P. Sköld. 2011. "Introduction." In *Indigenous Peoples and Demography: The Complex Relation between Identity and Statistics*, edited by P. Axelsson and P. Sköld, 1–14. New York and Oxford: Berghahn Books.

Clem, R. S. 1986. *Research Guide to Russian and Soviet Censuses*. Ithica and London: Cornell University Press.

Donahoe, B., J. O. Habeck, A. Halemba, and I. Santha. 2008. "Size and Place in the Construction of Indigeneity in the Russian Federation." *Current Anthropology* 49 (6): 993–1020.

Evjen, B., and L. I. Hansen. 2009. "One People – Many Names: On Different Designations for the Sami Population in the Norwegian County of Nordland through the Centuries." *Continuity and Change* 24 (2): 211–243.

First Alaskans Institute. 2014. "Alaska & Alaska Natives in the 2010 Census." Accessed 27 March 2019. https://firstalaskans.org/census-information-center/census-data/.

Gad, F. 1982. *The History of Greenland III: 1782–1808*. Montreal: Nyt Nordisk Forlag Arnold Busck.

Goldmann, G. J., and S. Delic. 2014. "Counting Aboriginal Peoples in Canada." In *Aboriginal Populations: Social, Demographic, and Epidemiological Perspectives*, edited by F. Trovato and A. Ramniuk, 65–71. Calgary, Alberta: University of Alberta Press.

Government of Norway. 2018. "Faglig analysegruppe for samisk statistikk (Academic analysis group for Sami statistics)." 13 August. Accessed 26 November 2019. www.regjeringen.no/no/tema/ urfolk-og-minoriteter/samepolitikk/midtspalte/faglig-analysegruppe-for-samisk-statisti/id5 35566/.

Hamilton, M. A., and K. Inwood. 2011. "The Aboriginal Population and the 1891 Census of Canada." In *Indigenous Peoples and Demography: The Complex Relation between Identity and Statistics*, edited by P. Axelsson and P. Sköld, 95–115. New York and Oxford: Berghahn Books.

Hansen, L. I., and B. Olsen. 2014. *Hunters in Transition: An Outline of Early Sámi History*. Leiden, Boston: Brill.

Heleniak, T., and H. R. Sigurjonsdottir. 2018. "Once Homogenous, Tiny Iceland Opens Its Doors to Immigrants." Migration Policy Institute, 18 April. Accessed 26 November 2019. www.migrationpolicy. org/article/once-homogenous-tiny-iceland-opens-its-doors-immigrants.

Hirsch, F. 1997. "The Soviet Union As a Work in Progress: Ethnographers and the Category Nationality in the 1926, 1937, and 1939 Censuses." *Slavic Review* 56 (2 (Summer)): 251–278.

Hirsch, F. 2005. *Empire of Nations: Ethnographic Knowledge and the Making of the Soviet Union*. Ithica and London: Cornell University Press.

Jóhannesson, G. T., G. T. Pétursson, and T. Björnsson. 2013. *Country Report: Iceland (Revised and updated January 2013)*. Florence, Italy: European University Institute, Robert Schuman Centre for Advanced Studies. Accessed 26 November 2019. http://globalcit.eu/.

Kaiser, R. J. 1994. *The Geography of Nationalism in Russia and the USSR*. Princeton, NJ: Princeton University Press.

Kent, N. 2014. *The Sami Peoples of the North: A Social and Cultural History*. London: Hurst and Company.

Langdon, S. 2020. "Alaska Native Marine Mammal Harvesting: The Marine Mammal Protection Act and the Crisis of Eligibility." In *Handbook of Arctic Indigenous Peoples*, edited by T. Koivurova et al., New York: Routledge.

Lehtola, V.-P. 2002. *The Sami People: Traditions in Transition*. Aanaar-Inari: Kustannus-Puntsi Publisher.

Lujan, C. C. 2014. "American Indians and Alaska Natives Count: The US Census Bureau's Efforts to Enumerate the Native Population." *American Indian Quarterly* 38 (3 (Summer)): 319–341.

Morning, A. 2015. "Ethnic Classification in Global Perspective: A Cross-National Survey of the 2000 Census Round." In *Social Statistics and Ethnic Diversity: Cross-national Perspectives in Classifications and Identity Politics*, edited by P. Simon, V. Piché, and A. A. Gagnon, 17–37. New York and London: Springer.

Norris, T., P. L. Vines, and E. M. Hoeffel. 2012. *The American Indian and Alaska Native Population: 2010*, 2010 Census Briefs. Washington, DC: US Census Bureau.

Petroff, I. 1882. *Report on the Population, Industries, and Resources of Alaska*. Washington, DC: US Census Bureau.

Sandberg, E. 2013. *A History of Alaska Population Settlement*. Juneau, Alaska: Alaska Department of Labor and Workforce Development.

Schweitzer, P., P. Sköld, and O. Ulturgasheva. 2014. "Culture and Identities." In *Arctic Human Development Report: Regional Processes and Global Linkages*, edited by J. N. Larsen and G. Fondhal, 104–150. Copenhagen: Nordic Council of Ministers.

Søbye, Espen. 2006. "A Dark Chapter in the History of Statistics?" *Statistics Norway*, 27 October. Accessed 11 March 2019. www.ssb.no/en/befolkning/artikler-og-publikasjoner/a-dark-chapter-in-the-history-of-statistics-1.

Sokolovsky, S. V. 2007. "Indigineity Contsruction in the Russian Census 2002." *Sibirica: Interdisiplinary journal of Siberian Studies* 6 (1 (Spring)): 59–94.

Sokolovsky, S. V. 2011. "Russian Legal Concepts and the Demography of Indigenous Peoples." In *Indigenous Peoples and Demography: The Complex Relation between Identity and Statistics*, edited by P. Axelsson and P. Sköld, 239–251. New York and Oxford: Berghahn Books.

Statistics Canada. 2005. "2006 Census Questionnaire." Accessed 12 February 2019. http://www23.stat can.gc.ca/imdb-bmdi/pub/instrument/3901_Q2_V3-eng.pdf.

Statistics Canada. 2017. "Census Profile. 2016 Census." Statistics Canada. https://www12.statcan.gc.ca/census-recensement/2016/dp-pd/prof/index.cfm?Lang=E.

Statistics Denmark. 2019. "People Born in Greenland and Living in Denmark 1 January by Time." Accessed 11 April 2019. www.statbank.dk/BEF5G.

Statistics Faroe Islands. 2019. "Statbank." Accessed 18 March 2019. www.hagstova.fo/en.

Statistics Finland. 2019. "Statistics Finland." Accessed 18 March 2019. on elections to Sami Parliament.

Statistics Greenland. 2019. "Statistics Greenland." Accessed 26 November 2019. www.stat.gl/.

Statistics Iceland. 1997. *Iceland Historical Statistics*. Reykjavik: Statistics Iceland.

Statistics Iceland. 2019. Statistics Iceland." Accessed 18 February 2019. www.statice.is/.

Statistics Norway. 2018. *Sami statistics 2018*. Oslo: Statistics Norway. Accessed 26 November 2019. www. ssb.no/en/befolkning/artikler-og-publikasjoner/sami-statistics-2018.

Statistics Sweden. 2019. "Statistics Sweden." Accessed 18 March 2019. www.scb.se/en/.

US Bureau of the Census. 1989. *200 Years of U.S. Census Taking: Population and Housing Questions 1790–1990*. Washington, DC: US Department of Commerce.

US Census Bureau. 2012. "Explore the Form: One of the Shortest Forms in History – 10 Questions in 10 Minutes." 1 March. Accessed 26 November 2019. www.census.gov/2010census/about/interactive-form.php.

US Department of Commerce, Bureau of the Census. 1943. *General Censuses and Vital Statistics in the Americas*. Washington, DC: United States Government Printing Office.

2

INDIGENIZING EDUCATION IN SÁPMI/NORWAY

Rights, interface and the pedagogies of discomfort and hope

Torjer A. Olsen

Abstract

In this chapter, I acknowledge the educational history of Norway regarding the Sámi, and I argue that there are three at times opposing main claims and issues for Sámi education today. These three also work to structure the chapter: First, Sámi students have the right to an education that ensures and develops Sámi languages, culture and society. Second, the educational system and implementation of Sámi rights seem to struggle to deal with Sámi diversity. Third, integrating and/or applying knowledge and perspectives of the Sámi and Indigenous peoples in the majority school is a complex matter and requires extensive competence and resources within school and teacher education. I look into the three claims separately, providing cases and analyses of curricula and textbooks, and discuss them also as connected to each other. In the discussion, I will bring to the table the approaches and concepts of cultural interface, Indigenous métissage, pedagogy of discomfort and pedagogy of hope. The goal is to see Sámi education in a bigger context both of pedagogical theorization and Indigenous studies globally.

Introduction

The history and situation of education in Sámi contexts is changing and diverse. Education and the educational systems have historically been part of a state strategy and practice of colonization and assimilation. Today, however, the tide seems to have turned for the educational system of Norway. When a new curriculum was made in 2017 and set to be implemented in the years following, the Sámi were explicitly recognized as an Indigenous people. The rights of Sámi students to a proper and Sámi-based education are recognized. The curriculum also states that all students of the Norwegian school need to have knowledge about the Sámi, their history, culture, language and status as an Indigenous people and an Indigenous perspective on democracy.

I realize that a curriculum and what a curriculum states is one thing; reality and practice are another (Goodlad 1979). At the same time, the curriculum is part of legislation and thus commits any teacher and school. Further, the curricula in past and present constitute an excellent source for (changing) state policy, for instance, when it comes to the Sámi. Implementation and realization will be part of the discussion here.

In this chapter, I acknowledge the educational history of Norway regarding the Sámi, and I argue that there are three at times opposing main claims and issues for Sámi education today. These three also work to structure the chapter: First, Sámi students have the right to an education that ensures and develops Sámi languages, culture and society. Second, the educational system and implementation of Sámi rights seem to struggle to deal with Sámi diversity. Third, integrating and/or applying knowledge and perspectives of the Sámi and Indigenous peoples in the majority school is a complex matter and requires extensive competence and resources within school and teacher education.

I look into the three claims separately, providing cases and analyses of curricula and textbooks, and discuss them also as connected to each other. In the discussion, I will bring to the table the approaches and concepts of cultural interface (Nakata 2007), Indigenous métissage (Donald 2009), pedagogy of discomfort (Røthing 2019) and pedagogy of hope (hooks 2003). The goal is to see Sámi education in a bigger context both of pedagogical theorization and Indigenous studies globally.

History: colonization, assimilation, marginalization, revitalization

In Norway, history has shown several great changes since the beginning of the public school. The School Act of 1739 made it compulsory for all children to go to school. A main rationale for this was making all citizens able to read – the Bible. This coincides with the intensified colonization policy towards the Sámi, the Indigenous people of the Nordic countries. A major part of colonization was Christian mission, with school as an important dimension. A century later, the school became the most efficient and important tool for the implementation of the assimilation (Norwegianization) policy that sought to make all minorities abandon their ethnic (and religiously dissident) identity and become Norwegian (and Christian). From the middle of the 19th century, the Sámi were one of several minorities who were more or less forbidden to speak their native tongue or learn about their own culture and history in school.

On a national level in Norway, assimilation policy turned Norway into more of a monocultural state through the attempt to get more or less rid of the "problematic" minorities. The Sámi became marginalized and exoticized, as shown in popular representations (cf. Olsen 2013) and textbooks (Ekeland 2017; Heimstad 2018; Mortensen-Buan 2016). The diversity of the Sámi communities had, in the eyes of the public, turned into a more or less monocultural reindeer-herding community. Within the national school, the Sámi seem to have had little or no place in the post–Second World War and post-Norwegianization era.

At the same time, there were forces within the Sámi communities working for the rights of the Sámi. The first wave of Sámi politics occurred in the beginning of the 20th century and crossed the borders of the states. The same happened with the second wave of Sámi politics, with the establishment of the Nordic Sámi Council in 1956. This was a factor when Sámi activists took part in the beginning of international Indigenism from the 1970s onwards. Norway changed its policy towards the Sámi from around 1980 onwards. Now, a policy of recognition was introduced, to a great extent as a result of the work of Sámi activists and politicians. Within the educational system, a preliminary peak was reached with the launch of the first Sámi national curricula in 1997. Since then, the national curricula have shown official recognition of the Sámi (cf. Olsen and Andreassen 2018). Still, it was not until 2017 that the Sámi were explicitly recognized as an Indigenous people in the general part of the curriculum. The development of the curricula has led to an extensive increase of the Sámi rights to education.

Sámi rights to education

The Sámi rights to education have several dimensions. When the Sámi were explicitly recognized as an Indigenous people in the new curriculum of the Norwegian school system, it was with reference to ILO-169. The curriculum further refers to the Constitution in stating that the State is obliged to enable the Sámi to ensure and further develop their language, culture and society. The school is ascribed an important role in this, both for education of Sámi students and for education about the Sámi for all students. The concept of "the Sámi school" is used for the education within the Norwegian school system that follows the Sámi parallel curriculum.

The Sámi curriculum, which has been a parallel part of the Norwegian curriculum since 1997, has been an important expression of recognition and (an apparent) indigenization. The Sámi curriculum is published in the Sámi languages and has in some subjects a particular curriculum different from what is in the national curriculum. The main parts of the Sámi curriculum, including most subjects, are the same, though, as in the national curriculum. Kajsa Kemi Gjerpe (2017), in her analysis of the Sámi curriculum, argues that it has been of huge importance but that the main importance may have been a symbolic one rather than one with actual impact. One reason for this is that not all Sámi students follow the Sámi curriculum at the same time, as the main effort towards Sámi education is put in the Sámi schools following the Sámi curriculum. Still, I would argue that the importance of the first national Sámi curriculum, be it symbolic or concrete, is real as it is the actual expression of Norway's move from assimilation, marginalization and multiculturalist tolerance to an explicit recognition of the needs of Sámi students and a Sámi community. Following closely the process of making a new curriculum (being a member of one of the curriculum committees), I would say that a potential reason for a slight lack of coherence in the previous curricula may be that the formal and overarching legislative and constitutional recognition of the Sámi has not been outspoken in the curricula.

A general challenge when it comes to the rights of Sámi students is that they are necessarily based on an idea of separate peoples. The national curriculum has, since 1997, used the concepts of "the Sámi school" and "the Sámi student" (Gjerpe 2017). At the same time as the concepts are useful and necessary signifiers of distinction, they also imply that the categories at hand are easily defined. This may not be the case. Or – at least – what are formal and legal categories may need more nuanced didactical and pedagogical practices related to them in order to avoid over-simplified claims and interpretations.

This curricular simplification is in line with central research literature on Sámi pedagogy and education. Asta Balto's work on the Sámi upbringing of children (1997) has become a monument within the discourse on Sámi pedagogy, child rearing and education. Here, a general perspective on Sámi pedagogy is expressed and articulated, even though it is based on a rather narrow part of Sámi society. Gjerpe (2018) argues that this can be seen as a strategic essentialism, similar to what is articulated or implied in many Indigenous contexts. In strategic essentialism, the differences between Indigenous and non-Indigenous communities, and the particularities of Indigenous communities, are highlighted and taken as starting points. Further, it creates a dominant discourse that leaves less room for Indigenous diversity and Indigenous students in mainstream schools. I have elsewhere (Olsen 2016, 31–32) argued that strict dichotomies – and what can be termed dichotomism – are at least problematic and cannot be seen as universal. The dichotomy between "the West" and "the Indigenous" can imply that internal variations and differences in both categories become blurry.

A similar, but from my point of view more problematic, kind of over-simplification, I would argue, is found in an often-quoted academic work on Sámi education. Hadi K. Lile (2011) has as a premise for his analysis of the knowledge status of students and teachers in Norwegian schools that the Sámi and the Norwegian peoples are inherently different and separated from one another based on national and international law and rights. When students and teachers participating in Lile's research state that, for them, in their life in the classroom, the picture may be more nuanced than a simple black-and-white one, he almost ridicules them (Lile 2011, 198, 290, 429–431). A different reading of what Lile's participants have said could conclude that there are different ways of being Sámi and different kinds of relationships between Norwegian and Sámi societies, individuals and identities. Further, as education scholars, we need to acknowledge and accept that legal and formal understandings of citizenship and definitions of Indigenous peoples are not necessarily the same as the understandings individual teachers and students may have. In addition to the different understandings of teachers and students, concepts such as *citizenship* and *Indigenous peoples* may be more relevant in some settings than in others. For instance, in local contexts, and dependent on issues and relations, people may refer to themselves as Sámi rather than as Indigenous. Sámi individuals may be and consider themselves citizens of both Norway and Sápmi. They are not wrong. People in strongly Norwegianized communities, such as the north-western coast of Sápmi and Norway, may even experience that they find themselves to be in-between citizenships, that they are "neither-nor" in some situations and "both-and" in others (Sollid and Olsen 2019). To add a slightly personal twist to this, I do also find myself having such an experience: I am in-between, I am a Norwegian and a Sámi at the same time.

Within educational research on Sámi issues, there is an approach that seems to be more open to the complexities of the Sámi schools, Sámi students and Sámi communities. Pigga Keskitalo's writing is part of a movement towards creating and articulating a Sámi pedagogy that is culture sensitive and takes the life and experiences of Sámi children as a starting point (Keskitalo et al. 2013).

Sámi diversity

Hence, Sámi society – or, more correctly, Sámi societies – are diverse. This goes for the language situations: The majority of the Sámi in Norway speak Norwegian. At the same time, the North Sámi, the Lule Sámi and the South Sámi languages are in different situations and under pressure in different ways. This goes for culture, ways of living, gender, sexualities and so on as well. Of course, this also goes for geography. There are Sámi living in different parts of Norway and Sápmi/Sábme/Saepmie. Some live in rural, others in urban areas. Some live in areas where Sámi languages are spoken; others do not. The consequences of diversity for Sámi education should be obvious.

Nonetheless, diversity seems to be a challenge in the field of Sámi education. The aforementioned dominant discourses have set their mark both in the making of the Sámi curriculum and in the representations of Sámi people and communities within educational contexts.

Perhaps the most striking aspect of Sámi diversity, and one that poses a serious challenge to the curricular and rights-based representation of "the Sámi student" and "the Sámi school", is the situation of the Lule Sámi and the South Sámi languages and communities. Where North Sámi is a language in a challenging situation, there are still language teachers and educational resources to be found – albeit scarce. Both Lule Sámi and South Sámi are languages that are threatened with extinction. The situation for teachers and educational resources on and in these two languages is strikingly different from the situation of the North Sámi (Gjerpe 2017).

A challenge of dealing with diversity shown from within is expressed through the afore-mentioned strategic essentialism. In writings on Sámi pedagogy and education, the hegemonic discourse is one where the Sámi community is a community of Inner Finnmark, the "heart-land" of the North Sámi language. This can be seen both in Sámi textbook representations in the previous curricula and in the educational resources made related to the curricula. Gjerpe (in a forthcoming work) refers to the Sápmi/Sábme/Saepmie found in these representations as "Textbook Sápmi", arguing that there is a particular kind of Sáminess, a particular Sápmi, that is portrayed. What is missing is other parts of Sápmi/Sábme/Saepmie, cultural, geographical and linguistic.

In addition, a similar kind of Textbook Sápmi is also found in representations of the Sámi coming from outside Sámi communities. Previous analyses of textbook representations of the Sámi in majority school contexts show that there is a tendency towards simplification and in some cases exotification. In my analysis of textbook representations of conflicts related to the Sámi, I argue that nuances and diversity are downplayed in favour of harmonization and sim-plification (Olsen 2017a).

Finally, a challenge in the general discourse on diversity is the place for gender and sexuali-ties. Diversity has to some extent become a concept that frames and includes many different issues and social categories. Ethnicity, identity, religion, (dis)ability, geography, class, gender and sexuality (and perhaps race) are all integral to or part of diversity. As diversity competence (cf. Røthing 2016) is earning an increasingly important status within educational contexts in Nor-way, knowledge about the potential aspects of diversity is equally important. Within Indigenous research, and especially within the field of Indigenous methodologies, diversity is not that much dealt with. A few examples: When dealing with Sámi diversity in educational contexts, there are few descriptions and presentations of issues related to gender and sexualities. Here, I would argue that this is connected to bigger tendencies. Within the literature on Indigenous method-ologies, gender is not given much attention or space, leaving different expressions of Indigenous feminism a side-note (Olsen 2017b). Within educational contexts, this coincides with the lim-ited space and attention given to Sámi and Indigenous issues as such. However, I would argue that the introduction and application of intersectional perspectives would be a useful and con-structive way of working with Sámi diversity also within education. Intersectional approaches imply the use of and basis in critical theories on power and normativity (Olsen 2018) – as does the field of diversity competence (Røthing 2016).

Sámi issues in the majority school

The 2020 curriculum emphasizes the importance of providing knowledge of and perspectives on Sámi and Indigenous issues to all students. This can be seen as a way of mainstreaming Indigenous issues, stating that the majority society could learn from and needs knowledge of Indigenous people and minorities.

In Aotearoa New Zealand, scholars have raised the potentially problematic issue of main-streaming related to Te Whāriki, the national curriculum for Early Childhood Education. Even though it has at least partial founding in Māori philosophy and concepts, the implementa-tion of the curriculum may be seen as happening on the principles of white majority society. This version of mainstreaming is termed "whitestreaming" (Ritchie and Skerrett 2014, 51). I acknowledge this potential whitestreaming pitfall in the analysis of Indigenous issues in the majority school.

Today's curriculum's emphasis on Sámi issues in the mainstream school is, in itself, despite its potential shortcomings and lack of implementation, an expression of changing Norwegian policy on the Sámi. From colonization, assimilation and marginalization, the educational system has moved via a more inclusive approach to one with certain aspects of decolonization and perhaps even indigenization (Olsen and Andreassen 2018; Gjerpe 2017).

From the introduction of the system of national curricula in 1974, Norway has had an interesting development when it comes to the curricular representation of the Sámi and of diversity. The concepts or strategies of politics of recognition and politics of integration, respectively, can be used to describe the curricula (Seland 2013). Norway's educational policy towards minorities and the Sámi seems to dwell between these two. The politics of recognition describe the first curriculum in 1974, with its idea of the school as an arena to ensure social equity. From the 1987 curriculum, the politics of recognition are more descriptive, with their multicultural roots and ideas about all cultures being equal. The 2020 curriculum seems to present a combination when it comes to Sámi issues. Group-based recognition has a central position. At the same time, there is an integrative dimension at play through the emphasis on an Indigenous perspective for all students (Olsen and Andreassen 2018, 14).

Within the same curriculum, *diversity* is a central and often-used concept. Through this, diversity becomes a key concept with clearly ideological dimensions. This means that diversity is presented and used both as a way of describing the social reality of the school and as a normative way of prescribing what the social reality of the school should be like. This requires some thinking and clarification for scholars, as we also attempt to use the concept of diversity as an analytical category. For the Sámi as an Indigenous people in Norway, this goes in two different directions: The explicit recognition as *Indigenous* provides a maintained proof of status. However, the focus on and extensive use of diversity may leave the Sámi as one of many groups belonging to the diverse side of society, that is, the side that is seen as different from the mainstream. From the work on making the 2020 curriculum, I can confirm that this was an ongoing debate: When should the different parts of the curriculum be explicit in mentioning the Sámi, and when would implicit mentions through use of "diversity" or "minority" suffice?

The curriculum mentions both knowledge and perspective. The students should, according to the curriculum, have knowledge about the Sámi and different aspects of their culture, history, society, ways of living and status as an Indigenous people. Further, when learning about democracy, they should receive or develop an Indigenous perspective. An Indigenous perspective requires a complex kind of competence: You need knowledge about Indigenous issues and situations. In addition, you need the ability and willingness to see from someone else's position and situation. This goes both for students from the majority population, who would need to be able to see from the position of the Indigenous minority, and for Sámi students, who would need to be able to see from the position of other Indigenous peoples. The ambitions are high when stating this. The combination of rather excessive knowledge and an Indigenous perspective implies the combination of different ways of teaching and learning. To teach and learn how to gain perspective require knowledge as well as the ability and attitude to see from someone else's point of view and situation. Safe to say, this will cause implementation challenges on local and national levels (Olsen and Andreassen 2018).

Paradoxes and challenges: a discussion on educational systems and (de)colonization

Schools – and entire educational systems – are excellent arenas for the implementation of state policy, especially when it comes to minorities and Indigenous peoples. Schools are the state's tools to provide its inhabitants with the knowledge and ways of getting knowledge that are defined as the most important. Thus, schools and education are tools for reproduction of ideology and for making policies come into reality. The experiences of Indigenous peoples include challenges both on local and global levels. Educational systems tend to be based on the majorities, on the mainstream. The experiences of Indigenous peoples worldwide clearly tell the story of school and education as main arenas for colonization, assimilation and the communication of states' monocultural ideologies. The paradox of education as decolonization or colonization can be better described, I would argue, in other terms that are connected and share some of the same potentials for meaning: cultural interface and Indigenous métissage.

The cultural interface

When discussing educational systems and colonization, a historical perspective is required. As colonial states have changed, either letting colonialism be a thing of the past or trying to move forward from explicit colonialism to a potentially implicit colonialism (choose which one you prefer), educational systems have changed, too. Even though there is still a need for decolonial criticism towards contemporary educational systems, there is no doubt that colonial states are making attempts to build educational systems that are more culturally responsive and less oppressive than the ones of the past. As such, the claim that the educational system is in itself colonial from top to bottom is no longer valid.

This creates the need for more nuance when understanding and analysing Indigenous issues in education. The educational system of Norway is an example of a system that has moved from colonization, assimilation and marginalization to inclusion and various expressions of indigenization (Olsen 2017a; Olsen and Andreassen 2018). In addition, both in Norway/Sápmi and other states with a more or less strong Indigenous presence, the situation for the Indigenous communities is complex and diverse. A "Sámi student" is not the same regardless of where s/he lives in Sápmi/Sábme/Saepmie. The legal distinction between two peoples, without deemphasizing the rights dimension, does not necessarily work as a pedagogical and didactic principle for use in classrooms.

The term "cultural interface" is coined and used by Martin Nakata (2007) to describe the complex situation both of Indigenous individuals and Indigenous communities. Cultural interface proposes an alternative to dichotomies and describes a space of relations that an individual person (and community) lives by and with. This space has numerous subject positions available, is multilayered and multidimensional and shapes how you speak of yourself and of others. Notions of continuity and discontinuity may provide good ways for understanding Indigenous people's relation both to other groups of people and to the past. Thus, cultural interface, and the idea of numerous subject positions, also seems a constructive alternative to simplistic dichotomies or dichotomism when speaking of Indigenous education.

A disclaimer is important to make: The theory of the cultural interface does not imply a deemphasis of the collective rights of Indigenous peoples. It is quite the opposite. The theory and premise of the cultural interface when understanding Indigenous communities makes it possible to have an eye also for the many living in or close to the periphery of their respective communities. In many Indigenous contexts, the boundaries can be blurry between who is Indigenous and who is not. Also, as many Indigenous children, regardless of geography, attend mainstream schools, a pure distinction between education for Indigenous peoples and education about Indigenous peoples and issues seems to be over-simplified. Thus, looking at and talking about Indigenous education as something that dwells and works in and through a cultural interface makes sense. It becomes a way of stating that Indigenous education, in practice, can have a lot of different variations and articulations and that different educational systems can be located on different parts of a continuum with assimilated and marginalized Indigenous students in mainstream schools on one end and Indigenous students going to Indigenous/indigenized schools within their Indigenous communities on the other end. Perhaps needless to say, most students, Indigenous or not, are found in between these two ends (Sollid and Olsen 2019).

Looking at the field of Sámi education as a continuum and/or an interface, there is a challenge related to how people are addressed and given status. The important, and now well-recognized legally, status of the Sámi as Indigenous – alongside the recognition of Norway being a state on the territory of two different peoples – has made it necessary to have two separate and parallel curricula for the two different peoples. However, there are still choices to be made when it comes to how this should play out and be put into practice. The apparent paradox needs to be considered in the implementation: The Sámi are an Indigenous people and, as such, a people of its own different from the Norwegians. At the same time, the boundaries and differences between Sámi and Norwegian are not at all clear cut. Many belong and see themselves in between (cf. Sollid and Olsen 2019). I argue that an application of cultural interface, with its idea of numerous subject positions available, is a valid and useful concept in trying to let the two national curricula, and their implementation, coincide and work together when needed.

Indigenous métissage

The encounter and potential synergy of the two national curricula in a Norwegian and Sámi context can be seen in two different ways depending on perspective. Having dichotomy and difference as premises is quite different from having the cultural interface as premise. I argue that the latter is both most useful and most reflective of the reality of Indigenous education in schools and teaching.

Connected to the situation of Indigenous education in Aotearoa New Zealand, Graham Smith puts it like this as he reflects on Kaupapa Māori as a way of transforming education: "education and schooling are also locations to be struggled over as they have the potential to be reformed as sites that can deliver Māori aspirations" (Smith 2017, 82). This is one of the many paradoxes related to Indigenous education. Smith terms the necessary actions to be taken from the Indigenous and decolonizing side as those of counter-hegemony. This means that, within an Aotearoa New Zealand context, Māori educators need to work towards a *critically conscientized education*. There is a movement in this line of argument from looking at education and schooling predominantly as arenas of colonization to looking at education and schooling in a quite different way. If they have the potential to be sites that can deliver Māori aspirations, educational practices need to be able to deal with the complex situation of students from Indigenous communities going to a diversity of schools. This calls for a way of taking the diverse situations of students as a starting point.

One way of addressing the issues of the cultural interface in an educational context is through the concept of *Indigenous métissage*. Dwayne Donald (2009; see also Solverson 2018) introduced the term in dealing with challenges in Canada related to Indigenous education. Here, the fort and the frontier had become key metaphors in teaching history, constructing and strengthening the differences and separations of people. Donald argues that there is a need for an "ethical space" between Aboriginal peoples and Canadians. As colonization is a shared condition, decolonization needs to be a shared endeavour. Indigenous métissage is introduced as a curriculum sensibility working to promote ethical relationality within curricular and pedagogical work (Donald 2009, 5–6). Thus, Aboriginal peoples and Canadians must be seen not as separate from each other but as related to each other through colonization. The term métissage refers to what has elsewhere been called hybridization. In a curricular context, métissage can connect personal and family stories to larger narratives of nations and identity. This can have provocative effects (Donald 2009, 8). *Indigenous* métissage, Donald argues, needs to take Indigenous perspectives as a starting point and be about particular places in Canada where different/differing stories are told. The goal is to have a curriculum and an educational practice that work to interpret mixed understandings and to have people reread and reframe understandings of Canadian history and communities as layered and relational in order to better acknowledge and recognize Indigenous presence and participation (Donald 2009, 10).

In my reading, this connects well both with the actual situation in Norway and Sápmi/Sábme/Saepmie and the theoretical concept of the cultural interface. A method of Indigenous métissage in a Norwegian/Sámi context would require the need for making – and making explicit – an ethical space between the Sámi and the Norwegians. Adding the cultural interface to this enables this space to be a space where there are many different subject positions available. Thus, students (and their families) can see themselves (be seen) as belonging to different positions between what is Norwegian and what is Sámi. Following Donald again, the Sámi and the Norwegians must be seen as related to – rather than separate from – each other through colonization. In Norway and Sápmi/Sábme/Saepmie, this is even more important and striking due to the process of assimilation (Norwegianization), which led many Sámi to change their ethnic identity. In many highly Norwegianized villages, the last three decades have shown a powerful revitalization of Sámi identity and language. At the same time, the question of who is Sámi and who is not is a complex one to answer in these same villages. In educational contexts, métissage may have the potential to connect these complex personal and family stories of Norwegianization and revitalization to larger narratives of nations and politics. To paraphrase Donald: The goal would be to have an educational practice that works to interpret and integrate mixed understandings of history and communities as relational – in order to acknowledge and recognize the Sámi presence and place in history and community.

Provocations, discomfort and hope

To address this further, I would want to connect to the potential provocative effects. The public debate on Sámi issues, as well as the established fact that Sámi in Norway suffer from harassment more than Norwegians do, shows that issues related to the Sámi may indeed have provocative effects. This can be resolved or seen through the lens of the approach of a pedagogy of discomfort (Røthing 2019). Despite the obvious discomfort of such a pedagogy, it may also be seen and used as a resource in educational settings. A key issue within the pedagogy of discomfort is the critique of any educational practice that is Othering and functions to reproduce stereotypes. For the pedagogical discourse, this means also the pedagogy of tolerance, which highlights knowledge about the Other as the main goal for education (Kumashiro 2002). Norm-critical pedagogies, like the pedagogy of discomfort, claim the need for critical reflections on power and injustice. The methods for doing so are to use and lift, instead of avoiding, situations of tension in the classroom (Røthing 2019).

Such an approach is quite different from, for instance, cultural-sensitive pedagogy and the pedagogy of tolerance. In the latter two, what is highlighted is cultural differences and the efforts made by the majority teacher to be cultural-sensitive or tolerant, respectively, in the encounter with cultural difference. An approach from Indigenous education that seems to be more in line with the pedagogy of discomfort than with the pedagogy of tolerance and cultural-sensitive pedagogy is culturally transformative pedagogy. Graham Smith, in his discussion of Kaupapa Māori theory and/in education (2017, 79–83), takes as a starting point that Māori are not homogenous in their educational aspirations and challenges any scholar and educator on the transformative dimension of their work: "I often challenge them: 'show me the blisters on your hands'" (Smith 2017, 79). A proper Indigenous education, even when theorized, needs to be hands on, metaphorically and in practice. The transformative dimension of education requires making space for Indigenous and minority cultures, protecting languages at risk, struggling for the minds to be educated out of false consciousness and hegemony – and recognizing the small victories along the way to transformation.

Taken over to the context of Sámi education, my question concerns the potential in a pedagogy of discomfort and/or transformation when dealing with Sámi and Indigenous issues. At the very least, it is important that the education about Sámi issues, in order to provide and enable a credible Indigenous perspective, not remain a pedagogy of tolerance. It cannot be a pedagogy where "we" (the majority undefined) must learn about the strange Others in order to tolerate differences. The pedagogy and the practice need to go into potentially complex and discomfortable areas of colonization, assimilation and oppression – without putting the blame on anyone in the classroom. Complicated? Absolutely. As it should be. An example from a Sámi context would be related to teaching about assimilation in a classroom where the students are of mixed heritage, some Sámi, some Norwegian, some located on different places along the cultural interface. In a local community, the Norwegians (some of whom would be descendants of assimilated Sámi) cannot be described or addressed as wrongdoers. At the same time, the diversity situation, on culture, language and identity, clearly show the effects of assimilation. As such, teaching about assimilation can be likely to cause discomfort – and cause blisters. However, it can have the power to create space for different kinds of culture and to struggle for a transformation of both mind and community.

Finally, and on purpose as I am getting close to the end of this chapter on such serious matters, I would like to see to the pedagogy of hope. bell hooks (2003) argues that the public school is a key in successful education about and democracy. In such an education, the pedagogy needs to be intersectional and lined with critical thinking. This means that the teachers need to move beyond mere knowledge and to see the interconnectedness of students from different backgrounds and identities.

Hope may at times seem hard to find and recognize in a field where assimilation, colonization and oppression have been dominant. Within the field of Indigenous education, suspicion and critical thinking from the side of the Indigenous towards educational systems are what can be expected. At the same time, it is striking – and an expression of hope – that Graham Smith (2017, 82) also talks about education and schooling having the potential to be part of transformative actions, as it is when Dwayne Donald talks about Indigenous métissage and the ethical space of Indigenous education. In such a line of thinking and practice, discomfort may be a necessary part of the practice in order to ensure that critical thinking and the diversity of the cultural interface remain and/or become integral part of Sámi education.

Conclusion

Ending on a hopeful note is somewhat controversial when writing about Indigenous education. I choose to do so, as there are clearly promising tendencies and movements within Norwegian Sámi education. One of these movements goes from having an educational system that was clearly an integral part of colonization to something that has at least elements of decolonization. Of course, you (and I) may ask if decolonization can come from above, through national curricula. I do not think it comes only from above, though. In Norway, the making of a national curriculum as well as educational practice is a joint effort wherein many actors have a role. The Sámi parliament has a key role in the entire process, ensuring that the official Sámi voice is heard. Still, hope – in order to prevail even as a thin-leaf version of hope – needs to live through and with suspicion and negotiation to be part of reconciliation.

In practice, this means that there is a need for the actors of the field to be able to work beyond dichotomist thinking. This goes both for teachers and school leaders as well as for politicians and bureaucrats – and scholars of the field. As it goes both for Sámi and Norwegian – and all those in between in the cultural interface. A disclaimer necessary to repeat: The Sámi and the Norwegian remain two peoples. At the same time, they are not distinct and separate in all manners and matters.

Such a relationship between majority society and Indigenous people is not unique on a global level. I have given examples from different areas, from Aotearoa New Zealand's Māori context, Australia's Aboriginal and Torres Strait Islander context and Canada's First Nations context, of different expressions of cultural interface, Indigenous métissage and complex diversity. The educational systems and the many educational practices of these varied communities are of course multifaceted and struggling with a number of challenges. A shared one is the history and experience of colonization, assimilation and oppression, which has led to the inevitable suspicion towards and alienation from national and/or federal educational systems and practices.

For the educational systems to succeed in the decolonization and making of more culturally responsive policies, curricula and practices, it is necessary to move beyond Indigenous suspicion and alienation. This is impossible without acknowledging state-led transgression and oppression. And here we are again at the potentially necessary pedagogy of discomfort. There are many encounters within the educational systems – between peoples, between individuals, between groups, between systems of knowledge, between systems and persons. And more. The making of good and sustainable education calls for understanding and considering the cultural interfaces through different versions of métissage. There are diverse ways of being Sámi – and of being Norwegian. Thus, Sámi education must move in diverse ways, too.

References

Balto, A. 1997. *Samisk barneoppdragelse i endring.* Oslo: Ad notam Gyldendal.

Donald, D. 2009. "Forts, Curriculum, and Indigenous Métissage: Imagining Decolonization of Aboriginal-Canadian Relations in Educational Contexts." *First Nations Perspectives* 2 (1): 1–24.

Ekeland, T. 2017. "Enactment of Sámi Past in School Textbooks: Towards Multiple Pasts for Future Making." *Scandinavian Journal of Educational Research* 61: 319–332.

Gjerpe, K. K. 2017. "Samisk læreplanverk – en symbolsk forpliktelse?" *Nordic Studies in Education* 3–4: 150–165.

Gjerpe, K. K. 2018. "From Indigenous Education to Indigenising Mainstream Education." *FLEKS – Scandinavian Journal of Intercultural Theory and Practice* 5 (1).

Goodlad, J. I. 1979. *Curriculum Inquiry: The Study of Curriculum Practice.* New York: McGraw-Hill Book Company.

Heimstad, J. B. 2018. "' . . . Halvparten skulle være om samene . . .' En komparativ analyse av norske- og samiskproduserte lærebøker for historiefaget i den videregående skolen etter Kunnskapsløftet 06 (k-06)." Master thesis, UiT The Arctic University of Norway.

hooks, b. 2003. *Teaching Community: A Pedagogy of Hope.* New York: Routledge.

Keskitalo, P., K. Määttä, and S. Uusiautti. 2013. *Sámi Education.* Frankfurt am Main: Peter Lang.

Kumashiro, K. 2002. *Troubling Education: Queer Activism and Antioppressive Education.* New York: Routledge.

Lile, H. K. 2011. "FNs barnekonvensjon artikkel 29 (1) om formålet med opplæring: En rettssosiologisk studie om hva barn lærer om det samiske folk." PhD diss., Oslo University.

Mortensen-Buan, A. 2016. "'Dette er en same.' Visuelle framstillinger av samer i et utvalg lærebøker i samfunnsfag." In *Folk uten land? Å gi stemme og status til urfolk og nasjonale minoriteter*, edited by N. Askeland and B. Aamotsbakken. Kristiansand: Portal.

Nakata, M. 2007. *Disciplining the Savages: Savaging the Disciplines.* Canberra: Aboriginal Studies Press.

Olsen, T. A. 2013. "Nordnorsk religiøsitet som begrep og fenomen." In *Hvor går Nord-Norge? Et institusjonelt perspektiv på folk og landsdel*, edited by S. Jentoft, K.-A. Røvik, and J. I. Nergård. Tromsø: Orkana.

Olsen, T. A. 2016. "Responsibility, Reciprocity and Respect: On the Ethics of (Self-) Representation and Advocacy in Indigenous Studies." In *Ethics in Indigenous Research: Past Experiences, Future Challenges*, edited by A.-L. Drugge. Umeå: The Centre for Sami Research-Umeå University.

Olsen, T. A. 2017a. "Gender and/in Indigenous Methodologies: On Trouble and Harmony in Indigenous Studies." *Ethnicities* 17 (4): 509–525.

Olsen, T. A. 2017b. "Privilege, Decentring and the Challenge of Being (Non-)Indigenous in the Study of Indigenous Issues." *Australian Journal of Indigenous Education* 47 (2): 206–215. doi:10.1017/jie.2017/16.

Olsen, T. A. 2018. "This Word Is (Not?) Very Exciting: Considering Intersectionality in Indigenous Studies." *NORA – Nordic Journal of Feminist and Gender Research* 26 (3): 182–196.

Olsen, T. A., and B.-O. Andreassen. 2018. "'Urfolk' og 'mangfold' i skolens læreplaner." *FLEKS Scandinavian Journal of Intercultural Theory and Practice* 5 (1).

Ritchie, J., and M. Skerrett. 2014. *Early Childhood Education in Aotearoa New Zealand: History, Pedagogy and Liberation.* New York: Palgrave Pivot.

Røthing, Å. 2016. *Mangfoldskompetanse. Perspektiver på undervisning i yrkesfag.* Oslo: Cappelen Damm akademisk.

Røthing, Å. 2019. "'Ubehagets pedagogikk' – en inngang til kritisk refleksjon og inkluderende undervisning?" *FLEKS – Scandinavian Journal of Intercultural Theory and Practice* 6 (1): 40–57.

Seland, I. 2013. "Fellesskap for utjevning – Norsk skolepolitikk for en flerreligiøs og flerspråklig elevmasse etter 1970." *Tidsskrift for samfunnsforskning* 54 (2): 188–214.

Smith, G. H. 2017. "Kaupapa Māori Theory: Indigenous Transforming of Education." In *Critical Conversations in Kaupapa Māori*, edited by T. K. Hoskins and A. Jones, 79–94. Wellington: Huia.

Sollid, H., and T. A. Olsen. 2019. "Indigenising Education: Scales, Interfaces and Acts of Citizenship in Sápmi." *Junctures: The Journal for Thematic Dialogues* 20: 29–42.

Solverson, E. 2018. "Education for Reconciliation. A Study of the Draft Curriculum for Mainstream Social Studies in Alberta, Canada." Master thesis, UiT The Arctic University of Norway. Accessed 22 May 2020. https://hdl.handle.net/10037/13077.

3

WHAT MAKES A GOOD POLITICAL LEADER?

Young people's perceptions from the republic of Sakha (Yakutia)

Aytalina Ivanova, Matrena Okorokova, Florian Stammler, and Emma Wilson

Abstract

This chapter analyses data from research carried out among young Indigenous people in the Republic of Sakha (Yakutia) to elicit what they perceive to be qualities of a good political leader. We investigate theoretical aspects of the formation of political youth culture and factors influencing the representation of political leadership. Empirical research results from two surveys, supported by qualitative fieldwork, in an urban and a rural setting, reveal young people's perceptions of what makes an ideal leader. We analyse similarities between urban and rural youth in their ideas about political leadership. Urban students emphasised the importance of the personal qualities of a good political leader, while rural Indigenous youth found professional qualities important. Using categories of political leadership identified by Hermann, we found that the preferred political leader for students studying in the regional capital fits the leadership type of 'leader-servant', whereas the rural young people prioritised what is called a 'leader-fireman'. All together, we find that ideas of political leadership among Indigenous youth follow the Russian mainstream political culture. We argue that this is because politics in the RSY is made within the Russian political system (albeit by local Indigenous politicians) and not by Indigenous politicians according to Indigenous traditional institutions of leadership.

Introduction

This chapter outlines ideas about political leadership in an Arctic region inhabited predominantly by Indigenous peoples. The focus was on young people under the age of 30, who will play a decisive role in the future of Russia's largest region, which is also its 'most Indigenous' and one of its most politically autonomous regions (Argounova 2004; Ivanova and Stammler 2017; Fondahl et al. 2019). The goal is to establish what kind of political leader is preferred by urban and rural young people, what the personal and professional qualities of an ideal leader are, and what problems such a leader is expected to address. This allows us to trace the process by which political values and political culture develop within the population. This is an area of political science that, to date, has rarely been based on research in Indigenous societies. The same is true for the concept of political leader, which has not yet been sufficiently explored within Indigenous societies.

The geographical focus of this study is the Republic of Sakha (Yakutia) (hereafter the RSY), a region almost as large as India (3 million sq km). The RSY is rich in natural resources and makes up a large part of the region officially known within the country as the Russian Arctic (even though much of the territory of the Republic lies outside the Arctic Circle). The capital of the Republic, Yakutsk, is the only big city in the RSY, and the entire political life of the Republic happens there. The city has 318,768 inhabitants and also hosts the region's only university, which is closely connected to the ethnic Sakha political intelligentsia.

The dominant ethnic group in the Republic are the Sakha (or Yakut), who make up 49% of the population (as of 2018).[1] The Sakha are recognised by the United Nations as being Indigenous. However, owing to their large population (478,085, according to the last census in 2010), they do not qualify to be recognised as 'small-numbered Indigenous peoples of the North' in Russian Federation law and therefore do not benefit from the rights that have been granted to such peoples. The Evenki (2%) and the Eveny (1.5%), as well as the smaller-numbered Dolgans, Chukchi, and Yukagir, are officially recognised as Indigenous in Russian law. Other ethnic groups that make up the population of the RSY include Russians (37.8%) and other non-Indigenous ethnic groups, such as Ukrainians and Tatars. Thus, roughly half of the population of the RSY is of Indigenous origin when internationally classified.

The city of Yakutsk has a much more diverse population than the rural areas, with ethnic Russians, Ukrainians, Tatars, Sakha, and other Indigenous groups. Due to post-Soviet outmigration of ethnic Russians (a decline from 50.3% of the population in 1989 to 37.8% in 2020), the share of the Indigenous population has increased in the city of Yakutsk as well as in most rural areas. The rural areas – such as our case study site in the south of the Republic – have a greater concentration of Indigenous populations, including the 'small-numbered' Indigenous groups, many of whom practice traditional livelihood activities. However, the areas that are being exploited for minerals, such as gold and diamonds, have populations made up mostly of Russians and other non-Indigenous ethnic groups, who migrated to the region for work in these industries.

The RSY has become well known for its political autonomy within Russia. The Republic has had its own political Indigenous intelligentsia since the early 20th century and since then has been navigating the tightrope between belonging to Russia and emphasizing its ethnic characteristics (Tichotsky 2000; Argounova 2004; Ivanova and Stammler 2017; Fondahl et al. 2019). The voting population of the RSY tends to be particularly active compared to other regions of the Russian Far East. Participation in presidential elections in March 2018 was 71% and resulted in the lowest share of votes cast for President Putin in the whole of Russia.[2]

According to statistical data, young people between the ages of 15 and 29 make up 23% of the total population of the Russian Federation and 26% of the population of the RSY. Thus, just over a quarter of the population of the Republic is made up of young people, over half of whom are Indigenous, according to the international definition.[3] This is a significant force, with considerable potential to influence political processes. In 2010, Yakutsk University was given a Russia-wide federal status, which meant much more funding but also more control by the Russian Ministry for Education. As a result, students there today are educated both in the spirit of the Indigenous political tradition and the Russian political system. It is important to be aware of this interplay when studying students' political perceptions.

The research for this chapter was carried out in two contrasting locations within the same region, in order to compare possible differences between urban and rural Indigenous youth. Based on the results of two surveys, one in the city and one in a rural municipality, supported by qualitative field work in both locales, we contribute empirically to classical theories of political leadership.

Theoretical framework

The sustainability of the political system in contemporary Russia depends on the involvement of a broad range of citizens in the political process. According to Easton's classic input-output model (2014), the active participation of civil society enhances the functioning of the political system. As a general rule, young people are an important component in the functioning of the political process and one of the most mobile.

The political culture of young people develops along the same lines as that of the wider population. The contemporary political culture in Russia is in a stage of transformation and is split into various components, including elements of paternalism and democratic values. Conceptually, the term 'political culture' as a functional approach began to be studied in the mid-20th century by Almond and Verba (1989). These authors imagined political culture as a certain collection of values, within which the political system operates – something along the lines of a historical-psychological backdrop, against which political events take place. We take this approach to our field in the RSY, which – as we shall demonstrate – has its own historical-psychological backdrop, which differs from that of central Russia and other Russian regions. Our research also explores the extent to which this backdrop differs within different parts of the RSY, between the urban capital and the rural areas.

In general, political culture can be defined as the totality of political knowledge, value orientations, and behavioural models produced by citizens and political and social groups within a given political system (Zimin 2008, 65). Almond and Verba (1989) identified three levels of political culture: 1) the level of world view, at which level individuals determine themselves within the realm of politics; 2) the citizen level, at which the attitudes of people towards power are formed; and 3) the political level, where the value relations of humans towards their own political events are formed and the roles of politicians in the lives of people are determined. Political culture and political socialisation are very closely interconnected, and it is important to take into account the way in which this interconnection between the two leads to the formation of political values. Malinova (2006) points to the need to develop alternative interpretations of the concept of political culture, drawing a boundary between scientific and politico-ideological discourse.

The concept of 'youth' itself, as a sociodemographic group, can be divided into various categories, according to interest and need, level of education and intellect, material circumstances, livelihoods, and lifestyle (student or professional, rural and urban) (Brader 2010; Sukarieh and Tannock 2014). In the process of political socialisation – that is, the process of integration into a political and politicised society – the most important youth category is 'students'. Having studied for a higher education degree and engaged in frequent scholarly and political debates, students are among all young people the group that is most oriented towards politics.

We divide political youth culture into three key components relating to 'intellect', 'values', and 'commitment' (Okorokova and Grigoriev 2019). The intellectual component is based on the accumulation of civic political knowledge, for instance, through the study of the humanities, including political science, sociology, and jurisprudence. The values-driven component of political youth culture is drawn from social and civic life itself; it is based on the system of citizen relations within which a young person has grown up and is also influenced by popular opinion on particular issues. This component is demonstrated in the activities of youth organisations and movements. For Indigenous people and any humans, youth is a period for accumulating knowledge and skills in various spheres of public life and for mastering the tactics and strategies of civil society aimed at reaching particular goals. These include taking part in elections, political rallies, and meetings organised by civil society organisations, and organising public actions. The values-driven component relates to both the operational and emotional aspects of political socialisation.

The component relating to 'commitment' also has both operational and emotional aspects and develops through participation in specific actions. The most obvious examples are the activities of young people's civil society organisations and student associations and student activities in support of greater self-determination. Political commitment is characterised by the willingness and ability of a young person to regulate her or his own behaviour and mobilise his or her own resources in the interests of the wider society. Policies oriented towards encouraging young people to take part in the resolution of issues that affect them and building a social partnership between young people and the state are especially pertinent in this regard. This allows the state to give a powerful stimulus to the resolution of issues related to the political socialisation of young people and make the development of wider society progressive and non-conflictual.

The process of political socialisation leads to the emergence of leaders, who possess certain leadership characteristics and charisma. Thus, it becomes pertinent to consider not only the 'perception politics' that the politicians themselves engage in, as Landtsheer et al. (2008) emphasize, but also the perceptions that young Indigenous people have about the image of a political leader. Differential psychology compares leaders' personal characteristics and assesses the influence of different leaders on political systems and their functioning (Simonton 2014). In political science, the concept of political leadership is understood as an institution endowed with power, based on uniting various groups around a programme promoted by the leader to address particular challenges. There are various approaches to the study of this phenomenon: philosophical reflection (Jaspers 1931; Russell 1938), the culturological approach (Tucker 1995), the behavioural approach, and psychological conceptions (Fromm 1942; Adorno 1973).

In political science, there are many typologies of political leadership, which have evolved from the work of Max Weber, who identified 'traditional', 'rational-legal' (bureaucratic), and 'charismatic' leadership (Weber 1922, 1926), which were later refined into 'interpersonal', 'charismatic', 'deliberative', 'creative', and 'neurotic' (Simonton 2009). Hermann identified four types of leader: the 'leader-flag-bearer' (the charismatic pioneer, such as Martin Luther King); the 'leader-servant' driven by what he or she believes is the will of the voters; the 'leader-trader', who can convince people to the extent that they 'buy in' to the leader's ideas; and the 'leader-firefighter', who excels at prompt reaction and decisions that the moment demands (Hermann 1991). In such a way, political leaders can be fitted into a typology according to their personal characteristics, the instruments they use to exercise power, and the specific context within which they act. Personal characteristics constitute leadership style, which can be analysed as a function of leaders' 'responsiveness to constraints, openness to information, and motivation' (Hermann 2010, 185). We shall apply Hermann's typology to our empirical research results from the RSY and thus contribute with new insights on how generally applicable such typologies are, including for an Indigenous context and in relation to a specific generational category – youth.

In the perceptions of young Indigenous people, the image of political leadership is formed in the course of elections, in the mass media, and by the surrounding community where people grow up. Coverage of a political leader in the mass media is essential for that leader to position him- or herself in the political realm. In the course of an election campaign, it is essential to consider certain popular stereotypes about politicians, which are formed by the impressions that are made during the election. This is how the image formed in the consciousness of the electorate is based not only on their reactions to the politician but also on their own level of political activity in the elections.

The literature suggests three broad groups of factors influencing the electorate's perceptions of a political leader (Egorova-Gantman 1994; Grishin 2007; Kozlovskaya 2012). These are: 1) factors relating to voters' informational environment and systems of communication, 2) factors that serve to motivate people (politically), and 3) existing political leader stereotypes (which are influenced in turn by voters' own sociopsychological characteristics). In addition, there are so-called 'background factors', which are assimilated from childhood and do not change much over time, as well as 'situational factors'. The former influence the image of a politician independently of specific conditions and events; the latter are activated only in certain situations, and in other circumstances they could be considered latent (Kozlovskaya 2012, 4).

The emergence and existence of effective leadership is determined by the coincidence of two factors: 1) the personal qualities inherent in a leader and 2) society's demand for the leader (Guseva 2011; Chuykov 2018; Pirogov and Rastimeshina 2018). Without both factors in play at the same time, you may have a formal leader, but it requires both to produce an effective leader. We shall consider the relevance of these factors in the specific context of our study, comparing a group of young Indigenous urban students with young people living in a rural municipality in the RSY.

Research, for example, surveys carried out in 2011 in the Chita Region of the Russian Far East, revealed that students had a particular perception of the image of a contemporary political leader (Rusanova 2011). This was based on the moral need for justice in political decisions; a pragmatic, utilitarian desire for security; and related factors such as historical stereotypes and political values, the (lack of) stability of the political regime, and perceived political corruption (ibid).

The political culture of young Indigenous people naturally correlates to the political culture of wider society. At the same time, it may also be revealed in three characteristic voting behaviours: 'activist', 'undecided', and 'absenteeist' (Okorokova and Grigoriev 2019, 148). The voting culture of young people includes the psychological orientation of individuals, their group, and their community in relation to the electoral process. However, the electoral culture – unlike other aspects of political culture – cannot exist outside of a legal framework and embraces only the public policy aspects of the electoral process (ibid). Young people can be inconsistent in their political affiliations, and according to the stereotype, they can be passive in the political process, owing to objective factors, such as difficult material conditions, a focus on career development, and day-to-day challenges, a tendency observed all over Europe (Pilkington and Pollock 2015).

Research of the political science department of the Ammosov North-Eastern Federal University, Yakutsk, between 2001 and 2016 aimed to elicit the political preferences of students from the university in the run-up to federal, regional, and municipal parliamentary elections (Petrov and Parfenova 2017). During the research, the students demonstrated a predominantly 'wavering' voting culture, to a certain degree overcoming this tendency in the parliamentary elections themselves. The research also revealed an increase in 'undecided' students from 27%

to 36% during that period and indicated that young voters were more active in federal elections than in regional and municipal elections (ibid).

Based on our understanding of the wider literature on political leadership and political preferences, we developed a research programme that sought to test out established theoretical frameworks and arguments in specific urban and rural contexts of the RSY. In particular, we sought to explore the general applicability of the categories of political leadership developed by Hermann (1991) for international political leaders and to what extent they can enhance understanding of young Indigenous people's political socialisation in a Russian Arctic context. Specifically, we sought to explore the following research questions:

1   What are the key factors influencing the way that people perceive politicians?
2   What are the key characteristics that make a good political leader in the minds of young people? and
3   What basic stereotypes of the ideal political leader have formed in the minds of young people?

Methodology

The empirical basis of the research comprised three main components:

1   A questionnaire survey of 500 university students in the regional capital of the RSY, Yakutsk, aimed at building a 'portrait' of the ideal political leader, based on their perceptions of the characteristics that make up a good political leader;
2a   A socioeconomic analysis of a sample rural municipality in order to identify key issues of importance to the rural electorate;
2b   A questionnaire survey of residents of the same villages to determine the factors influencing their own decision to vote for one or another candidate in municipal leadership elections. This was contextualised with some data from qualitative research by the authors between 2017 and 2019, based on questions to people about which local figures would meet their idea of a good politician.

Thus, the research was divided into an urban and a rural component, with the goal of comparing the perceptions of political leadership among young people in the contrasting urban and rural contexts.

The urban site was the city of Yakutsk, the capital of the Republic, with a population of 318,768, majority Indigenous (2019 data). Our rural sample site was the Belletskiy Evenki Natsionalnyi Nasleg (BENN), in Aldan District, in southern Yakutia, with a population of 1738 people in two villages: Khatystyr and Ugoyan.[4] Almost all of them are Indigenous, while the population in the surrounding gold extraction settlements of Nizhnyi Kurannakh and Aldan is mostly made up of Russians and other non-Indigenous ethnic groups.

Correspondingly, in both Yakutsk University (Efimova and Parnikova 2017) and our rural sample municipality, the absolute majority of the respondents were Indigenous young people. The difference between them is that in BENN, their connection to traditional livelihood activities is closer than in Yakutsk. The fact that the city students are mostly ethnic Sakha, while the BENN population is mostly Evenk, was not important for this research: the goal was not to separate Evenki from Sakha but study the perceptions of political leadership among Indigenous young people in the RSY. This makes our data meaningful in terms of possible non-Russian ethnic particularities of political culture in the Russian Arctic.

Part I: research in the regional capital

The first component comprised a questionnaire survey of randomly selected students studying at Ammosov North-Eastern Federal University in 2018. Five hundred students took part in the questionnaire survey, including 290 young women (58%) and 210 young men (42%). The respondents were aged between 17 and 26.

The students were first invited to choose their preferred options from a selection of basic characteristics that are meaningful for framing the portrait of a leader: gender, age, nationality, education, and profession. In the next step, respondents were invited to indicate at least three personal and three professional qualities which they considered particularly desirable in a politician and three personal and professional qualities that they considered unacceptable in a politician. This method was based on Shestopal's (2015) and L'vov and Trifonova's (2007) categorisations of psychological, political, and sociodemographic characteristics of politicians.

Part II: research in an Indigenous rural municipality

The objective of the second part of our research was to create a 'portrait' of a political leader, which would embody all the ideal qualities and values identified by the students, and compare this with the perceptions of the young rural villagers in the municipality of BENN in southern Yakutia. This allowed us to identify similarities and differences between urban and rural people, as well as to test the influence of concrete specifics of a place on general ideas. Moreover, this allowed us to test in a village context the significance of the categories of personal and professional qualities that the students identified in the city. BENN was interesting because this village council is a place where Indigenous peoples live and practice their traditional livelihoods, while the neighbouring larger settlements are mostly inhabited by Russians and other non-Indigenous ethnic groups.

To embed the survey on the ideal political leader locally in BENN and understand the responses, we needed to identify the basic specifics of the community there and analyse the sociodemographic, political situation and qualitative characteristics of the infrastructure of the municipality. Information gathering took place in the municipal administration offices and was supplemented by interviews and participant observation in BENN.

To form an idea of the perception of political leadership among the villagers, we carried out a questionnaire survey of the population of BENN. This covered the same areas and questions as the ones that were presented to the students in the city, which allows us to draw conclusions on rural and urban ideas about political leadership and the different agendas that people would like their leaders to pursue. For clarification, in addition to the questions from the student questionnaire, the respondents in the village were invited to formulate their own priorities and desires that they wished an ideal political leader to address.

A total of 50 people aged 18–29 years took part in the survey. Of these, 28 were female and 22 male. While the sample is 10 times smaller than the one conducted at the University in Yakutsk, our 50 people were approximately one-tenth of all youth in BENN, while our student sample of 500 was about 3% of the students enrolled at the University.

We amended the survey with more qualitative data from informal conversations in the main village of BENN, Khatystyr, during several field trips by several of the authors (Ivanova, Stammler, and a student of Okorokova) between 2016 and 2019. These informal conversations went deeper into the social dynamics within the community of the village and served as a sort of 'reality check' for the answers that the respondents gave in the survey as well. One of the specifics was detailed talks with and about political leadership figures, of which we present one subsequently, who illustrates best the idea of a political leader among the younger Indigenous villagers. This fieldwork-based background was essential to make the survey data meaningful.

Findings

We present the findings of our research, starting with the results of the student survey carried out in Yakutsk, followed by the two studies carried out in BENN: the sociodemographic assessment and the questionnaire survey carried out with the young villagers. The analysis section compares the results from the urban and rural components of the research to draw comparisons and wider insights.

Student perceptions of political leadership qualities

The results of the student survey allowed us not only to determine young people's knowledge of state leaders but also to assess their expectations from political leaders and to understand more about the formation and reproduction of the political elite, all of which are directly related to the future political course of the country.

Assembling the results of the student survey, we can propose a portrait of an 'ideal' politician as someone with the following traits: he is male (64%), aged between 35 and 50 (52%), has a higher education (72%), and is a political scientist (44%) or a lawyer (26%). Thirty percent of the surveyed students did not consider gender important, while only 6% of students preferred to see a female politician. This male-centred perception of Russian politics continues to predominate, despite the fact that, in late 2018, a female Sakha politician with no party affiliation – Sardana Avsentieva – was elected mayor of Yakutsk, becoming something of an Internet sensation, lauded across the country for her brave decisions on issues such as the economy and crime (MacFarquhar 2019).Thus, the mostly Indigenous respondents in our survey follow the overall Russian trend when they consider 'maleness' key in the image of a political leader, along with 'strong patriarchal values when it counts'. The ethnicity of a politician was not important for the majority (67%) of respondents. However, this has to be interpreted in light of the fact that Sakha politics has been strongly dominated by Indigenous Sakha politicians, and respondents therefore probably did not feel a particular need to underline that their candidates should be from their own people, because most of them are anyway.

Throughout the research process, it appeared that the students place more emphasis on personal and moral qualities, emotional characteristics, and judgements than professional qualities when forming their impressions of an ideal political leader.' Positive personal characteristics that our respondents considered important in a politician included 'honesty' (48.7%) and 'responsibility' (31.2%), followed less prominently by 'fairness' (18.5%) and 'being a hard worker' (18%). The least important positive characteristics were 'determination and the ability to achieve goals' (5.7%) and 'patriotism' (5%). Among negative personal characteristics not desirable in a political leader, the respondents found 'indifference' to be the most significant (31.4%). Less prominent among negative characteristics were 'egoism' (18%), 'greed' (16%), and 'dishonesty' (16%), and least important were 'laziness' (11%) and 'hypocrisy' (8.4%).

In regard to professional qualities, most of the surveyed young people identified the following as being important in a political leader: 'competence, decisiveness and responsibility' (37%), 'ability to achieve goals' (34%), 'ability to understand and assess a situation' (32%), and 'ability to convince people' (21.2%). Thus, a contemporary political leader must have superior organisational and managerial capabilities. The following positive qualities appeared not to be so decisive according to our respondents: 'the ability to acknowledge your mistakes' (7.3%), 'objectivity' (5%), and 'being demanding of oneself and others' (4%). An official position (48%) was a much more important feature of a future politician than age (22%) or ethnicity (20.1%). Young people are also more inclined to support politicians who promote care for the wellbeing of the population (45.3%), peace and stability in the country (29.8%), and guaranteeing and defending the rights and freedoms of people (21.3%).

Among contemporary political leaders, most students identified the current president Vladimir Putin (39.8%) as a notable political leader, which suggests that the image of unified power has formed in the mass consciousness of the young generation regardless of its ethnicity, concentrated in the hands of the president. Among regional political leaders, young people had most sympathy for the first leader of the RSY (then known as President), Mikhail Nikolaev (17.7%), and the current Head of the Republic (now known by the Indigenous term *Il Darkhan*), Aysen Nikolaev (12.8%). Among well-known historical political figures, they identified Vladimir Lenin (16%), Stalin (9%), Peter the Great (8%), and Mahatma Gandhi (6.2%) as being notable political leaders.

This student survey gave us an idea of the political orientations, values, and ideals among Indigenous city youth, which we then took to the village level for comparison

Socioeconomic characteristics of Belletskiy Evenki Natsionalnyi Nasleg

The socioeconomic research carried out in BENN helps us to understand the key issues that are important to local people and the problems that they expect political leaders to be able to address. This allows us to interpret their political priorities and understand from there the factors that shape their impressions of what makes a good political leader.

BENN municipality currently has two villages – Khatystyr (population 1385) and Ugoyan (population 375). Several other villages, including the former municipal centre Khapparastakh, were officially closed, and people were resettled to Khatystyr and Ugoyan in the 1950s, following the 'amalgamation' (*ukrupnenie*) policy that was being implemented all over the Soviet Arctic (Allemann 2020). The majority of BENN's population is Evenki, most of whom were relocated from the Amga River area. Only 40 inhabitants were registered as ethnically non-Indigenous according to the municipal records that we studied. The breakdown of the population of Khatystyr illustrates the importance of young people within the overall local population: 45% are under the age of 25.

By far the biggest employer in BENN municipality is the 'Khatastyr' Indigenous enterprise, a former Soviet collective farm engaged in reindeer herding and employing 57 people. Another 80 reindeer herders have 'self-organised into clan-based Indigenous communities (*rodovye obshchiny*), a form of non-commercial Indigenous unions that is granted special status under Russian federal law (Gray 2001; Stammler 2005; Fondahl 2014). The other important source of employment is gold mining (primarily with the company Aldanzoloto, a subsidiary of Polyus Gold); gold mining has a history of more than 100 years and is still the main source of income for the Aldan district (Stammler and Ivanova 2016; Fondahl et al. 2019). Another 130 people are employed by schools and kindergartens and 187 in various municipal services (housing, firefighters, culture, and healthcare). Some BENN residents have opened their own companies, mostly engaged in trading. A few local people work outside of BENN for the gold industry or on the East Siberia–Pacific Ocean (ESPO) and Power of Siberia pipelines. This indicates that the main productive occupation in BENN is reindeer herding, while all other employment is connected to state and municipal services and trading.

Both villages are currently classified as places of Indigenous occupation and traditional economic activity (*mesta traditsionnogo prozhivaniya i traditsionnoy khozyaystvennoy deyatel'nosti korennykh narodov*). In 2008, the joint council of the two villages decided to declare BENN a specially protected area of local significance for traditional nature use (*territoriia traditsionnogo prirodopol'zovaniia* or TTP).[5] Reindeer herding and hunting in the forest continue to be culturally and economically significant occupations for both Evenki and Sakha (Yakut) inhabitants of BENN. However, reindeer herding is facing several issues, including overall stagnation, threats from predators, and the low wages of herders and tent-workers. Herders are barely able to meet their own subsistence needs and those of their relatives. Since barium was banned for use as a poison for wolves in 2005, the increased wolf population has decimated the reindeer herds, and currently no meat is being produced; meat for subsistence comes mostly from hunted wild reindeer or other wild animals. Besides being the only occupation for these people, reindeer herding is also a crucial Indigenous identity marker. Therefore, an important goal for a local leader is to save herding from a complete collapse and restore reindeer numbers to a level where herders can produce something rather than just living from the subsidies that they get from federal and regional agricultural departments. For this purpose, one of the most hotly debated issues locally is support for wolf hunting or the possible re-allowing of barium as a wolf poison.

In the sphere of education, the complete lack of Evenk language teaching in the school curriculum is a big problem that contributes to the further gradual disappearance of the Evenk language in favour of Sakha (Varlamov et al. 2020). The school also lacks the staff to teach some other subjects. A key issue is the lack of children's transport, especially between the lower part of Khatystyr on the river bank and the elevated new part of the village where the school is located. A further pressing issue is the poor water supply in both villages. Moreover, many houses are old and run down, and some streets in the new part of Khatystyr still lack lighting. The absence of an Internet connection has also become increasingly relevant as a factor of structural disadvantage for the villages in BENN. Civil society in BENN is mainly shaped by a handicrafts club ('Gelen'), a grannies' club 'Dylacha', and the women's club 'Dalbar Khotun' – there are no youth organisations in the municipality. Even more importantly for BENN, the competition between two powerful clans – the Marfusalovs and the Dormidontovs – shapes social life in the municipality. Currently, nearly all governing organs are occupied by the Marfusalov clan.

The socioeconomic and cultural development of the Evenki and the conservation (or restoration) of the Evenk language and reindeer herding are often referred to as long-term goals among the population of BENN. In addition, local leaders need to have the practical capability to organise improvements in basic provision of infrastructure, such as water supply, transport, lighting, and the Internet. We would therefore assume that problem-solving characteristics are likely to feature strongly in the portrait of an ideal political leader.

Young villagers' perceptions of political leadership qualities: analysis of survey responses

Our survey of young villagers reveals a relatively low level of political activity or politicisation among respondents. Only one in four respondents stated that they even talk about politics, while hardly anybody was actually engaged in politics. Only one respondent stated that he had a sort of inner inclination for political activities; however, he did not feel that his relatives and the community would approve of him becoming more active in this sphere. One factor for low levels of political activism can be the belief that local politics would not change the socioeconomic situation much to the better. According to our survey responses, 42 of our 50 young respondents considered themselves as having an 'average' position within the local society with regard to status and wellbeing, while only 3 people positioned themselves at the lower end of society.

Despite a lack of political activism, only three respondents said they would definitely not vote in elections. More women than men stated that they would definitely vote, but mostly because either friends or family also do so and this is what every citizen should do. This shows that a high turnout in elections does not equate to high levels of political activity among citizens. The young people of BENN mostly trust the president of the country to solve political problems, a view expressed by 30 respondents (60%). The second most important politician was the head of the RSY, Aysen Nikolaev (14 respondents or 28%).

Out of 50 village respondents, 33 thought their ideal politician should be a man. Half of our respondents thought he should be between 35–45 years old, as at this age someone already has life experience but is still full of energy and without serious health problems. In the villages, ethnicity did not play an important role, but the kinship status of the politician did. This is not surprising, considering that two clans compete for power and prestige in BENN. Having a family and children is also a desired feature of a politician in BENN.

In terms of desired characteristics and values for a politician seen as ideal for BENN, the professional qualities were seen as more important than the personal qualities: the ratio was 30 to 18 respondents. The villagers tended to value professional more than personal qualities in their ideal politician. Among the personal qualities, respondents particularly valued 'honesty' (38 out of 50 people), 'responsibility' (22), 'fairness' (18), and 'determination and ability to achieve goals' (12). Less important were 'being a hard worker' (10) and 'patriotism' (7). Among negative personal characteristics not desirable in a political leader, the respondents emphasised 'indifference' (27). Far fewer respondents considered 'lying' (12), 'egoism' (9), 'laziness' (7), 'greed' (6), or 'hypocrisy' (4) to be particularly bad.

Among professional qualities, the most important were 'competence, decisiveness and responsibility' (23 people). Not quite as important, but still significant, were the 'ability to achieve goals' (14 respondents), 'ability to understand and evaluate a situation' (12 people), and 'ability to convince people' (11). Less important were the 'ability to admit their mistakes' (8 people), 'objectivity' (5 people), and 'being demanding of themselves and others' (6 people). According to the survey, the work of the politician in this rural area would mostly be judged by his success in dealing with concrete problems in the municipality, such as high food prices (20 respondents), environmental issues (water pollution, domestic and industrial waste) (11 respondents), and the bad road and transport conditions (10 respondents). Only 13 out of 50 people (26%) felt that caring for the wellbeing of the general population is important for a political leader, while only 9 out of 50 (18%) believed that protecting peoples' rights and freedoms is important, and only 7 people (14%) considered peace and stability in the country a priority.

In conversations with local people at the time of the questionnaire survey, it became clear that people in the village more readily commented and reflected on a specific person than choosing among abstract human qualities that researchers presented to them. We therefore started up informal conversations with local people about notable figures whom they knew and the extent to which these figures might fit their image of a respected political leader. In these informal conversations, several (mostly young) villagers indicated that a certain local figure would be a suitable leader. His name is Djulustan Vital'evich Sidorov, an Evenki in his 40s, a family father from a respected reindeer-herding clan with shamans among their ancestors. He is one of a few politically active men in BENN and was a candidate in the 2015 municipal elections,[6] where he gathered 28% of the vote (231 votes), supported by an influential entrepreneur from one of the dominant Evenki clans in BENN. In the meantime, Sidorov has made it to the chair of the RSY-side Union of Indigenous clan communities (*rodovye obschiny*).[7] Moreover, he is an active member of RAIPON, the Association of Indigenous small-numbered peoples of the North, Siberia, and the Far East of the Russian Federation.

His village members value that he is 'one of us', with his hobbies such as hunting, fishing, and gathering wild berries and mushrooms. Judging from conversations with some of his colleagues, people value his charisma, decisiveness, impassivity, courage, determination, confidence, and initiative. We can also hear pride about the fact that he has put the small municipality of BENN on a bigger map through his position as chair of the Union of Indigenous communities. This work experience has also enabled him to talk very confidently about the land rights legislation in the region. Sidorov has also made himself known through establishing his sound studio 'Tungus Records' that supports young Indigenous artists.[8] Many Indigenous people beyond BENN have come to value this initiative, as it brings the Indigenous culture movement to new audiences beyond village cultural centres.

The data from the 50 surveyed young villagers and the portrait of the currently most visible political activist in the main village give us an idea of how the younger Indigenous rural generation imagine a political leader. While we cannot say which data are more credible, the quantitative survey or the qualitative informal discussions, we believed that both sources of information were worthy of consideration in our analysis, taking into account the context and methods for the data gathering, including people's greater willingness to comment on specific local characters than abstract human qualities.

Analysis

Comparing the findings from the city students and the village youth, the most significant difference is that for city people, the personal qualities of their political leader of choice are more important, while for the rural youth, the professional qualities count more. The students in the city place more emphasis on general values that they want the politicians to find important. For instance, 45.3% of students believe that politicians should care for the wellbeing of the population, compared to 26% of village respondents; 29.8% of students believe that politicians should protect people's rights and freedoms, compared to 18% of village respondents, while 21% viewed peace and stability as a priority, compared to 14% in BENN.

However, the responses of the young people from BENN should not be read as an indication that they do not care about such values. Having lived during fieldwork in Khatystyr and talked to reindeer herders, their wives, hunters, municipal employees, and teachers, we can say rather that they do not think politicians can have a major role in guaranteeing these values. For example: for a reindeer herder, human well-being and freedom are in the forest with the herd. As long as he has the possibility to live that lifestyle, he has access to these values, and that is what counts most, regardless of the values that a politician in his municipality promotes. Moreover, like in many Russian rural regions, a lot of Indigenous people with land-based livelihoods are not very convinced that they can expect support from politicians, based on post-Soviet experience. This probably also explains the high level of political absenteeism stated by our respondents in BENN. It is therefore important to interpret the figures against the backdrop of people's life experience.

At the same time, the young people of BENN have a greater level of trust in the Russian president than students in the city, with 60% of village respondents choosing him as the most trusted leader to solve political problems, compared to 40% of students. The current head of the RSY, Aysen Nikolaev, was the second most trusted leader among BENN respondents (28%, compared to 12% of the students).

The BENN respondents had different priorities than those of the students. For instance, the gender division of labour was more significant in BENN, because of the natural resource-based livelihoods, than in the city of Yakutsk. The work of the rural politician is more likely to be judged on their ability to deal with practical problems, whereas in the city, the goals that students highlighted for politicians were more abstract. This shows the relevance of Kozlovskaya's (2012) emphasis on background and situational factors that influence the idea of a political leader. Therefore, in shaping their image of a political leader, BENN residents draw on their history and livelihood traditions, as well as their current situation. For them, the most important valued capacity is to be bold enough to make decisions, stand up for them, and react to problems as they come up. This resonates with the image that Yakutsk mayor Avsentieva stands for, who stated that she sees being a city mayor as being like leading an enormous household[9] – focusing on solving practical problems as they come up. In this light, we may wonder if students in the city start to favour a 'housekeeper type' over a 'human rights defender' as they grow older or as they consider options in future elections.

The difference in the significance of abstract values confirms the validity of earlier arguments that informational environments and systems of communication influence political values (Grishin 2007). Students in a big city are exposed to a different informational environment and systems of communication, compared to forest dwellers and villagers who do not even have a working Internet connection. Correspondingly, it is not surprising that the rural young people place their political interest more on concrete tasks such as improving transportation or restoring reindeer herding. Among respondents in the city, the link between general values and their concrete implementation may not be that straightforward. Therefore, such values may seem more abstract. The students have also had access to literature that takes examples from around the world, with authors theorising from these examples to create values. The lack of access to the Internet and big libraries in BENN creates a different intellectual environment, which may lead to more attention being paid to solving practical problems. Thus, Almond and Verba's (1989) idea of political culture as a collection of values is still relevant for both the city and the rural environment.

This shows that values are shaped by the cultural environment that people live in. It is therefore worth reflecting to what extent our material points to a specific Indigenous, or at least non-Russian, political orientation among our respondents in both the city and the village. Within the Russian Arctic, the RSY is one of the 'least Russian' or 'most Indigenous' ethnic territories. One might therefore assume that the political values of people in the RSY would correspondingly be linked to the tradition of its Indigenous residents, be they urban or rural. Given that there has been a revival of Indigenous worldviews in Yakutia (Stammler 2012; Jonutytė 2020; Vitebsky 2020), and given that worldview influences political culture, as outlined previously (Almond and Verba 1989), one could expect the ideas about politics and political leaders to be different from the mainstream in the country. However, the results from both the city and the rural area do not suggest such an expressed specificity. Rather, the most valued political leaders are the two heads of state, of the Russian and the Yakutian state (Putin and Nikolaev). Young people opted for the clearest representative of the mainstream Russian state system, in line with our finding that a responsible position (*dolzhnostnoe polozhenie*) is important for a political leader.

This emphasis on a responsible position also explains why people in the village in 2018–2019 accredited some of their valued traits of character for a political leader to Mr Sidorov. Besides really having the most charisma among active people in the village of Khatystyr, he is also the most experienced in rhetoric and has gathered fame beyond the borders of BENN with his sound studio and his position with the republic-wide Union of Indigenous communities. Nonetheless, he is still considered young enough to be perceived as progressive. This confirms what we found in the survey: that for people in the village, it is most important that they believe a political leader can actually get things done.

Applying these findings to the categories of political leadership coined by Hermann (1991), the village youth seem to value most the 'fireman' type of leader, as such a person is best at making confident and prompt decisions in response to problems as they arise. On the other hand, our informal conversations show that people in the village also valued the charisma of public figures such as Sidorov. So while the 'fireman' is important in the village, they are also receptive to the 'flag-bearer', for whom charisma is one of the most important traits. This view is shared by 30% of the city students, who believe a 'flag-bearer' would be able to inspire ideas among young people. The most popular leader category – among 40% of the city students – was the 'leader-servant', whom they believe is able to express and satisfy the interests of young people. Only 17% of the students opted for the 'fireman'.

Considering the importance of local and traditional livelihoods and awareness of respondents' non-Russian ethnicity in both settings, it might be surprising that such general theoretical models hold up in an Indigenous context. We suggest that this is because of the wider influences over Indigenous culture, for example, the presence of President Putin in all households, which is transported through Russian media to both the city and the villages, as well as the influence of international thinkers that students read on university courses.

Applying the classical categorisation of political leadership by Weber (1922, 1926) to our findings helps us see that for the village youth, traditional legitimisation such as belonging to a certain clan that enjoys local authority continues to play a role in the perception of political leadership. The low interest of young Indigenous villagers in formal political activity is not surprising but displays a rather general tendency (Pilkington and Pollock 2015). This low interest suggests that the belief in legal rational-bureaucratic political legitimacy as coined by Weber is not particularly highly valued. Even people's participation in the elections points more to a social *habitus* than to a belief in legalising political legitimacy through the state system. As was mentioned before: villagers do not vote in BENN because they believe that changes anything but because voting is what society expects. The importance of the 'leader-flagbearer' for both the city and the village points to the importance of charismatic leadership for both settings.

Conclusion

We can thus see how political culture and the values connected to it are shaped in this Indigenous population by the dominant state system, a system reproduced and supported at the local level by the Indigenous people themselves. This fact has enabled us to apply mainstream political leadership theory to an Indigenous context and to find that it has some relevance. A different way to approach this research would have been to start from the point of exploring Indigenous ideas of leadership, but this would have required a different methodology and produced different results (Ladner 2003). In this study, we focused on leaders operating within a political system of professional politicians. In traditional Indigenous societies in East Siberia, there were no professional politicians, and the leadership was with people who enjoyed authority because of their skills in a particular livelihood activity, such as reindeer herding, or because of their wealth. Therefore, even in a very non-Russian social and cultural environment, the idea of political leadership still resembles the Russian mainstream.

The expression of Indigenous values is, in this case, mainly visible in the importance that our young Indigenous respondents place on the regional state and its head, from where their sets of political values and ideas about leadership are partially inspired. However, although the RSY is a majority-Indigenous state with mainly Indigenous politicians, it is still a tightly integrated unit of the Russian Federation. Despite its strong history of political agency within Russia and all its talk of sovereignty and autonomy, it is still fully dependent on the Centre. Thus, however politics is practiced on Indigenous lands by Indigenous politicians, in Russia, this will still remain part of the all-Russian political system.

Acknowledgements

This research was carried out as part of a Russian-Finnish joint project 'Live? Work? Or leave? Youth wellbeing and the viability of (post) extractive Arctic industrial cities in Finland and Russia', jointly funded by the Academy of Finland (decision number 314471) and the Russian Fund for Basic Research (project number 18-59-11001).

Notes

1 Between detailed census, the Russian statistical office publishes yearly updates, some of which contain ethnic data. Their credibility is discussed in social networks and forums, for example, here on the city of Yakutsk: http://travel-ykt.ru/geografiya/naselennye-punkty/yakutsk.html

2 Итоги голосования в Республике Саха (Якутия) на выборах Президента РВ 18 марта 2018 г. [Электронный ресурс] Режим доступа: www.yakut.izbirkom.ru/arkhiv-vyborov-i-referendumov/2018/18.03.2018.php, accessed 22 April 2020.

3 We follow the international definition as in UNDRIP and ILO 169 for this chapter, whereby there is no numerical cap for defining who is Indigenous, as is the case in Russian law. For Yakutia, this means that the Sakha population is considered Indigenous.

4 As of 2018, official statistics: www.gks.ru/free_doc/doc_2018/bul_dr/mun_obr2018.rar, accessed 22 April 2020

5 Resolution (*rasporyazhenie*) of the Government of the Russian Federation of 08.05.2009, No. 631-p, 'areas of traditional living and traditional livelihoods of the Indigenous peoples of the Russian Federation and the list of types of traditional livelihoods activities of Indigenous and small-numbered peoples of the Russian Federation'. According to the articles 129, 209 and 215 of the RF Citizens' Code (*Grazhdanskii Kodeks*) TTP's can be owned by the municipality.

6 https://elections.istra-da.ru/person/188679/, accessed 22 April 2020.

7 http://yakutiakmns.org/archives/10530), accessed 22 April 2020.

8 https://sakhapress.ru/archives/199784, accessed 22 April 2020.

9 In a TV interview, https://youtu.be/iFlomYxiCRQ accessed 22 April 2020.

References

Adorno, T. W. 1973. *Studien zum autoritären Charakter [The Authoritarian Personality]*. Edited by L. von Friedeburg. Frankfurt and Berlin: Suhrkamp.

Allemann, L. 2020. "Soviet-Time Indigenous Displacement on the Kola Peninsula: An Extreme Case of a Common Practice." In *Handbook of Arctic Indigenous Peoples*, edited by T. Koivurova, E. G. Broderstad, D. Cambou, D. Dorough, and F. Stammler, Chapter 7. New York: Routledge.

Almond, G. A., and S. Verba. 1989. *The Civic Culture: Political Attitudes and Democracy in Five Nations*. Newbury Park, London and New Delhi: SAGE Publications.

Argounova, T. 2004. "Diamonds: A Contested Symbol in the Republic of Sakha (Yakutia)." In *Properties of Culture – Culture as Property: Pathways to Reform in Post-Soviet Siberia*, edited by E. Kasten, 257–265. Berlin: Dietrich Reimer Verlag (Pathways to Reform in Post-Soviet Siberia, 2).

Brader, A. 2010. *Youth Identities: Time, Space and Social Exclusion: Exploring Youth Policy and Practice in Sheffield, 1999–2003*. Saarbrücken: LAP Lambert Academic Publishing.

Chuykov, O. E. 2018. "Liderstvo v sisteme gosudarstvennogo i munitsipal'nogo upravleniya." [Leadership in the System of State and Municipal Government.] *Izvestiya Yugo-Zapadnogo Gosudarstvennogo Universiteta. Seriya Ekonomiya, Sotsioogiya, Menedzhment* 8 (3 (28)): 198–203.

Easton, D. 2014. "An Approach to the Analysis of Political Systems." In *Political System and Change: A World Politics Reader*, edited by I. Kabashima and L. T. W. III, 23–40. Princeton, NJ: Princeton University Press.

Efimova, S. K., and П. Parnikova. 2017. "Trudnosti i osobennosti studentov yazykovogo vuza pri obuchenii yaponskomu yazyku (na primere respubliki Sakha (Yakutia))." [The Difficulties and Peculiarities of Students' Experience in a Language University When Learning Japanese – The Case of the Republic of Sakha (Yakutia).] *Mir nauki, kul'tury, obrazovaniya Мир науки, культуры, образования* 5 (66). Accessed 18 April 2020. https://cyberleninka.ru/article/n/trudnosti-i-osobennosti-studentov-yazykovogo-vuza-pri-obuchenii-yaponskomu-yazyku-na-primere-saha-Yakutia.

Egorova-Gantman, E. V. 1994. *Imidzh lidera. Psikhologicheskoye posobiye dlya politikov [The Image of a Leader. A Psychological Handbook for Politicians]*. Moscow: Obshestvo 'Znanie'. Accessed 22 April 2020. http://nikkolom.ru/image-lidera/.

Fondahl, G. 2014. "Where Is Indigenous? Legal Productions of Indigenous Space in the Russian North." In *Nomadic and Indigenous Spaces: Productions and Cognitions*, edited by J. Miggelbrink, J. O. Habeck, and N. Mazzullo,77–91. Aldershot: Ashgate.

Fondahl, G., et al. 2019. "Niches of Agency: Managing State-Region Relations through Law in Russia." *Space and Polity* 23 (1): 49–66.

Fromm, E. 1942. *The Fear of Freedom*. London: Routledge.

Gray, P. 2001. "The Obshchina in Chukotka: Land, Property and Local Autonomy." *Max Planck Institute for Social Anthropology working paper* 29: 23.

Grishin, Ye. V. 2007. "Fenomen vospriyatiya i yego vliyaniye na elektoral'noye povedeniye Феномен восприятия и его влияние на электоральное поведение." [The Phenomenon of Perception and Its Influence on Electoral Behaviour]. *Vestnik Moskovskogo gosudarstvennogo oblastnogo universiteta. Seriya: Psikhologicheskiye nauki* 3: 199–210.

Guseva, N. I. 2011. "Sovremennaya model' liderstva v usloviyakh global'nogo konteksta." [The Modern Leadership Model in a Global Context]. *Izvestiya Baykal'skogo gosudarstvennogo universiteta Известия Байкальского государственного университета* 4. Accessed 22 April 2020. https://cyberleninka.ru/article/n/sovremennaya-model-liderstva-v-usloviyah-globalnogo-konteksta.

Hermann, M. G. 1991. "Stily liderstva v formirovanii vneshnei politiki." [Leadership Styles in the Shaping of Foreign Policy]. *Polis Politicheskie Issledovaniya* 1: 91.

Hermann, M. G. 2010. "Assessing Leadership Style: A Trait Analysis." In *The Psychological Assessment of Political Leaders: With Profiles of Saddam Hussein and Bill Clinton*, edited by J. M. Post, 178–214. Ann Arbor: University of Michigan Press.

Ivanova, A., and F. M. Stammler. 2017. "Mnogoobraziye upravlyayemosti prirodnymi resursami v Rossiyskoy Arktike." [The Diversity Natural Resource Governance in the Russian Arctic]. *Sibirskiye istoricheskiye issledovaniya* 5 (4): 210–225.

Jaspers, K. 1931. *Die geistige Situation der Zeit [Man in the Modern Age]*. Berlin and Leipzig: W. de Gruyter.

Jonutytė, K. 2020. "Shamanism, Sanity and Remoteness in Russia." *Anthropology Today* 36 (2): 3–7.

Kozlovskaya, N. V. 2012. "The Image of a Modern Politician in Students' Representations." *Psichologicheskie Issledovaniya* 1 (21): 4.

Ladner, K. L. 2003. "Governing Within an Ecological Context: Creating an AlterNative Understanding of Blackfoot Governance." *Studies in Political Economy* 70 (1): 125–152.

Landtsheer, C. D., P. De Vries, and D. Vertessen. 2008. "Political Impression Management: How Metaphors, Sound Bites, Appearance Effectiveness, and Personality Traits Can Win Elections." *Journal of Political Marketing* 7 (3–4): 217–238.

L'vov, S. V., and A. V. Trifonova. 2007. "Kachestva 'Ideal'nogo' politika: vzglyad naseleniya Качества." [Qualities of the 'Ideal' Politician: The Opinion of the People]. *Monitoring obshchestvennogo mneniya: ekonomicheskiye i sotsial'nyye peremeny Otkrytoe aktsionernoye obshchestvo Vserossiyskiy tsentr izucheniya obshchestvennogo mneniya* 3 (83).

MacFarquhar, N. 2019. "In Siberia, a First Female Mayor Builds a National Profile." *The New York Times*, 30 August. Accessed 18 April 2020. www.nytimes.com/2019/08/30/world/europe/siberia-mayor-yakutsk.html.

Malinova, O. Y. 2006. "'Politicheskaya kul'tura' v rossiyskom nauchnom i publichnom diskurse ['Political Culture' in Russian Scientific and Public Discourse]. *Polis Politicheskie Issledovaniya* 5 (5): 106–128.

Okorokova, M. P., and N. A. Grigoriev. 2019. "Obraz politicheskogo lidera v predstavleniyakh studencheskoy molodezhi Respubliki Sakha (Yakutia)." [The Image of a Political Leader According to Young Students in the Republic of Sakha (Yakutia)]. *Vlast'* 27 (5): 146–152.

Petrov, Y. D., and O. A. Parfenova. 2017. "Studencheskiye partiynyye predpochteniya kak otrazheniye molodezhnoy politicheskoy kul'tury sovremennoy Rossii (na primere Respubliki Sakha (Yakutia))." [Students' Political Party Preferences As a Reflection of Political Youth Culture in Contemporary Russia: The Case of the Republic of Sakha (Yakutia)]. *Obschestvo: politika, ekonomika, pravo* 4.

Pilkington, H., and G. Pollock. 2015. "'Politics are Bollocks': Youth, Politics and Activism in Contemporary Europe." *The Sociological Review*. Accessed 26 April 2020. http://journals.sagepub.com/doi/10.1111/1467-954X.12260.

Pirogov, A. I., and T. V. Rastimeshina. 2018. "Oborotnaya storona politicheskogo liderstva v sisteme vlasti" [The Flip Side of Political Leadership in the Power System]. *Ekonomicheskiye i sotsial'no-gumanitarnyye issledovaniya* 2 (18). Accessed 22 April 2020. https://cyberleninka.ru/article/n/oborotnaya-storona-politicheskogo-liderstva-v-sisteme-vlasti.

Rusanova, A. A. 2011. "Obraz sovremennogo politicheskogo lidera glazami studencheskoy molodezhi (na primere konkretnogo sotsiologicheskogo issledovaniya)" [The Image of a Modern Political Leader through the Eyes of Students (Based on a Sociological Case Study)]. *Vlast i upravlenie na vostoke Rossii* 4. Accessed 26 April 2020. https://cyberleninka.ru/article/n/obraz-sovremennogo-politicheskogo-lidera-glazami-studencheskoy-molodezhi-na-primere-konkretnogo-sotsiologicheskogo-issledovaniya/viewer.

Russell, B. 1938. *Power: A New Social Analysis*. London: Allen & Unwin. Accessed 22 April 2020. https://books.google.ru/books?id=Ql89zAEACAAJ&dq=5.%09Power:+A+New+Social+Analysis+by+Bertrand+Russell+(1st+imp.+London+1938,+Allen+%26+Unwin,+328+pp.&hl=en&sa=X&ved=0ahUKEwjd0bTxifzoAhWoIIsKHXqLDnIQ6AEIMDAB.

Shestopal, E. B. 2015. "Metody issledovaniya politicheskogo vospriyatiya v rossiyskoy politicheskoy psikhologii." In *Rossiyskaya politicheskaya nauka: Idei, kontseptsii, metody [Methods for Researching Political Perceptions in Russian Political Psychology]*, edited by L. V. Smorgunova, 76–94. Moscow: Aspekt Press.

Simonton, D. K. 2009. "Presidential Leadership Styles: How Do They Map onto Charismatic, Ideological, and Pragmatic Leadership?" In *Multi-Level Issues in Organizational Behavior and Leadership*, edited by F. J. Yammarino and F. Dansereau, 123–133. Emerald Group Publishing Limited (Research in Multi-Level Issues).

Simonton, D. K. 2014. "The Personal Characteristics of Political Leaders: Quantitative Multiple-Case Assessments." In *Conceptions of Leadership: Enduring Ideas and Emerging Insights*, edited by Goethals, G. R., et al, 53–69. New York: Palgrave Macmillan US (Jepson Studies in Leadership).

Stammler, F. 2005. "The Obshchina Movement in Yamal: Defending Territories to Build Identities?" In *Rebuilding Identities: Pathways to Reform in Postsoviet Siberia*, edited by E. Kasten, 109–134. Berlin: Reimer (Siberian Studies).

Stammler, F. 2012. "Dukhi gor vdol' Kolymskoy Trassy, ili kak zapolnit' dukhovnyy vacuum." *Nauchnyy vestnik YNAO Rossiiskii Sever i Severyane: Sreda-Ekologiya-Zdorov'e* 74(1): 53–56.

Stammler, F., and A. Ivanova. 2016. "Resources, Rights and Communities: Extractive Mega-Projects and Local People in the Russian Arctic." *Europe-Asia Studies* 68 (7): 1220–1244.

Sukarieh, M., and S. Tannock. 2014. *Youth Rising? The Politics of Youth in the Global Economy*. New York, London: Routledge.

Tichotsky, J. 2000. *Russia's Diamond Colony: The Republic of Sakha*. Amsterdam: Harwood Academic publishers.

Tucker, R. C. 1995. *Politics as Leadership*. Rev. ed. Columbia and London: University of Missouri Press.

Varlamov A., G. Keptuke, and A. Lavrillier. 2020. "Electronic Devices for Safeguarding Indigenous Languages and Cultures." In *Handbook of Arctic Indigenous Peoples*, edited by T. Koivurova, E. G. Broderstad, D. Cambou, D. Dorough, and F. Stammler, Chapter 5. New York: Routledge.

Vitebsky, P. 2020. "Indigenous Arctic Religions." In *Handbook of Arctic Indigenous Peoples*, edited by T. Koivurova, E. G. Broderstad, D. Cambou, D. Dorough, and F. Stammler, Chapter 8. New York: Routledge.

Weber, M. 1922. "Die drei reinen Typen der legitimen Herrschaft." [The Three Types of Legitimate Rule]. *Preussische Jahrbücher* 187 (1): 1–12.

Weber, M. 1926. *Politik als Beruf [Politics As a Vocation].* SSOAR Open Access Repository. Berlin: Duncker & Humblot. https://nbn-resolving.org/urn:nbn:de:0168- ssoar-59888–1.

Zimin, V. A. 2008. "Osobennosti formirovaniya politicheskoy kul'tury v sovremennoy Rossii." [Aspects of the Formation of Political Culture in Modern Russia]. *Vestnik VEGU* 3: 61–66.

4

ELECTRONIC DEVICES FOR SAFEGUARDING INDIGENOUS LANGUAGES AND CULTURES (EASTERN SIBERIA)

Aleksandr Varlamov, Galina Keptuke and Alexandra Lavrillier

Abstract

This chapter focuses on the electronic devices used to safeguard Siberian traditional cultures and languages. It explains the reasons for the modern cultural and linguistic decline (Soviet policy, globalisation, issues of native language teaching). It stresses the important interrelations between the preservation of the natural environment and the protection of culture and language. The chapter then examines Evenki community-driven experiences led by the first two authors in successful and unsuccessful initiatives to preserve Indigenous culture and language. It proves that tools of globalisation (new information technologies: social networks, smartphone applications, online databases and modern music) can create an e-learning environment that helps to protect and revitalise Evenki language and culture. It also stresses the crucial role played by Indigenous youth in urban environments. The chapter continues with recent recommendations from Evenki scholars to improve the teaching of their language and proposes comparisons with other experiences in the Arctic and elsewhere in the world. It ends with a discussion of the interactions between scholars and native communities in Siberia.

This chapter focuses on the use of electronic devices to safeguard Siberian cultures and languages on the basis of the example of the Evenki. The first two authors are Evenki scholars in oral literature and language and important actors in the cultural and linguistic revival (Varlamov 2009, 2013; Varlamova 2002, 2004, 2008, 2009; Keptuke 1991). The third author is a French social anthropologist with experience in applied anthropology and linguistic documentation. After explaining the reasons for the present situation of cultural and linguistic decline among the Evenki, the chapter shows the interrelation between the preservation of the natural environment and the protection of native culture and language. The authors present some community-driven Evenki initiatives for supporting cultural and linguistic preservation that are totally independent from the Decade for Indigenous Languages by the United Nations or the recent International Year of Indigenous Language. The chapter then focuses on how new technologies – like internet social networks, smartphone applications, online databases and modern music – are being used by the Evenki to safeguard their language and culture. The chapter continues with recent recommendations from the Evenki community for revitalising their language. It concludes with a discussion of some other Siberian, Arctic and international experiences with the development of e-learning tools and interactions between native/non-native scholars and native communities in Eastern Siberia.

Globalisation has affected Indigenous peoples worldwide, including those in Siberia. Today, many aspects of traditional Evenki culture are no longer a part of everyday life because of the shift of some of the population from nomadic economics (reindeer herding and hunting) to rural and urban ways of life, industrial development (mines, dams, urbanisation, etc.) and a subsequent decline of those portions of the Indigenous population engaged in nomadism.

In Russia, those places which safeguard Siberian Indigenous cultures are called 'places of compact Indigenous inhabitation': this refers to small, remote villages and nomadic areas with a population composed mainly of Indigenous peoples. These habitations can exist only in a wild landscape where people can follow the ancestral traditions of reindeer herding, hunting and fishing.[1]

Today, the Evenki people consist of three social groups: nomads, villagers and townspeople. Nomads and urban Evenki represent minority groups, while the rural population makes up the majority. While some individuals switch from one social group to another temporarily, each social group has its own specific activities, social organisations, ritual practices, anxieties, language situation and so on (Lavrillier 2005a). Although all Evenki are greatly concerned about the situation with reindeer herding or hunting, in terms of cultural loss, one of their biggest sources of anxiety is the decline in the use of their native language, which they consider the very basis of their culture and identity. Anthropologists and linguists have argued that culture is not built on language alone (Boas 1911; Sapir 1933; King 2011, among many others), but the situation is more complex for some regional Evenki groups. "Language means culture and identity" is a very widespread viewpoint in many places in Siberia, both among researchers (Filippova and Sokolovskii, 2019, for example) and Indigenous communities. As noted by the third author at the turn of the millennium, in internal debates and publications, urban Evenki often express the following sentiment: "I do feel myself to be Evenki, but in what way can I prove that I am an Evenki, since I do not know either my native language or the nomadic life style?" They have since resolved this concern by revitalising their language and creating collective rituals (see later in the chapter). We will also see that the first two authors link language preservation and herding conservation, just like the nomads do. For example, in the Sakha Republic (Yakutia) in Eastern Siberia (Russia), the Evenki language is still naturally practised (i.e. its use is not compelled by cultural or educational frameworks like schools, university exercises or cultural events) in one village, reindeer-herding camps and the forest, where it is still possible to hunt. In other areas of the Sakha Republic, most Evenki are still considered 'Evenki' only in terms of their worldview, because they have lost their native language. The third author remarks that Evenki native speakers consider and call these people "Yakol", that is, "Sakha". The Evenki honorific expression *tedie Evenki* ("a true Evenki") designates a person who is a reindeer herder/hunter and speaks Evenki (Lavrillier 2005a). The first two authors note that, unfortunately, entire groups of what was once a widely dispersed people have ceased to exist because of the destruction of the local ecology. Indeed, in the Mirnii, Vilyuisk and Lenskii regions of the Sakha Republic, the Evenki of the taiga have disappeared entirely.

Nowadays, the Evenki language is only known by 12.5 per cent of the Evenki of Russia, out of an entire population of just over 38,396 individuals. For example, in the Sakha Republic, where more than 22,000 Evenki live, the level of native language preservation is only 5–6 per cent (see also Grenoble and Whaley 1999).[2] Only 70–80 years ago, more than 60 per cent of the Evenki in the Sakha Republic practised the Evenki language. It must be noted that in the southern Sakha Republic, where most Evenki live, a significant process of 'yakutisation' has occurred over the last few centuries, resulting in a linguistic shift from Evenki to the Sakha language.[3] Consequently, despite the fact that the Evenki lifestyle and some elements of the lexicon were safeguarded in many places, these people now only speak Sakha (Maak 1886, 1887; Nikolaev 1964; Lavrillier 2005a).[4]

Nevertheless, it would be wrong to argue that the Evenki or Siberian non-native speakers have lost their ethnic self-identity. On the contrary, the younger generation of Indigenous peoples mostly identify themselves first of all as Evenki, Even, Yukaghir or Chukchi[5] and not only as citizens of Russia or the Sakha Republic. Citizenship is perceived locally as civic belonging, while feeling Evenki or Chukchi is about ethnic/cultural identity. Several researchers note that metis (including from Russian-Siberian or from different Indigenous Siberian parents) underline one or the other of their identities depending on the situation, location or purpose (Elwert 1997; King 2011, 19 et passim, among others); the third author has also often observed this in the field. Perhaps this positive tendency has been caused by what the first two authors call a specific 'process of ethnic defence'. To observe this, we can compare the new generation with the previous one. In contrast to today's youth, their parents, who possessed an important set of cultural features, were less motivated to maintain the culture of their people. The sociopolitical framework of the time also played a role. For example, there was a popular point of view in the USSR that the so-called 'primitive peoples' (i.e. Indigenous peoples of Siberia) needed to be educated so that they could partake in pan-Soviet culture. In addition, 'yakutisation' has had a calamitous effect on Evenki culture and, in particular, their language. Indeed, between the 1950s and 1970s under the Soviet regime, Evenki children in boarding schools were forbidden to speak their native language; they were forced to speak Russian or Sakha. Since this period, their native language has been taught in schools, but this has caused another problem: the Evenki taught is an artificially constituted literary language, very different from many of the mostly spoken dialects (see later in the chapter). As a result, this standard Evenki does not fulfil its role as a *lingua franca* for all Evenki, and most children (lost between the Evenki dialect spoken at home and incomprehensible literary Evenki) tend to forsake their language (Bulatova and Grenoble 1999, 3; Grenoble 2003; Lavrillier 2005b, 440–444). The Evenki are still divided between those "for the dialects", such as many village teachers with native-speaking Evenki families, and those "for a uniform Evenki language", such as the first two authors and most Evenki intellectuals. In recent years, the languages of Indigenous peoples in the Russian Federation have been taught at school for a maximum of two hours per week as a non-compulsory subject. If there are several Indigenous languages in one village, the first author noted that in the Sakha Republic, the school and parents will decide which language is taught. Thus, in Khatystyr village, Evenki, rather than Sakha, was compulsory. In order to have more hours for native language teaching, some teachers use those hours normally devoted to the school subject "Indigenous cultures". However, there is a lack of young native speakers and motivated teachers. More than 80 per cent, around 160 of 200 Evenki-language teachers who graduated from the three remaining Russian universities offering such a specialism, were non-native speakers. The urgent need to create systematic educational activity in order to preserve the language and culture of the Indigenous minorities of Siberia has been cited by many experts (Petrov 2013). An alternative for increasing children's contact with their native language is nomadic schools in Siberia (Lavrillier 2013; Terekhina 2019). Another acute problem is the lack of modern educational literature. While the Sakha Republic has a state policy favourable to the preservation of Indigenous cultures, only one set of textbooks for elementary grades has been issued with state funding in the past 20 years. All other classes continue to study obsolete books (Sharina 2015). This applies to most regions where Evenki is taught. In Siberia, few of the regions where the Indigenous peoples of Siberia live, such as the Sakha Republic, the Khanty-Mansi Autonomous Okrug, the Yamal-Nenets Autonomous Okrug and Sakhalin Oblast, are actively publishing schoolbooks that meet modern requirements (Glukhova 2011). The same is true for the publication of artistic literature: reprinting or publishing the works of Evenki authors is quite rare. For academic publications, Tungus[6] scholars resolve their difficulties through grants and fundraising, although the systematic support that existed during the USSR is no longer available.

According to the first two authors, ecological decline is related to language preservation. While the laws of the Russian Federation forbid the installation of pipelines near archaeological sites and monuments, it is possible to install pipelines or gold and diamond mines directly on reindeer-herding pastures, the territories of hunting clans and rivers which serve as breeding grounds for fish. Such places could also be considered ancient cultural sites for the Evenki. Thus, given how such industrial activities undermine traditional Indigenous economies, it is impossible to safeguard the cultures of the Evenki and other Indigenous peoples. Nevertheless, the Sakha Republic is known to have the most advanced legislation of all Russian regions, if not the Arctic as a whole, for industry cultural impact assessments (Stammler and Ivanova 2016). Thus, while the younger generation still learn about the culture of their people from the older generation, today they are finding out more from academic and literary sources.

It may sound paradoxical, but the Evenki youth see the solution to safeguarding their people in technological progress, such as the global use of sustainable energy sources (sun, wind, hydro-electricity and so on) and the development of artificial replacements for petrol, gas, diamonds, gold and so on when international demand falls:

> When people stop digging in the ground and taking from it things that are supposed to have value for them, when one stops destroying the taiga and learns to estimate the value of all surrounding living beings – only then will our people be able to live in harmony with themselves and with *Buga*.[7]
>
> (From a prediction of the shaman Keptuke,
> the father of the second author)

It is interesting to note that the ideas contained within this shaman's prediction are also present in the songs of the youth music group Mit evenkil ('We – the Evenki' in Evenki) from the village of Iengra in the southern Sakha Republic. These ballads use the Evenki language to express the hope that the natural environment can be safeguarded. The following verses once again prove that, for the Evenki (as for many other Indigenous peoples of Siberia and the Arctic; Nuttall 2009), the natural environment is their culture. As such, the preservation of their culture (including language) is strongly related to, and interdependent with, the preservation of the natural environment. This is referenced in anthropological debates on nature/culture. The idea that the destruction of the natural environment harms ethnic identity and culture was also argued by the Inuit (Nuttall 2009).

> They have dried out all rivers
> They have dug over all the mountains
> For the young generation
> Only stubs have been left
> Nobody was awaiting such guests
> They came when we slept

They did even not knock at the door
It will be hard to live:
We are forgetting our native language
Our land is burning and has gotten sick
What will happen after?
Nobody knows
If we lose our face
We won't be as they are!
From our thoughts the snow is melting
They will hear us!
May they not say
That we don't care!
We will survive everything – I know.
In the bed of the river there are huge stones
From our thoughts they are turned upside down
We are not afraid of the winter colds
Nor of the hot summer weather
Refrain:
We – the Evenki
Living in our own land
Herding reindeer!
It will be like that forever!
Odyo!

In this song, the line "if we lose our face" refers to the Evenki expression "to lose face" (evk. *dereve tekiv-da*, literally "to let one's own face fall"): this means "to lose dignity and pride", as well as self-confidence. It is often associated with the issue of language and culture preservation. During winter and spring native festivals, one can find the following phrase written on the stands: "Don't let your face fall: don't forget your language and way of life!" The ritual aspect of this youth song is marked by the final word "*Odyo!*", which means something like "let it be like this in reality!" (Lavrillier 2003, 2005b).

From the 1990s, Siberia has seen an impressive increase in Indigenous community-driven projects relating to culture/language survival or revitalisation. Today, the Evenki youth are working to safeguard their own culture. The funding assigned by the government for the safe-guarding of Indigenous languages is insufficient and is allotted mostly to scholars (as we will see), so most Indigenous projects depend on volunteering. They write new songs, organise and perform musical concerts, make recordings and sew native costumes, among many other activities. The older generation is very proud of their youth: elders hope that their efforts will not be wasted and that they will be able to keep Evenki culture alive until the arrival of better times. During 2014–15, as noted by the third author, the Evenki community, without the help of scholars or funds and with the support of their own record label, "Tunguz Rekord", created many songs in Evenki or Evenki/Russian that have circulated widely all over Russia and China (where the Evenki also live; Dumont 2018) through social networks (Vkontakt, Whatsapp and, more recently, Instagram). Some use traditional round-dances with ritual singing (like the *delehincho* recorded by the third author in the 1990s in the Tyanya village and distributed with other songs on a DVD); they have turned these into very successful techno songs while still keeping the old Evenki lyrics. Many similar examples can be found in the Arctic among the

Inuit or during the Sámi Grand Prix, where traditional Joik songs have been modernised and are very popular.

As already mentioned, the position of the Indigenous youth is very difficult because of the Soviet historical context and globalisation. The most important means for the preservation of the Evenki language is support from the state and the existence of a traditional framework for native language practice. Indeed, were an Evenki commission on native language preservation with similar authority to those in the Sakha Republic, Buryatia and other national territories to come into existence, the loss of language and other cultural elements would stop, and a revival would occur quickly through strong governmental support for language revitalisation. A brilliant example of this is the recovery of the previously lost Hebrew language after the creation of the Israeli state (according to the first author). Another success story is that of Kalaallisut, thanks to Greenlandic governmental measures (Grenoble 2016; Huctin 2016, 447–449) and institutions defending the language.[8] One can point to other international examples as well (Jones and Ogilvie 2013). The Sakha language also underwent successful revitalisation during the 1990s–2000s in the main city of Yakutsk, thanks to both republic-level measures and the urban migration of native speakers from the countryside (Lavrillier 2013, 110).

What can one do when the language of a numerically tiny Indigenous people is practised in only a few small, isolated places surrounded by majority populations that speak another language? The first two authors argue that the Evenki could lose hope, blame globalisation and place the Siberian Indigenous languages of the taiga and tundra in the historical archives of vanished tongues. However, it is also possible to oppose globalisation by using its own weapons against it. A similar idea was proposed by Hannerz (1992) and Appadurai (1996), among others. The Evenki community therefore consider it essential to employ modern information technologies. Such methods were successfully used by the Sámi (Riessler 2013).

Before developing the present theme further, the first two authors will share their previous experience with actions aimed at revitalising Evenki culture (including language). At the beginning of the 1990s, Evenki intellectuals and associations of the Sakha Republic were able to recreate the Evenki collective springtime shaman ritual *Ikenipke* (*Bakaldyn*) ('The Day of Playing', a meeting which was forbidden by the Soviet authorities from the 1930s) in Yakutsk. This ritual has taken place annually ever since, not only in Yakutsk but also in all the regions of Russia where Evenki live.[9] Since the early 1990s, many successful and unsuccessful projects were created, of which the first two authors are proud. Several Evenki dance groups were formed, and ancestral handicrafts were revitalised or more intensively practised thanks to the mobilisation of craftswomen. In addition, the first two authors created an ethnographic centre close to Yakutsk with small reindeer herds for the promotion of the Evenki culture and language to a wider public. For several years, it was the site of Evenki ritual performances, meetings, social events and tourism. Unfortunately, because of a lack of funding and several forest fires that destroyed the lichen cover, the centre had to be closed down, pushing all their efforts back to square one (see Figures 4.1 and 4.2).

In addition, in 2012 and 2019, the Evenki of Yakutsk took part in an important collective project focused on the history of the Evenki people. Within the framework of this project, an expedition was organised from Yakutsk to Lake Baikal. Both the older and younger generations participated. Lake Baikal is the very birthplace of the Evenki people, which is why the members of this trip set up a stone memorial with Evenki engravings (evk. *ityl*) (Varlamova 2009) (see Figures 4.3 and 4.4).

It was among all these initiatives for cultural recovery and language revitalisation that the first projects using new information technologies appeared. As mentioned previously, this was

Figure 4.1 Ritual practices in Bakaldyn: offerings of fat to the *sevek-moo* representing the three words of the Evenki universe

Source: © A. Varlamov.

a community-driven and independent initiative from non-native scholars. Many plans were successful thanks to the efforts of non-Evenki volunteers. For example, the website Evengus is very popular among the Evenki and specialists in the Evenki language (www.evengus.ru). It was created thanks to the enthusiasm of Rustam Yusupov from Tatarstan. This portal gathers general information and lectures on Evenki grammar that have been prepared on the basis of printed manuals. In addition, it offers a long list of useful links to many related websites. This website is also connected to a very rich online library about the Evenki people, culture, ethnography, literature, language teaching and so on (www.evenkiteka.ru). It is visited each day by several hundred people (including scholars) who wish to translate Russian into Evenki, learn about grammar and receive information. However, the third author would like to underline that the first two authors were not aware of many similar projects run by linguists from central Russia or outside Russia (like the impressive work of Kazakevich, 2014), a point that will be developed in the conclusion.

As previously noted, one of the reasons for the unfavourable situation in language practice is the decline or absence of opportunities to communicate with native speakers or to obtain continuous contact with natural and direct native speech. For these purposes, local native teachers like A. Kolesova sometimes spontaneously and voluntarily organise trips for children to nomadic camps (such as the project "Young reindeer herder", *yunyi olenevod* in Russian), but this is rare.[10] Instead of immersion in speaking communities, computing technologies allow one to partially create a framework for communication in Evenki and, more importantly, provide

Figure 4.2   Ritual practices in Bakaldyn: marking people with ash/cinder from camp fire

Source: © A. Varlamov.

opportunities to independently obtain linguistic knowledge. Given the critically low number of native speakers, such projects are extremely important for the population.

In 2011, the specialists of the Evenki Philological Chair at the Institute of Human Studies and Problems of the Indigenous Minority Peoples of the North (part of the Siberian branch of the Russian Academy of Sciences) created a free electronic phrase book (voiced by the second author) that one can find on the portal *evengus.ru*.[11] In 2012–13, these scholars started to realise an experimental project to create a framework for native-language practice with the support of the Department for the Affairs of Indigenous Peoples of the Sakha Republic (Yakutia). This project was designed to attract young people who neither know nor practise Evenki; it cost $10,000.

Between 2015 and 2016, the first two authors developed an experimental program for e-learning called 'Evenki for Beginners' (see Figure 4.5). As a methodological foundation for language teaching, they employed a framework that makes use of visual and audio perception of the language. The program is composed of around 30 lessons which cover the main spheres of knowledge about the Evenki language, as well as exercises, tests and additional documents like oral literature, literary texts in Evenki and modern music (hip-hop and rock).[12] This program has been distributed via the web, DVDs and USB drives. The main channels of diffusion are social networks, especially the group 'Evenki' on Vkontakte.[13]

This social networking group was created through volunteer work by Evenki and non-native enthusiasts from all over Russia. At present, it contains around 4,000 active members and is one of the most important inter-regional channels of communication. On this network, Evenki from different regions share and upload news about artistic projects, modern Evenki music and songs, videos, cartoons, links to interesting websites and oral literature recorded in the past; they

Figure 4.3 Installation of the stone memorial with engraved rules of Evenki good behaviours
Source: © A. Varlamov.

also discuss important topics in relation to Evenki culture and language and communicate in Evenki via the chat function. Most of the members do not know Evenki as a mother tongue and use this social network to enrich their practice. Representatives of other peoples, like scientists, artists or simply those interested in Evenki culture, are also members of the group. In 2013, with the help of computer specialists from Moscow, the first two authors made a mobile version of the Russian–Evenki e-phrase book and a Russian–Evenki translator; these apps can be installed on most modern phones (Evdik). Similar tools exist on Google Play for the Buryat language (Amitad, Amiskhaal).

The results of these initiatives show that the younger Evenki generation are very interested in using e-learning programs to acquire native languages. While only several dozens of the participants have gained a basic knowledge of the Evenki language, many of those who already possessed a foundational level of linguistic proficiency have significantly enriched their vocabulary: they use it during direct communication or in online chats conducted via social networks and smartphone apps (see Figure 4.6).

The experience of the first two authors has made it clear that there is a considerable opportunity to employ similar e-learning tools in kindergartens and during the first years of school. By presenting lessons in the form of games, children can receive basic knowledge about their native language and culture (see the following). Thus, in 2014, the first two authors started to develop a new e-learning program called The Speaking Alphabet (evk. *Dylgani Azbuka*).

Figure 4.4 Collective offerings in the stone memorial site
Source: © A. Varlamov.

This program is already used in kindergartens and schools and also in some Evenki households. This application can be downloaded for free (see Figure 4.7).[14]

The first author, along with his community, is planning to develop an English version of an Evenki language e-learning tool for a broader public. With this aim in mind, volunteers are undertaking ongoing work to translate materials into English. However, because of the lack of funding, Evenki youth are translating all the materials for free; this means that the work is not proceeding as quickly as desired. When this task is completed, they will distribute the results through English-language social networks, such as Facebook.

There has clearly been rapid progress in the use of information technology to safeguard Siberian and northern Indigenous cultures and languages. Of course, it will probably not be able to radically alter the current negative trend, but it may be able to curb the process. The first author emphasises that between 2016 and 2020, the Evenki scholarly community completed a great deal of work in assessing the language situation and developing methods for improving it. To improve the status of the Evenki language in places of compact residence, it is necessary to increase the number of teaching hours in schools, create a native language environment at school, train new teachers with a "full language immersion" methodology in native-speaking communities, publish textbooks in the eastern dialect of the Evenki language (spoken by more than 70% of Evenki in Russia; the literary Evenki taught is based on the western dialect) (see previously) and stimulate self-education in the mother tongue. Furthermore, electronic

Figure 4.5 Picture of the cover – Evenki language for beginners

Source: © A. Varlamov

software, like talking dictionaries with a translation function, mobile language applications and audio recordings of folklore and literature, should be created and distributed (Varlamov 2009, 2013; Varlamova-Keptuke 1991; Varlamova 2008, 2009, among many others).[15] However, all these measures require the direct participation of the state and considerable financial support.

For the moment, the scholarly and applied projects of many foreign researchers are the best hope. Thus, works on the revival of native languages in Scandinavia and other places in the

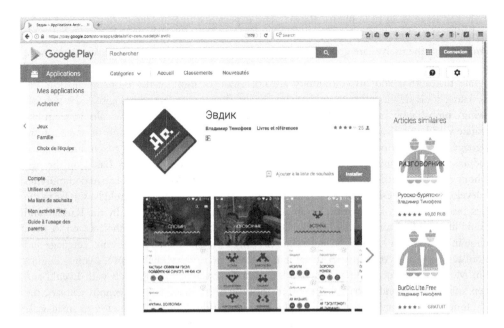

Figure 4.6   Google application – e-learning Evenki phrase book and dictionary

Figure 4.7   Youth working with e-learning tools

Source: © A. Varlamov

Arctic can serve as an example for Russia. The experience in Sámi territories shows that the language web technique is the most effective (Hinton 2002). The first two authors note that the ethnic groups themselves are gradually realising the need to preserve their language as the most important element of culture (Zamyatin et al. 2012).

The first author and his community are not aware of most of the e-documentation and e-learning tools developed by non-Evenki scholars (linguists or anthropologists) (see the following), probably because of insufficient communication or distribution. They almost seem like separate worlds. He also underlines that because calls for proposals are exclusively oriented toward language documentation, non-native scholars are probably not encouraged to produce applied products like e-learning tools that would really help revitalise the language. Based on the first author's experience, it is impossible to find funding institutions for Siberian native communities to develop their own e-learning tools. For him, useful projects by scholars should be similar to the Olympiad of the Evenki language, organised each year in the Amur region by the Russian and Evenki linguists O. Morozova and N. Bulatova with a committee of native scholars (and so, even online in the Covid-19 pandemic context). This truly increases the status of Evenki language and stimulates the youth. Another very useful initiative belongs to K. Mishchenkova, a young linguist from Moscow, who designed and funded a project for creating playing cards with Evenki and Even animal names and distributing them in native kindergartens. For Indigenous scholars, the only funding opportunities are grants by the president of the Russian Federation or small-scale assistance from the native mayors of Evenki villages, municipalities or mining enterprises.

The third author, based on her own experience of helping to implement an Indigenous project for a nomadic school, confirms that it is almost impossible to find, in academia in particular, money for such an applied anthropology project (Lavrillier 2013). As an anthropologist working in a language and cultural documentation project (Max Planck Institute EVA, Leipzig),[16] she has observed missing 'activation links' between the e-learning tools developed through years of effort by scholars and the local communities likely to use them (i.e. dozens of hours of video interviews subtitled in Even and Russian; the electronic trans-dialect dictionary of Even reindeer herding with several hundred entries, initiated and realised with K. Gernet, B. Pakendorf, D. Matic, N. Aralova and A. Lavrillier or the publication in Even and Russian of the ancient epic poetry of D. Osenina).[17] While the teams have shown these tools to the communities and distributed them, it seems that they are still not used in pedagogy. However, it is the decision of the Indigenous community whether they wish to use these. The team also observed the closure of one of the most generous funding sources for language projects, DOBES (Volkswagen Foundation), leaving only a few possibilities, like the Hans Rausing Endangered Language Project and Earth Action for projects in Siberia. E-documentation products for Siberian or Indigenous languages are numerous, like rich open-source publications in Koriak, Even and Itelmen (including language manuals) by the Kulturstiftung Siberien of the German anthropologist Erich Kasten;[18] the many publications in Indigenous languages from the same publisher (Odé 2016 on Yukaghir; Bulgakova 2016 on Nanai; Kasten et al. 2017) and the e-documentation of oral history along the Arctic shores by the Orhelia project (Dudeck et al. 2015; Kasten et al. 2017, 34 et passim, Stammler et al. 2017) or the DOBES database (see previously).

From Indigenous initiatives, several video games have been produced in Sámi (Outakoski 2013; Laiti et al. 2020, among others). There is also an impressive website, Giellatekno, from the combined efforts of Indigenous communities and native/non-native linguists, with different e-dictionaries of several Sámi and other Arctic languages, e-learning tools/games for the Sámi language like Oahpa and so on.[19] There are also successful experiences of collaboration between Skolt Sámi language activists and linguists in terms of the use of internet and digital devices for Kildin Sámi (Marten et al. 2015, 225; Riessler 2013). Some publications are also very good tools for linguists, anthropologists and native language activists with international experience (Grenoble and Whaley 2006; Grenoble 2018, among others), such as online dictionaries, the talking dictionaries developed by K. David Harrison (including one in the Tuva language),[20] an Evenki dictionary by O. Morozova from Amur State University or the 14 online dictionaries and smartphone applications for learning Kalaallisut.[21]

According to the third author, there are several reasons the "activation link" between non-Evenki scholars and Evenki scholars/activists in Eastern Siberia is under-utilised and native communities struggle to realise their own projects. Scholars may not consider it their task to "push" native communities to use the e-tools or e-documentation they have developed. Given that they have mostly been defined and developed by non-native scholars, these tools are perhaps not the most interesting to native communities. Equally, the scholars may not be interested in developing the tools desired by native communities. For instance, what can a schoolteacher do with a very technical reindeer-herding e-dictionary containing hundreds of words describing reindeer colour and organs? Furthermore, it may be the case that natives do want to develop community-driven and community-designed e-tools in order to decolonise scholarship and because the process of creating, designing, funding, realising and distributing an e-learning tool is pedagogically instructive and inspiring for the native youth. One can take the example of the Uummannaq Children's Home in Greenland, where native children themselves defined and realised projects in Kalaallisut (artistic and documentary films, songs, cartoons illustrating Inuit legends, etc.) (Huctin 2016, 558 et passim).[22] Thus, the third author's fieldwork shows that the e-tools developed by the Evenki themselves (like Evengus, the songs produced by Tunguz Record and social network discussions; see previously) are more frequently used and seem to be more effective for language revitalisation than any talking dictionary, subtitled interview or textbook. However, as noted previously, other examples in the Arctic show that a well-balanced combination of efforts between native/non-native scholars and communities can give rise to very powerful results. Regarding the problems of fundraising (a problem scientists also meet), Siberian natives are confronted by additional difficulties: there is insufficient funding for applied linguistic projects led by non-scholars or small local research institutions; they have little access to Western literature or workshops because of the exorbitant prices of many publications or trips and access to Western funding is almost impossible due to the 2012 "foreign agent" law, which forces Russian NGOs to go through a long and strict supervisory process before being allowed to receive foreign funding.[23] Additionally, the sanctions against Russia from 2014 have affected the ability to respond to international calls for proposals.

Thus, several elements could enhance the development and efficiency of e-learning tools in Eastern Siberia. Communication between non-native scholars, native academics and local communities could be developed toward more community-designed projects. Academic and funding institutions could better consider the status of concrete/applied projects for and by the community. Siberian communities could also benefit from a pan-Arctic fund especially created for such projects, allowing them to exchange information during special practical workshops with native Arctic scholars, youth and non-native scientists from other countries.

Notes

1 The expression 'compact (Indigenous) inhabitation' (ru. *kompaktnoe prozhivanie*) reflects precisely the contemporary situation of Evenki culture. In the past, the Evenki were masters of a huge taiga spreading from the Ural Mountains to the Okhotsk Sea, while today they live in small, isolated areas where

the traditional way of life can be maintained only because industrial development has not yet entirely destroyed the flora and fauna.

2 According to the 2010 Russian census: http://raipon.info/peoples/data-census-2010/data-census-2010.php (RAIPON 2010)

3 This 'yakutisation' is the result of the expansion of the Sakha (Yakut), a Siberian people speaking a Turkic language, who supposedly arrived in Siberia from Central Asia in the 13th century. Installed in Central Yakutia, Sakha cattle breeders started moving to the south across the Olëkma and Aldan Rivers from the 18th century onwards in order to expand their hunting territories, a step made necessary because of decline in fur game in their region. Their occupation of Evenki areas became more intensive with each passing century. During this process, local interethnic conflict (still present in the local oral history) and mixed marriages led to many Evenki communities adopting the Sakha language and some aspects of their culture (Nikolaev 1964, 12–20, 36–46; Lavrillier 2005a).

4 For instance, in the south Sakha Republic villages of Tyanya, Tokko, Kindigir, Khatystyyr and Khatymi, part of the hunting lexicon is still Evenki, while the spoken language is entirely Sakha. In contrast, in Iengra, the Evenki language is still practised, although the dialect contains many Sakha words, which the speakers consider an integral part of their Evenki dialect (Lavrillier 2005b).

5 In the Sakha Republic, there are four minority Indigenous peoples: the Evenki and the Even (from the Tungus-Manchu language group), the Chukchi (from the Chukotko-Kamchatkan group) and the Yukaghir (isolated language). On the complex Yupik language situation in Chukotka, see Schwalbe (2015).

6 Tungus is the previous term given by Russians for the Evenki and Even people. For further details, see Lavrillier et al. (2018).

7 *Buga* simultaneously designates the biophysical natural environment, the spirits inhabiting it and a global spiritual entity governing the natural environment (see also Varlamov 2013).

8 https://oqaasileriffik.gl/ and https://oqaasileriffik.gl/language/ (information from J.M. Huctin).

9 For the history of this collective ritual in the past and its revitalisation in the 1990s, see Lavrillier (2003, 2005a, 2005b).

10 www.youtube.com/watch?v=dGF4m6z7eLE&pbjreload=106; www.youtube.com/watch?v=fy7DK5P8qG8

11 www.evengus.ru/razgovornik/index.html

12 www.evengus.ru/prilozheniya/windows-samouchitel

13 www.vk.com/evenkil

14 https://yadi.sk/d/o6zCJopxjYiL8

15 These recommendations can be compared with Hirvonen's book gathering the thoughts of Sámi teachers about effective ways of teaching Sámi at school (Hirvonen 2004).

16 Project "Documentation of Dialectal and Cultural Diversity among Evens in Siberia": https://dobes.mpi.nl/projects/even/

17 See Lavrillier and Matic (2013).

18 www.kulturstiftung-sibirien.de/materialien_E.html

19 http://giellatekno.uit.no/Giellatekno.eng.html

20 http://talkingdictionary.swarthmore.edu/

21 Information from J.-M. Huctin: https://play.google.com/store/apps/details?id=emp.hbu.daka&hl=da, www.ilinniusiorfik.gl/oqaatsit/daka; https://learngreenlandic.com/; www.oqaatsit.gl/

22 https://bhjumq.com/UummannaqChildrensHome.htm

23 The "Russian Federation Law on Foreign Agents" (*Federal'nyi zakon 121 "Ob inostrannykh agentakh"* in July 2012). It complicated many foreign NGO activities in Russia and in some cases ended them altogether (Lavrillier field note).

References

Appadurai, A. 1996. *Modernity at Large: Cultural Dimensions of Globalization*. Minneapolis: University of Minnesota Press.

Boas, F. 1911. "Introduction." In *Handbook of American Indian Languages*, edited by F. Boas, vol. I. Bureau of American Ethnology Bulletin 40, 1–76. Washington, DC: Government Printing Office.

Bulatova, N., and L. Grenoble. 1999. *Evenki*. München: Lincom Europa (Languages of the world materials 140).

Bulgakova, T. 2016. *Kamlaniia nanaiskikh shamanov [Shamanic Rituals of the Nanai]*. Fürstenberg: Kulturstiftung Sibirien, Languages and Cultures of the Russian Far East.

Dudeck, S., F. Stammler, L. Allemann, N. Mazullo, N. Meshtyb, and R. Laptander. 2015. *Nomadic Memories: People and Oral History in the 20th Century Along the Shores of the Arctic Ocean*. Audio-visual publication on DVD and USB flash drive. Rovaniemi: Anthropology Research Team, Arctic Centre, University of Lapland.

Dumont, A. 2018. "Are the Evenki Reindeer Herders Still Nomads? The Alternate Use of Different Types of Spaces in Inner Mongolia, China." *Études mongoles & sibériennes, centrasiatiques & tibétaines* 49. Accessed 20 December 2018. http://journals.openedition.org/emscat/3398.

Elwert, G. 1997. "Boundaries, Cohesion and Switching. On We-Groups in Ethnic, National and Religious Forms." In *Rethinking Nationalism and Ethnicity: The Struggle for Meaning and the Order in Europe*, edited by H. R. Wicker, 251–271. Oxford and New York: Berg.

Filippova, E., and S. Sokolovskii. 2019. *Smert' yazyka – smert' naroda? Yazykovye situatsii, yazykovye prava v Rossii I sopredel'nykh gosudarstvakh [Does the Death of the Language Mean the Death of the People? Language Situations and Language Rights in Russia and Neighboring States]*. Moskva: Rossiiskaia Akademiia nauk Institut etnologii I antropologii.

Glukhova, V. A. 2011. "Sovremennyy repertuar izdaniy literatury narodov Severa: razrusheniye traditsiy" [The Modern Repertoire of Publications of the Literature of the Peoples of the North: The Destruction of Traditions]. *Vestnik Moskovskogo Gosudarstvennogo Universiteta Pechati* 6: 320–340.

Grenoble, L. A. 2003. *Language Policy in the Former Soviet Union*. Dordrecht: Kluwer Academic Press.

Grenoble, L. A. 2016. "Kalaallisut, the Language of Greenland." In *Minority Languages: Threatened to Compete*, edited by R. Higgins and P. Rewi, 273–289. Honolulu: University of Hawaii Press.

Grenoble, L. A. 2018. "Arctic Indigenous Languages: Vitality and Revitalization." In *The Routledge Handbook of Language Revitalization*, edited by L. Hinton, L. Huss, and G. Roche, 345–354. New York: Routledge.

Grenoble, L. A., and L. J. Whaley. 1999. "Language Policy and the Loss of Tungusic Languages." *Language and Communication* 19 (4): 373–386.

Grenoble, L. A., and L. J. Whaley. 2006. *Saving Languages. An Introduction to Language Revitalization*. Cambridge: Cambridge University Press.

Hannerz, U. 1992. *Cultural Complexity*. New York: Columbia University Press.

Hinton, L. 2002. *How to Keep Your Language Alive: A Commonsense Approach to One-on-One Language Learning*. Berkeley: Heyday Books.

Hirvonen, V. 2004. *Sámi Culture and the School. Reflections by Sámi Teachers and the Realization of the Sámi School. An Evaluation Study of Reform 97*. Karasjok: CalliiidLagadus.

Huctin, J.-M. 2016. "Maltraitance et bientraitance des jeunes au Groenland: de l'éducation traditionnelle inuit (XVIIe-XXe siècles) à l'actuelle maison d'enfants d'Uummannaq." PhD thesis, University of Paris Diderot Paris 7.

Jones, M. C., and S. Ogilvie. 2013. *Keeping Languages Alive: Documentation, Pedagogy and Revitalization*. Cambridge: Cambridge University Press.

Marten, H. F., M. Riessler, J. Saarikivi, and R. Toivanen. 2015. *Cultural and Linguistic Minorities in the Russian Federation and the European Union Comparative Studies on Equality and Diversity*. Cham: Springer.

Kasten, E., K. Roller, and J. Wilbur. 2017. *Oral History Meets Linguistics*. Fürstenberg and Havel: Kulturstiftung Sibirien.

Kazakevich, O. A., and S. F. Chlenova. 2014. "Polveka laboratorii avtomatizirovannykh leksikograficheskikh sistem NIVTS MGU im. M. V. Lomonosova." [Half a Century of the Laboratory of Automated Lexicographic Systems NIVC at M. V. Lomonosov MSU]. *Vestnik RGGU. Seriya YAzykoznaniye/Moskovskiy lingvisticheskiy zhurnal*, Tom 16, 8 (130): 28–39.

Keptuke, G. I. 1991. *Malen'kaya Amerika: povest' i rasskazy [Little America: A Novel and Stories]*. Moskva, Sovremennik.

King, A. 2011. *Living with Koryak Traditions: Playing with Culture in Siberia*. Lincoln: University of Nebraska Press.

Laiti, O., S. Harrer, S. Uusiautti, and A. Kultima. 2020. "Sustaining Intangible Heritage through Video Game Storytelling – The Case of the Sámi Game Jam." *International Journal of Heritage Studies*. doi:10.1080/13527258.2020.1747103.

Lavrillier, A. 2003. "De l'oubli à la reconstruction d'un rituel collectif. L'*Ikènipkè* des Évenks." *Slovo, Sibérie. Paroles et mémoires* 28–29: 169–191.

Lavrillier, A. 2005a. "Nomadisme et adaptations sédentaires chez les Évenk de Sibérie postsoviétique: 'Jouer' pour vivre avec et sans chamanes." PhD diss., Paris: École Pratique des Hautes Etudes.

Lavrillier, A. 2005b. "Dialectes et norme écrite en évenk contemporain [Dialects and Written Norm in Contemporary Évenk Language]." In *Les langues ouraliennes aujourd'hui. Approches linguistiques et cognitives*, edited by M. M. J. Fernandez-Vest, 433–446. Paris: Bibliothèque de l'École des Hautes Étude.

Lavrillier, A. 2013. "A Nomadic School in Siberia Among Evenk Reindeer Herders." In *Keeping Languages Alive: Documentation, Pedagogy and Revitalization*, edited by M. C. Jones and S. Ogilvie, 140–154. Cambridge: Cambridge University Press.

Lavrillier, A., and M. Dejan. 2013. *Even tales of Dar'iia Mikhailovna Osenina – Dar'ja Mikhajlovna Osenina nimkarni – Evenskie nimkany Dar'i Mikhajlovna Osenina*. Fürstenberg, Germany: Kulturstiftung Sibirien.

Lavrillier, A., A. Dumont, and D. Brandišauskas. 2018. "Human-Nature Relationships in the Tungus Societies of Siberia and Northeast China." *Études mongoles et sibériennes, centrasiatiques et tibétaines*. Accessed 13 May 2020. http://journals.openedition.org/emscat/3088.

Maak, R. K. 1886. *Viljujskij okrug [Vilyui Okrug]*. Tipogafiya i khromolotografiya A. Transhelia, Stremiannnaya N°12, St-Petersburg.

Maak, R. K. 1887. *Viljujskij okrug Jakutskoj oblasti [Vilyui Okrug of Yakut Oblast']*. I, II and III. Tipogafiya i khromolotografiya A. Transhelia, Stremiannnaya N°12, St-Petersburg.

Nikolaev, S. I. 1964. *Eveny i evenki yugo-vostochnoi Yakutii [The Even and Evenki of Southeastern Yakutia]*. Yakutsk: Yakutskoe knizhnoe izdatelstvo, 201.

Nuttall, M. 2009. "Living in a World of Movement: Human Resilience to Environment Instability in Greenland." In *Anthropology and Climate Change: From Encounters to Actions*, edited by S. A. Crate and M. Nuttall, 292–310. Walnut Creek: Left Coast Press.

Odé, C. 2016. *Il'ia Kurilov: My Life, Songs*. Fürstenberg: Kulturstiftung Sibirien.

Outakoski, H. 2013. "Teaching an Endangered Language in Virtual Reality." In *Keeping Languages Alive: Documentation, Pedagogy and Revitalization*, edited by M. C. Jones and S. Ogilvie, 128–139. Cambridge: Cambridge University Press.

Petrov, A. A. 2013. "Etnolingvistika v pedagogicheskom obrazovanii: istoriya i kul'tura (na materiale tunguso-man'chzhurskikh yazykov) [Ethnolinguistics in Pedagogical Education: History and Culture (Based on the Material of the Tungus-Manchu Languages)]." *Istoriya pedagogiki estestvoznania [History and Pedagogy of Natural Sciences]* 3: 55–59.

Riessler, M. 2013. "Towards a Digital Infrastructure for Kildin Saami." In *Sustaining Indigenous Knowledge: Learning Tools and Community Initiatives on Preserving Endangered Languages and Local Cultural Heritage*, edited by E. Kasten and T. de Graaf, 195–218. Fürstenberg: SEC Publications, Verlag der Kulturstiftung Sibirien.

Russian Association of Indigenous Peoples of the North (RAIPON). Russian Census. 2010. Accessed 18 May 2020. http://raipon.info/peoples/data-census-2010/data-census-2010.php.

Sapir, E. 1933. *Language: An Introduction to the Study of Speech*. New York: Harcourt Brace Jovanovich.

Schwalbe (Morgounova), D. 2015. "Language Ideologies at Work. Economies of Yupik Language, Maintenance and Loss." *Sibirica* 14 (3): 1–27.

Sharina, S. I. 2015. "Evenskiy yazyk: problemy sokhraneniya [The Even Language: Conservation Issues]." *Arktika. XXI vek. Gumanitarnye nauki [Arctic. 21th century. Human Sciences]* 3 (6): 93–100.

Stammler, F., and A. Ivanova. 2016. "Resources, Rights and Communities: Extractive Mega-Projects and Local People in the Russian Arctic." *Europe-Asia studies* 68 (7): 1220–1244.

Stammler, F., A. Ivanova, and L. Sidorova. 2017. "The Ethnography of Memory in East Siberia: Do Life Histories from the Arctic Coast Matter?" *Arctic Anthropology* 54 (2): 1–23.

Terekhina, A. N. 2019. "Kochevoye obrazovaniye: chto vy khoteli znat', no boyalis' sprosit [Nomadic Education: What You Want to Know, but Were Afraid to Ask]." *Arktik Go*, 17 September. Accessed 4 April 2020. https://goarctic.ru/live/kochevoe-obrazovanie-chto-vy-khoteli-znat-no-boyalis-sprosit/.

Varlamov, A. N. 2009. *Istoricheskie obrazy v evenkiiskom fol'klore [Historical Images in Evenki Oral Literature]*. Novossibirsk: Nauka.

Varlamov, A. N. 2013. *Buga ganalchil. Stikhi. Perevod na evenkiiskii Galiny Kepuke [What Must Be Taken from Buga. Poems. Translation in Evenki Language of Galiny Kepuke]*. Yakutsk: Ofset.

Varlamova, G. I. 2002. *Epicheskie i obryadovye zhanry evenkiiskogo fol'klora [The Epic and Ritual Genres of Evenki Oral Literature]*. Novossibirsk: Nauka.

Varlamova, G. I. 2004. *Mirovzzrenie evenkov: otrazhenie v fol'klore [The Evenki Worldview: Its Reflection in Oral Literature]*. Novossibirsk: Nauka.

Varlamova, G. I. 2008. *Zhenskaya ispolnitelskaya tradiciya evenkov [The Traditional Female Performance of Oral Literature]*. Novossibirsk: Nauka.

Varlamova, G. I. 2009. "Ity i odyo [Ity and Odyo]." In *Slovar' Dzheltulakskogo govora evenkov Amurskoi oblasti*, 553–562. Blagoveshchshensk: OAO PKI "Zeya".

Varlamova-Keptuke, G. I. 1991. *Dvunogii da poperechnoglaznyi, chernogolovyi chelovek i ego zemlya Dulin Buga [The Two-Legged, Cross-Eyed, Black-Headed Man and His Land, the Middle Universe]*. Yakutsk: Yakutskoe knizhnoe izdatelstvo.

Zamyatin, K., S. Pasanen, and A. Janne. 2012. *Kak i začem soxranjat jazyki narodov Rossii? [How Are the Languages of the Peoples of Russia to Be Maintained and Why?]*. Helsinki: Vammalan kirjapaino.

5

VOICES OF THE FORESTS, VOICES OF THE STREETS

Popular music and modernist transformation in Sakha (Yakutia), Northeast Siberia

Eleanor Peers, Aimar Ventsel, and Lena Sidorova

Abstract

Music, and especially song, have been the means by which Sakha communities in north-eastern Siberia have interacted with their environment over the centuries. And this environment has incorporated an enormous pantheon of deities, area spirits, ancestors, ghosts, and demons, particularly in the years before Soviet-era modernisation began in earnest. These entities and their relationships with Sakha communities were and are voiced through sung Olongkho epics, algys prayers, chabyrghakh chants, and Ohuokhai choral dances. Sakha men and women praised or petitioned deities and spirits through these musical genres. However, modernisation and urbanisation have radically changed Sakha peoples' relationships with their environment, in transformations replicated throughout the Circumpolar North. During the mid-twentieth century, Sakha people moved first into Russian-style villages and then urban settlements – and in particular to Yakutsk, their Republic's capital. Modernised farming and industry have taken root in the Republic of Sakha (Yakutia), bringing their associated environmental challenges. And with modernisation and urbanisation have come a plethora of new musical genres, often emerging out of the adaptation of mainstream Russian or global musical forms. Soviet-era Estrada music has given way to Sakha-language rap, pop, and rock. In this chapter, we chart the Sakha people's changing interrelation with their environment, through a history of twentieth- and twenty-first-century popular music. In doing so, we show how Sakha people have incorporated music into the articulation of new identities and relationships, in addition to ways of combating the negative impact of modernising change.

Introduction

The inhabitants of Yakutsk, the capital of the Republic of Sakha (Yakutia) (RSY hereafter[1]), are not short of things to do in the evening, even in the depths of a north Siberian winter. The temperature outside may be below minus 40 degrees Celsius, but Yakutsk's concert venues and nightclubs still seethe with activity: there is plenty of music to choose from, ranging from folk to rock, rap and dance, mainly performed by artists from RSY's largest Indigenous community, the Sakha people. This chapter outlines the history of this Sakha popular music. As it explains, Sakha popular music has emerged over the twentieth and twenty-first centuries in tandem with radical transformations in Sakha ways of life, brought on by Soviet-era modernisation and the subsequent urbanisation of the Sakha population. These transformations have generated the urbanised, educated Sakha population we see today – but they have also fractured older interrelations between the Sakha people and their natural environment. The Sakha popular music tradition reveals the paradoxical vulnerability and persistence of these interrelations, while it enables them to endure. RSY's popular music industry appears to be a typical example of global commercialised music – and yet it has kept alive the Sakha people's intimate relationship with their land.[2]

The history of the nearly five hundred thousand-strong Sakha population, most of whom live in the Republic of Sakha Yakutia, is fairly representative of that of Siberia's Indigenous communities and indeed of northern Indigenous communities in general. Russian Cossacks first encountered Sakha settlements during the 1620s (Gogolev 2000, 30). Sakha territories were quickly subsumed into the Russian Empire, although sustained attempts to convert Sakha populations to Russian religion and lifestyle only began in the late eighteenth century (Dmitrieva 2013, 35; Forsyth 1992, 69). The lives of Sakha people changed dramatically over the nineteenth century, as the infrastructure of the Tsarist state strengthened: many were converted to Russian Orthodox Christianity and a largely settled form of livestock herding (Middendorf 1878). These changes were highly uneven, given the territory's vast size (over three million square kilometres) and small population.[3]

And yet at the beginning of the twentieth century, it was possible to distinguish even the most russianised Sakha individuals from incomers, through characteristic assumptions and activities. These assumptions and activities emerged out of an experience of environmental setting that differed fundamentally from that of the educated European Russians who ran Tsarist institutions. The latter's worldview is more recognisable to those of us who have attended the schools and universities of the late twentieth and twenty-first centuries: like ours, it has roots in the European Enlightenment's separation of man from the natural world, mental from material, and cultural from natural (Blaser 2013). As the ethnographic work testifies, these boundaries were more fluid for Sakha communities (Popov 1910; Sieroszewski 1993). The Sakha people's experience of life seems to have been typical of the northern phenomenologies set out by Tim Ingold (2000). A wide variety of persons – both human and non-human – existed in and through their interrelation with each other. It was impossible to demarcate self-sufficient human persons clearly from their environment. Instead, life was characterised by the perpetual re-alignment of beings in a constantly moving 'meshwork' (Ingold 2000). Hence, both humans and their livestock could be stricken by destructive forces called *abaahi*, which would bring about their illness and death – while they could also call on a huge pantheon of deities and the benevolent beings known as *aiyy*. The landscape was alive with beings dwelling in trees, mountains, animals, or streams, who were entirely capable of both rejoicing in and taking offence at human behaviour. The Sakha tradition has been classified in the European and Russian ethnographic literature as 'shamanic'; Sakha ethnographers and practitioners have also adopted the term 'shaman', even though the Sakha word often used as an equivalent (*oyuun*) is not an exact translation (Sieroszewski 1993; Kondakov 2012).

The genres of Sakha music that were widespread at the start of the twentieth century reflect this interrelation of persons, and in fact much of this music was concerned with managing this interrelation. The ritual specialists (*oyuun*s or *udaghan*s) at the forefront of this mediation relied on a variety of vocal techniques both to address *abaahy*, *aiyy*, and the deities and to make these beings present to Sakha people – such as the *kylyhaakh*, a one-note trill made from flexing the throat muscles. The Sakha epic the Olongkho incorporates the widest variety of these vocal forms: it was generally performed by a single bard, who would become in turn *abaahy, aiyy*, and the Sakha epic heroes through his or her use of techniques such as *d'ieretii yrya*, which combined elongated tones with decorative *kylyhaakh* trills (Sheikin 2002, 456). Drums were important instruments, as were jaw harps called *khomus*, and string instruments, the *kyrympa*. Music was experienced as a force with the power to produce concrete change, precisely because it mediated semi-hidden forces in the landscape. *Oyuun*s could heal through sung poem-prayers called *algys*; less careful singers could cause harm (Sieroszewski 1993, 574). Music could also be entertaining – such as the *chabyrdakh* chant, a stream of fast, rhythmic speech, which could be humorous (Spiridonova 2019; Sheikin 1996) – and it could incorporate dance, like the Ohou-hai circle dance. Russian music was also exerting an influence: Sakha-language melodic songs had started to appear (Ergis and Emel'yanov 1976, 5; Zhirkov 1981).

As the following section explains, the Soviet administration attempted radically to change Sakha ways of life and cultural forms in its effort to create a modernised, socialist Sakha society. Within this modernisation, the Soviet administration fostered a genre of entertainment known in Russian as *Estrada*. The second and third sections show how professional Soviet *Estrada* and its concomitant amateur music-making generated the range of popular music styles currently on offer in Yakutsk – as Sakha people grappled with the shifts in their political setting and the dislocation from the natural world these shifts caused. Two of this chapter's authors – Aimar Ventsel and Eleanor Peers – have been observing these changes since 2000 and 2004, respectively, as frequent visitors to RSY. Over the years, they have attended as many concerts as they could, in addition to speaking to both singers and producers; for a short time in the early 2000s, Ventsel also worked as a DJ with various event promotors. Lena Sidorova has witnessed the emergence of Sakha popular music at first hand and indeed helped to document it during the early 2000s through her involvement with *Trilisnik*, a supplement of the literary magazine *Ilin*. She was born in the Churapcha region and has spent most of her life in Yakutsk.[4]

Sovietisation and the emergence of *Estrada*

The Bolshevik party took some time to establish its hold on what was known from 1922 as the Yakut Autonomous Soviet Socialist Republic (YaASSR, or Yakutia), as was the case across Siberia. In assuming the governance of Sakha territories – and indeed the entire Tsarist Empire – the Bolshevik party had set itself an enormous task. It had to convert a highly varied population, occupying a vast and in parts inaccessible territory, to the project of building a communist society. As this section describes, the Bolshevik endeavour included the renovation of popular culture. For popular music, this renovation consisted of the generation of *Estrada* performance as the appropriate form of entertainment for the entire Soviet population (Rothstein 1980).

The aspiration towards communism, along with the entire body of theory from which it emerges, is based on the European intellectual assumptions about the separation of person and environment mentioned previously – assumptions which large numbers of Sakha people did not entirely share, in common with the rest of the Soviet population. This difference in mindset was understood by Bolshevik activists as exhibiting the Sakha people's 'primitive' level of development (Hirsch 2007). Their task, as they saw it, was to re-form Sakha communities into developed, modernist societies, capable of understanding the importance of a socialist economy and the broader journey towards communism. Soviet policy was to generate new, Soviet Sakha men and women to join all the other national incarnations of Soviet Man across the Soviet Union, whether Asian, Caucasian, or European (Hirsch 1997; Johansen 1996; Roy 2000; Slezkine 1991).

In common with the entire Soviet population, Sakha communities soon became the targets of an avalanche of initiatives and resources aiming to transform their lives and beliefs. From the late 1920s, Sakha families were moved from small, scattered farmsteads in distant forest clearings (*alaas*) into villages serving collective farms (Argounov 1985; Safronov 2000, 247–248). A concerted effort meanwhile was made to provide schools to educate both children and adults into becoming committed members of the Soviet citizenry and a network of hospitals to maintain their health. This emphasis on biomedical healing was one strand of the Soviet administration's effort to convert the entire population to materialist atheism. In itself, it demonstrated the assertions made by the various atheist campaigns of the Soviet period: all religions are lies; the gods and spirits they claim to address do not exist; life's problems can be solved through human technologies, like the biomedicine that cures disease.

The assertion of power that underpinned this particular manifestation of Soviet ideology lay, whether consciously or unconsciously, in the unreflective imposition of standardised models of 'religion', 'spirit', 'god', and 'person' on all the practices and people that did not refer to a materialist universe. Christian saints, Buddhist Bodhisattvas, and shamanic deities were apparently equally preposterous, despite considerable differences in their relationships with humans. As far as Soviet ideology was concerned, the universe consisted of human beings existing separately from a material world, itself seen either as a backdrop for human activity or a resource (Ingold 2000). Human beings were complex organisms incorporating a material body and a spirit (*dukh*, Russian) and rational mind (*um*, Russian). This perception of the universe underpinned all the institutions the Soviet Union built – whether educational, economic, or cultural (e.g., see Hoffman 2003). These institutions worked to absorb the Sakha population into this materialist experience of life in the world and in so doing remove the environment they had inhabited hitherto, along with its gods, saints, and spirits (Popov 2009, 54). Life was to proceed without dealings with these entities, and hence the Sakha musical genres that mediated these interactions – such as *algys* prayers or Olongkho epic recital – had either to be reconstituted or dispatched.

As this suggests, cultural expression was understood to be a vital aspect of the transformation the Soviet administration hoped to achieve: cultural workers were regarded as 'fighters on the invisible front' (Alekseyev 2007; Ammosov 2007). Specific forms of theatre, art, music, and literature were both pedagogical tools and the correct attributes of the new Soviet person. During the 1930s high culture – *kul'tura*, in Russian – was promoted as the right aspiration of an appropriately educated, developed person (Volkov 2000). The adjective *kul'turniy* was a positive descriptor for people, while emerging Soviet towns struggled to build the theatres, concert halls, and opera houses that would establish their status as civilised settlements. Yakutsk, like other older towns, already had a theatre from the Tsarist era – reflecting the continuities between Tsarist and Soviet *kul'tura* that became increasingly evident after the temporary wave of avant-garde experimentation just after the Civil War (Slezkine 1994). There were also strong continuities between Tsarist and Soviet popular entertainment – even if policymakers understood their task to be the generation of wholesome, edifying entertainment for the new Soviet population.

The genre of light entertainment called *Estrada* in Russian was no exception. The term comes from the French vaudeville tradition; it included a similarly wide range of forms of entertainment (MacFadyen 2002, 3). Soviet *Estrada* performances included circus acrobats, clowns and magicians, ballet, Flamenco dancing, operatic arias, and stand-up comedy (Ventsel 2016, 73–74).[5] *Estrada* music is correspondingly difficult to define because of the breadth of this variety; in addition, the music has changed greatly over the decades. One common denominator is that *Estrada* singers tended to perform with large orchestras and a high standard of musicianship: perfection and quality were and are very important for the genre. *Estrada* performers were professionals, who had undergone lengthy training within the Soviet Union's extensive network of circus, theatrical, and musical educational establishments. They conducted their professional lives within the framework of performance events and spaces provided by the state – which included a system of determining performers' salaries, according to their maturity and conformity to the accepted professional standard (Dmitriev et al. 2004). This process was called *tarifikatsiya*, in Russian, and was introduced in 1927 (Dmitriev et al. 2004, 685). Performers would agree on their seasonal repertoire in advance with the committee in charge of *Estrada* in each region.

The establishment of *Estrada* as a state institution therefore generated a prominent group of professionals, providing both high-quality recreational entertainment and a good example for those interested in amateur performance (*samodeyatel'nost'*, in Russian) (Ergis 1974, 390). Amateur performance was highly encouraged and supported by the Houses of Culture that appeared in every settlement across the Soviet Union (Donahoe and Habeck 2011; Habeck, 2011). In the materialist Soviet universe, human achievement and well-being were the central aspirations and hence foci of value. Good Soviet citizens strove to better themselves and their communities through improving their education and their achievements in sports or cultural performance.

Although the Soviet administration of Yakutia promoted their Republic's developing network of theatres, training colleges, and houses of culture as hard as they could (Androsov 2019), they tended at first to rely on *Estrada* performers brought in from other parts of the Soviet Union. The technical possibilities open to performers, whether amateur or professional, were limited both by the size and remoteness of Yakutia's territory and the considerable differences between earlier Sakha and Russian music. Sakha performers had not necessarily been introduced to European scales, musical notation, or melodies: older genres of Sakha music tend to emphasise differences in rhythm, pitch, and vocal technique closely incorporated into a song's words rather than melody and tone. Towards the end of the 1930s, a Bureau for *Estrada* Concerts (*kontsertno-estradnoe biuro*, in Russian) was established, which would employ performers from outside Yakutia to tour the region for a couple of seasons at a time (Fillipov 2014). These performers co-existed with Cultural Enlightenment workers (*kul'tprosvetrabotniki*, Russian) trained at a school set up in Yakutsk in 1937; this school was eventually to become RSY's current College of Culture.[6] The first duty of Cultural Enlightenment workers was to teach people Russian and to read and write – however, they also taught politics and culture, encouraging amateur performances in the cultural genres that were deemed appropriate (Basharin et al. 1968, 8).

As Sakha people were moved into villages and collective farms, therefore, they were offered new forms of recreational performance that carried a heavy emphasis on European-style melodic song. The expanding network of government institutions reaching out to absorb them into the state's workforce, education, political project, and worldview included performance spaces in the form of Houses of Culture and a body of cultural workers ready to disseminate the skills and activities that should be pursued within them (Parnikov 2015, 27; Spiridonova 2018). These cultural activities were predicated on European perceptions of person and environment, in common with all the other Soviet institutions. Music in these genres was therefore not experienced as an enhanced communication with unseen beings in a live environment but an exclusively human artistic achievement. Music and song were seen to express a person's deeper feelings and to engineer or manifest a particular bond between the people participating in a given performance; this bond did not extend, however, to deities or area spirits. The 1930s, 40s, and 50s saw a rapid expansion of melodic singing and European folk dancing throughout Sakha settlements. It seems that Sakha people were adopting the forms of artistic expression and recreation that suited their new village settings and the corresponding transformation of their interaction with the natural world. This transformation was conditioned by a number of factors – importantly, the concentration of Sakha populations in settlements, the introduction of mass education, and the powerful and determined atheist propaganda.

The widespread opinion among both artists and public in RSY is that the first Sakha *Estrada* artist was Khristofor Maksimov, a singer, culture-house worker, and actor who started to perform in 1937. Maksimov's music continues to be popular: the 100th anniversary of his birthday was celebrated with large concerts in 2017. His style presaged the mainstream Sakha *Estrada* music of later decades and indeed subsequent genres of Sakha popular music (Sidorov 2011, 72). He accompanied himself on an accordion, singing songs composed in Sakha that broadly conformed to mainstream Russian popular music – even if his melodies and vocal technique displayed an influence from older Sakha genres of song (Tumanov 2001, 3). This influence corroborates Mark Zhirkov's claim that the first *Estrada* composers developed popular Sakha songs using the melodies he transcribed during expeditions to Sakha villages in the 1940s and 50s (Zhirkov 1981). The themes of Maksimov's songs also echoed recognisably Sakha preoccupations. In particular, he often sang of the home *alaas* – the sub-Arctic forest clearing from which a Sakha family had been moved into a new Soviet village. Many Sakha people retained an attachment to the *alaas* from which they came – an emotion that reflected their continuing interrelationship with a live natural environment (Bravina 2005, 24). Some continued surreptitiously to offer libations to area spirits, for example, even if it was no longer possible or desirable to conduct lengthy *algys* prayers or healing sessions (Crate et al. 2017; Mészáros 2012).

As this suggests, the Sakha adoption of the Soviet mindset was uneven and gradual. There were plenty of Sakha people who were happy to pursue the Soviet project, especially if they were young enough to have been educated in Soviet schools (Yakovlev-Dalan 2003). If older Sakha genres like the Olongkho fell out of use as the twentieth century progressed, it was as much because Sakha people themselves regarded them as obsolete as it was the consequence of pressure from Soviet authorities (Illarionov 2006). Post-war Sakha generations were as eager to reach forward towards the communist future and the opportunities for education and recreation it offered as anyone else (Chakars 2014). And yet it was difficult to deny or ignore the habits and relationships with the land that had been integral to Sakha life for centuries. This tension was manifested in the hybrid versions of mainstream Soviet cultural forms produced by performers such as Maksimov (Sidorov 2011). Performances at the new Houses of Culture springing up in Sakha villages showed a similar inventiveness in adapting new performative conventions to older Sakha habits. For example, the Olongkho was performed onstage by several actors as a kind of play (Illarionov 2006; Alekseyev 1994).

Sakha people were adapting to their new political setting, as they had done throughout the Tsarist era (Forsyth 1992). As part of this, their music changed to suit their new circumstances – but not entirely. It reflected a continuing ambivalence about the atheist, materialist outlook their new state was promoting, in addition to the distinction between humankind and environment this outlook presupposed (Ingold 2000). The uneven progress of sovietisation was complicated further by the Second World War, which brought further upheaval, along with a massive loss of life. The emergence of professional Sakha popular music was stalled, but only temporarily.

The rise of Sakha *Estrada*

Yakutia's administration quickly returned to its active development of Sakha *Estrada* performance after the defeat of Nazi Germany in 1945: a music and art college was established in 1948. As the 1950s progressed, Yakutia's Bureau for *Estrada* concerts started to include Sakha performers in its programmes. These performers had either trained at Yakutsk's musical school or were amateurs; their mastery of *Estrada* performance indicates the extent to which the cultural genres introduced by the Soviet administration were catching on (Fillipov 2014). Ol'ga Ivanova was one of these performers and remained a famous and well-respected singer and composer. She was one of the first prominent Sakha singers to adopt a distinctly Russian folk vocal style. Like Maksimov, her songs suited accompaniment from an accordion. Individual amateur singers could easily learn and perform these songs themselves, and they have become part of the large fund of Sakha melodic songs that continue to be performed by both professionals and amateurs (Sidorov 2011).

1968 is now regarded as the year that saw the birth of professional Sakha *Estrada* – even if the first professional Sakha *Estrada* performers, Yurii Platonov and Natal'ia Trapeznikova, received their government sponsorship in 1967 (Fillipov 2014). Trapeznikova was sent to train at the All-Russian Creative Workshop of the Art of *Estrada* in Moscow (*Vserossiiskaia masterskaia estradnogo iskusstva*, in Russian, or *VTMEI*) (Fillipov 2014). In 1969, the first Sakha *Estrada* singers received their official authorisation (*attestatsia*, in Russian) from the Soviet Union's Ministry of Culture. In 1970, a group of fifteen promising young performers was sent to study at *VTMEI*. They graduated in 1972 and became known as *Yakutskaia Estrada-72* – the first cohort of professional Sakha *Estrada* performers, who along with predecessors like Maksimov, Ivanova, Platonov, and Trapeznikova were to lay the foundations of professional Sakha popular music. Most of these performers were and are singers, such as Vladimir Zabolotskii and Elizaveta Romanova, but the group included dancers, comedians, and circus artists, reflecting the variety of genre encompassed by the *Estrada* performance of the time (Sidorov 2011).

The emergence of Sakha *Estrada* performance during the late 1950s, 60s, and 70s coincided with the retreat of Stalinist repression in 1956 and the advent of Nikita Khruschev's thaw. Most importantly for non-Russian populations like the Sakha, the political regimes under Khruschev and later Brezhnev allowed and could even encourage a greater level of non-Russian national expression – even if this expression still had to conform to the most essential Soviet conventions, such as the presumption of an atheist worldview. From the late 1940s, Sakha singers increasingly began to appear in what had come to be regarded as Sakha national costume – or to incorporate older Sakha vocal techniques into their songs. These songs continued to conform to Europeanised Soviet conventions and expectations; they consisted of several verses interspersed with a repeated chorus, sung to a melodic tune. Sakha singers could introduce a distinctively Sakha sound by elongating a note into a *kylyhakh* trill, for example.

Sakha performances were either in Sakha or Russian; if they were in Russian, they could draw attention to Yakutia and its people in their words. One example is the Russian-language song 'Little Woman from Yakutia' (*Yakutianochka*), by the well-known composer Arkadii Alekseyev – a typical Soviet song about a pretty woman who dances but who in this case is from Yakutia. The Sakha-language original was called 'Sakha Dance' (*Sakha Ünküüte*) and was based on verses by Semen Danilov, a Sakha poet, which evoke the movement and strength of Sakha dances. The Russian-language version contains a number of themes often associated with Yakutia: extreme cold, glistening white snow, reindeer, diamonds, and an Indigenous Siberian people with an ancient connection to the earth. These themes had in fact become Yakutia's brand within the wider Soviet Union. In doing so, they both manifested and reproduced an understanding of Yakutia as one component of a broader Soviet state – a component that had its distinguishing characteristics but was nonetheless a typical representative of the family of modernised, socialist Soviet nation-states. The extent to which *Yakutianochka* and its tropes idealised Sakha communities is demonstrated by its factual inaccuracy: the Sakha people's heritage is primarily connected to horses, rather than reindeer (Sieroszewski 1993). And yet *Yakutianochka* was a popular song in Yakutia. People were happy to accept it as an identification and promotion of their community within the Soviet space – to the extent that in 2016 the Russian rapper L'One, who was brought up in RSY, produced a second version to promote RSY's flagship sports tournament, the Children of Asia. This version and its promotional video reproduce all the stereotypes of the original song – and yet it was circulated widely on Sakha social media after its release and generally as an object of pride.

The popular acceptance of *Yakutianochka* indicates the changes in relationships to state and environmental setting Sakha people were experiencing. The status of Yakutia as a component of the Soviet Union had become important to the new generations of Soviet-educated Sakha people. Many were glad to have their region and community displayed in a positive way to Soviet society in general, even if the representation itself was idealised – if not exoticised – to the point of inaccuracy. Sakha people could also take pride in the fact that their community had produced *Estrada* singers of a sufficient professional standard not only to perform in other parts of the Soviet Union but to win prizes in Soviet-wide competitions (Fillipov 2014). It seems many Sakha people wanted their region to be respected and accepted as a representative modernised state in Soviet terms, populated by appropriately educated, aspirational, and accomplished people (Forsyth 1992, 259, 319; Vitebsky 2005; Chakars 2014; Mowat 1970). And if Sakha people were framing these hopes and intentions within the corresponding materialist Soviet universe, the educated and accomplished population they desired also had to be atheist. Large parts of the Sakha population therefore were being drawn closer into the Soviet state, along with its aspirations – but their incorporation into the Soviet project precluded their older relationships with a lived environment. As numerous Soviet performances show, the natural world was very often praised and adored for its beauty – for example, *Yakutianochka* reminds us of Yakutia's glorious, shining snow – however, it was certainly not treated as a community of living entities with their own purposes, who were capable of answering back.

But Sakha *Estrada* songs retained a preoccupation with the Republic's landscape throughout the Soviet period, including its *alaas*. Several themes recur again and again as the central focus of these songs: the greatness of the Sakha country; a person's attachment to their home village, *alaas*, and district; love songs; songs in praise of the beauty of the local landscape; and a beloved, caring mother figure. Ivanova's most famous songs are about the Amga river, reflecting her own provenance in the region that takes its name from this river: her songs are acknowledged to be unmistakeably Sakha, despite her Russian vocal style (Ivanova-Sidorkevich 2003). Her work also exemplifies the close associations Sakha singers had and continue to have with their home regions and communities. Another instance of this association is the widespread habit of inviting pop singers from one's home region to perform at large family events, such as weddings or important birthdays.

Thus, longstanding, loving relationships between people and their home landscape were and continue to be central to the practice and content of *Estrada* music. These relationships add a distinctive tone to the praise of the natural world occurring in Sakha songs, whether Soviet or contemporary: the natural world has its own presence, as an active member of the network of relationships that forms a regional community. The conventions of Soviet-era *Estrada* in fact had provided Sakha people with an opportunity to express the connection they felt with their natural setting. Sakha singers adapted the praise of natural beauty that appeared so often in Soviet-era art and performance into the expression of their distinctive experience of the land (Sidorov 2011). They were able to acknowledge the importance of the land in their lives – but in the medium of a newly fashionable musical genre, as conventional, modernised Soviet citizens.

Perestroika and Popular Music in the Republic of Sakha (Yakutia)

The Soviet state continued to develop Sakha *Estrada* through the 1970s and 80s, sending more artists to be trained at the *VTMEI* in Moscow (Fillipov 2014). Yakutia's first 'vocal-instrumental ensemble' (*vokal'no-instrumental'niy ansambl'*, or *VIA*), a group of seventeen professionally trained singers and musicians called Choroon, was brought together in 1974 (Fillipov 2014). *VIA*s, like *Estrada* music, are another distinctively Soviet musical phenomenon. Their performance was similarly conditioned by the felt need to authorise and control musical expression while making it meaningful and appealing to audiences who continued to have interests and attractions that were not necessarily endorsed by the state. From the 1970s onwards many people in the Soviet Union became increasingly drawn to western rock music, which had become available through the black market (Cushman 1995; Yurchak 2006, 207–237). Those in charge of the Soviet Union's popular music did not feel able to endorse this attraction by authorising the Soviet Union's very own rock bands, instead sponsoring the carefully named *VIA*s as a compromise (Grabowsky 2012: 31). These were relatively large groups of highly accomplished professional musicians who incorporated a level of musical experimentation into their performance, without the explicit rejection of musical convention that was so characteristic of Western rock. And as the state-authorised Sakha musical establishment generated their own *VIA*, other Sakha people were forming a parallel amateur rock scene, inspired by the Beatles, Pink Floyd, and Deep Purple: the fashions sweeping through the Soviet Union had also engulfed Yakutia. Soviet amateur music-making had taken a turn the Soviet authorities had not expected.

Sakha rock emerged as a distinctive musical force at the end of the 1970s. Several bands – a second incarnation of Choroon, Serge, Hardyy, and Ai-Tal – claim the honour of being the first Sakha rock group. They invented what is today known as Sakha shamanic rock. As this name suggests, Sakha rock was shaped by the tensions sovietisation engendered, in common with Sakha *Estrada*. Many Sakha people were torn between the desire for education and technological development and the sense that these very education and development were destroying their cultural tradition – including the close relationship with the natural world that had been mediated through shamanic performance (Ambros'ev 2000). Ironically, the process of sovietisation had also generated a powerful awareness of Sakha national identity, along with a simultaneous consciousness that the Sakha people were the subjects of a broader state that was sucking oil, natural gas, gold, and diamonds out of their territory; polluting the land; and keeping the profits (Crate 2002). The people who came of age during the late Soviet period often express a yearning for the traditions and practices they feel have been lost, within a broader search for the meaning of a Sakha identity in the contemporary world (Iakovleva 2010).

Sakha shamanic rock therefore combined progressive rock with motifs from Sakha traditional music, adding the Sakha mouth harp (*khomus*, in Sakha), shamanic drum, and traditional string instruments to conventional rock instruments (Ventsel 2004a). These bands mixed into their music elements of the older Sakha genres that fostered intercommunication with deities and area spirits, for example, by performing *algys* prayers. The lyrics of shamanic rock groups focused on mysterious, esoteric themes, in contrast to those of *Estrada* artists. Most of the songs were about Sakha history and especially about mythical ancient warriors and shamans;

they engaged with the contradictions and tensions of Sakha experience while emphasising the spiritual bond Sakha people have with their natural environment. This focus on the natural environment reflects the fact that most rock musicians came from villages or small towns rather than Yakutia's urban centres. As this suggests, the younger generations of rural Sakha communities had retained an awareness of a live natural environment into the 1970s, 80s, and 90s – an awareness that the growing influence of Yakutsk's urban lifestyle was increasingly to obscure (Kalininskai and Kalininskaia 2012).

The second wave of shamanic rock groups appeared in the mid-1980s, fronted by Cholbon, the most famous and successful band of the genre. The 1980s was a particularly tumultuous decade. Gorbachev's *perestroika* initiative in 1986 further liberalised the music landscape – while the mass urbanisation of the Sakha people was just beginning, as increasing numbers moved to work or study in Yakutsk, then a town with a predominantly non-Sakha population (Drobizheva 1998). This urbanisation was the consecutive step of the Soviet Union's modernisation: more and more Sakha people aspired and were qualified to receive the higher education and professional work Yakutsk offered. There were considerable tensions between the different ethnic communities in Yakutsk throughout the 1980s, 90s, and into the 2000s. During the 1980s, Sakha people were strongly discouraged, if not forbidden, to speak Sakha in public (Argounova-Low 2007, 2012; Ferguson 2013). As a result, the first large generation of Sakha people to be born and brought up in Yakutsk were unable to speak Sakha, creating an awkward disconnect with relatives who had remained in Sakha-speaking villages (Argounova-Low 2007). This urbanised Sakha generation also had difficulty relating to the thousands of Sakha students who flocked to Yakutsk during the 1990s and 2000s. Urban Sakha people found rural incomers aggressive, while these immigrants were disgusted to find that urban Sakha people could only speak Russian (Habeck, Pirie, Eckert and Ventsel 2005). Sakha people saw the disappearance of their culture in the disappearance of their language – and, with it, the dissipation of their close relationship with the natural environment into urbanised preoccupations and habits. This convoluted and increasingly overt resentment at Russian domination was articulated in political and intellectual circles as a desire first for a more autonomous Sakha state and second for the revival of the Sakha traditions suppressed during the Soviet period (Vinokurova 1994, 1995, 2002). In this, Sakha activists were joining the wave of non-Russian national revival movements that swept the Soviet Union over the late 1980s and which ended in the collapse of the Soviet state in 1991.

Shamanic rock groups became the ambassadors of Sakha culture outside Yakutia's post-Soviet incarnation, the Republic of Sakha (Yakutia), established in 1992 (Troitskii 1992, 2007). Sakha rock reached its pinnacle of fame during a decade that saw simultaneously a new and short-lived autonomy for the Sakha Republic and financial chaos throughout Russia as a whole. The famous Soviet rock journalist Artemii Troitskii travelled to Yakutsk to visit the newly established Sakha rock festival Tabyk in 1989. As a result, he aired Ai-Tal in his iconic music programme "Programma A", which was popular across the former Soviet space (Ventsel 2004a, 2004b, 2012, 2014). From the early 1990s to the early 2000s, Sakha shamanic rock groups performed outside Russia at a variety of world music festivals while attracting a following within the wider Russian Federation.

As Sakha rock flourished, professional Sakha *Estrada* also found a new institutional stability in the foundation of a specialised college for *Estrada* performers in 1992.[7] *Estrada* performance in fact split into circus performance and *Estrada* music (Fillipov 2014). It is this *Estrada* music, along with Sakha shamanic rock, that has formed the basis of the Sakha popular music we see today – in part due to the emergence of a longstanding musical establishment whose members continue to direct and support younger artists.

However, over the longer term, the disappearance of the Soviet Union meant that the generous state support for the arts also collapsed. Musicians and singers enjoyed a new freedom of expression and, correspondingly, new possibilities for making money out of their music – but they lost the regular living wage the Soviet state had provided. Salaries paid by the *Estrada* Theatre have dwindled to the point where they barely contribute to covering performers' costs. Popular music artists have had to acquire an entrepreneurial attitude towards their music and its distribution (Ventsel 2006). New platforms for local music sprang up during the 1990s and 2000s in the form of private radio stations, bars, and nightclubs. In 1998, a rock musician, Büökke Bötürüöp, established Duoraan Records – which for a long period was the central record company for Sakha music and continues to exert an influence. Artists would also tour extensively throughout RSY in order to earn money. Very often, in order to cut costs, they performed in smaller places only with a playback track or a synthesiser player. The need to earn money, in combination with the appearance of new recording technologies, has focused musical artistry on vocal performance: the vast majority of artists sing and dance to a backing track rather than playing musical instruments live (Fillipov 2014). These artists have also had to compete with a deluge of Russian and foreign popular music for their audiences' attention. In response, popular Sakha performers have adopted a wide variety of foreign or Russian musical styles – from hard rock or rap to the simple, poppy Russian genre of *popsa* (Ventsel 2006). A small hardcore punk scene was established during the early 2010s. Electronic dance music combining samples of traditional music with techno or house beats has become increasingly popular; influences from K-pop have been introduced by the prominent singer Künnei.

Sakha people have continued relocating to Yakutsk, which has become a city of gleaming blocks of flats, inhabited largely by Sakha people: the majority of the Sakha population now lives in Yakutsk rather than the regions. Sakha people are distinctly conscious of a difference between the members of their community who were brought up in a village and those who were brought up in Yakutsk (Ferguson 2019). People frequently express this difference in terms of an attachment to or detachment from the natural world. The urban lifestyle affords a much more limited exposure to the natural world than village life – even if many of Yakutsk's inhabitants carefully maintain their connections with their home regions. The attractions and distractions of the post-Soviet, globalised world in fact pose as much of a threat to the Sakha experience of a lived environment as Soviet-era atheism – just as the vitality of this lived environment suffers as acutely from capitalist industrial exploitation as it did from Soviet.

The uncertainty about Sakha identity that has been the corollary to the gradual disintegration of Sakha relationships with their environment has continued, along with its influence over Sakha popular music. For example, Sakha-language rap artists, started off by Dzhida in 2003, have consciously articulated new, urban Sakha identities, asserting the right of younger, urbanised generations to speak for the Sakha people (Ventsel et al. 2017; Ferguson 2019). They articulate a growing feeling among younger urban Sakha people that they should not be made to feel ashamed of their interest in global cultural forms or their inability to speak Sakha. As their Soviet-educated grandparents found their orientations shifting towards the wider Soviet Union and away from their immediate environmental setting, so younger Sakha people are drawn increasingly not only to the wider Russian Federation but also to the world beyond. Tensions caused by the erosion of rural Sakha communities are also displayed in a division within the Sakha musical community, which emerged during 2008 and 2009. The Sakha music scene became divided into groups performing either *Estrada-popsa* or ethno-rock. Rock musicians perceive that *popsa* performers have sold themselves to the attraction of making cheap, profitable imitations of Western music while disregarding their heritage (Fillipov 2014). *Estrada* performers meanwhile believe that ethno-rock musicians are simply unable to keep up with changing contemporary tastes and technologies (Ventsel 2014).

Contemporary Sakha popular music thus incorporates a wide variety of styles and musical intentions. It is now the music of a largely urbanised community who are accustomed to sharing tracks and celebrity gossip via social media. And yet younger artists are continuing to reference a living natural environment in their work – if not also actively to engage with it through their music. One prominent example is the reggae artist and rapper Kit Dzha, who has reached a Russia-wide audience after winning a prize on the Russian television programme *Novaia Zvezda* (*New Star*) in 2019. Kit Dzha's appearance frames him as a rebel who challenges the establishment. Simultaneously he has produced various tracks which are in fact shamanic prayers: his song *Khotoiton Khotoi* (Eagle from the Eagle) goes, 'I ask the higher *aiyy*, [the god] Urun Aar Toion/holding out my hands in prayer to my older sister [a female spirit]/ . . . / defend the Sakha people, who have weakened/give them fire'.[8] Kit Dzha is paralleled by the *Estrada* singer Vitalii Ochirov: Ochirov also incorporates shamanic prayers into his lyrics while combining *Estrada* singing with his practice as an *algyschyt*, or speaker of prayers to the deities of the natural world. And, in fact, plenty of Sakha people – both performers and listeners – do report unexpected sensations during musical performances by particular artists. People have described 'energy' (*energetika*, in Russian) coursing through their bodies at concerts, particularly when a singer uses a traditional Sakha vocal technique. This is why the figure of the combined pop star and traditional shamanic practitioner is not a contradiction in RSY: music and song are still felt to instantiate the presences of the natural world. The lives of Sakha people may have changed beyond recognition, but the natural environment has retained a potent presence in popular music.

Conclusion

Sakha people have continued to adapt to their changing economic and political circumstances, as the twentieth and twenty-first centuries progressed. Key to this adaptation have been the skill and inventiveness of individual Sakha people in adopting new cultural forms to suit the relationships and practices they sought to preserve. The emergence of Sakha popular music shows this particularly clearly: the music Sakha people listen to today would have been barely recognisable to the Sakha people living a hundred years ago, and yet music is still the medium through which a personal interrelationship with a live environment is both expressed and maintained. Sakha people can engage with the highly charged words and sounds of artists such as Kit Dzha or Vitaliy Ochirov as they listen to music on the bus, at a concert, or in a night-club. As they do so, they can also engage with the mysterious, fleeting perceptions and presences that are associated with the land – or at least with the possibility that these presences still exist and are intimately connected to their histories and lives. Music offers people the opportunity both to access these felt presences and to acknowledge their existence and validity. The identification of this music as distinctively Sakha enables people to associate their more mysterious experiences of it with their cultural tradition – thus providing a language with which to articulate and authorise these experiences.

Sakha people, then, continue to devote their time and resources to popular music because of its capacity to re-open old interconnections between people and land, releasing a source of refreshment and strength in Yakutsk's busy and stressful urban setting. Many Sakha people's answer to the benevolence they intuit in nature takes the form of a determined resistance to projects that potentially damage the environment. For example, street protests in 2016 were significant enough to stave off the building of a chemical processing plant on the banks of the Lena, the river that runs alongside Yakutsk. Music therefore works to sustain not only the Sakha people's harmonious relationship with their land but also the environment itself. It does this in the face of constant and increasing pressure on the natural world from both commercial industrialisation and global climate change. And as we indicated in the Introduction, the Sakha people's history, circumstances, and responses are repeated all over the far north. For example, Inuit and Saami communities are also producing their own forms of global popular music, as are Buryat communities in the neighbouring Republic of Buryatia, just south of RSY.[9] The Sakha case indicates the power of popular music to ameliorate the breakdown in relationships that accompanies modernising and globalising change – including the breakdown of relationships within the ecosystem.

Acknowledgements

We would like to thank all the performers, producers, and specialists who have helped with our research over the years.

Notes

1 We refer to the Republic of Sakha (Yakutia) as RSY for simplicity's sake and use this term for the administrative unit of the Russian Federation established in 1992. Before that, in historical times, the more generic term Yakutia as a name for the region is more appropriate.
2 The word for Sakha in Russian is Yakut. The translated words in this piece are all in Sakha, unless specified otherwise.
3 According to the 2010 census, 478,100 Sakha people live in Russia, while 466,500 live in RSY. https://gks.ru/free_doc/new_site/perepis2010/croc/perepis_itogi1612.htm, accessed October 2019.
4 There is very little published literature on Sakha popular music – with the exception of Aimar Ventsel's articles; books about individual singers published in RSY; a university diploma dissertation and a small booklet by Vasilii Fillipov, formerly director of the Sakha *Estrada* Theatre; and Jenanne Ferguson's more recent piece (Ventsel 2004a, 2004b, 2006; Ventsel et al. 2017; Androsov 2019; Fillipov 2009, 2014; Ferguson 2019).
5 See also РУССКАЯ СОВЕТСКАЯ ЭСТРАДА, www.ruscircus.ru/progstrada, accessed September 2019.
6 http://yakkii.ru/kolledzh/kolledzh.html, accessed September 2019.
7 http://yakkii.ru/kolledzh/kolledzh.html, accessed October 2019.
8 www.youtube.com/watch?v=odoiOjOdOFU, accessed October 2019.
9 The following YouTube link shows a collaboration between Dzhida and the Buryat rapper Alarui in 2019: www.youtube.com/watch?v=8Hrl7FdR8C0, accessed 22 March 2020.

References

Ambros'ev, Aleksei. 2000. "Tabyk, Myzyka Kornei." *Trilisnik* 2.
Ammosov, M. K. 2007. *V Gusche Sobytii: Stat'i, pis'ma, telegrammy, rechi, besedy, 1920–1928.* Edited by L. E. Vinokurova and V. N. Ivanov. Yakutsk: Institut Gumanitarkykh Issledovanii AN RS(Ya).
Androsov, V. T. 2019. *Khomuian Ongordo O. I. Mestnikova.* Yakutsk: NEFU University Press.
Alekseyev, E. E. 2007. *Natsional'nyi Vopros v Yakutii (1917–1972).* Yakutsk: Bichik.
Alekseyev, G. G. 1994. *Ot Fol'klora do Professional'noi Muzyki.* Yakutsk: Bichik.
Argounov, I. A. 1985. *Sotsial'noe Razvitie Yakutskogo Naroda (Istoriko-Sotsiologicheskoe Issledovanie Obraza Zhizni).* Novosibirsk: Nauka.
Argounova-Low, T. 2007. "Close Relatives and Outsiders: Village People in the City of Yakutsk, Siberia." *Arctic Anthropology* 44 (1): 51–61.
Argounova-Low, T. 2012. *The Politics of Nationalism in the Republic of Sakha (Northeastern Siberia) 1900–2000.* Lewiston: Edwin Mellen.

Basharin, G. P., V. N. Chemezov, and L. T. Ivanova. 1968. *Kul'turnaia Revoliutsiia v Yakutii (1917–1937 gg.) Sbornik dokumentov i materialov*. Yakutsk: Yakutskoe Knizhnoe Izdatel'stvo.

Blaser, M. 2013. "Ontological Conflicts and the Stories of Peoples in Spite of Europe: Toward a Conversation on Political Ontology." *Current Anthropology* 54 (5): 547–568.

Bravina, B. 2005. *Conceptzia jizni i smerti v kulture etnosa: na materiale traditsii sakha*. Novosibirsk: Nauka.

Chakars, M. 2014. *The Socialist Way of Life in Siberia: Transformation in Buryatia*. Budapest: Central European University Press.

Crate, S. 2002. "Co-option in Yakutia: The Case of Diamonds and the Vilyuy Sakha." *Polar Geography* 26 (4): 418–435.

Crate, S., M. Ulrich, J. O. Habeck, A. R. Desyatkin, R. V. Desyatkin, A. F. Fedorov, T. Hiyama, Y. Iijima, S. Ksenofontov, C. Mészáros, and H. Takakura. 2017. "Permafrost Livelihoods: A Transdisciplinary Review and Analysis of Thermokarst-Based Systems of Indigenous Land Use." *Anthropocene* 18 (18): 89–104.

Cushman, T. 1995. *Notes from Underground: Rock Music Counterculture in Russia*. Albany: State University of New York.

Dmitriev, O. A., et al. 2004. *Estrada v Rossii. XX Vek. Entsiklopediia*. Moscow: Olma-Press.

Dmitrieva, I. A. 2013. *Na Sever Skvoz' Veka: Istoriia Pravoslaviia v Yakutii. Chast' 1. Seredina XVII – pervaia polovina XIX vekov*. Yakutsk: Izdatel'skii Otdel Yakutskoi Eparkhii Russkoi Pravoslavnoi Tserkvi Moskovskogo Patriarkhata.

Donahoe, B., and J. O. Habeck. 2011. *Reconstructing the House of Culture. Community, Self, and the Makings of Culture in Russia and Beyond*. New York and Oxford: Berghahn.

Drobizheva, L. M. 1998. *Sotsial'naia i kul'turnaiia distantsiia. Opyt mnogonatsional'noi Rossii*. Edited by L. M. Dobrizheva. Vol. 1, Etnicheskie i administrativnye granitsy: faktory stabil'nosti i konfliktnosti. Moskva: Izdatel'stvo Instituta sotsiologii RAN.

Ergis, G. U. 1974. *Ocherki po Yakutskomu Fol'kloru*. Moscow: Izdatel'svo Nauka.

Ergis, G. U., and N. V. Emel'yanov. 1976. *Sakha Narodnai Yryalara*. Yakutsk: Yakutskoe Knizhnoe Izdatel'stvo.

Ferguson, J. 2013. "Khanna Bardyng? Where Are You Going? Rural-Urban Connections and Fluidity of Communicative Practices among Sakha-Russian Speakers." PhD diss., University of Aberdeen, Scotland.

Ferguson, J. 2019. "Rapping in Sakha. Indigenous Languages." *Anthropology News*, 19 September. Accessed 25 March 2020. www.anthropology-news.org/index.php/2019/09/19/rapping-in-sakha/.

Fillipov, V. V. 2009. *40-let yakutskoi Estrade, otvet*. Yakutsk: Ministerstvo Kul'tury i Dukhovnogo Razvitiia.

Fillipov, Vasilii Vasil'evich. 2014. "Yakutskoe Estradnoe Isskusstvo: Etapy i Perspektivy Razvitiia." Diploma diss., North-Eastern Federal University in Yakutsk. Accessed 25 March 2020. www.bibliofond.ru/view.aspx?id=813893.

Forsyth, J. 1992. *A History of the Peoples of Siberia: Russia's North Asian Colony*. Cambridge: Cambridge University Press.

Gogolev, A. I. 2000. *Istoriia Yakutia (Obzor istoricheskikh sobytii do nachala XX v.)*. Yakutsk: Izdatel'stvo Yakutskogo Gosuniversiteta.

Grabowsky, I. 2012. "Motor der Verwestlichung. Das sowjetische Estrada-Lied 1950–1975." *Osteuropa* 4: 21–35.

Habeck, J. O. 2011. "Introduction: Cultivation, Collective, and the Self." In *Reconstructing the House of Culture*, edited by B. Donahoe and J. O. Habeck, 1–28. New York and Oxford: Berghahn.

Habeck, J. O., F. Pirie, J. Eckert, and A. Ventsel. 2005. *What It Takes to Be a Man: Constructions of Masculinity*. Report 2004/2005: 35–40. Halle (Saale): Max Planck Institute for Social Anthropology.

Hirsch, F. 1997. "The Soviet Union As a Work-in-Progress: Ethnographers and the Category Nationality in the 1926, 1937, and 1939 Censuses." *Slavic Review* 56 (2): 250–278.

Hirsch, F. 2007. "State and Evolution: Ethnographic Knowledge, Economic Expediency, and the Making of the USSR, 1917–1924." In *Russian Empire. Space, People, Power, 1700–1930*, edited by J. Burbank, M. V. Hagen, and A. Remnev. Bloomington and Indianapolis: Indiana University Press.

Hoffman, D. L. 2003. *Stalinist Values: The Cultural Norms of Soviet Modernity, 1917–1941*. Ithaca: Cornell University Press.

Iakovleva, A. 2010. "Andrei Borisov: marginaly vsekh stran, soediniaites!" *Yakutsk Vechernii*, 20 August.

Illarionov, V. V. 2006. *Yakutskoe Skazitel'stvo i problem vozrozhdeniia Olonkho*. Novosibirsk: Nauka.

Ingold, T. 2000. *The Perception of the Environment: Essays on Livelihood, Dwelling and Skill*. London: Routledge.

Ivanova-Sidorkevich, O. P. 2003. *Ologhum Chechirdere: Stranitsy Moei Zhizni.* Yakutsk: Bichik.

Johansen, U. 1996. "Die Ethnologen und die Ideologien. Das Beispiel der estnischen Ethnographen in der Sowjetzeit." *Zeitschrift für Ethnologie* 121: 181–202.

Kalininskai, A. K., and V. I. Kalininskaia. 2012. *Sakha Yryata Kem-Kerdii Tetiminen.* Yakutsk: Kemeul.

Kondakov, V. A. 2012. *Religiia Aar Aiyy: Osnovye Svedeniia.* Yakutsk: OOO SMIK-Master. Poligrafiia.

MacFadyen, D. 2002. *Estrada?! Grand Narratives and the Philosophy of the Russian Popular Songs since Perestroika.* Montreal, Kingston, London and Ithaka: McGill-Queen's University Press.

Mészáros, C. 2012. "The Alaas: The Interplay between Environment and Sakhas in Central-Yakutia." *Max Planck Institute for Social Anthropology Working Papers*: 137.

Middendorf, A. 1878. *Puteschestvie na severe i vostok Sibiri.* Saint Petersburg.

Mowat, F. 1970. *The Siberians.* London: Heinemann.

Parnikov, V. S. 2015. *Sakha kien tuttar dzhono. Khomuian ongordular.* Yakutsk: Bichik.

Popov, G. M. 1910. *V Yakutskoi Glushi: Sbornik Statei iz Prikhodskoi Zhizni Yakutskoi Eparkhii.* Irkutsk: Parovaia tipo-litografiia N. I. Makushina I V. Moscow: Tsosokhipa.

Popov, N. S. 2009. *Ekologicheskaia kul'tura Yakutov: monografia.* Sankt Peterburg: Asterion.

Rothstein, R. A. 1980. "The Quiet Rehabilitation of the Brick Factory: Early Soviet Popular Music and Its Critics." *Slavic Review* 39 (3): 373–388.

Roy, O. 2000. *The New Central Asia: The Creation of Nations.* London: I.B.Tauris.

Safronov, F. G. 2000. *Entsiklopediia Iakutii.* Vol. 1. Moskva: OOO "Sakha Sirin Entsiklopediiata".

Sheikin, Y. I. 1996. *Muzykal'naia Kul'tura Narodov Severnoi Azii.* Yakutsk: Respublikanskii Dom Narodnogo Tvorchestva.

Sheikin, Y. I. 2002. *Istoriia Muzykal'noi Kul'tury Narodov Sibiri.* Moscow: Izdatel'skaia firma 'Vostochnaia literatura' RAN.

Sidorov, O. 2011. *Ot Alekseia Kulakovskogo do Nickolaia Luginova: shtrikhi k istorii yakutskoi kul'tury.* Yakutsk: Bichik.

Sieroszewski, W. 1993 [1896]. *Yakuty: Opyt etnicheskogo issledovaniya.* Moscow: Rosspen.

Slezkine, Y. 1991. "The Fall of Soviet Ethnography, 1928–38." *Current Anthropology* 32 (4): 476–484.

Slezkine, Y. 1994. "The USSR As a Communal Apartment, or How a Socialist State Promoted Ethnic Particularism." *Slavic Review* 53 (2): 414–452.

Spiridonova, A. E. 2018. *Luboviu k zhizni zemnoi.* Yakutsk: Saidam.

Spiridonova, S. E. 2019. "Chabyrghakh – Tugen, Sytyy Tyllaakh Aiymnyy." *Ilin*: 16–25.

Troitskii, A. 1992. *Liner Notes for An Album "Prokliatyi Kamen".* Cholbon: SP 'Aprelevka-Sound Inc.'.

Troitskii, A. 2007. *Back in the USSR.* Sankt-Peterburg: Amfora.

Tumanov, V. 2001. *Kuiregeiim, illaa . . . (Irannik). Taata uluusun melodistarin tymsyyte.* Yakutsk: Kuduk.

Ventsel, A. 2004a. "Sakha Pop Music and Ethnicity." In *Properties of Culture – Culture as Property. Pathways to Reform in Post-Soviet Siberia*, edited by E. Kasten, 67–86. Berlin: Dietrich Reimer Verlag.

Ventsel, A. 2004b. "Stars without Money: Sakha Ethnic Music Business, Upward Mobility and Friendship." *Sibirica* 4 (1): 88–103.

Ventsel, A. 2006. "Sakha Pop Music – A Celebration of Consuming." *The Anthropology of East Europe Review* 24 (2): 68–86.

Ventsel, A. 2012. "Religion and Ethnic Identity: Sakha Shamanic Rock." *Forschungen zur Anthropologie und Religionsgeschichte* 42: 247–259.

Ventsel, A. 2014. "World Music Routes: The Modification of the Sakha Musical Tradition." *InterDisciplines* (1): 187–209.

Ventsel, A. 2016. "Estonian Invasion as Western Ersatz-pop." In *Popular Music in Eastern Europe. Breaking the Cold War Paradigm*, edited by E. Mazierska, 69–88. London: Palgrave Macmillan.

Ventsel, A., E. Peers, and Z. Tarasova. 2017. "Rep kak sredstvo vyrazheniya sotsial'nykh izmenenii na severo-vostoke Sibiri: Khip-khop, urbanizatsiya i etnichnost' Yakutov (Sakha)." *Etnograficheskoye Obozreniye* 1: 168–182.

Vinokurova, L. I. 1994. *Skaz o narode sakha.* Yakutsk: Bichik.

Vinokurova, L. I. 1995. "The Ethnopolitical Situation in the Republic of Sakha (Yakutia)." *Anthropology and Archeology of Eurasia* 34 (1): 60–78.

Vinokurova, L. I. 2002. "Aboriginy Yakutii i gosudarsctvo: istoricheskaia retroperspektiva vzaimootnoshenii." In *Tezisy vserossiiskoi konferentsii "Dukhovnaia kul'tura narodov Severa i Arktiki v nachale tret'ego tysiacheletiia"*. 52–53. Yakutsk: Severoved.

Vitebsky, P. 2005. *Reindeer People: Living with Animals and Spirits in Siberia.* London: Harper Collins.

Volkov, V. 2000. "The Concept of Kultur'nost': Notes on the Stalinist Civilising Process." In *Stalinism: New Directions*, edited by S. Fitzpatrick, 210–230. London: Routledge.

Yakovlev, V. S.-D. 2003. *Zhizn' i sud'ba moya: Roman-esse*. Yakutsk: Bichik.

Yurchak, A. 2006. *Everything Was Forever, Until It Was No More: The Last Soviet Generation*. Princeton, NJ: Princeton University Press.

Zhirkov, M. N. 1981. *Yakutskaia Narodnaia Muzyka*. Yakutsk: Knizhnoe Izdatel'stvo.

6

SOVIET-TIME INDIGENOUS DISPLACEMENT ON THE KOLA PENINSULA

An extreme case of a common practice

Lukas Allemann

Abstract

Sedentarisation and urbanisation, displacement and relocation among Arctic Indigenous populations have been common traits of all modern Arctic nation states in the past century, with heavily traumatic consequences everywhere. This chapter focuses on the Kola Peninsula, Northwest Russia, as an extreme and yet exemplary case of such social engineering in its Soviet variety. I look into Indigenous people's displacement and its consequences through the theoretical framework of social engineering and using oral history and archival data. The chapter shows the reasons and motivations of the state behind its efforts to remake the social assembly of its Indigenous backyards, as well as the circumstances that were faced by the people concerned. Considerable space is given for quoting oral testimonies, in order to lay bare the heavily traumatic events.

Introduction

Displacement, relocation and urbanisation among Arctic Indigenous populations has been a common trait of *all* modern Arctic nation states in the past century, with amazingly similar consequences: in North America and in the Nordic countries (Armstrong et al. 1978; Wenzel 1991; Fogel-Chance 1993; Hamilton and Seyfrit 1994; Tester and Kulchyski 1994; Csonka 1995; Kohlhoff 1995; Marcus 1995; Hamilton et al. 1996; Einarsson et al. 2004; Lantto 2014), as well as in the Soviet Union, as will be shown in this chapter. This chapter shall focus on the case of the Kola Peninsula, northwest Russia, looking into Indigenous people's displacement during Soviet times, on the basis of oral history and archival data and using the concept of social engineering (Scott 1998).

The Kola Peninsula's Indigenous minority are the Sámi, an ethnic group whose population spans Norway, Sweden, Finland and Russia. There are estimates that seventy to 80 per cent of all Russian Sámi in the twentieth century had to change their place of residence due to state measures. Being displaced in one's life more than once was not uncommon (own field materials;

Bogdanov 2000). The displacement of the Indigenous minority Kola relocations should be seen within global tendencies of urbanisation and industrial rationalisation, as "more people were involuntarily displaced in the twentieth century than in any other in recorded history" (Oliver-Smith 2009, 3). On the other hand, urbanisation was also "a process of increasing access to new career paths, as well as marriage possibilities offered by the rapidly emerging towns" (Konstantinov 2015, 167). At any rate, displacement policies almost never went the way originally intended, with consequences that were beyond the imagination of both the relocated people and the social engineers.

Between the 1930s and the 1970s, the Indigenous population of the Kola Peninsula – as in the rest of the Soviet Arctic – was concerned by different forms of displacement. The reasons included policies aiming directly at the concerned communities (collectivisation, sedentarisation and economic rationalisation), as well as outer constraints (the needs of industry, infrastructure development and the military) (Gutsol et al. 2007; Afanasyeva 2013; Allemann 2013, 2020). We can find the only other region with a similarly high share of relocated Indigenous people at the opposite end of the country: Chukotka. Krupnik and Chlenov (2007, 74) estimate that two-thirds of the Yupik, the Indigenous population of the region, found themselves relocated, and many families had to change their place of residence three to four times.

By the end of the Soviet period, the Murmansk Region was the most industrialised, most militarised and most populated area in the entire Arctic (Hønneland and Jørgensen 1999; Fedorov 2009; Overland and Berg-Nordlie 2012; Hønneland 2013; Josephson 2014). Throughout the twentieth century, the incoming population of the Murmansk Region grew exponentially. While the share of the Indigenous population had been 22.3% in 1895 (Shavrov 1898), it had become 0.2% by 1989, with an overall population in the Murmansk Region of almost one million (Vsesoiuznaia perepis' naseleniia 1989 goda 2018). Besides that, since the nineteenth century, the Sámi had experienced pressure by immigration from the East of a group of Komi reindeer-herding families who settled down on the Kola Peninsula, founded several villages and competed for pastures (Bruno 2016; Mankova 2018). Throughout all these decades, the Sámi population remained relatively stable at around 1600 to 1900 people. These numerical prerequisites are crucial for understanding the above-average strength of Soviet social engineering towards its Indigenous minorities on the Kola Peninsula.

Both before and after the Second World War, a repeated reason for village closures was industrial development, which necessarily led also to the growth of infrastructure and a steady population influx (Fedorov 2009, 149–366). Already by the 1940s, the entire western part of the Murmansk Region had been cleared from rural and Indigenous livelihoods and become the main arena for extractive and other industries. In the 1960s, the increased need for electricity led to the construction of two dams, and several Indigenous settlements were flooded (Gutsol et al. 2007).

While sedentarisation and collectivisation took place already before the Second World War, the most incisive relocations of Indigenous communities took place under Khrushchev's policy of consolidation (*ukrupnenie*) from the 1960s onwards. This policy was a rushed-through, centralised economic rationalisation of the whole agricultural sector of the Soviet economy. As a result, 120,000 villages were identified as "viable" (*perspektivnye*) and thus deemed worthy of investment and development. The remaining 580,000 [*sic!*] were deemed "non-viable" (*neperspektivnye*) and scheduled for liquidation (Melvin 2003, 64), a ratio of one surviving village to five villages wiped out. In the country's Arctic regions, consolidation amounted to the goal of finishing the project of sedentarisation, which had not been completed before the war. In the

Figure 6.1   Upper picture: Voron'e, a Sámi settlement, in the late 1950s or early 1960s. The settlement was flooded in 1967; lower picture: Lovozero, on the day of the yearly Festival of the North, probably 1986

Source: Private archive of Apollinariia Golykh, Lovozero.

case of the Indigenous, non-industrialised part of the Murmansk Region, Lovozero had been chosen as "viable" and was thus destined to become a the new Indigenous "hub" (Krupnik and Chlenov 2007, 62), to which most people from the "non-viable" and subsequently closed-down Sámi settlements were relocated.

In addition to the consolidation policy, another policy played a crucial role in drastically reducing the number of settlements in the Indigenous North: the 1957 resolution "On the Measures for Further Economic and Cultural Development of the Peoples of the North" (Vakhtin 1992, 18 f.). This decree provides a formal explanation for the ethnic disbalance of village closures on the Kola Peninsula: most of the predominantly Pomor villages along the White Sea coast and the predominantly Komi villages of Krasnoshchel'e and Kanevka survived consolidation despite the same level of remoteness and disconnectedness. Also, Lovozero, the chosen new Indigenous "capital", was a village dominated by the Komi. In the Soviet hierarchy of nations (Slezkine 1994a, 445 f.), Pomors and Komi did not belong the 'small' Northern native peoples, and thus they were not covered by this decree.

Another decisive factor was the state border proximity of the Kola Peninsula, a commonality shared with Chukotka. That village closures and relocations happened to the most extreme extent in the country's westernmost and easternmost Arctic regions is connected to their strategically important location. In both places, military and border guards wanted to clear the territories from civilians. Due to the inaccessibility of according archival materials, it is hard to provide hard proof for this claim. However, among elder informants, there is still a lot of talk about this (own field data).

Konstantinov's (2015, 147) poignant remark subsumes well the pervasiveness of village closures and population displacement during the 1960s: "The principal rationale of agglomeration [*ukrupnenie*], in the local application of this programme, can be likened to the macabre adage, attributed to Stalin: 'No person – no problem' (*Net cheloveka, net problemy*). In other words: 'No village – no problem'."

Population displacement as social engineering

The Kola Peninsula is not only the most densely populated region of the circumpolar North, but within the Soviet North, it had already been since the end of the 1960s the region where sedentarisation had been carried out most consistently. As stated in a formerly confidential sociological report, "in the Murmansk Region the sedentarisation of the former nomadic population has been accomplished. . . . Similar processes can be observed among other Northern peoples, but among most of them they did not go that far" (Bogoiavlenskii 1985, 92–93). The Kola Peninsula is proudly presented here as a place where all the 'problems' of the local Indigenous population have been solved in an exemplary manner by the Soviet planners:

> Currently the Indigenous people of the North [in the Murmansk Region] live in comfortable villages and settlements with modern conveniences: Lovozero, Revda [etc.], where the social infrastructure is well-developed: there are apartments with modern facilities, schools and kindergartens, shops, canteens, hospitals, health centres, Houses of Culture [*doma kul'tury*] and clubs. *This guarantees the high level of adaptation* [*prizhivaemosti*] of the Indigenous population in the extreme conditions of the Kola North.
>
> (Balakshin 1985, 6, my emphasis)

In the logic of high-modernist social engineering in the Arctic, it was not just innovation that Soviet power brought to the North. It goes deeper: it is the proper *adaptation to* the North that Soviet power aspired to bring to the local Northerners, implying that until that moment they led a maladaptive life. To central planning from outside, the fine-tuned, organically grown ways of *habitation in* the North (Ingold 2011, 2019) looked simply too messy, intransparent, irrational to be included into its initial design. Local knowledge was rationalised away under the pretence of scientific planning, while in fact these were simplifications of reality, fanciful dreams about mechanising intricate social and environmental adaptations (Scott 1998, 253). Remembering the village meetings before the relocation and the persuading orations of the bureaucrats, one elderly witness of the Voron'e relocation shows scant respect for the self-declared wisdom of officials:

> In '63 they already started dismantling everything. . . . They came, they started explaining, those, well, those folks who think they're smart. They grasped things worse than

we did, but they posed as smart. "We will transport you" [to the new village], and "you will live in such and such houses", and "it will be all for free."

(AI 2013)[1]

If anything, the sociological report quoted previously and this testimony show us well what Scott (1998, 343) put in a nutshell: "The progenitors of such plans regarded themselves as far smarter and far-seeing than they really were and, at the same time, regarded their subjects as far more stupid and incompetent than they really were."

Legibility

The emphasis on sedentarisation and concentration in a few settlements was connected to deep convictions by Soviet planners about the benefits for the local population. These included providing services and goods, most prominently in the spheres of education, healthcare, hygiene and 'civilised' living conditions. Additionally, the belief in the eventual fusion (*sliianie*) of all participants of the Soviet experiment into a Soviet nation (Slezkine 1994b, 343 f.) served as an ideological basis. However, benign intentions and ideological convictions justifying displacement and sedentarisation represent only one side of the medal. Undeclared goals and benefits for the state form the other side. Scott (1998) subsumes such concealed effects under the umbrella term of *legibility*.

Several aspects of legibility pertain to Indigenous displacement on the Kola Peninsula. Collectivisation, sedentarisation, concentration in one settlement and reorientation towards monoculture are supposed to make the population more legible to the state: the actions and movements of people, domesticated animals and all their belongings become more easily controllable and countable; education, healthcare, logistics, taxation and many more duties and ambitions of state power become more easily reachable.

In social engineering, there is a correlation between radically simplified designs for social organization and for natural environments (Scott 1998, 7). Failures of monocultures or mono-industries correlate with the failures of collective farms and/or planned settlement. Conversely, there is a correlation between social and natural diversity as a source of resilience. Concentrating people in one big settlement (Lovozero) and elevating reindeer herding to a state of monoculture form an example for both: we can see evident failures, but we can also see how diverse practices persisted and emerged as 'grey' resilience under the surface. These forms of resilience included, on the one hand, a wide array of tundra activities not legible to the state (meaning those that are not part of the planned economy and thus marginalised or even illegalised, such as hunting and fishing) and, on the other hand, accommodating and making the best of what the new urban life can offer.

According to Scott's analysis of the Ujamaa relocation campaign in Tanzania in the 1970s, "administrative convenience, not ecological considerations, governed the selection of sites; they were often far from fuelwood and water; and their population often exceeded the carrying capacity of the land" (1998, 235). The commonalities of high-modernist social engineering are confirmed in the comparison with the Kola Peninsula. Examples can be found already in the early stage relocations before the war, when the migration between summer and winter settlements was abolished for the sake of sedentarisation. An informant remembers:

> Later [at the beginning of the winter], the Sámi – my parents among them – moved to a winter village. . . . We moved into the forest where we could find more firewood and better moss for our reindeer. . . . When the Communists came to power, as early as

the 1930s they created collective farms and started converting the Sámi to a sedentary lifestyle. That's why the village of Semiostrov'e was closed. . . . Because there were no roads leading to it, you could use reindeer to get there in winter, but you couldn't do that in summer. That's why they decided to create a permanent settlement on the Barents Sea coast. This is my native village ((shows a photo)). It was a summer village. . . . That's where I was born. It was called Varzino. . . . You ask why Sámi didn't live there before [all year round]? Because it was a very cold place. . . . One couldn't survive there in winter, that's why Sámi didn't live there. But the Soviet authorities were building a new life and hadn't a clue how the settlers would survive in a location where there was no firewood and no trees. There was a forest [in Semiostrov'e], so one could go and fell a couple of birches and cut them up for firewood. . . . That's why life was so difficult there [in Varzino] in winter.

(NE 2008)

In this case, the failure of the Soviet planners to take local knowledge into account meant that in winter, the locals experienced a deficit of firewood. Additionally, they were exposed to a windy and humid climate during wintertime, while the hinterland would have offered better conditions. From a centralised state's vantage point, however, this was not simply thoughtlessness but a matter of priorities: sedentarising people in one place, and connecting them to a year-round transport infrastructure (in this case, the never-freezing Barents Sea), meant control over the production of goods and the movements of people in this sensitive border zone area. In short, it meant increased legibility for the state's eyes.

During the Cold War confrontation, the Kola Peninsula became one of the most heavily militarised areas in the world (Hønneland and Jørgensen 1999; Fedorov 2009; Heininen 2010; Hønneland 2013). During this time, most remaining civilian settlements along the Barents Sea coast were eliminated and some transformed into military ones. No coastal Indigenous settlement survived. While there is very limited written, accessible proof about any direct link between the resettlements of Indigenous communities and military activity, many accounts, events and circumstances in their sum point at such a causal connection. Mentioning the strong military presence on Chukotka, Holzlehner (2011, 1967) acknowledges that "direct evidence of military induced resettlements is rather limited". The declared reasons for village closures were always other than military ones. It is, however, hardly a coincidence that exactly in the two Arctic regions of the Soviet Union directly bordering to the NATO, the coast was most consistently cleared of civilians. Oral testimonies of relocation witnesses reflect this opinion in many instances:

It was not so much the border zone restriction that mattered but the nuclear submarine fleet. There in Drozdovka were all those never freezing inlets. . . . That's one of the most important reasons for the closure of Varzino. Developing the nuclear fleet meant freeing up certain territories. They closed the coastline, not only Varzino but the whole coast. Iokan'ga, Varzino, Kharlovka – they closed down everything up to Teriberka.

(NE 2013a)

A report of the village assembly of Varzino, a coastal settlement closed down in 1966, forms a rare instance of written evidence about this secretive topic. The report states that civilians and border guards were bothering each other: "The village population has been informed about the border zone rules. At the same time, there are still shortcomings. . . . There are instances, when people take a boat or walk into the tundra without informing the border guards" (f.302 op.1 d.134 l.4, 1966). As a consequence, in the same report, the collective farm chairman is asked to inform the border guards unit every time a village inhabitant plans to leave the village – an utterly impossible task. Shortly after that, the issue of troubling civilians was solved by simply removing the settlement altogether.

Trying to control the movement of every single individual outside of the village, as stipulated by the border control regulations and, as a result, their confinement to a concentrated settlement, amounts to the apogee of legibility – while it was a difficult task to control the moves of a border zone village population, it would have been an altogether impossible task if the intransparent holism of semi-nomadic family life and work in a fluid multitude of places had been kept.

Mono-settlement, however, was not only the answer to security concerns – such concerns were important, but they were a territorially limited peculiarity of some parts of the Kola Peninsula and some other areas of strategically outstanding importance, such as the Chukchi Peninsula. Far more representative for the countrywide rationalisation of rural settlement is the relation of mono-settlement to mono-industry and its rural equivalent, monoculture. For many of the Arctic Indigenous people who used reindeer as one of their resources, this meant a reorientation dictated from above towards reindeer herding as a monoculture. After the collective farms of the closed villages had been merged into only two collectives in Lovozero and Krasnoshchel'e, fishing and hunting were dropped altogether. Cows were kept for local production of dairy products, but reindeer herding remained as the only large-scale activity. The tundra-connected population of the Kola Peninsula had been assigned its small but clearly defined place in the large mosaic of Soviet planned economy. This is something that elderly informants still meet with the most indignation: it does not seem rational to many that especially the profitable salmon fishing had been dropped (own field materials).

There is, however, a rationale behind these transformations. From a high-modernist socio-economic planners' point of view, it lies in creating a canon of monocultures as the presumed key to maximum efficiency: designated entities – be that whole republics or single villages – were assigned clearly defined and limited roles within the Soviet economy, the most prominent large-scale examples being probably the Uzbek cotton and the Ukrainian wheat and corn monocultures. On the Murmansk regional level, sea fishery at industrial scale was based in Murmansk, with a 10 per cent share of the countrywide fish production by the end of the Soviet times (Fedorov 2009); salmon fishery as a specialisation had been assigned to the White Sea coast Pomor collective farms, which did not interfere with military activity and were not closed down (according entries in Agarkova (ed) 2016), and the remaining central sovkhozes in Lovozero and Krasnoshchel'e were responsible for reindeer herding. In short, collectivisation and consolidation were one large "program designed to establish central control over food production" in a worldwide trend towards monoculture (Bruno 2016, 142, 157). Thus, on the Kola Peninsula, Lovozero became the epitome of rural legibility by creating compact settlement in the village and reindeer monoculture around it.

Pull factors and push factors

I will focus here on how a major displacement wave on the Kola Peninsula happened in the 1960s, trying to identify the push factors (constraints) and pull factors (incentives) of moving. However, a clear distinction between push and pull factors is not always possible, as I will show in more detail subsequently. Changes, leading to large-scale displacement, occur on several levels: In a *first stage*, decisions at the top tend to be taken quickly. This is the case especially in a one-party system with a strong political leadership as the Soviet Union. I already discussed this first stage, in which the countrywide consolidation policy was adopted, drawing up a plan within one year to close down over half a million villages.

The *second stage* consists of announcing the planned changes to the population and "manufacturing consent" (Herman and Chomsky 1988). In the case of the Soviet Union, there were no institutionalised civil society mechanisms at hand as a strong counteracting negotiating party. Also in such a setting, however, forms of negotiation do take place. They have to deal with a decision already taken at the top, or they are post-factum adaptations, in the grey zone of legality or altogether illegal (Scott 1989; Konstantinov 2015). Several interviewees described the negotiations around the imminent relocation:

> And so there was this directive from above, decision by the Murmansk Region Executive Committee, this and that date, this and that year, that the village of Varzino, as a non-viable settlement, shall be closed down. That's it! And when this decision came, where was it first discussed? In the collective farm board. The board first agrees, the board members . . . discuss, and they start working with the people. You see, in any settlement there are always are those who say: "Oh, great, let's go, it will be better there!". But there were others who were against it, that's why they had to work with the population, in order to prevent too much noise ((chuckling)).
>
> (NE 2013b)

> Q: All in all, in your village, do you think more people were for or against the relocation to Lovozero?
> A: God knows. Well, I think about fifty-fifty. Yes. (AI 2013)

The quotations show that, at the stage of announcement and negotiations in village meetings, the main goal of officials was to present the new place as an improvement compared to the old place of habitation. Pull factors were highlighted in order to create consent. Later many of these pull factors turned out not to correspond to the promised reality. Among such overstated pull factors, the most prominent is housing: electricity, gas and running water looked to many people like real and important benefits. However, there was an immense gap between the promises and the reality. Another pull factor was education: parents from remote villages had to send their children to the boarding school in Lovozero. Living in the same village was supposed to terminate this separation and was a significant pull factor. However, here, too, reality lagged behind the promises: children of relocated families were kept at the boarding school despite having their parents in the same village, among other things because of the catastrophic housing conditions met by the relocatees (Allemann 2018). Road access was a fulfilled promise and a pull factor for some – but not for everyone, as the contemporary 'happy roadlessness' of the remote Kola villages shows (Konstantinov 2009). Improved sanitation was an undoubted and uncontested advantage. Finally, yet importantly, the new career and marriage opportunities that a much bigger settlement could offer were certainly pull factors, too. The highly gendered differences in marriage and career patterns after the relocations clearly shows that women profited much more from these pull factors (Vitebsky and Wolfe 2001; Povoroznyuk et al. 2010; Vitebsky 2010).

Overstating these pull factors during village meetings was not intentional lying, but it concealed "important goals of appropriation, security, and political hegemony" (Scott 1998, 191). As a result, all relocations were formally supported by a majority of the concerned villagers. In the case of Varzino, for instance, at the collective farm meeting – the collective farm was the only employer in the village – twenty people voted for the relocation, four were against it and five abstained (Gutsol et al. 2007, 42).

The *third stage* consists of the actions following the manufacture of consent: due to strong vertical power structures in the bureaucratic apparatus, often combined with an emphasis on strong personal responsibility, officials are urged to show results swiftly. Interviews and archival materials show that with each relocation, less time and attention were given for negotiating, informing and manufacturing consent. For example, a look through the village council protocols of Varzino shows business as usual until the very last moment before the relocation in 1968, with no signs of an imminent village liquidation (f.302 op.1 d.145, 1967–1968). Interviewees draw a similar picture:

> Here in Varzino we all had good houses. . . . We had to leave like refugees, we left everything behind. . . . My parents could bring quite some goods and chattels, which they brought with the reindeer. But those who had no reindeer, they just packed a bag.
>
> (EK 2014)

> They sent the tractors and said: "Three days to get ready. . . . Pack your belongings, you have three days to get ready, here is the transportation so you can get to Lovozero." That's how they relocated us.
>
> (AD 2013)

The same pattern has been described, for instance, about Chukotka (Krupnik and Chlenov 2007, 69) and the Ujamaa campaign (Scott 1998, 234). Thus, hastiness as an external push factor seems to be another basic trait of many social engineering projects, with the goal to curtail negotiation power and resistance opportunities and with the result that there is no proper relocation programme.

On the Kola Peninsula, there were nominal relocation programmes for every village closure but to varying degrees and with inconsistent implementation. The construction of pull factors (as shown previously) but also of endemic push factors (problems of non-viability supposedly not connected to the world outside of the community, such as remoteness or poor economic performance) could serve as a means of justification for offering minimal relocation programmes. The following testimony illustrates this:

> It was winter, she had to give birth soon, but the infirmary had been already closed down, the shop too. So why did the people begin to leave themselves? I mean, they created the conditions [to push out the people]. That's how it was.
>
> (AD 2013)

In planned displacement, both push and pull factors are presented to the population as reasons for moving. They are often created or exaggerated by the people who stand behind the initial idea to move a settlement. For example, *before* the actual relocation, the village's slow death is induced by removing, one by one, its vital organs: shop, school, infirmary and other infrastructure is closed down, thus exacerbating the non-viability that in village meetings is presented to the population as a reason to move. From the state's point of view, such constructed non-viability is likely to make more people agree that moving away is the only option (push factors are created) and that the new, prospective settlement is comparatively a more attractive option (pull factors are created).

Manufactured consent, in its turn, offers the advantage of minimising logistic support and compensation, as the relocation can be turned in such a way that it is at least partially wanted by the relocated people themselves. For instance, the relocations of Voron'e and Varzino are commonly acknowledged as the most "slapdash and unprepared" (Vakhtin 1992, 19) relocations. While some support was available for the relocated people, many families, especially those with children, could not benefit from it because circumstances forced them to leave before the official relocation programme would start: services were cut or completely closed down before the collective farm and thus the village had been formally liquidated (see also Gutsol et al. 2007, 52).

Filling up a viable village and its single enterprise with people from a number of non-viable villages required, with each relocation, an increased effort to provide housing and meaningful work. As one informant put it, Lovozero "wasn't made of rubber" (NE 2013a), meaning that expansion had its limitations. However, local administrators had to somehow make the dreams of distant planners come true, by hook or crook. The easiest solution, from that point of view, would be then to turn the tables on the non-viable villages and put their people and farms as petitioners: it should be *their* wish to move because *their* village became non-viable and this is why *they* should bear the burden.

The last relocation of a Sámi village, Varzino, illustrates this best. In the light of the non-viability verdict from Moscow, the collective farm *Bol'shevik* members in Varzino voted for a merger with the Lovozero collective farm *Tundra* and relocation (Gutsol et al. 2007, 42). The Lovozero collective farm, which by that time had already absorbed several other smaller collective farms, was concerned about the limited pastures, but they had to agree, half-heartedly, as there was no way around the formal fulfilment of the consolidation directives from above. However, Lovozero, caught between the centralised consolidation planning and local constraints, set conditions that were probably far from what the original consolidation planners in Moscow would have imagined: that not the Varzino relocatees but the *Tundra* people would be given priority to move into new housing and that the relocation costs would not be paid. As a consequence, the Varzino relocatees met the most difficult situation among all relocated groups.

When looking at the course of the relocations, we can clearly identify a tendency towards fewer benefits and higher coerciveness. There were enormous differences in how Chudz'iavr (in 1959), Voron'e (in 1963) and Varzino (in 1968) were relocated to Lovozero. Block houses from Chudz'iavr were disassembled, moved and reassembled at the state's expenses. The Voron'e peoples' houses were not moved, but new housing was promised, though given with a five-year delay. For Varzino, there was nothing the like (Gutsol et al. 2007, 42, 50, 53). For instance, Vatonena (1988), herself from Lovozero, remembers how eleven relatives from Varzino moved into her home. There were all together sixteen people living in two rooms for five years, until they received a flat.

The inhabitants of Voron'e and Varzino have been clearly the most deeply traumatised by relocation (own field materials). Discussing the psychosocial consequences of the post-displacement period goes beyond the scope of this chapter. It may suffice to cite a few figures. Between 1968 and 1988, the share of "natural" death (not accident, intoxication, suicide, homicide) among the relocated people from Voron'e was at an extraordinarily low 1.8 per cent and from Varzino at 4.3 per cent, while for those from Chudz'iavr, who had not met the same housing difficulties, the according value was around 50 per cent (Gutsol et al. 2007, 54 f.). More generally, during two decades following the relocations, violent death (accident, intoxication, suicide, homicide) among the relocated groups rose to a constant rate of about 50 per cent, and among men of working age even around 80 per cent, while the birth rate and the male marriage rate among the Sámi were among the lowest across all Indigenous peoples of the Soviet North (Bogoiavlenskii 1985; Kozlov and Bogoiavlenskii 2008). The Murmansk region led not only in terms of displacement and sedentarisation of Arctic Indigenous people but also in terms of the negative consequences of these policies.

Concluding his analysis of the Ujamaa relocation campaign, Scott said: "Positive inducements were, apparently, more typical for the early, voluntary phase of villagization than the later, compulsory phase" (Scott 1998, 236). Also on the Kola Peninsula, growing de-facto coercion meant fewer consultations, fewer promises, fewer benefits and a more hasty, sudden, order-like resettlement. In an apparent paradox, while *de-facto* coercion was growing, *de-iure* voluntariness was emphasised.

Conclusion: the Kola Peninsula, an extreme case or an exemplary case?

I will answer this question straight away: in terms of Arctic Indigenous population displacement, the Kola Peninsula in north-west Russia is both an extreme *and* an exemplary case. An extreme case can help us to better see and understand phenomena which may be less visible yet widespread elsewhere (Maxwell and Chmiel 2014). While the Kola Peninsula is one of the Arctic regions – if not *the* Arctic region – where social engineering (Scott 1998) has been most thoroughly implemented, many other regions in the Soviet Union and in the other circumpolar countries have gone through similar developments.

We have seen that the Kola Peninsula features all aspects of Soviet social engineering in the Indigenous Arctic implemented *ad extremis*. The social and spatial reorganisation of Indigenous livelihoods was due to policies aimed directly at the concerned communities (collectivisation, sedentarisation and consolidation) as well as outer constraints (the needs of industry, infrastructure development and the military). As an additional, concealed benefit for the state, I discussed the increased legibility (Scott 1998) of the population that that those measures implicated. With all these dimensions in one place, thus, the Kola Peninsula became the Arctic pinnacle of Soviet rural socioeconomic transformations.

Relocations meant an attempt to terminate the scattered ways of dwelling and the multiple resource-use of fine-tuned Indigenous adaptations, which looked so messy to the outsider gaze of the arrived bureaucrat. Transforming them into radically simplified patterns of concentrated living and monoculture assumed an increase of efficiency and legibility. In practice this meant cutting off and closing down Indigenous communities and moving their people to a larger settlement, which in the case of the Kola Peninsula was Lovozero.

This simplification – from a state's perspective – had a twofold nature: on the one hand, openly stated goals grounded in the spread of the state's ideology: first, improving the lives of "primitive" people, supposedly poorly adapted to the harsh North, by providing all the blessings of modern civilisation, and, second, allocating to clearly defined communities clearly defined roles in the planned Soviet economy. On the other hand, undeclared goals about increasing the population's legibility played an important role: simplified structures were supposed to increase control over the production of goods and state revenue and over the movements of people in a strategically highly sensitive part of the country.

While implementing the formally not coercive relocations, the state applied various strategies in order to increase the attractivity of moving. This was done by overstating pull factors (gas, electricity, road connection, schools) and creating push factors (by closing down services and disrupting supply chains). An increasing hastiness in preparing and carrying out relocations can be seen as another stratagem: it served to pre-empt resistance. Additionally, during the course of the relocation campaign, we can see a tendency towards decreasing the support offered to the relocated communities and shifting the burden of relocation onto them, with the result that there was no proper relocation programme, especially for the later relocations.

While many people were truly convinced by the argument in favour of moving, most of the people, once they arrived to Lovozero, did not find ready conditions to *dwell* but just to *be*, we may say inspired by Ingold (2000). Hence, in the new post-relocation environment, it became very challenging (but not impossible) to build up a satisfying life, with devastating social consequences. This seems to be one of the most common traits of development-induced displacement (many examples in Oliver-Smith 2009).

As a conclusion from these observations, I suggest that, terminologically, *displacement* should be preferred to *resettlement* or *relocation*. Displacement is broader: there is often displacement without relocation, or without proper relocation, if we define the latter as a dedicated programme offered and organised by the development project, which initially made it necessary to move a settlement (Oliver-Smith 2009, 8). In this sense, displacement is not only broader, but it depicts the reality of the Arctic village closedowns more accurately.

Note

1 All quoted interview sections have been anonymised.

References

Afanasyeva, A. 2013. "Forced Relocations of the Kola Sámi People: Background and Consequences." Master thesis, University of Tromsø.

Agarkova, T. V. (ed.). 2016. *Kol'skaia Entsiklopediia*. Apatity, Murmansk and Sankt-Peterburg: IS, KNTs RAN.

Allemann, L. 2013. *The Sami of the Kola Peninsula: About the Life of an Ethnic Minority in the Soviet Union*. Tromsø: Septentrio Academic Publishing.

Allemann, L. 2018. "'I Do Not Know if Mum Knew What Was Going on': Social Reproduction in Boarding Schools in Soviet Lapland." *Acta Borealia* 35 (2): 1–28.

Allemann, L. 2020. "The Experience of Displacement and Social Engineering in Kola Saami Oral Histories." PhD diss., University of Lapland, Rovaniemi.

Armstrong, T., G. Rogers, and G. Rowley. 1978. *The Circumpolar North*. New York: Methuen.

Balakshin, Y. Z. 1985. "Perevod korennykh zhitelei na osedlyi obraz zhizni – aktual'naia zadacha partiinykh, sovetskikh i khoziaistvennykh organov." In *Sotsial'no-ekonomicheskie i demograficheskie problemy zaversheniia perevoda kochevogo naseleniia na osedlyi obraz zhizni*, edited by B. M. Levin, S. N. Batulin, and F. S. Donskoi, 3–6. Apatity: Akademiia nauk SSSR/Murmanskii oblastnoi ispolnitel'nyi komitet.

Bogdanov, N. B. 2000. "Protsess urbanizatsii korennogo naseleniia Kol'skogo poluostrova: Saami v XX veke." *Lovozerskaia pravda*, 18 November.

Bogoiavlenskii, D. D. 1985. "Vliianie etnicheskikh protsessov na dinamiku chislennosti narodnostei Severa." In *Sotsial'no-ekonomicheskie i demograficheskie problemy zaversheniia perevoda kochevogo naseleniia na osedlyi obraz zhizni*, edited by B. M. Levin, S. N. Batulin, and F. S. Donskoi, 92–93. Apatity: Akademiia nauk SSSR/Murmanskii oblastnoi ispolnitel'nyi komitet.

Bruno, A. 2016. *The Nature of Soviet Power: An Arctic Environmental History*. New York, NY: Cambridge University Press.

Csonka, Y. 1995. *Les Ahiarmiut: à l'écart des Inuit Caribous*. Neuchâtel: Éditions V. Attinger.

Eìnarsson, N., J. Nymand Larsen, A. Nilsson, and O. R. Young. 2004. *Arctic Human Development Report*. Akureyri: Stefansson Arctic Institute.

Fedorov, P. V. 2009. *Severnyi vektor v Rossiiskoi istorii. Tsentr i Kol'skoe zapoliar'e v XVI-XX vv*. Murmansk: Murmanskii Gosudarstvennyi Pedagogicheskii Universitet.

Fogel-Chance, N. 1993. "Living in Both Worlds: 'Modernity' and 'Tradition' among North Slope Inupiaq Women in Anchorage." *Arctic Anthropology* 30 (1): 94–108.

Gutsol, N. N., S. N. Vinogradova, and A. G. Samorukova. 2007. *Pereselennye gruppy Kol'skikh saamov.* Apatity: Kol'skii nauchnyi tsentr RAN.

Hamilton, L. C., R. O. Rasmussen, N. E. Flanders, and C. L. Seyfrit. 1996. "Outmigration and Gender Balance in Greenland." *Arctic Anthropology* 33 (1): 89–97.

Hamilton, L. C., and C. L. Seyfrit. 1994. "Coming out of the Country: Community Size and Gender Balance among Alaskan Natives." *Arctic Anthropology* 31 (1): 16–25.

Heininen, L. 2010. "Globalization and Security in the Circumpolar North." In *Globalization and the Circumpolar Nort*, edited by L. Heininen and C. Southcott, 221–264. Fairbanks: University of Alaska Press.

Herman, E. S., and N. Chomsky. 1988. *Manufacturing Consent: The Political Economy of the Mass Media.* New York: Pantheon Books.

Holzlehner, T. 2011. "Engineering Socialism: A History of Village Relocations in Chukotka, Russia." In *Engineering Earth*, edited by S. D. Brunn, 1957–1973. Dordrecht: Springer.

Hønneland, G. 2013. *Borderland Russians: Identity, Narrative and International Relations.* New York: Palgrave Macmillan.

Hønneland, G., and A.-K. Jørgensen. 1999. *Integration vs. Autonomy: Civil-Military Relations on the Kola Peninsula.* Aldershot: Ashgate.

Ingold, T. 2000. *The Perception of the Environment: Essays on Livelihood, Dwelling and Skill.* London and New York: Routledge.

Ingold, T. 2011. *Being Alive: Essays on Movement, Knowledge and Description.* London and New York: Routledge.

Ingold, T. 2019. "The North Is Everywhere." In *Knowing from the Indigenous North: Sámi Approaches to History, Politics and Belonging*, edited by T. H. Eriksen, S. Valkonen, and J. Valkonen, 108–119. Abingdon, Oxon and New York: Routledge.

Josephson, P. R. 2014. *The Conquest of the Russian Arctic.* Cambridge: Harvard University Press.

Kohlhoff, D. 1995. *When the Wind Was a River: Aleut Evacuation in World War II.* Seattle: University of Washington Press.

Konstantinov, Y. 2009. "Roadlessness and the Person: Modes of Travel in the Reindeer Herding Part of the Kola Peninsula." *Acta Borealia* 26 (1): 27–49.

Konstantinov, Y. 2015. *Conversations with Power: Soviet and Post-Soviet Developments in the Reindeer Husbandry Part of the Kola Peninsula.* Uppsala: Acta Universitatis Upsaliensis.

Kozlov, A. I., and D. D. Bogoiavlenskii. 2008. "Smertnost' ot vneshnikh prichin i otkloniaiushcheesia povedenie." In *Kol'skie saamy v meniaiushchemsia mire*, edited by A. I. Kozlov, D. V. Lisitsyn, and M. A. Kozlova, 78–85. Moskva: Institut naslediia.

Krupnik, I., and M. Chlenov. 2007. "The End of 'Eskimo Land': Yupik Relocation in Chukotka, 1958–1959." *Études/Inuit/Studies* 31 (1/2): 59–81.

Lantto, P. 2014. "The Consequences of State Intervention: Forced Relocations and Sámi Rights in Sweden, 1919–2012." *Journal of Ethnology and Folkloristics* 8 (2): 53–73.

Mankova, P. 2018. "The Komi of the Kola Peninsula within Ethnographic Descriptions and State Policies." *Nationalities Papers* 46 (1): 34–51.

Marcus, A. R. 1995. *Relocating Eden: The Image and Politics of Inuit Exile in the Canadian Arctic.* Hanover: University Press of New England.

Maxwell, J. A., and M. Chmiel. 2014. "Generalization in and from Qualitative Analysis." In *The SAGE Handbook of Qualitative Data Analysis*, edited by U. Flick, 540–553. Thousand Oaks, CA: SAGE Publications.

Melvin, N. 2003. *Soviet Power and the Countryside: Policy Innovation and Institutional Decay.* Houndmills, Basingstoke, Hampshire, and New York: Palgrave Macmillan in association with St. Antony's College, Oxford.

Oliver-Smith, A. 2009. "Introduction: Development-Forced Displacement and Resettlement: A Global Human Rights Crisis." In *Development and Dispossession: The Crisis of Forced Displacement and Resettlement*, edited by A. Oliver-Smith, 3–23. Santa Fe, NM: School for Advanced Research Press.

Overland, I., and M. Berg-Nordlie. 2012. *Bridging Divides: Ethno-Political Leadership among the Russian Sami.* Berghahn Books.

Povoroznyuk, O., J. O. Habeck, and V. Vaté. 2010. "Introduction: On the Definition, Theory, and Practice of Gender Shift in the North of Russia." *Anthropology of East Europe Review* 28 (2): 1–37.

Scott, J. C. 1989. "Everyday Forms of Resistance." In *Everyday Forms of Peasant Resistance*, edited by F. D. Colburn, 1–34. Armonk, NY: Sharpe.

Scott, J. C. 1998. *Seeing Like a State: How Certain Schemes to Improve the Human Condition Have Failed.* New Haven: Yale University Press.

Shavrov, N. A. 1898. *Kolonizatsiia, ee sovremennoe polozhenie i mery dlia russkogo zaseleniia Murmana.* Sankt-Peterburg: Tipografiia I. Gol'dberga.

Slezkine, Y. 1994a. "The USSR as a Communal Apartment, or How a Socialist State Promoted Ethnic Particularism." *Slavic Review* 53 (2): 414–452.

Slezkine, Y. 1994b. *Arctic Mirrors: Russia and the Small Peoples of the North.* Ithaca, NY: Cornell University Press.

Tester, F., and P. Kulchyski. 1994. *Tammarniit (Mistakes): Inuit Relocation in the Eastern Arctic, 1939–63.* Vancouver, BC: UBC Press.

Vakhtin, N. 1992. *Native Peoples of the Russian Far North.* London: Minority Rights Group International.

Vatonena, L. 1988. "Est' takie problemy." *Lovozerskaia pravda*, 11 May, 3.

Vitebsky, P. 2010. "From Materfamilias to Dinner-Lady: The Administrative Destruction of the Reindeer Herder's Family Life." *Anthropology of East Europe Review* 28 (2): 38–50.

Vitebsky, P., and S. Wolfe. 2001. "The Separation of the Sexes Among Siberian Reindeer Herders." In *Women As 'Sacred Custodians' of the Earth*, edited by S. Tremayne and A. Low, 81–94. Oxford: Berg.

Vsesoiuznaia perepis' naseleniia 1989 goda [online]. 2018. Accessed 24 Apr 2018. www.demoscope.ru/weekly/ssp/rus_nac_89.php.

Wenzel, G. W. 1991. *Animal Rights, Human Rights: Ecology, Economy and Ideology in the Canadian Arctic.* Toronto: University of Toronto Press.

Other documents

f.302 op.1 d.134 l.4, 1966. *Protokol No. 8 zasedaniia Varzinskogo sel'skogo Soveta deputatov trudiashchikhsia.*

f.302 op.1 d.145, 1967. *Protokoly zasedanii Varzinskogo sel'skogo Soveta deputatov trudiashchikhsia.*

Interviews

AD. 2013. Female, born in the early 1960s, Murmansk.

AI. 2013. Female, born in the early 1930s, Lovozero.

EK. 2014. Female, born in the early 1940s, Lovozero.

NE. 2008. Female, born in the late 1930s, Murmansk.

NE. 2013a. Female, born in the late 1930s, Murmansk.

NE. 2013b. Female, born in the late 1930s, Murmansk.

7

INDIGENOUS ARCTIC RELIGIONS

Piers Vitebsky and Anatoly Alekseyev

Abstract

Arctic Indigenous religions tend to share distinctive common features: an environment imbued with spirits, respect toward the bodies and souls of animals, reincarnation of animals and humans, and a vertical cosmology with shamanic trance and soul-flight. Hunting is surrounded by many taboos. Animals are often controlled by a spirit owner or manager and have human-like consciousness. The moral tension between guilt over killing and the necessity of eating is mediated through ideas of respect and gratitude, with careful treatment of the animal's remains so that it will be reincarnated in new flesh. Humans may have several kinds of soul, or be reincarnated from ancestors. Shamans in addition often have animals as helper spirits. Colonialism was generally hostile to Indigenous religions throughout the Arctic, introducing writing and schooling, with Christian or Marxist ideologies. Christianity is now the dominant religion, leading to varying forms of rejection or adaptation of earlier theologies and religious sensibilities. Today, there are many revival movements involving Indigenous artists, scholars, and activists. Traditional religious ideas may be deployed politically, acquiring a new validation as "Indigenous knowledge" in the fight against mineral extraction and the destruction of sacred sites and using modern media such as newspapers, websites, and social media.

Introduction

The Arctic contains diverse ecologies, peoples, language families, histories, and political regimes. Yet Indigenous religions throughout the region share striking common features: an environment imbued with spirits, elaborate shows of respect toward the bodies and souls of animals, reincarnation of animals and humans, and a vertical cosmology, with shamanic trance as the core technique for operating the entire system. Archaeological evidence suggests that this nexus is very ancient (Arutyunov 1998; Mulk and Bayliss-Smith 2006; Jordan 2011; Bäckmann and Hultkrantz 1985; Hilton et al. 2014). Some of the closest similarities run from west to east, throughout the Inuit (Eskimo) peoples (Damas 1984) or among the very different peoples around the Bering Strait (Fitzhugh and Crowell 1988); other shared traits extend southwards to sub-Arctic America (Helm 1981) or link Siberia with Mongolia and Central Asia (Hamayon 1990; Basilov 1984; Pedersen 2001).

In every Arctic language, there are various names and classifications for the entities that fill the environment with forces and agency and which European languages, with their Christian heritage, call "spirits". Spirits represent the essence or inner being of things and are also the causes of events and experiences in human life. Animals, mountains, and rivers can have a consciousness which is similar to human consciousness and makes them act as they do. They can love humans and thus feel compassion and cooperate with us. They also have needs, hunger, jealousy, and pride and so can attack people and eat them or drive them mad. Separate body parts and even tools and weapons can have their own souls and intentions which make them function well or badly, a reflection of a philosophical tendency to decentralise the world, giving a special importance to the role of fortune (Oskal 2000; Brandišauskas 2017).

Traditional Arctic religions have no written texts and are inspirational rather than doctrinal. Hunting, and even dreaming, take place within a field of shared consciousness that encompasses other humans and animals alike. The introduction of Christian or Marxist hegemonic texts, and of modern individualism, marks a major shift and leads to complicated tensions and compromises. Forms of Christianity, in particular, have permeated all Arctic populations without necessarily eliminating the un-Christian idea that animals have souls similar to the souls of humans. Though this chapter is written mostly in the present tense, available sources do not always allow us to know how far older ideas or practices continue today.

Space, time, and movement

Religion (sometimes called cosmology or worldview) underlies traditional Indigenous Arctic perceptions of what others might call "nature" or "environment" through an animistic perception of the world as alive and conscious. For example, Inuit have two metaphysical principles: *inu(a)*, the "essence" of a thing or person, and *sila*, meaning "the world", "air", "weather", and "mind" or "intelligence" (Fitzhugh and Kaplan 1982; Saladin d'Anglure 1994).

Topography, routes, hunting territories, dwellings, and burial sites are all integrated into a landscape where material and spiritual aspects, economy, and religion cannot be understood separately from one another (Jordan 2011; Brody 1982; Hallendy 2001; Ahlbäck and Bergman 1991). This world is the middle layer between various upper and lower worlds, each with its own spirits and other inhabitants (Eliade 1964). The ability to traverse levels is a key shamanic skill. Arctic life is highly seasonal (Turner 1996; MacDonald 1998; Lowenstein 1993; Vitebsky 2005), often with major rituals at midwinter (Fienup-Riordan 1994 for bladder festival) or midsummer (Romanova and Ignat'yeva 2011; Alekseyev 1993, where each person flies symbolically on a reindeer to the sun for a blessing). Myth and storytelling remain a vital way of relating to the world (Lowenstein 1992, 1993 in northern Alaska; van Deusen 1999; Dolitsky 2019 in Chukotka; Cruikshank 2000 in sub-Arctic Canada).

The landscape itself may listen and observe us (Cruikshank 2005; Nelson 1983). Spirits may be everywhere, but they are concentrated more intensely in certain significant places and become apparent at particular moments. Focal points for sacrifice or gift exchange with local spirits become sacred sites (Jordan 2003; Mulk and Bayliss-Smith 2006; Saladin d'Anglure 2004; Gemuyev and Sagalayev 1986), which demand or respond to human attention. Eveny reindeer herders renew their relationship with the spirit of each seasonal site by feeding the campfire there with offerings and leaving it to fall dormant again as they move on; their predictive dreams are fulfilled a year later when returning to the same site (Vitebsky 2005). Human presence or action may leave a permanent spiritual trace or residue (Lavrillier 2012), and anxiety about disturbing the dead is a serious issue for Indigenous people who become archaeologists. Among Sami, ghosts of abandoned babies can sometimes be heard crying in the mountains (Pentikäinen 1968; Nergård 2006); in Siberia, some graves are so dangerous that reindeer herders may change their yearly migration route to avoid them (Vitebsky 2005); Indigenous people living on old gulag sites see ghosts of blonde Russian prisoners (Ulturgasheva 2012).

Some deities, like Num (Nenets), Torum (Khanty), or Hövki (Eveny: Alekseyev 1993), made the world and now reside in the sky. But unlike the Christian God, these are not set within a fundamental dualism between good and evil. Instead, spiritual entities are associated with power or agency, and their morality may be more ambiguous. In North America, the world was generally created by the trickster Raven, who brought us daylight (Nelson 1983; Lowenstein 1992); among Sami, by the sun (Pentikäinen 2000); in northeast Siberia, by a mammoth who drew up mud from the water with its tusks.

Animals and humans, spirits and souls

Creatures can change outward form, and humans echo this shape-shifting every day as they wear animal skin transformed by women's needlework into clothes (Chaussonet 1988). Both animals and humans have souls which are continually reincarnated in new bodies. While in Hindu and Buddhist *karma*, beings are reincarnated up and down a scale of value according to merit (Obeyesekere 1994), in the Arctic, each being is generally reincarnated in the same species. In animals, this ensures the supply of meat through species continuity; in humans, it brings back important aspects of a named person.

To ensure the reincarnation of animals, their flesh, bones, and souls must be treated with special procedures of respect and be protected from being gnawed by dogs or trodden underfoot. Eveny in Siberia may lay out an elk's head facing east so that it will return like the rising sun (Vitebsky 2005, 263); many Inuit give the thirsty soul of a dead seal an offering of fresh water and believe that the soul resides in the bladder, which they use as harpoon float. In the old midwinter bladder festival of the Alaskan Yupik, inflated bladders were returned to the sea through a hole in the ice with a request to grow into a new body and come back next year (Fienup-Riordan 1990, 1994). Like the sacrificial bear cub in Siberia (see subsequently), the seals would tell other spirits of their good treatment by humans.

Hunting is highly ritualised, with many taboos. Hunters speak of their plans and achievements in code, as to utter the terminating word "kill" could end an animal's cycle of reincarnation forever. While her husband is away at sea, an Iñupiat wife keeps a welcoming hearth for the whale and avoids actions which resemble the breaking of his harpoon line or escaping of his prey (Bodenhorn 1990; cf. Rethmann 2001, 64–65). Animals give themselves willingly to a respectful hunter, in a collaborative reciprocity which is the basic cosmic scheme of Arctic existence but only on strict conditions, as they watch and judge us for correct thoughts and behaviour. The relationship between hunter (usually male) and prey is often depicted as seductive or erotic, with the moment of killing as a sexual consummation, and hunters must refrain from other female contact before setting out (Willerslev 2007, 110–114; Tanner 1979; Brightman 1993).

Animals often have a spirit owner and can be understood as incarnations, manifestations, or refractions of such figures, who instruct them to give themselves to the hunter – or to withhold themselves. Among Eveny and Yukaghir in northeast Siberia, this is an old man, an elemental force who pervades animals and landscape alike, and the hunter's erotic relationship is with the old man's daughter (Vitebsky 2005; Willerslev 2007). Throughout the Inuit world, animals are similarly controlled by a woman at the bottom of the sea (called Sedna or other names; Laugrand and Oosten 2010), originally a girl whose fingers were cut off by her father in a dispute about her unsuitable husband (a dog or a stormy petrel). Each finger grew into a species of sea mammal so that in order to survive, humans must eat her flesh. Animals and their owners in many areas particularly dislike individualism or boasting. Meat is not the hunter's personal property but is bestowed by spirits and must be generously shared. For all our dependence on them, relations with spirits are inherently dangerous. Their love may become fatal, and a hunter who is given too many animals will pay for it with his own life (Vitebsky 2005, 295–298; Willerslev 2007, 42–49).

The bear is a particularly powerful figure (Hallowell 1926 for a classic survey), and there are many taboos about killing or eating it. For Inuit, a powerful shaman may turn into a polar bear, moving freely between land and water (Laugrand and Oosten 2014). In Siberia, the brown bear is the lord of the forest. In a rite now partially revived, the Khanty in western Siberia hunted an adult bear and sang an elaborate ritual over its body (Jordan 2003, 115–124; Siikala and Ulyashev 2011; Wiget and Balalaeva 2011; https://arcticanthropology.org/2016/04/24/the-khanty-bear-feast-revisited/). The peoples around the Amur Gulf in the Siberian Far East used to capture a wild bear cub and rear it, before sacrificing it and sending it to the sky (Batchelor 1967 [1909]; Kitagawa 1961). Like the seals of Alaska, it would report how it had been well treated by humans, and like the consecrated reindeer (see subsequently), it colluded in its own death (Willerslev et al. 2015).

Animals are human-like, so the moral tension between hunger and guilt can never be resolved. As a shaman in Iglulik told Rasmussen (1929, 56), "human food consists entirely of souls, which must be propitiated lest they should revenge themselves on us for taking away their bodies". The domestication of reindeer some 2–3,000 years ago (which never happened in North America) transformed the economies and cosmologies of the European and Asian Arctic. Unlike wild animals, domesticated reindeer can be sacrificed and are identified closely with their human owners. Ingold (1986) argues that domestication represents a domination by humans and contrasts this with a relationship of trust between hunter and prey. Willerslev et al. (2015) argue by contrast that it is from domestication that trust and cooperation emerge and that the sacrifice of a willing tame reindeer or a hand-reared bear cub symbolically neutralises the violence of the hunt. Domestic animals can be intensely loyal to humans. Many Eveny have a consecrated reindeer that will deliberately shield its owner from danger, dying as their substitute (Vitebsky 2005). By contrast, even though wild animals supposedly give their bodies willingly, hunters still have to deceive them with secret language. They do not say "I'm going hunting" but "I'm going to look around" (Vitebsky 2005, 268–270; Willerslev 2007, 100–101); bears are mentioned as "grandfather", "the mossy one", or "the barefooted one". Animals are further deceived by seductive songs, smells, and sexual imagery and are caught by hidden traps (Jochelson 1926, 378–382; Bogoras 1904–9, 138–147).

Though animals are like non-human persons, they are not much more than an exemplar of their species. Human personhood is more complex. A Yukaghir person's soul is both helper and deceiver, which must be appeased with sacrifices and can be carried around as a wooden figurine (Willerslev 2007, 57–62; cf. Hultkrantz 1953, 357–358 for Native America). Among Inuit and related groups, the person is composed of elements which can be translated as shade, breath, mind, and name (Fienup-Riordan 1994, 211–250 for Yupik; Flora 2019 for Greenland). These are often described as multiple souls. But though most Arctic people are now also Christian, the term "soul" (in any European language) is inadequate to translate such Indigenous concepts.

Arctic peoples vary in just what aspects of a human they say are reincarnated. After the separation and dissolution of death, the return of the deceased is a complex process. A dead person's return may be suggested by relatives' dreams or by a new baby's habits or body markings (Mills and Slobodin 1994), ideas that extend south into Native America and Mongolia and also to Iceland (Heijnen 2014). The baby's naming ceremony confirms the identification, and adults may then address the baby, literally, as "grandmother" or "uncle". Even if we understand the name as a kind of soul, what is reincarnated is not so much an individual as an identity (Sharp 1997, 99–103), a nexus in a web of social relations: "The who of the name . . . not only implies identity but also sociality" (Bodenhorn 2006, 139); "there is no point in returning unless you are longed for" (Flora 2019, 5).

The subtleties of Arctic metaphysics are further revealed in the Inuit concept of mind (Fienup-Riordan 1991) and in linguistic studies, for example, of shamanic speech (Therrien 1987). In Siberia, Eveny travellers emanate *djuluchen*, an intention which arrives at their destination before the physical person (Ulturgasheva 2012) and can be perceived predictively through messages crackled by campfires (Vitebsky and Alekseyev 2015a). Spiritual forces constantly affect people's daily lives and long-term destinies. In a dangerous landscape, it is hard to stay alive, and dreams and omens often concern the personal safety of oneself or others (Vitebsky 2005, 285–310). An immature or "open" soul or body may remain close to the dead and be especially vulnerable to spirits (Ulturgasheva 2012, 43–55). Suicide, a serious problem throughout the Arctic, is widely said to happen at the beckoning of previous victims, a reasoning not understood by medicalised social services. Suicide need not be the end of the person, who in Lutheran Greenland is buried in a churchyard and reincarnated like anyone else. Here, a final, irreversible death is experienced only by the *qivittoq*, a person who rejects society and disappears out on the ice to become a sort of zombie (Flora 2019) – perhaps equivalent to an animal whose reincarnation has been blocked by a hunter's disrespectful treatment.

Shamans and their performances

Shamans, male or female, are specialists who can perceive the presence of spirits and influence their actions (Balzer 1990; Yamada and Irimoto 1998; Vitebsky 2005; Ahlbäck 1987; Siikala 1978). The word "shaman" was introduced from the Eveny and Evenki languages of eastern Siberia (Flaherty 1992; Znamenski 2007; Hutton 2001), and each Arctic language has its own words for similar figures. It may be that spirits have made a patient sick by stealing their soul or that animals are not giving themselves to hunters. In a state of trance, the shaman's soul leaves their body and travels to other realms of the cosmos to debate and persuade or do battle with the spirits who are causing the problem. In a famous example from Arctic Canada (Rasmussen 1929, 124–129), an Iglulik shaman descends to the seabed to negotiate with the woman who has been withholding sea mammals from a starving community. After overcoming various deadly dangers, he finds her with the sea creatures swimming around her and facing angrily away from her blubber-oil lamp. Her hair is matted and filthy from the wrong behaviour of the community, such as incorrect treatment of animals' remains and improperly managed miscarriages. Since her fingers were cut off to make the species of animals they eat, she cannot comb her own hair. Gently the shaman turns her toward the light, combs her hair, and pleads for the community's survival. Mollified, she releases the animals. Above ground, the audience, who have been following events the other side of a curtain, begin to confess their secret sins. In a new mood of relief and purification, the hunters return to the sea, presumably to be successful.

To underpin a religion, such performances must be dramatically, sociologically, and psychologically persuasive. If shamans do not have captive clientele, as in an Evenki clan, they may compete for clients through spectacular magical tricks (Pedersen 2001). They may perform openly in front of their audience or out of sight, where an invisible performance makes the audience more sensitive to aural cues (Stépanoff 2019). Among many sub-Arctic American peoples, the audience may sit outside a tent which shakes violently as shaman and spirits converse inside (Feit 1994, 1997). All of these are ways of revealing a hidden reality which underlies the surface phenomena and processes of the world, like the *inkonze* (knowledge of reality) of the Dene and related peoples (Smith 1997; Sharp 1997) or the *anguakua* (inner fire, associated with inner sight) of the Inuit shaman's initiation (Merkur 1992).

Unlike the format of prayer (Nelson 1983; and in most Christian ritual), or the "thank you" murmured when returning a seal's bladder to the sea, shamanic rituals of this type move from problem to resolution. But this resolution can never be final, as the world is forever shifting and uncertain, like knowledge itself, or the weather (*sila*). Such elaborate religious procedures are also practical, helping humans to live together with other humans, animals, and the environment. They enjoin morality by working with the diversity of phenomena rather than by integrating them hierarchically under a single monotheistic principle. Theologically, they are more immanent (with spiritual forces embedded throughout the world) than transcendent (with a divinity concentrated somewhere beyond this world, as in Christianity).

While any ordinary person may have several souls or aspects, a shaman's personhood is even more complex, augmented or enhanced during initiation by acquiring helper spirits, such as a bear for strength, or an eagle to reach the sky, or a mouse to penetrate to the underworld. In their intense intimacy, it is debateable how far the shaman controls the spirits (Eliade 1964) or the other way round. A hunter must think himself into an animal's mind, but a shaman may go further and turn into an animal to gain an inside understanding of a species. One man among the Iñupiat was abducted by migrating whales and spent some time in a whale's body. He later used this shamanic initiation to divert their course to feed his friends and starve his community's enemies (Lowenstein 1992, 90–94). Initiation is usually through shock or serious illness and a period of withdrawal from society, often with the appearance of madness and the experience of extreme torture by spirits, before returning to the community (Merkur 1992). One woman in Iglulik was struck by a ball of fire and felt lit up from within by a spirit which was half human and half polar bear (Rasmussen 1929, 123–124). Siberian shamans are typically initiated by being dismembered by spirits, just as game animals are dismembered by hunters, and boiled in a cauldron. Then, just as all an animal's bones are saved so that it can be reincarnated, the spirits reassemble the apprentice shaman into a new body, but now with extra bones and enhanced powers of sight and understanding: "[the blacksmith spirit] pierced my ears with his iron finger and told me, 'You will be able to hear and understand the speech of plants'" (Popov 1936, 84ff; discussed in Basilov 1984, 59–63; Vitebsky 1995, 58–61).

Attacks, conversions, and responses

Our knowledge of Arctic religions comes through various cultural, political, and intellectual filters, each inflecting this abstract realm with a particular tone and interpretation. There are virtually no written records until recent centuries of colonial incursion. Early European travellers were soon followed by Christian missionaries, who were intensely interested in Indigenous religions but usually in order to combat them. Russian scientific expeditions to Siberia in the 18th century created a European fascination with shamanism (Flaherty 1992; Hutton 2001;

Znamenski 2007). In the later 19th century, Jewish and Polish political exiles in Siberia wrote works which remain classics in anthropology and comparative religion (Jochelson 1908, 1926; Bogoras 1904–9, 1910; Sieroszewski 1896). Bogoras and Jochelson participated around 1900 in expeditions directed by Boas (1888) which studied the entire North Pacific as an integrated culture area, a fieldwork perspective which became politically impossible from 1917 until the late 1980s (Krupnik and Fitzhugh 2001; Fitzhugh and Crowell 1988; Vitebsky and Alekseyev 2015b). Rasmussen, himself half-Greenlandic, produced a monumental classic (1929), but generally, apart from a few exceptions like Laestadius (2002[?1845]) and Turi (2011[1910]), it is only more recently that Indigenous people can be heard speaking about religion for themselves.

By the time a community is written about, it has usually already been touched by Christianity (Laugrand and Oosten 2009) and in Russia after 1917 by Communism (Slezkine 1994; Basilov 1984). These totalising ideologies introduced literacy but were also carried by invasive populations and promoted by powerful governments, generally hostile to Indigenous religious forms and practitioners. Even the most sympathetic Christian missionaries (Fienup-Riordan 1991) or Soviet scholars were ultimately caught up in a wider project to suppress Indigenous religions and even execute their shamans (Slezkine 1994; cf. Botheim 1999) or to condemn them as fraudulent or mad, an attitude which remained common until the later 20th century when global counter-culture re-branded shamans as exceptionally wise (Merkur 1992; Znamenski 2007; Hutton 2001).

Many Arctic people themselves became enthusiastic Christians (or in Soviet Russia, Communists) as they converted, not so much *to* a new religion as *from* an old religion which no longer fitted their colonial situation. Christianity is now the dominant religion across the Arctic, whether Lutheran in Scandinavia and Greenland, Catholic and Protestant in North American, or Orthodox in Russia (though recently with infiltration of evangelical Protestant sects, Vallikivi forthcoming; Vaté 2009; Wiget and Balalaeva 2011). Where conversion is already centuries old, historical documents were written by churchmen, portraying encounter and conflict but revealing little about local people's actual experience. Christians, like Communists, may start by seeing Indigenous religions as primitive, superstitious, and simply wrong. As the focus of spirituality shifts from the environment to the sky and intimacy with spirits is replaced by devotion to God and Jesus, or as spirituality itself is replaced by atheist materialism, communities must drastically reconfigure their sense of personhood and social relations. Newly converted communities may vehemently reject their ancestral spirits and identify them with barbarity (Slezkine 1994) or the devil (Vallikivi 2009), often burning drums and idols (Rydving 1993; Ahlbäck and Bergman 1991; Vallikivi 2011; Dombrowski 2001). But on a longer timescale, through these militant ideological filters one can perceive a complex process of persistence, transformation, and compromise, with important continuities in healing, land use, kinship, and morality. The impact of missionisation produced many prophets (Laestadius 2002[?1845]) and social disturbances, as it still does today where it is new (Vallikivi 2009; Plattet 2013). The complexities of this are explored in many historical studies (Fienup-Riordan 1991; Dombrowski 2001; Znamenski 1999; Petterson 2014 on how Indigenous catechists in Greenland became dependent on the new Danish monetary economy; Kleivan 1999, where a woman in the 1780s claimed to have borne Jesus' child).

The persecution of Sami shamans by church and state in the 17th–18th centuries (Rydving 1993; Botheim 1999) eventually subsided into the compromise of Laestadianism (Norderval and Nesset 2000; Minde 1998). Some Sami still see spirits and consult sacred stones (Joy 2018), but their healers rescue abandoned babies' ghosts by baptising them and exorcise haunted houses with Bible readings (Nergård et al. 2019). Eveny in Siberia top off their graves with an ensemble of sacrificed reindeer skins, Orthodox crosses, and Communist red stars (Vitebsky 2005). Sacred objects in museum exhibits may receive offerings of coins, alcohol, or cloth (Liarskaya 2011). Fienup-Riordan (1991) gives a particularly subtle study of Indigenous/Christian negotiation. Though governments and churches everywhere almost totally wiped out the old shamans, much remains of the surrounding worldview and sensitivities: shamanism even without shamans.

The terms "myth" or "folktale" miss the depth of Indigenous narrative forms. Collections of stories (Holtved 1951; Millman 2004; Dolitsky 2019; van Deusen 1999; Saladin d'Anglure 2006) show a world of warfare, sex, magic, tricksters, animal helpers, animal-human marriages, reincarnations, transformations, journeys to the sun and moon, and dangerous monsters. But a text on a page risks losing performativity, context, and religious impact. These narratives contain the germplasm of religion. As they unfurl, they explore fundamental dilemmas of subsistence, intimacy, social life, and morality (Cruikshank 2000; Ridington 1988).

Indigenous scholars write academic or popular accounts of their own religious traditions, particularly in Russia (Alekseyev 1993; Lar 2003). Indigenous literary writing strongly echoes older oral styles (Patsuaq 1970; Vaschenko and Clayton Smith 2010 for samples; Langgård and Thisted 2011 for a regional survey), just as modern understandings of historical events tend to follow mythic narrative templates (Saladin d'Anglure 2006). Indigenous novelists are very aware that they are bridging genres and worlds. Many explore the cruelties of Christianisation or Sovietisation and the boarding schools they both imposed. The family in Nappaaluk (2014) is caught between rival Christian groups as well as their own spirits. Ipellie (1993), likewise set in a world of spirits and missionaries, combines magical realism with fierce political cartoons. Soviet writers had to tread a particularly delicate line, creating a world saturated with Indigenous religion without explicitly endorsing it (Rytkheu 2006, 2011), though by 1982 a Nenets author was able to portray a sophisticated city girl returning home to the *chum* (tepee), being given the family idols, and realising that "she has just received the soul of her father, mother, grandfather and everybody who had lived before her" (Anna Nerkagi in Slezkine 1994, 369–371). Indigenous authors and outsiders alike make their Indigenous characters intuitive, observant, and open to spirits (Arsenyev 2000 [1923]). When they are investigators or detectives (Hoeg 1992; MacDonald and MacDonald 2002), they may also use their spiritual powers to right historical injustices.

Possible futures

At the moment, it is not forbidden or dangerous anywhere in the Arctic to revive or adapt traditional religious forms. Russia has been swept by a new wave of religiosity since the late 1980s as dreams and visions, an interest in family history, the reprinting of old anthropological classics, and international New Age influences all allow Indigenous artists, activists, and scholars to promote updated developments of old religious forms. The Russian government is supportive of Orthodox Christianity and hostile to Protestant and evangelical sects, but many new churches of all types have sprung up (Plattet 2013; Vallikivi 2009; Vaté 2009). In Scandinavia and North America, there is a new tolerance and pluralism in many communities which were defensively Protestant a few decades ago. People may modernise their previous practices of healing (Nergård 2006; Miller 2007) or burial (Siikala and Ulyashev 2011) or give their old actions a new rationale: young Chukchi continue to look after the carcasses of whales, but this is no longer in order to reincarnate them but for "ecological" reasons. They name their babies after ancestors, not as reincarnations but "out of respect" (Krupnik and Vakhtin 1997). Traditional religious ideas may be deployed politically, for example, to defend Inuit sealing and whaling against animal-rights lobbyists, seen as urban dwellers alienated from nature (Wenzel 1991; Lynge 1993). Iñupiat hunters in the 1990s used their superior knowledge of whales (Lowenstein 1992, 1993; Turner 1996) to overturn restrictions from the International Whaling Commission. Religious imagery can also serve identity politics, for example, to demand ethnically based administrative territories. Among larger groups like the Sakha (Yakut), these can become significant political movements. In the 1990s, the Sakha regional government taught shamanism in schools, and Sakha society now contains vigorous religious movements with a nationalist agenda far from their traditional horse- and cattle-herding origins (Kondakov 2012; Peers and Kolodeznikova 2015).

Researchers, too, interpret old religious styles in a new, more political light. Reincarnation among the endangered Yukaghir may now also stand for reversing the possible extinction of an entire people (Vallikivi and Sidorova 2017) and the ghosts of crying babies in Norway for generations of grieving Sami children incarcerated in boarding schools (Nergård 2006; Nergård et al. 2019); Eveny children unconsciously echo the healing voyage of the shaman in their hope of going to the city for education and then returning home to rescue their community from poverty and depression (Ulturgasheva 2012). Our own fieldwork shows how Siberian shamans' children who were forbidden to practise their ancestral calling became doctors or musicians, thereby developing separate healing or performative strands of the old religious vocation in a secular, permitted idiom. Religion underlies new forms of popular music and dance (King 2011; Ventsel 2006; Peers et al. this volume), perhaps with more depth than the folkloristic ethnic cabaret of Soviet times. Performers may emphasise environment and land claims (Ikuta 2011; Koester 2002) and also make hidden deliberate mistakes to protect the inner religious core.

One of the most effective areas of Indigenous activism concerns sacred sites, a concept which governments can understand in terms of their own monuments (Vorren and Eriksen 1993; Vinokurova 2014; Lar 2003; Pratt 2009; Heinämäki and Herrmann 2017, all for detailed documentation; Goulet 2008 for Canadian land claims made in terms of the "Creator" spirit). Mineral extraction makes it especially urgent to develop a legal framework for their protection in Russia (Kharyuchi 2004; RAIPON 2000), even if they were previously secret (Gemuyev and Sagalayev 1986). Earlier, these were vulnerable to destruction by Christians or Communists (Golovnev and Osherenko 1999), but now Russian oil and gas workers may even feed local Indigenous spirits at wayside shrines (Stammler et al. 2017).

Old ideas about animal souls and landscape spirits are acquiring a new validation as "Indigenous knowledge", wisdom about sustainability and stewardship of nature. At first sight, this looks like empowerment. However, there is often an avoidance of explicit religion talk in Indigenous peoples' self-presentation to the outside world, as these ideas are discreetly filtered and stripped of their deep religiosity when brought to the cosmopolitan table of scientific environmental discourse. The use of Inuit and Cree "Traditional Ecological Knowledge" (TEK) can supposedly allow "data" to be "independently verified" and "integrated into the cumulative effects assessment" (*Northern Perspectives* 1992, 15–16). However, this is quite unlike the Inuit or Cree of Lagraund and Oosten or Feit. Homogenising everyone's knowledge or spiritual experience is just how Arctic people have *not* traditionally approached the world. When the publishers of Lynge, himself a Greenlander, write (1993: jacket blurb) that "the native perspective [on whaling] is entirely consistent with international conservation strategies and global environmental concerns", we see the metaphysics of Arctic animal knowledge subordinated to an alien agenda of rational resource use.

Many Indigenous people today live in northern cities or southern capitals, where they are exposed to new social and economic pressures and are offered new solutions by new religions. Young people, both urban and rural, are now closely integrated into global youth culture and the Internet (Habeck 2019). Spirituality becomes more individualised, though there are also moves toward a more doctrinal religious style (Kondakov 2012 is a rulebook of the new Sakha religion; Vallikivi forthcoming shows a craving for the rigidity of extreme Baptism). Some Indigenous people are simultaneously exploring their own traditions and exotic religions such as Buddhism or Bahai'i (Peers and Kolodeznikova 2015; Hakkarainen 2009). Patients afflicted by spirits are treated by psychiatrists who do not believe in them; crowded apartments, job insecurity, and political intrigue lead to sorcery accusations (Zorbas 2015), and people turn to psychic healers and neo-shamans (Kharitonova 2006, 2015; Balzer 2011) who have no space for drums, fire, and animal sacrifices. Communities in Arctic Russia under Ukrainian or Korean missionary influence today are going through similar dilemmas to communities in Arctic Europe, Greenland, and North America generations ago (Vallikivi 2009; Plattet 2013; Vaté 2009). Arctic people's present-day religious energy is very apparent in websites, pamphlets, social media, local newspapers, meetings, and political and cultural movements. Yet such developments are largely unexplored by researchers, who either remain fascinated by archaic survivals or neglect religion altogether for better-supported studies of environment and ecology, as if these can be understood separately. There is a pressing need for Arctic research to catch up with current religious reality.

Acknowledgements

Thanks for comments or references to the following, who are not responsible for any remaining faults: Barbara Bodenhorn, Julie Cruikshank, Janne Flora, David Koester, Roza Laptander, Dmitriy Oparin, Eleanor Peers, Anastasia Piliavsky, Florian Stammler, Nikolay Vakhtin, Virginie Vaté, Laur Vallikivi.

A note about the references

Centuries of travel, government, missionising, and research have generated a huge literature on religion in every part of the Arctic. The sources quoted here are necessarily very selective and oriented toward the English reader. They also include some examples showing continuities with the sub-Arctic. Main sources are sometimes hard to find, but the same authors often cover similar ground in other articles and book chapters. Older classic works are also often recycled or paraphrased, more or less accurately, by other authors in secondary or popular literature. Most primary sources on Greenland and the Sami are in Scandinavian languages and those on Siberia in Russian, though few of these are given here. With very few exceptions (Diószegi 1998; Balzer 2011), modern studies of the Russian Arctic in English became possible only after foreign anthropologists were allowed to do fieldwork there more freely in the late 1980s (Vitebsky 2005; Brandišauskas 2017; Willerslev 2007; Wiget and Balalaeva 2011), though Russian and Western researchers still pursue very different questions (Vitebsky and Alekseyev 2015b).

As well as in very local publications, such as *Recherches amérindiennes au Québec* or *Samefolket*, articles on religion frequently appear in journals such as *Études/Inuit/Studies*; *Folk: Dansk Etnografisk Tidskrift*; *Arctic Anthropology*; and *Acta Borealia*. For Russia, see also *Sibirica: Journal of Siberian Studies*; *Folklore* (Estonia); *Soviet Anthropology and Archeology* (now *Anthropology and Archeology of Eurasia*); *Archaeology, Ethnology and Anthropology of Eurasia*; *Études mongoles et sibériennes, centrasiatiques et tibétaines*; and *Sovetskaya Etnografiya* (now *Etnograficheskoye Obozreniye*).

References

Ahlbäck, T. 1987. *Saami Religion*. Stockholm: Almqvist and Wiksell.

Ahlbäck, T., and J. Bergman. 1991. *The Saami Shaman Drum*. Åbo: Donner Institute.

Alekseyev, A. 1993. *Zabytyy mir predkov [The Forgotten World of the Ancestors]*. Yakutsk: Sitim.

Arsenyev, V. 2000 [1923]. *Dersu the Trapper*. Kingston, NY: Macpherson (various editions, also called *Dersu Uzala*).

Arutyunov, S. 1998. "Twenty-Five Centuries of Stability and Continuity in the Shamanistic and Animistic Beliefs of the Bering Strait Eskimos." In *Circumpolar Animism and Shamanism*, edited by T. Yamada and T. Irimoto, 185–192. Sapporo: Hokkaido University Press.

Bäckmann, L., and A. Hultkrantz. 1985. *Saami Pre-Christian Religion; Studies on the Oldest Traces of Religion among the Saamis*. Stockholm: Almqvist and Wiksell.

Balzer, M. M. 1990. *Shamanism: Soviet Studies of Traditional Religion in Siberia and Central Asia*. Armonk, NY: M.E. Sharpe.

Balzer, M. M. 2011. *Shamans, Spirituality, and Cultural Revitalization: Explorations in Siberia and Beyond*. New York: Palgrave Macmillan.

Basilov, V. N. 1984. *Izbranniki dukhov [Chosen by the Spirits]*. Moscow: Politizdat.

Batchelor, J. 1967[1909]. "The Ainu Bear Sacrifice." In *From Primitives to Zen: A Thematic Sourcebook on the History of Religions*, edited by M. Eliade, 206–211. London: Collins.

Boas, F. 1888. *The Central Eskimo, Bureau of American Ethnology, 6th Annual Report*, 399–670. Reprinted University of Nebraska 1964.

Bodenhorn, B. 1990. "I'm Not the Great Hunter, My Wife Is: Iñupiat and Anthropological Models of Gender." *Études/Inuit/Studies* 14(1–2): 55–74.

Bodenhorn, B. 2006. "Calling into Being: Naming and Speaking Names on Alaska's North Slope." In *An Anthropology of Names and Naming*, edited by G. vom Bruck and B. Bodenhorn, 139–156. Cambridge: Cambridge University Press.

Bogoras, W. 1904–9. *The Chukchee*. Jesup North Pacific Expedition, Vol. 7. Leiden: Brill.

Bogoras, W. 1910. *Chukchee Mythology*. Jesup North Pacific Expedition, Vol. 8. Leiden: Brill.

Botheim, R. 1999. *Trolldomsprosessane i Bergenhus en 1566–1700: hovedoppgave i historie [Witchcraft Trials in Bergenhus 1566–1700: An Exercise in History]*. Bergen: Universitet i Bergen.

Brandišauskas, D. 2017. *Leaving Footprints in the Taiga: Luck, Spirits and Ambivalence among the Siberian Orochen Reindeer Herders and Hunters*. New York and Oxford: Berghahn.

Brightman, R. A. 1993. *Grateful Prey: Rock Cree Human-Animal Relationships*. Berkeley: University of California.

Brody, H. 1982. *Maps and Dreams: Indians and the British Columbia Frontier*. London: Faber.

Chaussonet, V. 1988. "Animals and Needles: Women's Magic." In *Crossroads of Continents . . . of Siberia and Alaska*, edited by W. FitzHugh and A. Crowell, 209–226. Washington, DC: Smithsonian Institution Press.

Cruikshank, J. 2000. *The Social Life of Stories: Narrative and Knowledge in the Yukon Territory*. Vancouver: University of British Columbia.

Cruikshank, J. 2005. *Do Glaciers Listen? Local Knowledge, Colonial Encounters and Social Imagination*. Vancouver: University of British Columbia.

Damas, D. 1984. *Handbook of North American Indians*. Vol. 5. Arctic. Washington, DC: Smithsonian.

Diószegi, V. 1998. *Shamanism: Selected Writings of Vilmos Diószegi*. Budapest: Akadémiai Kiadó.

Dolitsky, A. 2019. *Ancient Tales of Chukotka*. Juneau: Alaska-Siberia Research Center.

Dombrowski, K. 2001. *Against Culture: Development, Politics, and Religion in Indian Alaska*. Lincoln: University of Nebraska.

Eliade, M. 1964. *Shamanism: Archaic Techniques of Ecstasy*. New York: Pantheon.

Feit, H. 1994. "Dreaming of Animals: The Waswanipi Cree Shaking Tent Ceremony in Relation to Environment, Hunting and Missionization." In *Circumpolar Religion and Ecology: An Anthropology of the North*, edited by T. Irimoto and T. Yamada, 289–316. Tokyo: Tokyo University Press.

Feit, H. 1997. "Spiritual Power and Everyday Lives: James Bay Cree Shaking Tent Performers and Their Audiences." In *Circumpolar Animism and Shamanism*, edited by T. Yamada and T. Irimoto, 121–150. Sapporo: Hokkaido University Press.

Fienup-Riordan, A. 1990. *Eskimo Essays: Yup'ik Lives and How We See Them*. New Brunswick and London: Rutgers University Press.

Fienup-Riordan, A. 1991. *The Real People and the Children of Thunder: The Yup'ik Eskimo Encounter with Moravian Missionaries John and Edith Kilbuck*. Norman: University of Oklahoma.

Fienup-Riordan, A. 1994. *Boundaries and Passages: Rule and Ritual in Yup'ik Eskimo Oral Tradition*. Norman and London: University of Oklahoma.

Fitzhugh, W., and A. Crowell. 1988. *Crossroads of the Continents: Cultures of Siberia and Alaska*. Washington, DC: Smithsonian.

Fitzhugh, W., and S. Kaplan. 1982. *Inua: Spirit World of the Bering Sea Eskimo*. Washington, DC: Smithsonian.

Flaherty, G. 1992. *Shamanism and the Eighteenth Century*. Princeton, NJ: Princeton University Press.

Flora, J. 2019. *Wandering Spirits: Loneliness and Longing in Greenland*. Chicago: University of Chicago Press.

Gemuyev, I., and A. Sagalayev. 1986. *Religiya naroda Mansi: kul'tovyye mesta, XIX – nachalo XX v [The Religion of the Mansi People: Sacred Sites, 19th–Early 20th Century]*. Novosibirsk: Nauka.

Golovnev, A., and G. Osherenko. 1999. *Siberian Survival: The Nenets and Their Story*. Ithaca: Cornell University Press.

Goulet, J.-G. 2008. "La dimension religieuse des revendications autochtones au Canada [The Religious Dimension of Aboriginal Claims in Canada]." *Recherches amérindiennes au Québec* 38 (2–3): 83–93.

Habeck, O. 2019. *Lifestyle in the Russian North*. Cambridge: Open Book.

Hakkarainen, M. 2009. "Tibetan Landscapes in Chukotka: Consumption of Esoteric Mass Production in Indigenous Community of Markovo Village (Chukotka)." *Folklore* (Estonia) 41: 97–118.

Hallendy, N. 2001. *Inuksuit: Silent Messengers of the Arctic*. Vancouver: Douglas and McIntyre.

Hallowell, I. 1926. "Bear Ceremonialism in the Northern Hemisphere." *American Anthropologist* 28 (1): 1–175.

Hamayon, R. 1990. *La chasse à l'âme: esquisse d'une théorie du chamanisme sibérien [Hunting the Soul: Sketch of a Theory of Siberian Shamanism]*. Nanterre: Société d'ethnologie.

Heijnen, A. 2014. *The Social Life of Dreams: A Thousand Years of Negotiated Meanings in Iceland*. Berlin: LIT Verlag.

Heinämäki, L. 2017. "From Knowledge to Action: How to Protect Sacred Sites of Indigenous Peoples in the North?" In *Experiencing and Protecting Sacred Natural Sites of Sámi and other Indigenous Peoples: The Sacred Arctic*, edited by L. Heinämäki and T. M. Herrmann, 181–192. Berlin: Springer.

Heinämäki, L., and T. M. Herrmann. 2017. *Experiencing and Protecting Sacred Natural Sites of Sámi and other Indigenous Peoples: The Sacred Arctic*. Berlin: Springer, Springer Polar Science).

Helm, J. 1981. *Handbook of North American Indians*. Vol. 6. Subarctic. Washington, DC: Smithsonian.

Herrmann, T. M. 2017. "Experiencing and Safeguarding the Sacred in the Arctic: Sacred Natural Sites, Cultural Landscapes and Indigenous Peoples' Rights." In *Experiencing and Protecting Sacred Natural Sites of Sámi and other Indigenous Peoples: The Sacred Arctic*, edited by L. Heinämäki and T. M. Herrmann, 1–8. Berlin: Springer.

Hilton, C. E., M. Auerbach, and L. W. Cowgill. 2014. *The Foragers of Point Hope: The Biology and Archaeology of Humans on the Edge of the Alaskan Arctic*. Cambridge: Cambridge University Press.

Hoeg, P. 1992. *Miss Smilla's Feeling for Snow*. London: HarperCollins.

Holtved, E. 1951. "The Polar Eskimo: Language and Folklore." *Meddelelser om Grønland* 152 (1–2).

Hultkrantz, A. 1953. *Conceptions of the Soul among North American Indians: A Study in Religious Ethnology*. Stockholm: Ethnographical Museum of Sweden.

Hutton, R. 2001. *Shamans: Siberian Spirituality and the Western Imagination*. London: Hambledon Continuum.

Ikuta, H. 2011. "Embodied Knowledge, Relations with the Environment, and Political Negotiation: St Lawrence Island Yupik and Iñupiaq Dance in Alaska." *Arctic Anthropology* 48 (1): 54–65.

Ingold, T. 1986. "Hunting, Sacrifice and the Domestication of Animals." In *The Appropriation of Nature: Essays on Human Ecology and Social Relations*, edited by T. Ingold, 242–276. Manchester: University Press.

Ipellie, A. 1993. *Arctic Dreams and Nightmares*. Pentikton, BC: Theytus Books.

Jochelson, W. 1908. *The Koryak*. New York: American Museum of Natural History.

Jochelson, W. 1926. *The Yukaghir and the Yukaghirized Tungus.* New York: American Museum of Natural History.

Jordan, P. 2003. *Material Culture and Sacred Landscape: The Anthropology of the Siberian Khanty.* Walnut Creek: AltaMira.

Jordan, P. 2011. *Landscape and Culture in Northern Eurasia.* Walnut Creek: Left Coast Press.

Joy, F. 2018. *Sámi Shamanism, Cosmology and Art as Systems of Embedded Knowledge.* Rovaniemi: Lapland University Press.

Kharitonova, V. 2006. *Feniks iz pepla? Sibirskiy shamanizm na rubezhe tysiachiletii [A Phoenix from the Ashes? Siberian Shamanism at the Turn of the Millennium].* Moscow: Nauka.

Kharitonova, V. 2015. "Revived Shamanism in the Social Life of Russia." *Folklore* (Estonia) 62: 37–54.

Kharyuchi, G. 2004. "Nenets Sacred Sites as Ethnographic Landscape." In *Northern Ethnographic Landscapes: Perspectives from Circumpolar Nations,* edited by I. Krupnik, R. Mason, and T. Horton. Fairbanks: University of Alaska.

King, A. 2011. *Living with Koryak Traditions: Playing with Culture in Siberia.* Lincoln: University of Nebraska.

Kitagawa, J. 1961. "Ainu Bear Festival (Iyomante)." *History of Religions* 1: 95–151.

Kleivan, I. 1999. "Habakuk og andre profeter i Vestgrønland [Habakuk and Other Prophets in West Greenland]." In *Inuit, kultur og samfund. En grundbog i eskimologi,* edited by J. Lorentzen, E. Jensen, and H. C. Gulløv. Århus: Forlaget Systime.

Koester, D. 2002. "When the Fat Raven Sings: Mimesis and Environmental Alterity in Kamchatka's Environmentalist Age." In *People and the Land: Pathways to Reform in Post-Soviet Siberia,* edited by E. Kasten, 45–62. Berlin: Dietrich Reimer Verlag.

Kondakov, V. A. 2012. *Religiia Aar Aiyy: osnovye svedeniia [The Aar Aiyy Religion: Basic Information].* Yakutsk: OOO SMIK-Master Poligrafiya.

Krupnik, I., and W. Fitzhugh. 2001. *Gateways: Exploring the Legacy of the Jesup North Pacific Expedition, 1897–1902.* Washington, DC: Smithsonian.

Krupnik, I., and N. Vakhtin. 1997. "Indigenous Knowledge in Modern Culture: Siberian Yupik Ecological Legacy in Transition." *Arctic Anthropology* 34 (1): 236–252.

Laestadius, L. 2002 [manuscript around 1845]. J. Pentikäinen. *Fragments of Lappish Mythology.* Beaverton, ON: Aspasia.

Langgård, K., and K. Thisted. 2011. *From Oral Tradition to Rap: Literatures of the Polar North.* Nuuk: Ilismatusarfik/Forlaget Atuagkat.

Lar, L. 2003. *Kul'tovye pamyatniki Yamala: Khebidiya ya [Cult Sites of Yamal: Khebidiya Ya].* Tyumen: Institut problem osvoyeniya Severa SO RAN.

Laugrand, F., and J. Oosten. 2009. *Inuit Shamanism and Christianity: Transitions and Transformations in the XXth Century.* Montreal and Kingston: McGill-Queen's University Press.

Laugrand, F., and J. Oosten. 2010. *The Sea Woman: Sedna in Inuit Shamanism and Art in the Eastern Arctic.* Chicago: University of Chicago Press for University of Alaska Press.

Laugrand, F., and J. Oosten. 2014. *Hunters, Predators and Prey: Inuit Perceptions of Animals.* Oxford and New York: Berghahn.

Lavrillier, A. 2012. "'Spirit-Charged' Animals in Siberia." In *Animism in Rainforest and Tundra: Personhood, Animals, Plants and Things in Contemporary Amazonia and Siberia,* edited by M. Brightman, V. E. Grotti, and O. Ulturgasheva, 113–129. Oxford and New York: Berghahn.

Liarskaya, L. 2011. "A Working Model of a Sacred Place: Exhibits Appearing in Dreams and Other Miracles in a Small Museum at the Edge of the World." *Sibirica* 10 (2): 1–25.

Lowenstein, T. 1992. *The Things That Were Said of Them: Shaman Stories and Oral Histories of the Tikiġaq People.* Berkeley: University of California.

Lowenstein, T. 1993. *Ancient Land, Sacred Whale: The Inuit Hunt and Its Rituals.* London: Bloomsbury.

Lynge, F. 1993. *Arctic Wars, Animal Rights, Endangered Peoples.* Dartmouth: Dartmouth College Press.

Macdonald, G., and A. Macdonald. 2002. *Shaman or Sherlock? The Native American Detective.* Westport, CT: Greenwood.

MacDonald, J. 1998. *The Arctic Sky: Inuit Astronomy, Star Lore, and Legend.* Toronto: Royal Ontario Museum/Nunavut Research Institute.

Merkur, D. 1992. *Becoming Half Hidden: Shamanism and Initiation among the Inuit.* New York and London: Garland.

Miller, B. 2007. *Connecting and Correcting: A Case Study of Sámi Healers in Porsanger.* Leiden: CNWS.

Millman, L. 2004. *A Kayak Full of Ghosts: Eskimo Folk Tales.* Northampton, MA: Interlink.

Mills, A., and R. Slobodin. 1994. *Amerindian Rebirth: Reincarnation Belief among North American Indians and Inuit*. Toronto: University of Toronto Press.

Minde, H. 1998. "Constructing 'Laestadianism': A Case for Sami Survival?" *Acta Borealia* 1: 5–25.

Mulk, I. M., and T. Bayliss-Smith. 2006. *Rock Art and Sami Sacred Geography in Badjelánnda, Laponia, Sweden: Sailing Boats, Anthropomorphs and Reindeer*. Umeå: University of Umeå.

Nappaaluk, M. 2014. *Sanaaq: An Inuit Novel*. Winnipeg: University of Manitoba.

Nelson, R. K. 1983. *Make Prayers to the Raven: A Koyukon View of the Northern Forest*. Chicago: University of Chicago Press.

Nergård, J. I. 2006. *Den levende erfaring: en studie i samisk kunnskapstradisjon [The Living Experience: A Study in a Sami Knowledge Tradition]*. Oslo: Kappelen.

Nergård, J. I., P. Vitebsky, S. Nergård, and S. Wolfe. 2019. "Eahpáraš: nyfortolkning av fortellingene om fornekter barndom [A Reinterpretation of Stories of Denying Childhood]." In *Kulturen som patient: uvanlige møter for vanlige folk [The Culture As a Patient: Unusual Meetings for Ordinary People]*, edited by J. I. Nergård, and P. Vitebsky, 22–34. Oslo: Universitetsforlaget.

Norderval, Ø., and S. Nesset. 2000. *Vekkelse og vitenskap. Lars Levi Laestadius 200 år [Revival and Science: Lars Levi Laestadius at 200 Years]*. Tromsø: University of Tromsø.

Obeyesekere, G. 1994. "Foreword: Reincarnation Eschatologies and the Comparative Study of Religions." In *Amerindian Rebirth: Reincarnation Belief among North American Indians and Inuit*, edited by A. C. Mills and R, xi–xxiv. Toronto: University of Toronto Press.

Oskal, N. 2000. "On Nature and Reindeer Luck." *Rangifer* 20 (2–3): 175–180.

Patsuaq, M. (Markoosie). 1970. *Harpoon of the Hunter*. Montreal: McGill Queen's University.

Pedersen, M. A. 2001. "Totemism, Animism and North Asian Indigenous Ontologies." *Journal of the Royal Anthropological Institute* 7: 411–427.

Peers, E., and L. Kolodeznikova. 2015. "The Post-Colonial Ecology of Siberian Shamanic Revivalism: How Do Area Spirits Influence Identity Politics?" *Worldviews* 19: 245–264.

Pentikäinen, J. 1968. *The Nordic Dead-Child Tradition: Nordic Dead-Child Beings: A Study in Comparative Religion*. Helsinki: Suomalainen Tiedeakademia.

Pentikäinen, J. 2000. *Sami Folkloristics*. Turku: Nordic Network of Folklore.

Petterson, C. 2014. *The Missionary, the Catechist and the Hunter: Foucault, Protestantism and Colonialism*. Leiden: Brill.

Plattet, P. 2013. "Sick of Shamanizing." *Civilisations: revue internationale d'anthropologie et de sciences humaines*: 69–88.

Popov, A. A. 1936. "Tavgitsy [The Tavgy]." In *Trudy Instituta Antropologii i Etnografii*. Vol. 1, No. 5. Moscow and Leningrad: Akademiya Nauk SSSR.

Pratt, K. 2009. *Chasing the Dark: Perspectives on Place, History and Alaska Native Land Claims*. Anchorage: Department of the Interior.

RAIPON (Russian Association of Indigenous Peoples of the North). 2000. *The Conservation Value of Sacred Sites of Indigenous Peoples of the Arctic: A Case Study in Northern Russia: Report on the State of Sacred Sites and Sanctuaries*. Akureyri: Conservation of Arctic Flora and Fauna.

Rasmussen, K. 1929. *Intellectual Culture of the Iglulik Eskimo. Report of the Fifth Thule Expedition, 1921–1924*. Vol. 7. Copenhagen: Gyldendalske Boghandel, Nordisk Forlag.

Rethmann, P. 2001. *Tundra Passages: History and Gender in the Russian Far East*. Philadelphia: Pennsylvania State University.

Ridington, R. 1988. *Trail to Heaven: Knowledge and Narrative in a Northern Native Community*. Iowa City: University of Iowa.

Romanova, Ye. N., and V. B. Ignat'yeva. 2011. "Yakutskiy natsional'nyy prazdnik ysyakh v situatsii perekhoda: istoricheskiy mif, etnokul'turnyy obraz i sovremennyy prazdnichnyy narrativ [The Yakut National Festival of Ysyakh in Transition: Historical Myth, Ethnocultural Image and Modern Festive Narrative]." *Etnograficheskoye Obozreniye* 4: 29–40.

Rydving, H. 1993. *The End of Drum Time: Religious Change among the Lule Saami, 1670's-1740's*. Stockholm: Almqvist and Wicksell.

Rytkheu, Yu. 2006. *A Dream in Polar Fog*. New York: Archipelago Books.

Rytkheu, Yu. 2011. *The Chukchi Bible*. South Royalton, VT: Steerforth.

Saladin d'Anglure, B. 1994. "Brother Sun (Taqqiq), Sister Sun (Siqiniq), and the Direction of the World (Sila): From Arctic Cosmography to Inuit Cosmology." In *Circumpolar Religion and Ecology: An Anthropology of the North*, edited by T. Irimoto and T. Yamada. Tokyo: University of Tokyo.

Saladin d'Anglure, B. 2004. "La toponymie religieuse et l'appropriation symbolique du territoire par les Inuit du Nunavik et du Nunavut [Religious Toponymy and Symbolic Appropriation of Territory by the Inuit of Nunavik and Nunavut]." *Études/Inuit/Studies* 28 (2): 107–131.

Saladin d'Anglure, B. 2006. *Être et renaître Inuit: homme, femme ou chamane [To Be and to Be Reborn as Inuit: Man, Woman or Shaman]*. Paris: Gallimard.

Sharp, H. 1997. "Non-Directional Time and the Dene Life-Cycle." In *Circumpolar Animism and Shamanism*, edited by T. Yamada and T. Irimoto, 93–104. Sapporo: Hokkaido University Press.

Sieroszewski, W. (Seroshevskiy, V). 1896. *Yakuty: opyt etnograficheskogo issledovaniya [The Yakut: An Experiment in Ethnographic Research]*. St. Petersburg: Imperial Russian Geographical Society.

Siikala, A.-L. 1978. *The Rite Technique of the Siberian Shaman*. Helsinki: Suomalainen Tiedeakatemia.

Siikala, A.-L., and O. Ulyashev. 2011. *Hidden Rituals and Public Performances: Traditions and Belonging among the Post-Soviet Khanty, Komi and Udmurts*. Helsinki: Studia Fennica Folkloristica No 19.

Slezkine, Y. 1994. *Arctic Mirrors: Russia and the Small Peoples of the North*. Ithaca, NY: Cornell University Press.

Smith, D. M. 1997. "World As Event: Aspects of Chipewyan Ontology." In *Circumpolar Animism and Shamanism*, edited by T. Yamada and T. Irimoto, 67–91. Sapporo: Hokkaido University Press.

Stammler, F., S. R. Nystø, and A. Ivanova. 2017. *Taking Ethical Guidelines into the Field for Evaluation by Indigenous Stakeholders*. Drag, Norway: Arran Lule Sami Institute.

Stépanoff, C. 2019. "Dark Tent and Light Tent: Two Ways of Travelling in the Invisible." *Inner Asia* 21 (2): 199–215.

Tanner, A. 1979. *Bringing Home Animals: Religious Ideology and Mode of Production of the Mistassini Cree Hunters*. St John's: Memorial University of Newfoundland.

Therrien, M. 1987. *Le corps Inuit (Québec Arctique) [The Inuit Body (Arctic Quebec)]*. Bordeaux: Presses Universitaires de Bordeaux.

Turi, J. 2011 [1910]. *An Account of the Sámi*. Chicago: Nordic Studies Press.

Turner, E. 1996. *The Hands Feel It: Healing and Spirit Presence among a Northern Alaskan People*. DeKalb, IL: Northern Illinois University Press.

Ulturgasheva, O. 2012. *Narrating the Future in Siberia: Childhood, Adolescence and Autobiography among the Eveny*. Oxford and New York: Berghahn.

Vallikivi, L. 2009. "Christianization of Words and Selves: Nenets Reindeer Herders Joining the State through Conversion." In *Christian Conversion after Socialism: Disruptions, Modernisms and the Technologies of Faith*, edited by M. Pelkmans, 59–73. Oxford and New York: Berghahn.

Vallikivi L. 2011. "What does Matter? Idols and Icons in the Nenets Tundra." *Journal of Ethnology and Folkloristics* 5 (1): 75–95.

Vallikivi, L., and L. Sidorova. 2017. "The Rebirth of a People: Reincarnation Cosmology among the Tundra Yukaghir of the Lower Kolyma, Northeast Siberia." *Arctic Anthropology* 54 (2): 24–39.

Vallikivi L. forthcoming. *Words and Silences: an Encounter between Nenets Reindeer Herders and Russian Evangelical Missionaries in the Post-Soviet Arctic*. Ithaca: Cornell University Press.

Van Deusen, K. 1999. *Raven and the Rock: Storytelling in Chukotka*. Seattle and London: University of Washington Press/Edmonton: Canadian Circumpolar Institute.

Vaschenko, A., and C. Clayton Smith. 2010. *The Way of Kinship: An Anthology of Native Siberian Literature*. Minneapolis: University of Minnesota.

Vaté, V. 2009. "Redefining Chukchi Practices in Contexts of Conversion to Pentecostalism (Russian North)." In *Christian Conversion after Socialism: Disruptions, Modernisms and the Technologies of Faith*, edited by M. Pelkmans, 39–57. Oxford and New York: Berghahn.

Ventsel, A. 2006. "Sakha Pop Music: A Celebration of Consuming." *The Anthropology of East Europe Review* 24 (2): 68–86.

Vinokurova, U. 2014. *Zapovednye etnokul'turnye landshafty Arktiki i Evrazii: sbornik nauchnykh trudov [Preserved Ethnocultural Landscapes of the Arctic and Eurasia: A Collection of Scientific Works]*. Yakutsk: Arctic State Institute of Arts and Culture.

Vitebsky, P. 1995. *The Shaman: Voyages of the Soul, Trance, Ecstasy and Healing from Siberia to the Amazon*. London: Duncan Baird (also published as *Shamanism*, Norman: University of Oklahoma 2001).

Vitebsky, P. 2005. *Reindeer People: Living with Animals and Spirits in Siberia*. Boston: Houghton Mifflin and London: HarperCollins.

Vitebsky, P., and A. Alekseyev. 2015a. "Casting Timeshadows: Pleasure and Sadness of Moving among Nomadic Reindeer Herders in Northeast Siberia." *Mobilities* 10 (4): 518–530.

Vitebsky, P., and A. Alekseyev. 2015b. "Siberia." *Annual Review of Anthropology* 44: 439–455.

Vorren, Ø., and H. Eriksen. 1993. *Samiske offerplasser i Varanger [Sami Offering Sites in Varanger].* Tromsø: Tromsø Museums Skrifter.

Wenzel, G. 1991. *Animal Rights, Human Rights: Ecology, Economy and Ideology in the Canadian Arctic.* Toronto: University of Toronto.

Wiget, A., and O. Balalaeva. 2011. *Khanty, People of the Taiga: Surviving the 20th Century.* Chicago: University of Chicago for University of Alaska Press.

Willerslev, R. 2007. *Soul Hunters: Hunting, Animism and Personhood among the Siberian Yukaghirs.* Berkeley: University of California.

Willerslev, R., P. Vitebsky, and A. Alekseyev. 2015. "Sacrifice as the Ideal Hunt: A Cosmological Explanation for the Origin of Reindeer Domestication." *Journal of the Royal Anthropological Institute* 21 (1): 1–31.

Yamada, T., and T. Irimoto. 1998. *Circumpolar Animism and Shamanism.* Sapporo: Hokkaido University Press.

Znamenski, A. 1999. *Shamanism and Christianity: Native Encounters with Russian Orthodox Mission in Siberia and Alaska, 1820–1917.* Westport, CT: Greenwood.

Znamenski, A. 2007. *The Beauty of the Primitive: Shamanism and the Western Imagination.* Oxford: Oxford University Press.

Zorbas, K. 2015. "The Origins and Reinvention of Shamanic Retaliation in a Siberian City." *Journal of Anthropological Research* 71: 401–422.

SECTION 2

The fundamental importance of land, territories and resources

Else Grete Broderstad and Dalee Dorough

The Arctic is characterized by the emergence of new diverse governance arrangements, as well as long-standing governance practices deeply rooted in cultural traditions (Arctic Human Development Reports [AHDR] 2004, 2014). In this section on sustainable land and resource development, the first four chapters address different governance changes and challenges, while the last three chapters in particular focus on the role of Indigenous knowledge, a paramount focus, as states are increasingly obliged to recognize and respect Indigenous knowledge systems in land- and marine-based decision-making.

The development of governance arrangements and practices has unfolded within different legal and political systems, reflecting demographic, geographic and political variations, as well as differences in colonial histories as expressed through overseas and internal colonization and nation-building processes. Significant drivers of the current debate on governance are the interacting forces of climate change and globalization, affecting different Indigenous realities in the Arctic. Furthermore, Indigenous peoples living in the Arctic insist on the right to self-determination and self-government. The prerequisite of the right of self-determination, including recognition of Indigenous knowledge in support for decision-making and regional and local claims of effective participation, have resulted in an array of Arctic resource governance arrangements.

The first four chapters of this section (8–11) present challenges for Indigenous land and resource governance, although in different ways. While the rights of Indigenous peoples across the Arctic over the last three to four decades have been recognized and strengthened through legislation, including protective measures for Indigenous traditional land and resource use, this development is far from straightforward. One of the emerging trends in the Arctic is devolution and tension with centralization. While a devolution of authority and an upsurge in Indigenous governance bodies have taken place over two decades, including within the Russian Federation in the 1990s, this process in Russia has been caught up in a broader entrenchment of central state power (Poelzer et al. 2015, 187). In Chapter 8, Fondahl et al. verify this trend in their examination of evolving federal legislation related to Indigenous territorial rights, creating numerous challenges for Indigenous peoples in Russia due to narrow interpretations of the laws and lack of comprehensive implementation. Though achievements in legislation characterize some regions of Russia, a trend of unifying regional legislation in accordance with the provisions of federal laws poses a threat to Indigenous rights, resulting in the diminishment of these essential rights.

While Chapter 8 relates to a trend of devolution and recentralization, the development of Sámi law (Chapter 9) and institutionalization (Chapter 10) in the Sámi-Norwegian context can be seen as a reflection of a second trend in the Arctic, of increasing Indigenous political and legal empowerment. In Chapter 9, Ravna describes the development of Sámi law and Sámi legal protections over the past 30 years and summarizes the latest advances in international law related to the Sámi. Simultaneously, Sámi law encounters difficulties in being recognized and accepted as a source of law in Norwegian courts. Ravna addresses challenges in relation to international law obligations applying to the legal interpretation of international law incorporation into the Finnmark Act and to the accompanying processes of surveying and recognizing existing rights in Finnmark.

Founded on the Finnmark Act, a management agency – the Finnmark Estate (FeFo) – was established. As with other land and resource management arrangements in the Arctic, FeFo is a result of Indigenous land claims, implying that particular values, norms and principles are constitutive for and underpin FeFo as an institution. Broderstad and Josefsen in Chapter 10 discuss the role of the board of FeFo and the lack of a joint understanding of the two appointing bodies of the FeFo board – the Sámi Parliament and the Finnmark County Council – on how to address these principles in order to strengthen the focus of FeFo governance, illustrating challenges related to public support and residents' conflicting views on Indigenous lands claims (Broderstad et al. 2020).

Whereas FeFo is a result of political processes resulting in the adoption of the Finnmark Act in 2005 by the Norwegian Parliament, the political processes related to Sámi rights are different in Sweden (Josefsen et al. 2014). The most recent development illustrating that the political system in Sweden has failed to clarify Sámi property rights is the Girjas case in northern Sweden. A 1993 revision of law by the state allowed everybody to hunt and fish in the high mountain area. In 2009, the Swedish National Union of Reindeer Herders and the Girjas Sámi community turned to the judicial system to have their specific rights enforced. Brännström in Chapter 11 outlines with reference to the historical context in case law the legal assessments of the District court and the Court of Appeal in the Girjas Case. In January 2020, the Supreme Court of Sweden ruled that the Girjas Sámi community does in fact hold exclusive rights to decide who gets to hunt and fish within this area. In addition to the legal implication of the Girjas verdict, the Sámi right to decide on hunting and fishing will expectedly add to the array of Arctic governance arrangements established on Indigenous use of lands and resources. A key issue of contention in the Girjas Case was "how Sámi utilized the landscape and the resources during historic times." The Sámi referred to the importance of historical Indigenous land use as evidence and topography, seasonal changes and weather conditions. Sophisticated harvesting strategies and a profound, long-standing relationship to this area were emphasized.

Settlement patterns and prerequisites for mobility as referred to by the Sámi in the Girjas Case demonstrate that for centuries, Indigenous peoples have developed social practices ensuring sustainable use of key renewable resources and resilience and adaptability to major changes in the biophysical system with which they interact. Many of these practices remain relevant today and are as critical as the example of Inuit use of marine areas depicted by Aporta and Watt in Chapter 12, which illustrates the counter-mapping of Inuit routes illuminating the extraordinary complexity of the interrelationships between Inuit, animals, land and ocean. As pointed out by Aporta and Watt, understanding Inuit history and mobility patterns in marine regions is paramount in times of increasing interest by multiple outside stakeholders. These routes and the mapping itself can potentially be "interpreted as political statements towards a recognition of the existence of an Inuit homeland across (or even beyond) national jurisdictions."

The debate on multilevel Arctic governance and resource governance includes land claim agreements, co-management arrangements, the role of Indigenous knowledge and question of eligibility and the distinct nature of Indigenous rights holders. In Chapter 13, Langdon presents information on the history and current status of marine mammal hunting by Indigenous peoples in Alaska, the legal context of that hunting and the implications of definition based on recently acquired demographic data. He problematizes eligibility in the case of ANCSA and the definition of Alaska Natives, which has two elements. The first relates to descent, and the second provides that if a person does not meet the first standard, they are to be deemed Alaska Native if they are so regarded by their collective membership. However, management agencies have chosen not to operationalize the second standard. The question of eligibility illustrates challenges related to how states have interfered with the way Indigenous peoples identify themselves. The practices of definition are framed by three broad characteristics, descent, cultural requirements and self-definition (de Costa 2016, 26), are recognizable in Langdon's discussion. Demographic changes have resulted in increasing numbers of Alaska Native peoples becoming ineligible to hunt marine mammals, resulting in a contest of Indigeneity between those who hold and claim Indigenous status.

Indigenous knowledge systems are diverse. In Chapter 14, Dahl and Tejsner explore the various knowledge concepts and their interrelations to better understand the distinctions. Despite an increased acknowledgement of the importance of using Indigenous knowledge alongside science in research, it is still unevenly used. Dahl and Tejsner explain this with a lack of agreed definitions – "complicating the utilization of knowledge is that there are no single agreed definitions of what each knowledge concept means." The knowledge of Indigenous peoples is linked to their distinct and collective rights, to power relations and to governance structures. They argue that IK is a legitimate knowledge concept that emphasizes identity and provides legal leverage for the empowerment of Indigenous peoples.

Barnhardt and Kawagley (2005, 10–11) point out Indigenous knowledge systems are themselves diverse and constantly adapting and changing in response to new conditions. As the chapters of this section illustrate, Arctic regions have a history of innovative responses to difficult problems. When political solutions like the Finnmark land governance arrangement (Chapters 9–10) are unattainable, litigation as a strategy is turned to, as the Indigenous (Evenki) from the village of Iengra in Sakha Republic (Yakutia) and the Girjas Sámi community in northern Sweden reveal (Chapters 8 and 11). However, despite legal recognition, new challenges of legal interpretation will likely assert themselves (Chapter 9). Indigenous agency is reflected through the participation in ethnological expertise procedures and efforts at establishing community-based monitoring programs (Chapter 8), as well as through counter-mapping and documenting traditional and Indigenous knowledge (Chapter 12). The current emphasis on utilizing the concept of Indigenous Knowledge of Indigenous peoples highlights the distinct status and collective rights of Indigenous peoples (Chapter 14). Still, despite a strong desire by Alaskan Natives to determine Indigenous eligibility, the legacy of state or colonial interference with the way Indigenous peoples identify themselves manifests itself (Chapter 13). Meanwhile, as de Costa (2016, 55) points out, "In the aspirations set out in UNDRIP and endorsed by most states, there would seem to be little role for the state in defining who was or was not an Indigenous person."

References

Arctic Human Development Report (AHDR). 2004. Edited by N. Einarsson, J. Nymand Larsen, A. Nilsson, and O. R. Young. Akureyri: Stefansson Arctic Institute.

Arctic Human Development Report (AHDR): Regional Processes and Global Linkages. 2014. Edited by J. Nymand Larsen and G. Fondahl. Copenhagen: Nordic Council of Ministers.

Barnhardt, R., and A. O. Kawagley. 2005. "Indigenous Knowledge Systems and Alaska Native Ways of Knowing." *Anthropology and Education Quarterly* 36: 8–23.

Broderstad, E. G., V. Hausner, E. Josefsen, and S. U. Søreng. 2020. "Local Support among Arctic Residents to a Land Tenure Reform in Finnmark, Norway." *Land Use Policy* 91(1): 12. doi.org/10.1016/j.landusepol.2019.104326.

de Costa, R. 2016. "States' Definitions of Indigenous Peoples:A Survey of Practices." In *Indigenous Politics: Institutions, Representation, Mobilisation*, edited by M. Berg-Nordlie et al., 25–60. Colchester, UK: ECPR Press.

Josefsen, E., U. Mörkenstam, and J. Saglie. 2014: "Different Institutions within Similar States: The Norwegian and Swedish Sámediggis." *Ethnopolitics: Formerly Global Review of Ethnopolitics* 1 (14): 32–51. doi: 10.1080/17449057.2014.926611.

Poelzer, G., G. N. Wilson, E. G. Broderstad, D. Hirshberg, M. Kimmel, and A. Sleptsov. 2015. "Governance in the Arctic: Political Systems and Geopolitics." In *Arctic Human Development: Report Regional Processes and Global Linkages*, 185–220. Copenhagen: Nordic Council of Ministers.

8

CHANGING INDIGENOUS TERRITORIAL RIGHTS IN THE RUSSIAN NORTH

Gail Fondahl, Viktoriya Filippova, Antonina Savvinova, and Vyacheslav Shadrin

Abstract

Russia's legal approach to Indigenous territorial rights differs significantly from that of other Arctic states. Key Russian laws acknowledge the importance of territorial rights to the cultural and physical survival of Indigenous northerners, focusing on the vulnerability and need for protection of their traditional activities and life ways, while not recognizing inherent Indigenous title to/ownership of ancestral lands. Thus, early laws focused on the protection of lands necessary to the traditional activities and on granting Indigenous peoples use rights, free-of-charge and in perpetuity, to these lands. However, since the turn of the millennium, legal protection of Indigenous territorial rights has eroded in favour of removing impediments to the extractive activities on which Russia's economy depends, and introducing rents and in cases competitive bidding for lands for traditional activities. Moreover, federal laws conflict with each other on these matters. In this chapter, we review the evolving federal legislation on/affecting Indigenous territorial rights, focusing especially on the past 15 years. We conclude with observations on what is needed to improve the situation of Indigenous territorial rights in the Russian North.

Introduction

Territorial rights (rights to land and the resources on those lands) are critical to the survival of all Indigenous peoples, both physically and culturally.[1] In the Russian Arctic, the Indigenous numerically-small Peoples of the North (*korennye malochislennye narody Severa*; hereafter, KMNS[2]) remain highly dependent on "traditional" activities that require the use of extensive territories, such as reindeer herding and hunting. Approximately two-thirds of these people live in rural areas; for many, herding, hunting, fishing, and plant-gathering provide both subsistence needs and some income generation (Sulandziga and Berezhkov 2017). Moreover, these activities are often cited as critical to cultural vitality and identity: "Without reindeer, we are not Evenki" (or Eveny, Nentsy, Chukchi, etc.) is a familiar and recurrent assertion.

Russia's legal approach to Indigenous territorial rights differs significantly from that of other Arctic states. Inherent rights to (including but not limited to ownership, past or current, of) ancestral lands are not legally recognized. Yet key Russian laws have directly acknowledged the importance of territorial rights to the cultural and physical survival of KMNS, focusing on the vulnerability of the traditional activities and life ways of the KMNS and their need for protection. The extensive territorial requirements of the traditional activities necessitate the protection of lands for these activities.

However, whereas at the turn of the millennium, legal attention focused on the protection and enabling of traditional KMNS activities, more recently, we observe a shift toward assisting extractive activities. The homelands of the KMNS contain a vast amount of Russia's natural resources, including 100% of its diamonds and platinum, over 90% of its natural gas and nickel, over 75% of its oil, and over 60% of its gold (Sulandziga and Berezhkov 2017; Sulandziga and Sulandziga, this volume). Russia's economy depends heavily on these resources: in the past 15 years, removing impediments to their development has taken precedence over the legal protection of Indigenous territorial rights. In practice, the development of such resources has removed significant tracts of land from traditional activities, including both directly and due to pollution of lands and waters. As one author noted, KMNS's "ancestral lands and pastures are by one means or another alienated under grandiose projects of developing gas, oil and other sub-surface resources, and they are left with crumbs and a 'folklore-exhibition' model of existence" (Tranin 2010, 22).

This chapter reviews the evolving legislation that addresses Indigenous rights to land and resources in the Russian Arctic, focusing especially on the past 15 years. We admittedly write from a statist perspective, not discussing the current status and roles of customary (Indigenous) law in governing territorial rights. We describe briefly key federal legislation on Indigenous territorial rights in the early post-Soviet period. The greater part of the chapter documents changes to the federal laws that have mostly eroded protections initially provided, as well as insufficiencies in the legislation that have resulted in opportunities for alienation of land. Although KMNS's ancestral lands (*iskonnaya sreda obitaniya*) and traditional way of life are the joint responsibility of the federal government and the governments of Russia's federal 'subjects' (republics, territories, autonomous districts, etc.) according to Russia's Constitution[3] and thus influenced by both federal and regional law, we focus mostly on federal law due to space limitations. In providing examples, we often choose to include those from Sakha Republic (Yakutia), which is not only home to three of the authors but is considered at the vanguard of Indigenous rights in the Russian Federation (both in terms of legislation and its implementation). We conclude with some observations on what is needed to improve the situation of Indigenous territorial rights in this part of the Arctic.

Early legislative initiatives regarding Indigenous territorial rights in post-Soviet Russia

Numerous English-language articles[4] discuss in some detail the legislation on territorial rights during the first decade or so of the post-Soviet period (Diatchkova 2001; Fondahl et al. 2001; Fondahl and Poelzer 2003; Garipov 2013; Kryazhkov 2013a, 2015; Osherenko 2001; Vakhtin 1998; Xanthaki 2004); thus, we will limit our recapitulation. An early (1992) Presidential Edict[5] ordered the federal government "to determine, in places of [KMNS's] habitation and economic activities, territories of traditional nature use which are the inalienable property of these peoples and without their consent are not subject to alienation for industrial development." These

Territories of Traditional Nature Use, discussed subsequently, have become a key spatial means of protecting Indigenous rights. The Russian Constitution,[6] adopted in 1993, guarantees the rights of Indigenous small numbered peoples in agreement with international treaties and norms (Article 69); much discussion has ensued on what this means regarding the recognition of Indigenous territorial rights. Perhaps more importantly, its Article 72 assigns the issue of "protection of ancestral lands and traditional ways of life of numerically small ethnic communities" as the joint responsibility of federal government and its subjects, thus recognizing the need for such. Ensuing sectoral laws, such as those on the faunal world (1995) and on the continental shelf (1995), included specific clauses on priority Indigenous rights to natural resources necessary for the pursuit of traditional activities.[7] However, new versions of these laws have often omitted such clauses. When such clauses remain, they often are impossible to apply, in that they require sub-legal acts to direct procedures for realization (e.g., decrees, orders), and these are not forthcoming. And other laws have also eliminated some of the protections provided by the earlier laws.

The turn of the millennium is often identified as the watershed in legal protection of Indigenous rights in the Russian Federation. Three framework federal laws were adopted: "On the Guarantee of Rights of the Indigenous Numerically-Small Peoples of the Russian Federation" (1999),[8] "On General Principles of the Organization of Obshchinas[9] of the Indigenous Numerically-Small Peoples of the North, Siberia and Far East of the Russian Federation" (2000),[10] and "On Territories of Traditional Nature Use of the Indigenous Numerically-Small Peoples of the North, Siberia and Far East of the Russian Federation" (2001).[11] In terms of territorial rights, the 1999 law "On Guarantees" provided Indigenous peoples with free-of-charge use of land for their traditional activities. The 2000 law on *Obshchinas* enabled the creation of such Indigenous organizations, which could petition for a land base on which to pursue traditional activities. Land allocated to obshchinas was originally done so free of charge and without term, with the right to inheritance. The 2001 law on Territories of Traditional Nature Use (*territorii traditionnogo prirodopol'zovanie* or TTPs) finally (partially) addressed the directive of the 1992 Presidential Edict in providing the basis for the establishment of TTPs – territories for Indigenous traditional activities protected from industrial encroachment.

However, none of these laws were complete in themselves: they were 'framework' laws that needed to be backed by further legislative acts (by-laws, decrees, orders, etc.) to provide the specific mechanisms by which the provisions of each could be operationalized (see Sulandziga and Sulandziga, this volume). Some of this was provided at the regional level: 'subjects' of the Federation – the republics, territories, autonomous districts, and so on – passed their own legislation that often fleshed out procedures and mechanisms. Indeed, many subjects passed legislation on territorial rights well before the turn of the millennium (see Kryazhkov 2013b, Volume 2, for a comprehensive compilation). These regional legal acts sometimes duplicate federal law and sometimes contradict it. For example, the 1998 Sakha Republic's law, "On Suktul of the Yukaghir People" (regarding self-government), had to be amended numerous times to remove provisions that conflicted with new federal laws regarding jurisdiction as these laws were adopted, and by 2005, it was deemed necessary to adopt a whole new edition of the republican law (Astakhova and Shadrin 2019). The use of different terminology across legislation provides a further complication to interpretation and implementation. The corpus of federal law itself is replete with ambiguities and contradictions.

Most importantly, changes in both the political and economic situation in Russia – the centralization of power and the increasing focus on the extraction of non-renewable resources, which strengthened the lobbying power of industries opposing legal reforms that would hinder their activities – impeded further development of the legal acts needed to realize the provisions of these three framework laws. Nevertheless, the laws provide notable *opportunities* for protection of Indigenous territorial rights in Russia. Clear, coherent legislation that elaborates the stated objectives of these laws, and provides mechanisms for their realization, could result in significant protection of Indigenous rights in the Russian Federation.

Legal erosions of Indigenous territorial rights: the last 15 years

"Land is a layered pie. And on this layered pie there can be several possible users" (Indigenous leader, Yakutsk, March 2019). That is, a specific area of land comprises strata of territorial rights: different persons or judicial entities (such as obshchinas or companies) may hold the legal rights to different resources on the same land-base, such as to reindeer pasture, species of hunted animals, timber resources, mammoth tusks, various sub-surface mineral resources, and so on. Thus, a whole range of laws can impinge on Indigenous rights on a given territory. Land used by Indigenous persons and collectives for traditional activities is most commonly 'owned' by the federal government, while legislation provides for joint oversight and protection of traditional activities by the federal and regional governments. Subsequently, we focus on the key legal changes that have affected Indigenous territorial rights over the past 15 years, including by enabling the different 'layers' to be allocated to different users.

Obshchina (clan community) lands

The 2000 federal law on Obshchinas allowed Indigenous collectives to organize and to petition for a land base on which to pursue traditional activities.[12] Lands would be available to obshchinas free of charge, for perpetual (without term) use (though the lands would not be transferred as property; that is, the obshchinas did not receive title to the land). However, even prior to this federal law, changes in other laws challenged the existence of obshchinas that had been created in the wake of regional laws. While the Tax Codex required that all organizations be registered in accordance with its provisions, 'obshchina' was not a category within Russia's Civil Code.[13] Many obshchinas thus re-registered as 'agricultural cooperatives' (*sel'skokhozyaystvennye proizvodstvennye kooperativy*, SPK/SKhPK) to meet the requirements of the Tax and Civil codes. However, SPKs/SKhPKs were categorized as a 'commercial' form of organization, which technically triggered the loss of rights to land provided to Indigenous obshchinas for non-commercial, 'subsistence' purposes, a fact unclear to many obshchina leaders. Moreover, in some areas, like Yakutia, the Indigenous SPKs/SKhPKs continued to enjoy free-of-charge use of land.

Two other changes brought the issue of land payments to the fore. Most of the land plots allocated to obshchinas are recognized as federal property. The Forest Code of the Russian Federation (2006)[14] (paragraph 1 of Art. 8) establishes forest lands as federal property. The legal regime of the forest fund prohibits the emergence of private property on the lands of the forest fund, thus excluding the possibility of issuing the right of permanent (in perpetuity) right to obshchinas. Ongoing centralization policies, increasingly requiring coherence with federal statutes, challenged and eroded the rights of regions to regulate issues regarding land use, thus removing protections that the more generous regions (in terms of Indigenous territorial rights), such as Yakutia, provided to KMNS.

In 2010, changes to the Civil Code finally legally introduced the concept of 'obshchina of Indigenous Numerically Small Peoples' as a non-commercial form of organization. This meant that former obshchinas, which for the most part had reconstituted as SPKs/SKhPKs, now could, and indeed had to, change back to (non-commercial) obshchinas if they wanted to maintain their privileges regarding land use. However, to do so, they had to liquidate the SPK/ SKhPK and form a 'new' obshchinas (rather than reverting to their former status). And as new entities, under the law, they were not eligible for permanent rights to the hunting and fishing grounds and reindeer pastures that they had originally enjoyed (these rights were only 'grandfathered' for those obshchinas that had existed as such without re-registering as SPKs/SKhPKs). They were now subject to the competition for such lands, as described in the next section.

In Yakutia, for instance, this issue affects 27.2 million hectares of land transferred to the use of 174 obshchinas. Whereas previously those obshchinas received 'free-of-charge, in perpetuity use' of land, they now receive use rights to allotments for a set number of years (often 25, but up to 45), after which they will need to re-register. They must pay rent for the lands, a concept antithetical to many Indigenous people's understandings of proper relationships with the land on/with which they coexist.[15] One head of an obshchina noted: "If we use land and water and pay for it – then we immediately alienate Indigenous peoples from their land" (Interview, Yakutsk, March 2019).

Beyond this existential challenge is a lesser but nevertheless significant financial one: although rents can be very low per hectare, obshchina territories often encompass hundreds of thousands of hectares, as required for traditional activities of reindeer husbandry and hunting (see subsequently). Thus, the requirement of rent has frequently prevented obshchinas from retaining their full land base. Lands released by clan communities due to their inability to pay rent, though still critical for the traditional activities, become available to others (non-KMNS) and can end up being leased or even sold to others whose land use is detrimental to traditional activities.

The registration of hunting grounds and reindeer pasture requires special documents and application procedures. Obshchinas (and other collectives) pursuing hunting must submit scientifically based plans for their hunting territories. If during the Soviet period such planning was carried out for the state farms (*sovkhozy* and *gospromkhozy*) by specialized and state-financed expeditions, the responsibility has now devolved to the obshchina which must allocate additional funds for the cadastral registration of its lands and for establishing a hunting management plan.

Obshchinas are also faced with an ever-increasing amount of reporting regarding the lands and resources they now rent. This provides further disincentives to petitioning for land and further means by which land can be alienated (if paperwork is not submitted as required) (authors' field notes, May 2017).

Changes to territorial rights regarding hunting and fishing

Fishing and hunting still serve as the basis of existence for many KMNS. Thus, their access to hunting and fishing territories and terrestrial and aquatic bioresources are of particular importance. Federal law protects Indigenous peoples' rights to fish, as part of their traditional way of life, without permits and with no limits to their takes. Indigenous persons nevertheless frequently face serious difficulties in exercising their rights to these resources. They frequently do not enjoy the right to determine *where* and *when* they fish.

Hunting

A 1995 Law on the Faunal World[16] provided priority use rights for Indigenous peoples to harvest for subsistence purposes. Revisions to laws, based on the principles of payment for the use of land and resources, did not include clear, specific exclusions for KMNS. Thus, in 2008,

KMNS found themselves faced with a new system by which the rights to hunting lands and fishing sites were now offered via auction.[17] This has had a profoundly negative effect on KMNS access to hunting and fishing resources. "We have become poachers on our own lands" is a repeated complaint (authors' field notes, August 2016, May 2017, March 2018).

Difficulties regarding access to hunting and fishing resources have been compounded by narrow interpretations of the laws regarding who can harvest resources. KMNS retain the right to hunt without payment and without permits when pursuing 'traditional' (subsistence) hunting. However, at least two problems arise here. Some law enforcers have interpreted this law very narrowly, citing hunters who have been found to hunt 'beyond their own needs', where they were also hunting for community members who could not hunt themselves. Thus, an elderly person or invalid, who is unable to personally hunt, has no way to access her/his country foods. Currently this is being challenged in court (see, e.g., Kryazhkov 2016). In one case, the plaintiff, who heads an obshchina and also serves as chair of the regional association of KMNS of the Taymyr Dolgan-Nenets Municipal District, Gennady Shchukin, has argued that this practice is not only unreasonable but contravenes Russia's Constitution. The case made it all the way to the Constitutional Court of the Russian Federation in April 2019 (Constitutional Court 2019).

Another critical problem is the current lack of legal means to determine who is a member of a KMNS and thus entitled to free hunting (and fishing and use of reindeer pasture), since the abolishment of the indication of 'ethnicity' in Russian passports. Following over a half-decade of debate on how best to deal with this problem, in early 2020, the Russian government adopted a law to establish a registry of Indigenous persons.[18] This law stipulates that such a Registry will be functional in two years (by February 2022). Numerous challenges remain to be resolved regarding limitations on criteria for a person to be recorded in the registry, burden of proof required, and control over the process (Sulandziga and Sulandziga, this volume; Fondahl et al. 2020).

A 2017 revision to the Hunting Law regulates the adoption of Hunting Management Agreements for Indigenous hunters who do have long-term licenses for the right to use the faunal resources, outside of the auction process.[19] Such agreements enable such hunters to better regulate hunting on their lands and include the right to pursue regulation of predator species, to develop hunting infrastructure, and to draw up indictments against illegal hunters on their lands.

Fishing

Indigenous persons and collectives experience similar problems with accessing fishing grounds. The introduction of the distribution of fishing grounds by auction at the end of 2008[20] has led to some obshchinas losing their sites. For instance, in northern Yakutia, in the summer of 2011, the local communities of Russkoe Ustye and Allaykha lost access to key fishing places. The sites 'belonged' to the Lena State Basin Administration for Waterways and Navigation (LGBUVPS). While villagers assumed that the LGBUVPS would conclude catch agreements with them, the administration decided to deploy its own fishers and ordered others to leave the fishing sites. In protest, local residents staged a two-day picket on the central square of the county's administrative centre, Chokurdakh. Then-President of the Republic V.A. Shtyrov intervened, demanding that LGBUVPS enter into agreements with the local obshchinas.

Problems arise in the implementation of the laws regarding traditional fisheries. In accordance with the current version of Article 25 of the Federal Law "On Fisheries", to protect the traditional way of life of KMNS, fishing can be carried out by KMNS members and their obshchinas on allotted fishing grounds (as well as anywhere they wish for subsistence needs).[21] The Federal Law "On Fisheries" also provides a mechanism for allocating catch quotas to ensure support for Indigenous peoples' 'traditional ways of life'. 'Subjects' of the Russian Federation (republics, territories, etc.) receive total allowable catches of various aquatic species (Art. 30, 31), the distribution of which they are responsible for. Subjects also receive a quota for *traditional* fishing, to be distributed to KMNS members and their obshchinas.[22] By Order of the Federal Ministry of Agriculture, the boundaries of fishing grounds are determined by the subjects' authorized executive organs of power.[23] Yet the subjects' regulatory legal acts often fail to lay out the procedures to do this, hindering KMNS in their fishing. Practice shows that without allocation of fishing grounds and permits, Indigenous peoples often cannot enjoy their rights as envisaged by this quota allocation process.

In 2015, Yakutia's Constitutional Court considered the question of the legality of the declarative principle in the implementation of traditional fishing by KMNS. It resolved that persons belonging to the KMNS who have no allocated fishing grounds are not required to submit an application for allocation of a quota in order to pursue traditional fishing. Thus, starting in 2016, KMNS and their obshchinas were able to carry out traditional fishing in their traditional places without submitting applications, which greatly simplified the lives of KMNS. In other subjects, the situation has not been nearly so positive: for instance, in perhaps a worst-case scenario, the Amur Oblast tried to award Indigenous norms on a *daily* basis, amounting to an allocation of 12 grams of fish per person per day.

Protection of reindeer pastures

Reindeer pasture occupies 335.2 million hectares, or 19.6% of Russia's territory, although the growth of industrial development of the Arctic is reducing this figure. Like hunting lands, most reindeer pasture (>80%) is categorized as 'forested land' and thus under federal domain and guided by related laws. At the same time, there is actually no defined legal category of 'reindeer pasture' in Russian law, despite that term appearing in a number of laws.

Most pasture has not been officially assigned to reindeer-herding enterprises. In many cases, the same land is used by KMNS and their obshchinas for hunting and pasturing reindeer.

Law provides for the protection of land occupied by reindeer pastures in the regions of the Far North, including from privatization.[24] However, as for hunting lands, reindeer pasture is now subject to rent, though without competitive bidding (auction). Indeed, two federal legislative acts come into conflict – the 1999 Federal Law "On Guarantees" and the 2004 Land Code.[25] The Land Code introduced rent for reindeer pasture (and hunting lands, as noted previously), from 1 January 2018. The requirement applies to those obshchinas that had not registered their land in the federal land cadastre (an expensive proposition), as well as for any obshchinas formed after the Land Code's adoption. As explained previously, many obshchinas that did register later had to convert to another status and then converted back to obshchinas only after the Land Code went into effect. Reindeer pasture now is allocated for no more than ten years.

Indigenous persons who herd find the charging of rent for reindeer pasture even more insidious than rent for hunting lands, in that the lands required for herding are vast:

> we use tens of thousands of hectares for reindeer husbandry – reindeer husbandry covers a huge part of Russia – so we (KMNS) would have to pay for all of Russia! Well, half of it. I alone have almost 300,000 hectares, which means I would have to pay almost 300,000 roubles annually for my pasture. The Council of the Federation just doesn't think about this.
>
> (Head of obshchina, Yakutsk, March 2019)

In some places, the cost of renting pastures is much less (in the north of Zabaykalskiy Kray in 2018, it was reported to 5 kopecks per hectare; authors' field notes, March 2018). Nevertheless, such charges can be difficult to pay, as reindeer herding often provides little income. Moreover, Indigenous people wonder why they must pay to use their ancestral lands.

A draft law on reindeer husbandry has been in the works since at least 1994 (author's field notes, May 1994). Various versions have at times proposed a variety of tenure arrangements, including free-of-charge, in perpetuity use rights, as well as leasing pastureland on a competitive basis (e.g. Federal'nyy 2003; Krivoshapkin 2014). Discussion of such a law appears to be dormant at present.

Changes to Territories of Traditional Nature Use

The 2001 law on TTPs mandated the protection of lands needed for traditional activities from industrial development through the creation of such territories and the subsequent limitation of other activities damaging to traditional activities within their boundaries. Even prior to the 2001 law, as early as 1995, TTPs were mentioned as one form of "Specially Protected Nature Territories" (*osobo okhranyaemye prirodnye territorii*, or OOPTs), in a section determining the use regimes of OOPTs.[26] That section provided for the use of specific parts of OOPTS for traditional economic activities when such territories were organized on the ancestral lands of KMNS.

A main goal of the law on TTPs was to enable the setting aside of lands of adequate size for the protection of traditional activities, as needed by current and future generations of those obshchinas practicing such activities. That is, they would be excluded from property transfer (via buying-selling, inheritance, lease, etc.; Tranin 2010, 12). A secondary objective is to ensure that in cases where industrial development is 'unavoidable', those obshchinas affected by it receive compensation (including monetary compensation and the allocation of other lands for their activities).

TTPs are created 'from the bottom up': Indigenous people have to petition to have these created (that is, they have to take the initiative). TTPs can (ostensibly) be created at the local, regional, or federal level. To date, none have been created at the federal level, despite numerous attempts. Their establishment is impeded by the lack of further legal acts to clarify the legal procedures for forming and using TTPs (such as provisions [*polozheniya*] or orders [*rasporyazheniya*]).

However, large numbers of TTPs exist in some federal subjects, which have passed their own, often more detailed, legislation on TTPs. These TTPs serve a variety of purposes. For instance, the majority of about 500 TTPs established within the Khanty-Mansi Autonomous Okrug lie within hydrocarbon extraction areas; rather than protecting lands for traditional activities from industrial encroachment, these TTPs enable KMNS to receive compensation for the damages done to their homelands (Wiget and Balalaeva 2011). In Sakha Republic (Yakutia), 59 TTPs provide opportunities both for protection and compensation against new industrial encroachment. There, the status of TTP invokes the need for carrying out 'ethnological expertise' (a cultural impact assessment of potential impacts of development on Indigenous peoples and their cultures) as well as environmental impact assessment for any prospective industrial development. Since the republican Law on Ethnological Expertise (2010) came into effect, 13 such processes have been carried out. The federal government is now drafting a similar law.[27]

Most recently, TTPs have been created to protect lands used for traditional Indigenous activities against the threat of alienation for other purposes, notably to Far East Hectares (Fondahl et al. 2019). Faced with a federal program allowing any Russian citizen to request a hectare of land in the Far East Federal District (of which Yakutia is a part) for development purposes, the government of Yakutia first proposed amendments to the law, during its drafting state, to remove TTPs from lands available for allocation to Far East Hectares. It then pursued a vigorous campaign to register existing TTPs and to create numerous new ones.

Revisions to the law on Specially Protected Natural Territories in 2013 removed TTPs from the status of OOPTs.[28] This has diminished their ability to provide protection from industrial encroachment. Notably, they are no longer covered by Article 27 of the Land Codex (part 5), which forbids land within OOPTs becoming private land or the object of transactions (Yakel' 2014). Indeed, some industrial developers have taken the loss of OOPT status to mean that compensation payments for damage done on TTPs are no longer necessary (interview with legal expert on Indigenous rights, Moscow, March 2019; see also Stammler and Ivanova 2016; Sulandziga and Sulandziga, this volume). However, not all Indigenous leaders feel that the TTPs losing the status of OOPT was negative: some note that the OOPT's focus on protecting nature, including animals, can lead to prohibitions on hunting, which conflicts with the TTP's status of protecting traditional activities, including hunting (interview, Indigenous leader, Yakutsk, March 2019). Moreover, the majority of TTPs are located on forested lands, that is, federal lands. Rather subtle changes to the OOPT law could remove this impediment to Indigenous rights.

The legal status of TTPs formed at the regional and local level remains precarious. Most are situated on federal forest lands, and thus the authority of local or regional authorities to unilaterally establish these is uncertain and so far untested. Of specific concern to many Indigenous leaders are suggested changes to the Law on TTPs that propose that all TTPs should be federal level. What this would mean for the hundreds of extant TTPs, created at the regional and local levels, is unclear. Some fear that such a revision to the law could force current TTPs to be 're-registered' at the federal level, and in the process, many might be refused, thus removing the protection they provide (author's field notes, March 2019; Zakonoproekt 2017).

At present, TTPs ostensibly provide some protection from industrial development and, in cases where such development does occur, enable affected KMNS to receive compensation for losses sustained to traditional activities. However, as of yet, no methods have been developed to calculate such losses, thus undermining this provision.

Over the past half-decade, several groups have worked on new versions of the TTP law. Most recently, the Federal Agency for Nationality Affairs and the Russian Association of Indigenous Peoples of the North (RAIPON) have taken lead roles in this process. The process has involved significant contention among the interested parties, including the Ministry of Energy, the Ministry of Natural Resources, representatives from Gasprom, Rosneft, and Rosgidro (a major Russian hydroelectric company): resource developers have lobbied against any provisions that might restrict resource extraction or make it more costly and have even suggested that no law is needed and that agreements between Indigenous groups and companies could suffice to address concerns of the former. Indigenous groups disagree with this assessment, seeing the law as providing the minimum needed protection.

Other challenges

Challenges to Indigenous territorial rights are numerous; subsequently we outline two critical ones. First, the regional legislation of numerous federal subjects provides better protection of Indigenous rights than does the corresponding federal law. However, federal law supersedes regional law; thus, moves to adopt federal laws often threaten a reduction in protection of territorial rights. Second, we note a relatively new threat, that of the creation of 'Territories of Priority Development,' in which interests of natural resource extraction (and sometimes processing) threaten to override territorial interests and rights of KMNS.

The potential erosion of regional law

Regional ('subject') laws on Indigenous rights to land vary across Russia. Some subjects, such as Yakutia and Khanty-Mansi Autonomous Okrug, have been at the vanguard of developing regional laws. In Yakutia, such laws are more often implemented (a major complaint of many Indigenous leaders is not the lack of laws but the lack of their implementation). For example, Yakutia's 2018 law "On Responsible Subsurface Resource Use"[29] enshrines legal mechanisms that have no parallel in federal legislation, including the stipulation that the interests of KMNS must be taken into account when developing such resources. These legal mechanisms include 1) the assessment of social and environmental impacts of subsurface resource projects and informing the interested public on the results of such assessment; and 2) monitoring subsurface resource use and providing an annual rating of such users. The law also defines the required forms of interaction between subsurface resource developers and KMNS – negotiations, consultations, agreements, mutual participation of subsurface resource users and obshchinas in meetings, and so on.

As another example, Yakutia's Law "On the legal status of Indigenous numerically small peoples, minorities of the North (new edition)"[30] provides that in places inhabited, and used for traditional activities, by KMNS, their interests must be taken into account by enterprises and organizations, which can operate only in accordance with state programs of natural resource development and environmental protection. Environmental assessment and ethno-ecological expertise (cultural assessment) are mandatory (Art. 22). At the same time, this law enshrines the right of Indigenous peoples to participate in such assessments (Art. 8). Similar provisions are not found in the relevant federal law (on TTPs).

However, in recent years, there has been a dangerous trend of unifying regional legislation in accordance with the provisions of federal laws, which often results in lowering protection. As a result, Indigenous peoples living in places more progressive in terms of Indigenous rights, such as Yakutia, may lose the benefits of these regional laws. Moreover, given that forested lands, where most of the lands used by KMNS to pursue their traditional activities (hunting lands and reindeer pastures) and waters are under federal domain, regional legislation has limited ability to address the current situation in terms of improving KMNS access.

Territories of Priority Development

Territories of Priority Development (TOR/TOSER)[31] pose a new threat to the territories of KMNS. These special economic regions, enabled by a 2014 law,[32] create advantages for industrial enterprises in order to promote such activity, including processing of raw materials. Advantages include not only tax breaks and simplified administrative procedures but also provision of land for such activities, which can be potentially dangerous for Indigenous communities. In northern areas, such enterprises are most often interested in maximizing the extraction of natural resources for subsequent sale and commercial profit. The participation of companies in the development of local areas in which they operate is minimal. Federal legislation insufficiently addresses the norms and rules governing the interaction of businesses and the population of such territories. If larger industrial enterprises practice some level of corporate social responsibility, any such activities are not clearly regulated by current legislation. Under these conditions, the status of TOR (TOSER) completely releases companies from social obligations, including those to KMNS.

What is to be done? Addressing deficiencies in Indigenous territorial rights in Russia

In lieu of a conclusion, in this final section, we briefly identify needed changes in the law to ensure Indigenous territorial rights, note the important role that courts and prosecutors' offices can play in protecting these rights, and comment on the nascent promise of, and continued need for, increased Indigenous activism regarding territorial rights.

Changing the law

Indigenous leaders are currently working to restore the rights of Indigenous peoples in the Russian North to receive land in order to carry out traditional activities on a free-of-charge, in-perpetuity basis. Key issues that they feel need to be addressed include changes to several laws, including the law on TTPs, the Land and Forest Codes, the Law on Subsurface Resources, and the Law on Specially Protected Nature Territories. Some also promote the codification of Russia's various laws on Indigenous rights (authors' field notes August 2016, February May 2017, March 2018, March 2019; see also Mazurov et al. 2018, 126).

Restoring Indigenous people's right to free-of-charge, in-perpetuity use of land for their traditional activities is of foremost importance. This will require changes to the Land Code.

A new version of the TTP law needs to clarify how these can be established at the federal level while retaining the ability to establish these at lower (regional, local) levels. Revisions should also set out how TTPs should be managed and monitored. The implementation of the law should better protect from industrial encroachment rather than depending heavily on compensation for such.

Laws on hunting and fishing need to clearly recognize priority rights of Indigenous northerners to the faunal resources on which they depend for their physical and cultural survival. The implementation of these should ensure that Indigenous peoples receive access to hunting and fishing grounds outside of the auction process and prior to it. Indigenous persons and obshchinas that hunt and herd should receive unfettered use rights to both pasture and faunal resources on their lands.

Also needed is the introduction of the concept of Free, Prior and Informed Consent to the Law on subsurface resources. This law needs to strengthen requirements for timely consultation with KMNS when planning development projects that would remove land from use by Indigenous peoples or damage it in ways that undermine its ability to support traditional activities. The procedures for doing so and methods for calculating and providing compensation for damage caused by industrial activity need to be clearly detailed.

All revisions to the laws must involve substantive opportunities for input from Indigenous persons and careful consideration of recommendations. Indigenous peoples should not fear that new versions of any of the federal laws would weaken their rights.

Using the courts to protect/achieve rights

Courts and prosecutors' offices provide important if underused instruments for the realization of the rights of KMNS. Unfortunately, by virtue of the established practices, KMNS often perceive these bodies as bodies that infringe on, rather than protect, their rights. Yet these bodies monitor the implementation of laws, including laws on KMNS rights. And they can play a positive role in protecting these rights.

In 2016, at the suggestion of the Deputy Prosecutor General of the Russian Federation Yury Gulyagin, the Federal Fisheries Agency took measures to eliminate legal violations of Indigenous rights, identified as a result of verification activities by the North-Eastern Territorial Department of the Federal Agency for Fisheries.[33] Specifically, the Kamchadal obshchina "Kalan" had had its catch reduced without cause. In 2017, upon appeal by the prosecutor's office of the Chukotka Autonomous Region, the Anadyr City Court cancelled what were burdensome conditions of catching anadromous fish species that had been imposed on KMNS without any legislative basis.[34] These rulings provide some optimism about the support role prosecution authorities can provide in ensuring the implementation of the legislation coordinating the activities of economic entities with the local population.

Increasing Indigenous activism

Indigenous individuals and their associations have increased activities in recent years, protesting against the violation of their territorial rights. Scientists frequently support such efforts. We offer one example, from southern Yakutia. Indigenous (Evenki) residents from the village of Iengra (Neryungri County) were concerned with the barbarous behaviour of the "Novaya" gold mining guild, which operated within the TTP "Iengrinskoe". They began to voice their frustration using social media, distributing videos that showed pollution of the local Amunakta River by placer mining. This was picked up by national radio and newspapers. It became emblematic of the way small extractive companies working on the ancestral lands of KMNS often completely ignore and violate environmental, forestry, and other legislation.

The publicity created by social media incited the Government of Sakha Republic (Yakutia) to take a role. To resolve this issue, it called a meeting in the village of Inengra in November 2018, which allowed villagers to express their views without having to travel long distances to the 'county' centre or the Republic's capital. The conflict then went to the courts: in April 2018, the Supreme Court of Sakha Republic (Yakutia) ruled that the Novaya guild had to carry out an ethnological expertise, given its location within a TTP. Here the teamwork of local activists and government authorities demonstrated a resolve to assert and protect KMNS rights, as guaranteed by law.

Associations of KMNS are currently working to increase legal literacy in defending KMNS rights. This includes conducting seminars, round tables, and consultations. In 2018, Yakutia's KMNS held such in 31 settlements in 14 counties of the republic, which covered more than 1,000 people. Further, methodological manuals on how to organize the work of TTPs have been published and efforts made to increase the participation of the Indigenous people in ethnological expertise procedures. We also see early efforts at establishing community-based monitoring programs that help to educate Indigenous communities in documenting their traditional knowledge and traditional environmental resources and in introducing co-management principles to working with industrial companies (Isakov 2018).

Acknowledgements

This research was supported by SSRHC grant 435-2016-0702 and NORRUS grant 257644/H30.

Notes

1 UN Declaration on the Rights of Indigenous Peoples (2007) UN Doc A/61/L.67.
2 The Russian Federation recognizes 40 'Indigenous numerically-small peoples of the north, Siberia and the Far East' [KMNS]; Perechen' korennykh malochislennykh narodov Severa, Sibiri i Dal'nego Vostoka [List of Indigenous numerically small peoples of the North, Siberian and Far East of the Russian Federation], Order of the Government of the RF no.536-r, 17 April 2006.
3 Article 72, part 1, paragraph "m" of the Constitution: *Konstitutsiya Rossiyskoy Federatsii* [Constitution of the Russian Federation]. 1993. Moscow: Juridicheskaya literatura.
4 Of course, the Russian-language literature is much more extensive and includes numerous books as well as a myriad of articles. Most notable are Vladimir Kryazhov's compendia of the relevant laws (Kryazhov 1995, 1999, 2005, 2013) and his book-length analysis of these laws (Kryazhov 2010). The journal *Mir korennykh narodov/Zhivaya arktika* (1999–2017; www.raipon.info/info/magazines/) regularly ran articles discussing legislative initiatives and lacunae; its successor *V mire korennykh narodov* (2017–2018) continues to do so.
5 O neotlozhnykh merakh po zashchite mest prozhivaniya i khozyaystvennoy deyatel'nosti malochislennykh narodov Severa [On urgent measures to defend the places of habitation and economic activity of the numerically small peoples of the North]. Edict 397 of the President of the Russian Federation, 22 April 1992.
6 Article 72, part 1, paragraph "m". *Konstitutsiya Rossiyskoy Federatsii* [Constitution of the Russian Federation]. 1993. Moscow: Juridicheskaya Literatura.
7 O zhivotnom mire [On the Faunal World]. Federal Law No. 52, 24 April 1995; O kontinental'nom shelf'e Rossiyskoy Federatsii [On the continental shelf of the Russian Federation]. Federal Law No. 187, 30 November 1995.
8 O garantiyakh prav korennykh malochislennykh narodov Rossiyskoy Federatsii [On guarantees of the rights of Indigenous numerically-small peoples of the Russian Federation]. Federal Law No. 82, 30 April 1999.
9 Obshchina is translated as 'clan community', 'tribal community', 'clan commune', and a number of similar variants. Here we maintain the Russian word.
10 Ob obshchikh printsipakh organizatsii obshchin korennykh malochislennykh narodov Severa, Sibiri i Dal'nego Vostoka Rossiyskoy Federatsii [On general principles for the organization of *obshchina*s of Indigenous Numerically-small peoples of the North, Siberia and the Far East of the Russian Federation], Federal Law No. 104, 20 July 2000.
11 O territorii traditsionnogo priorodopol'zovaniya korennykh malochislennykh narodov Severa, Sibiri i Dal'nego Vostoka Rossiyskoy Federatsii [On territories of traditional nature use of Indigenous numerically small peoples of the North, Siberia and the Far East of the Russian Federation] Federal Law No. 49, 7 May 2001.
12 Earlier 'subject' (regional) laws resulted in the creation of many clan communities prior to the passage of the federal law; see Kryazhkov 1994, 1999.
13 Nalogovyy kodeks Rossyskoy Federatsii [Tax code of the Russian Federation], Part 1, No. 146, 31 July 1998; Part 2, No. 117, 5 August.
14 Lesnoy kodeks Rossiyskoy Federatsii [Forest Code of the Russian Federation], No. 200, 4 December 2006; see Article 8, paragraph 1.
15 Many Indigenous people find the concept of both ownership (*sobstvennost'*) of land and the idea of renting land (*arenda*) contrary their worldviews and their very understanding of appropriate relationships with other parts of what 'Westerners' term 'nature' or 'the environment'.
16 O zhivotnom mire [On the Faunal World]. Federal Law No. 52, 24 April 1995.
17 O provedenii konkursa na pravo zaklyucheniya dogovoroa o predostavlenii rybopromyslovogo uchastka dlya osushchestvleniya rybolovstva v tselyakh obespecheniya vedeniya traditsionnogo khozyaystvennoy deyatel'nosti korennykh malochislennykh narodov Severa, Sibiri i Dal'nego Vostoka Rossiyskoy Federatsii i o zaklyuchneii takogo dogovora [On holding a tender for the right to conclude an agreement on the provision of a fishing site for fishing in order to ensure the conduct of traditional economic activities of the Indigenous peoples of the North, Siberia and Dal'nego East of the Russian Federation and on the conclusion of such an agreement]. Decree of the Government of the Russian Federation No. 986, 24 December 2008.

18 O vnesenii izmeneniy v Federal'nyy zakon "O garantiyakh prav korennykh malochislennykh naro-dov Rossiysoy Federatsii" v chasti ustanovleniya poryadka ucheta lits, otnosyashchikhsya k koren-nym malochislennym narodam" ["On introducing changes to the Federal Law "On guarantees of the rights of Indigenous numerically small peoples of the Russian Federation," regarding the rules for the registration of persons belonging to Indigenous numerically small peoples]. Federal Law No. 11, 6 February 2020.

19 O vnesenii izmeniniya v stat'yu 71 Federal'nogo zakona "Ob okhote i sokhranenii okhotnich'ikh resursov i o vnesenii izmeneniy v otdel 'nye zakonodatel'nye akty Rossiyskoy Federatsii [On amend-ments to Article 71 of the Federal Law "On Hunting and on the Conservation of Hunting Resources and on Amending Certain Legislative Acts of the Russian Federation"] Federal Law No. 224, 21 July 2017.

20 See footnote 16.

21 O rybolovstve i sokhraninii vodnykh biologicheskikh resursov [On fisheries and the preservation of aquatic biological resources], Federal Law No. 166, 7 February 2003; see Article 25.

22 Pravila raspredeleniya organami ispolnitel'noy vlasti sub"yektov Rossiyskoy Federatsii kvot dobychi (vylova) vodnykh biologicheskikh resursove v tselakh obespecheniya traditsionnogo obraza zhizni i osushchestvleniya traditionnoy khozyaystvennoy deyatel'nosti korennykh malochislennyh narodov Severa, Sibiri i Dal'nego Vostoka Rossiyskoy Federatsii [Rules of distribution of quotes for the har-vest (catch) of aquatic biological resources by executive bodies of subjects of the Russian Federation, with the goal of providing for a traditional way of life and the realization of traditional economic activities of the Indigenous numerically small peoples of the North, Siberia and the Far East of the Russian Federation], Decree of the Government of the Russian Federation No. 558, 5 May 2018; see point 5.

23 "Ob utverzhdenii Poryadka opredelniya granits rybopromyslovykh uchastkov [On approval of the procedure for determining the boundaries of fishing grounds], Order of the Minister of Agriculture of the Russian Federation No. 143, 22 March 2013.

24 Ob oborote zemel' sel'skokhoyzystvennogo znacheniya [On the turnover of land of agricultural sig-nificance], Federal Law No. 101, 24 July 2002, Article 1, point and Article 10, points 5, 6; Zemel'nyy kodeks Rossiyskoy Federatsii [Land Code of the Russian Federation]. No. 136, 23 October 2001, Article 39.10, point 13.

25 Law on Guarantees (see footnote 8); Land Code (see footnote 24).

26 Ob osobo okhranyemykh prirodnykh territoriyakh [On Specially Protected Nature Territories]. Fed-eral Law No. 33, 14 March 1995. The Land Codex (2001) also listed TTPs as one form of OOPT (Article 95).

27 Ob otsenke vozdeystviya na iskonnuyu sredu obitaniya, traditsionnyy obraz zhizni i traditsionnoe prirodopol'zovanie korennykh malochislennykh narodov Rossiyskoy Federatsii. Draft Law. See Ledkov et al. 2017, 218–232 for drafts and explanatory notes.

28 O vnesenii izmeneniy v Federal'nyy zakon "Ob osobo okhranyayemykh prirodnykh territoriyakh" i otdel'nyye zakonodatel'nyye akty Rossiyskoy Federatsii" [On the introduction of Changes to the Federal Law 'On Specially Protected Natural Territories' and Other Legal Acts of the Russian Federation, Federal Law No. 406, 28 December 2013; especially articles 5, 6; see also O vnesenii izmeneniy v Federal'nyy zakon "Ob osobo okhranyayemykh prirodnykh territoriyakh" i otdel'nyye zakonodatel'nyye akty Rossiyskoy Federatsii, Federal Law No. 321, 3 August 2018.

29 Ob otvetstvennom nedropol'zovanii na territorii Respubliki Sakha (Yakutia) [On responsible subsur-face resource use] Sakha Republic (Yakutia) Law No. 1645-V, 3 July 2018.

30 O pravovom statuse korennykh malochislennykh narodov Severa (novaya redaktsiya) [On the legal status of Indigenous small peoples of the North (new edition)". Sakha Republic (Yakutia) Draft Law, accepted at first reading, 25 April 2019. (http://iltumen.ru/content/v-pervom-chtenii-prinyat-proekt-zakona-o-vnesenii-izmeneniy-v-pravovoy-status-korennykh).

31 *Territorii operezhayushchego razvitya* (TOR); in full, Territories of Priority Social-Economic Develop-ment; *territorii operezhayushchego sotsial'no-ekonomicheskogo razvitya* (TOSER).

32 O territoriyakh operezhayushchego sotsial'no-ekonomicheskogo razvitya v Rossiykoy Federatsii [On territories of priority social-economic development in the Russian Federation], Federal Law No. 473, 29 December 2014.

33 Po protestu zamestitelya General'nogo prokurora Rossiyskoy Federatsii Yuriya Gulyagina otmenen pravovoy akt regional'nogo upravleniya Rosrybolovstva [By protest of the Deputy Prosecutor General of the Russian Federation Yuri Gulyagin, the legal act of the regional administration of the Federal

Fisheries Agency] www.genproc.gov.ru/special/smi/news/news-1455766/; https://kamchatinfo.com/news/society/detail/17352/.

34 Prokuratora Chukotskogo avtonomnogo okruga (17 October 2018) Po isku prokurora okruga sud priznal vozlozheniye dopolnitel'nykh obyazannostey na predstaviteley korennykh malochislennykh narodov, osushchestvlyayushchikh traditsionnoye rybolovstvo [According to the suit of the district prosecutor, the court recognized the imposition of additional duties on the representatives of the Indigenous peoples engaged in traditional fishing]. https://prokuror.chukotka.ru/news/po-isku-prokurora-okruga-sud-priznal-nezakonnym-vozlozhenie-dopolnitelnyh-obyazannostey-na-predstaviteley-korennyh-malochislennyh-narodov-osuschestvlyayuschih-traditsionnoe-rybolovstvo

References

Astakhova, I. S., and V. I. Shadrin. 2019. "Zakon RS(Ya) 'O Suktule Yukagirskogo Naroda' kak otrazhenie evolyutsii regional'nogo zakonodatel'stva v sfere etnicheskogo mestnogo samoupravleniya v postsovet-skogo period [The Law of the RS(Ya) on Suktul of the Yukaghir People as a Reflection of the Evolution of Regional Legislation in the Sphere of Ethnic Local Self-Government in the Post-Soviet Era]." *Gumanitarnye issledovaniya v Vostochnoy Sibiri i na Dal'nego Vostoke* 1: 16–24.

Constitutional Court of the Russian Federation. 2019. "Meeting of the Constitutional Court of the Russian Federation." 16 April 2019. Accessed 15 May 2020. www.ksrf.ru/ru/Sessions/Pages/ViewItem.aspx?ParamId=189.

Diatchkova, G. I. 2001. "Indigenous Peoples of Russia and Political History." *The Canadian Journal of Native Studies* 21 (2): 217–233.

Federal'nyy. 2003. "Federal'nyy zakon o severnom olenevodstve (proekt) [Federal Law on Reindeer Husbandry (Draft)]." In *Olen' vsegda prav. Issledovaniya po yuridicheskoy antropologii*, edited by N. Novikova, 2210223. Moscow: Strategiya.

Fondahl, G., V. Filippova, and A. Savvinova. 2020. "Introducing a Registry of Indigenous Persons in Russia: Rationales and Challenges." *Espace/Populations/Sociétés* 2020 (1/2). https://journals.openedition.org/eps/9572.

Fondahl, G., V. Filippova, A. Savvinova, A. Ivanova, F. Stammler, and G. Hoogensen Gjørv. 2019. "Niches of Agency: Managing State-Region Relations through Law in Russia." *Space and Polity* 1 (23): 49–66.

Fondahl, G., O. Lazebnik, G. Poelzer, and V. Robbek. 2001. "Native 'Land Claims', Russian Style." *The Canadian Geographer* 45 (4): 545–561.

Fondahl, G., and G. Poelzer. 2003. "Aboriginal Land Rights in Russia at the Beginning of the Twenty-First Century." *Polar Record* 39 (309): 111–122.

Garipov, R. 2013. "Resource Extraction from Territories of Indigenous Minority Peoples in the Russian North: International Legal and Domestic Regulation." *Arctic Review on Law and Politics* 4 (2): 4–20.

Isakov, A. 2018. "Na territorii Zhiganskogo i Olenekskogo rayonov Yakutii nachal rabotat' proekt po dokumentirovaniyu traditsionnykh znanii." *V Mire korennykh narodov. Al'manakh* 2017–2018: 145–149.

Krivoshapkin, A. 2014. "Zakon o domashnem olenevodstve (A Law on Reindeer Husbandry)." *Mir Koren-nykh Narodov/Zhivaya Arktika* 30: 34–35.

Kryazhkov, V. A. compiler. 1994. *Status korennykh malochislennykh narodov Rossii. Pravovye akty i Dokumenty [The Status of the Indigenous Numerically-Small Peoples of Russia. Legal Acts and Documents]*. Moscow: Yuridicheskaya literatura.

Kryazhkov, V. A. compiler. 1999. *Status korennykh malochislennykh narodov Rossii. Pravovye akty. Kniga Vtoraya [The Status of the Indigenous Numerically-Small Peoples of Russia. Legal Acts. Book 2]*. Moscow: Izdanie g-na Tikhomirova M.Yu.

Kryazhkov, V. A. 2013a. "Development of Russian Legislation on Northern Indigenous Peoples." *Arctic Review on Law and Politics* 4 (2): 140–155.

Kryazhkov, V. A., compiler. 2013b. *Status korennykh malochislennykh narodov Rossii. Mezhdunarodnye pravovye akty i rossiyksoe zakonodatel'stvo. [The Status of the Indigenous Numerically-Small Peoples of Russia. International Legal Acts and Russian Legislation]*. 2 vols. Moscow-Salekhard: Russian Association of the Indigenous Peoples of the North, Siberia and Far East.

Kryazhkov, V. A. 2015. "Legal Regulation of the Relationships between Indigenous Small-Numbered Peoples of the North and Sub-Soil Users in the Russian Federation." *The Northern Review* 39: 66–87.

Kryazhkov, V. A. 2016. Pravo korennykh malochislennykh narodov na traditsionnoe priroropol'zovanie (na primeri okhoty) [The Rights of Indigenous Numerically-Small Peoples to Traditional Resource Use (the Example of Hunting)]. *Gosudarstvo i pravo* 11: 32–42.

Ledkov, G., A. Permyakov, and V. Istomin, V. 2017. *Sbornik normativno-pravovykh actov, regiliryushchikh otnosheniya s ychastiem korennykh malochislennykh narodov [Collection of Normative-Legal Acts Regulating Relations of Participation of Indigenous Numerically Small Peoples].* Moscow.

Mazurov, Y. L., R. V. Sulyandziga, and N. V. Vronskiy. 2018. *Tseli ustoychivogo razvitiya-2030 i korennye malochislennye narodoy Severa, Sibiri i Dal'nego Vostoka: ot informirovannosti k uchastiyu. "Nikogo ne ostavit pozadi" [2030 Sustainable Development Goals and Indigenous Numerically Small Peoples of the North, Siberia and the Far East: From Awareness to Participation. "Leave No One Behind"].* Moscow: TsS KMNS.

Osherenko, G. 2001. "Indigenous Rights in Russia: Is Title to Land Essential for Cultural Survival?" *The Georgetown International Environmental Law Review* 13 (3): 695–734.

Stammler, F., and A. Ivanova. 2016. "Resources, Rights and Communities: Extractive Mega-Projects and Local People in the Russian Arctic." *Europe-Asia Studies* 68(7): 1220–1244. doi:10.1080/09668136.2016.1222605.

Sulandziga, R., and R. Berezhkov. 2017. "Reflections on the Influence of the Current Political Development in Russia on Indigenous Peoples' Land Rights." In *Indigenous Peoples' Rights and Unreported Struggles: Conflict and Peace,* edited by E. Stamatopoulou, 80–95. New York: ISHR, Columbia University.

Sulandziga, R., and L. Sulandziga. In this volume. "Indigenous Self-Determination and Disempowerment in the Russian North." In *Handbook of Arctic Indigenous Peoples,* edited by T. Koivurova et al. New York: Routledge.

Tranin, A. A. 2010. *Territorii traditsionnogo prirodopol'zovaniya korennykh malochislennykh narodov Rossiyskogo Severa (Problemy i Perspektivy) [Territories of Traditional Nature Use of Indigenous Numerically Small Peoples of the Russian North (Problems and Prospects)].* Moscow: Institute of Government and Law, Russian Academy of Sciences.

Vakhtin, N. B. 1998. "Indigenous People of the Russian Far North: Land Rights and the Environment." *Polar Geography* 22 (2): 79–104.

Wiget, A., and O. Balalaeva. 2011. *Khanty. People of the Taiga. Surviving the 20th Century.* Fairbanks: University of Alaska Press.

Xanthaki, A. 2004. "Indigenous Rights in the Russian Federation: The Case of Numerically Small Peoples of the Russian North, Siberia and the Far East." *Human Rights Quarterly* 26: 74–105.

Yakel', Y. 2014. "Byt ili ne byt? [To Be or Not to Be?]." *Mir korennykh narodov/Zhivaya Arktika* 30: 46–49.

Zakonoproekt. 2017. "Zakonoproekt o territoriyakh traditsionnogo prirodopol'zovaniya protivorechet interesam korennykh malochislennykh narodov Severa [The Draft Law on Territories of Traditional Nature Use Contradicts the Interests of Indigenous Numerically Small Peoples of the North]." 14 June 2017. Accessed 15 May 2020. http://iltumen.ru/content/zakonoproekt-o-territoriyakh-traditsion nogo-prirodopolzovaniya-protivorechit-interesam-koren.

9

SÁMI LAW AND RIGHTS IN NORWAY – WITH A FOCUS ON RECENT DEVELOPMENTS

Øyvind Ravna

Abstract

This chapter addresses the legal protection of the Sámi as Indigenous peoples in Norway and thus the laws that protect Sámi language, culture and way of life or, in other words, *Sámi law*. The chapter describes the development of Sámi law and Sámi legal protections over the past 30 years, with a primary focus on recent developments and current legal status. Through this review, the chapter summarises the latest advances in international law related to the Sámi, the process of surveying and recognising land rights in Finnmark (the most central Sámi area in Norway) and the latest amendments to the Reindeer Husbandry Act. Questions about the right to the usage, management and control of natural resources, or "land and water", in the Sámi area of Norway therefore occupy a central place in this examination.

Introduction

Over the past four decades, an extensive development of Sámi rights and legal protections for Sámi culture has occurred in the Nordic countries. Norway is considered a leading power in this expansion, which makes its progress an interesting object of study.

Due to the controversial Alta case from the 1970s and beginning of the 1980s, the Norwegian government was pushed to establish the Sámi Rights Committee and grant it a mandate to examine the legal position of the Sámi and propose legislative amendments. This committee, which became an engine of progress in ensuring Indigenous rights in Norway, spurred the Sámi Act (1987), a constitutional amendment protecting Sámi language, culture and way of life (1988), and the Sámi Parliament (1989). In 1990, Norway was the first country in the world to ratify ILO Convention No. 169 on Indigenous and Tribal Peoples in Independent States (ILO 169).

Thus, Norway has assumed significant legal obligations for the protection of Sámi language and culture, which include the right of the Sámi to be consulted in legislative or administrative measures which may affect them directly, and participate in decision-making processes related to natural resources pertaining to their lands. This progress continued into the first decade of the 21st century, as Norway adopted the Human Rights Act (1999), which incorporated the International

Covenant on Civil and Political Rights (ICCPR) as Norwegian law, with precedence over internal legislation. In 2005, the Finnmark Act was adopted, which completed the work of the Sámi Rights Committee (I) when it comes to the area of Finnmark. In 2001, the Sámi Rights Committee (II) was reappointed to investigate the situation in the Sámi areas south of Finnmark. At the same time, Norway was at the forefront of the adoption of the UN Declaration on the Rights of Indigenous Peoples (2007) by the UN General Assembly. This declaration presumably strengthened the legal protection of the Sámi culture as well.

In this chapter, the laws that protect the Sámi culture, language and way of life in Norway are examined. The purpose of this study is to discuss how Norway legally protects Sámi culture and language, including how the country complies with its international obligations and what place customary Sámi law holds in Norwegian law. Queries about the right to "land and water", or the material basis of Sámi culture, consequently occupy a significant place in the analyses. Particular attention is paid to development connected to the Finnmark Act and the latest Supreme Court cases related to Sámi issues.

Until 2005, the doctrine held that the Norwegian state was the owner of unceded or unsold land in most of the Sámi areas in Norway, without any responsibilities for use rights belonging to others beyond those evidenced by statutory laws or deeds. This has since shifted, and the area in focus today largely belongs to the Finnmark Estate, a new ownership body that represents the rights that have been transferred from the state to the people in Finnmark. On this land, a process is ongoing to clarify and recognise usage and property rights. To date, six investigative reports and two Supreme Court judgements have emerged from this recognition process in the Sámi areas of Finnmark (Finnmarkskommisjonen, Rapport Felt 1 to 6, HR-2016–2030-A, HR-2018–456-P).

The regulation of Sámi reindeer husbandry is another controversial issue, as are other intervention cases related to the Sámi use of land and water. Also, although the Parliament has adopted fishery law amendments, disputes concerning rights to marine resources in coastal areas outside of Finnmark are not yet fully resolved; however, coastal rights are not addressed in this chapter.

Background of Sámi Law in Norway

Historically, the first treaty regulating the use of land by the Sámi was The Lapp Codicil, an annex to the border treaty of 1751 between Sweden and Denmark – Norway. The codicil confirmed the Sámi right to annually cross the border with their reindeer, based on "old customs". The codicil is generally characterised by a positive spirit and a desire to facilitate the continued existence of the Sámi nation. The pending Nordic Sámi Convention can be regarded as an extension of the codicil (The Draft Nordic Sámi Convention 2017).

From the mid-19th century, the Sámi were subjected to a targeted assimilation policy, while the state's ownership of Sámi areas was ascertained through the state's land doctrine and the adoption of the Land Sales Act of 1863. Although the assimilation policy gradually declined in use after the Second World War, the state's land doctrine was, in principle, in force until 2006, when the Finnmark Act replaced the Land Sales Act of 1965.

The decision to construct a hydropower plant on the Alta-Kautokeino watercourse in 1978 forced the debate on the Sámi's rights to land and water to the forefront. To the extent that positive attitudes towards the Sámi had developed during the post-war period, the hydropower plant issue revealed how little this entailed practically in politics. Although the Alta river was ultimately dammed, the case became a turning point in Norwegian Sámi policy; in the fall of 1980, the Sámi Rights Committee was founded, which spurred the legislative push that culminated in the Finnmark Act 25 years later.

In the following three sections, I outline how Sámi uses, rights and customary law are safeguarded and regulated in Norway by both internal law and the international law to which Norway is bound, what impact Sámi law exerts and how these issues affect possibilities for the Sámi to participate in managing natural resources in Sápmi.

National legislation

Article 108 of the Constitution

The Norwegian Constitution, which is the national source of the highest law, includes a "Sámi Article", which reads as follows:

> The authorities of the state shall create conditions enabling the Sámi people to pre-
> serve and develop its language, culture and way of life.

This provision establishes a general protection for Sámi culture. In 1988, it was adopted as Article 110a, based on a proposal from the Sámi Rights Committee. During the constitutional revision in 2014, it was updated, linguistically and chronologically, as Article 108. The article outlines legal obligations primarily addressing the state authorities, which was emphasised by the Sámi Rights Committee when they drafted the provision (NOU 1984: 18, 433) and later reinforced by executive authority (St. meld. nr. 28 [2007–2008], 29) and the Supreme Court (HR-2018–456-P, para. 91). Article 108 is modelled after ICCPR Article 27, which, according to the Sámi Rights Committee II, means that it must be interpreted in accordance with contin-uously applicable international law demands on Norwegian authorities (NOU 2007: 13, 191).

Little case law on Article 108 exists. However, it exerted significant impact when the Inner Finnmark District Court, a district court with Sámi-speaking judges possessing specific knowl-edge in Sámi customs and customary law, was established in 2004. A public investigation pre-ceding the establishment of the Court concluded that Article 110a requires that judges possess expertise and insight into Sámi language, law and culture. This was crucial for the establishment of the Court (NOU 1999: 19, 20, NOU 1999: 22, 72).

The Sámi Act and the Reindeer Husbandry Act

The Sámi Act (1987) and the Reindeer Husbandry Act (2007) are both intended to safeguard Sámi culture. The latter, however, also provides authority for interventions in reindeer herders' livelihoods.

The Sámi Act aims "to enable the Sámi people in Norway to safeguard and develop their language, culture and way of life" (cf. Section 1–1), complementing Article 108 of the Consti-tution. Section 1–2 authorises the establishment of the Sámi Parliament, which is further regu-lated in Chapter 2. The preparatory work states that the purpose of the Act cannot be achieved without special measures towards the Sámi; "the formal equality in a narrow sense" must give way to equality in the wider sense or, in other words, "a real equality between Sámi culture and other culture in Norwegian society" (Ot.prp. nr. 33 [1986–1987], 23).

Chapter 3 deals with the Sámi language. Section 3–2 states that "[s]tatutes and regulations of particular interest to all or parts of the Sámi population shall be translated into Sámi". The chapter ensures extended rights to use the Sámi language in the courts (cf. Section 3–4) and the health sector (cf. Section 3–5). In 2014, the government appointed a committee to study leg-islation and measures to protect the Sámi language in Norway. This committee has suggested, amongst other things, that all municipalities create plans for strengthening Sámi language and culture in their service offerings and that a language area replace the current Sámi language administrative area, wherein relevant municipalities are divided into categories with differentiated obligations and responsibilities (NOU 2016:18, 95, 99 129).

The Reindeer Husbandry Act (2007) aims at facilitating ecologically, economically and culturally sustainable reindeer husbandry. The Act emphasises the importance of Sámi traditions and customs, as demonstrated by the Sámi reindeer husbandry community, or *siida*, gaining a place in the act. The government, however, recently proposed that the ecological objective take precedence over the cultural and economic objectives (Meld. St. 32 [2016–2017]), so that ecological considerations may be emphasised more heavily than Sámi culture and requirements for economic operations when interpreting the Reindeer Husbandry Act. After consultations with the Sámi Parliament, the proposal was taken out of the bill and not adopted by the Parliament (Prop. 90 L [2018–2019]). Rules on making reindeer numbers available and mandatory individual marking of reindeer were however, adopted.

Section 3 states that the Act shall be applied in accordance with international law on Indigenous peoples and minorities. The rule on the burden of proof, established in Section 4, para. 2, entails that in case of dispute, landowners must prove that no reindeer-herding rights exist on their lands within the Sámi reindeer pasture area (Nrt 2001, 769 [786–788]). Section 4, para. 1 codifies legal rules that have emerged through case law, ensuring that reindeer husbandry's legal basis stems from immemorial or historical usage.

The Reindeer Husbandry Act was substantially revised in 2014. The revision meant that the regional administrations, which had reindeer herders on their boards, were discontinued and their duties transferred to the county governor (Fylkesmannen). The amendment was met with criticism from the Sámi Parliament and reindeer husbandry organisations, as it was considered, amongst other issues, an intervention against Sámi self-determination and self-management (Prop. 89 L [2012–2013], 9–10).

In recent years, the act has been criticised for its strict rules in regulating and reducing the number of reindeer, as was made very clear in Supreme Court Case HR-2017-2428-A (*Sara*). The appellant, Jovsset Ante Sara, has acquired significant support, even beyond the Sámi community, and the Supreme Court decision has been sent as an individual complaint to the UN's Human Rights Committee (HRC).

The Finnmark Act, the Finnmark Estate and the process of the survey land rights

The Finnmark Act (2005) resulted from inquiries into Sámi rights to land and water which were accentuated during the Alta case and which the Sámi Rights Committee was mandated to investigate.[1] It is also a result of the commitment Norway undertook by ratifying ILO 169. Although an expert subgroup of the Sámi Rights Committee concluded that the Norwegian state owned the previously unsold or unceded land in Finnmark (NOU 1993: 34, 266), the Committee proposed abolishing this ownership (NOU 1997: 4). The rationale for this decision lays in requirements in international law and the absence of a legal basis for the state's ownership. The durability of the legal group's conclusions has recently been questioned (Ravna 2020).

Due to the Finnmark Act, the Norwegian state transferred the title of 95% of the county's area (46,000 km²) to a new legal entity, the Finnmark Estate. A primary objective of the Act, according to preparatory works, was "to replace uncertainty and dispute over the right to land and water in Finnmark with security and predictability regarding the natural basis for Sámi culture, [and] for the inhabitants' use of the outlaying fields . . . based on a sustainable utilisation of resources" (Ot.prp. nr 53 [2002–2003], 7).

In the same preparatory works, the Ministry of Justice heralded "a significant historical shift to local governance and a clear statement of confidence to all the people of Finnmark, regardless of ethnic or cultural background". This historical shift, however, has not been as significant as predicted by the Ministry. This may be because many old management traditions have been continued by former employees of Statskog, who gained positions in the new body via statutory requirement. Furthermore, none of the local management structures proposed by the Sámi Rights Committee have been adopted.

The Finnmark Act's purpose is to facilitate the management of land and natural resources in the county of Finnmark in a balanced and ecologically sustainable manner for the benefit of the residents of the county and particularly as a basis for Sámi culture, reindeer husbandry, the use of non-cultivated areas, commercial activity and social life (cf. Section 1). Section 3 states that the law "shall apply with the limitations that follow from ILO Convention No. 169". This is a sectorial incorporation (Innst. O. nr. 80 [2004–2005], 33). In addition, "The Act shall be applied in compliance with the provisions of international law concerning Indigenous peoples and minorities and with the provisions of agreements with foreign states concerning fishing in transboundary watercourses". Beyond treaties concerning Indigenous peoples and minorities, this sentence references an agreement about fishing in the border rivers of Tana and Neiden (between Finland and Norway). Fishing in the Tana River has, since 2014, been regulated in a separate Tana Act (20 June 2014 no. 51) aimed at securing the particular rights of local people to fish in the Tana watercourse on the basis of law, immemorial usage and local custom. Section 3 of the Tana Act is worded similarly to Article 3 of the Finnmark Act.

This section (of the Finnmark Act) assumes that ILO 169 applies to the process of surveying and recognising existing rights, regulated under Chapter 5 of the Finnmark Act. The Supreme Court has, however, interpreted the scope of the incorporation more narrowly, stating that ILO 169 takes precedence only over the Finnmark Act's own provisions (Ravna 2019, 173–178). I return to that interpretation subsequently.

Of major interest is Section 5, para. 1, which states,

> Through prolonged use of land and water areas, the Sámi have collectively and individually acquired rights to land in Finnmark.

Para. 2 states that the Act does not interfere with collective or individual rights acquired by Sámi and other people through prescription or immemorial usage, while para. 3 establishes the process of surveying and recognising land rights. It reads,

> In order to establish the scope and content of the rights held by Sámi and other people on the basis of prescription or immemorial usage or on some other basis, a commission shall be established to investigate rights to land and water in Finnmark and a special court to settle disputes concerning such rights, cf. chapter 5.

Section 29 authorises the establishment of the Finnmark Commission, "which, on the basis of current national law, shall investigate rights of use and ownership to the land to be taken over by Finnmark Estate". The preparatory works explain that the wording of "current national law" is used to emphasise that Sámi customs and legal opinions are part of the law (Innst. O. nr. 80 [2004–2005], 18–19). The establishment of the Finnmark Commission and the Uncultivated

Land Tribunal for Finnmark for Finnmark is anchored in ILO 169 Article 14 (Innst. O. nr. 80 [2004–2005], 28).

The Finnmark Commission began its work in the spring of 2008. As of March 2020, it has completed six fields of investigation. In the first five reports, the Commission found no areas of collective rights, Sámi property rights or use rights beyond what was previously settled in the Finnmark Act (except for the peculiar case of Gulgofjord/Vuođavuotna in Field 6; Finnmark Commission 2015, 170). The justification for these findings lies not in a lack of actual use from the concerned people or other requirements directly related to the rules of prescription or immemorial usage but rather the state's, or the government's, dispositions and activities as resource manager and public authority (Ravna 2019, 483–485). In its latest report, *Field 4 Karasjok*, delivered in December 2019, the Commission has placed significantly less emphasis on governmental dispositions, and it also assesses the local population's use in a different manner than in previous reports. As a result, the Commission concludes that the local population in the Karasjok municipality has property rights to areas that are currently managed by the Finnmark Estate. Whether the result shall remain standing is uncertain, as the conclusion was made with dissent (3–2).

The Uncultivated Land Tribunal for Finnmark, which was established in 2014 with a mandate to settle disputes arising from investigations by the Finnmark Commission, has heard 16 cases as of January 2020. An initial initiative to consider Norway's obligations under international law vis-à-vis the Sámi – by recognising the local Sámi people's right to control their rights of use in a January 2017 judgement concerning *Nesseby* (UTMA-2014–164739) – was reversed through the *Gulgofjord* judgement in the fall of 2018. Here, efforts by the Finnmark Commission to recognise ownership rights of a coastal Sámi population were strictly rejected (UTMA-2017–62459).

Legally speaking, however, there is no reason to criticise the Finnmark Commission's application of the law, as its general opinions and application of law were confirmed by the Supreme Court in the *Nesseby* case (HR-2018–456-P). Furthermore, the Land Tribunal has anchored its *Gulgofjord* judgement on this Supreme Court case.

The Finnmark Act has been internationally upheld as an example (Anaya 2011, para 44). As the results of the surveying process have become visible, the international response has been stunning. In 2016, the UN Special Rapporteur of the Rights of Indigenous Peoples, Victoria Tauli-Corpuz, noted that "the Commission has almost exclusively found no grounds for recognising Sámi individual or collective ownership or usage rights beyond usage rights already granted to all inhabitants in Finnmark", after which she added that "[s]uch conclusions seem to have been motivated by the State's active and extensive disposition of land and resources in the investigated fields which is seen to have precluded property or usage rights for the local population" (Tauli-Corpuz 2016, para 23). The Special Rapporteur then stated,

> the State's earlier dispositions as the claimant of property rights in Finnmark cannot be considered to create law in order to support its continued ownership of land. The importance of that point can be further underscored by the fact that in many cases, the Sámi communities' severed connection to their lands and resources is a result of earlier government policies and assimilation efforts towards the Sámi. A starting point for any measures to identify and recognize Indigenous peoples' land and resource rights should be their own customary use and tenure systems.
>
> (Tauli-Corpuz 2016, para 24)

Her point of view, however, stands in direct contrast to the Norwegian Supreme Court, which has emphasised the state's dispositions as crucial for the Finnmark Estate's mandate to govern local people's rights to their traditional lands.

International obligations

The significance of international law for Norway's commitments to the Sámi

The Norwegian legal system comprises a dualistic structure, in which international law must be implemented through an act of the Parliament to become national law. At the same time, Norway has placed considerable weight on adhering to international standards and obligations, as demonstrated by the *presumption principle* being part of Norwegian law. The principle entails that, unless otherwise expressly stated, Norwegian law shall be interpreted in accordance with the rules of international law (Ruud and Ulfstein 2018, 68–69). This applies to both basic human rights as prohibition of torture, racial discrimination and attacks on freedom of expression, as well as specific human rights aimed at preserving the culture and languages of Indigenous peoples and minorities. Norway's emphasis on international law is shown by its ratification of ILO 169 and incorporation of the most significant human rights conventions through the Human Rights Act of 1999. Norway's active work for the UNs adoption of the Declaration on the Rights of Indigenous Peoples also demonstrates this intention. The most relevant international obligations when it comes to protecting Indigenous culture and livelihood in Norway are International Labour Organization (ILO) Convention Concerning Indigenous and Tribal Peoples in Independent Countries (ILO 169) and International Covenant on Civil and Political (ICCPR). But the United Nations Declaration on the Rights of Indigenous Peoples (UNDRIP) is also gaining more significance, which I will return to in short.

The ICCPR is, as demonstrated previously, a part of Norwegian law. Article 27 is particularly relevant to Indigenous people and ethnic minorities. The provision establishes a right for persons bevinging to minorities not to be denied the right to enjoy their own culture, profess and practise their own religion or use their own language. Statements by the UN's HRC, which is the Covenant's monitoring body, shows that Article 27 not only provides minorities with protection against being denied the right to exercise their culture but also imposes "positive measures by States . . . to protect the identity of a minority and the rights of its members to enjoy and develop their culture and language" (HRC 1994, para 6.2).

The HRC has made clear through both general statements and assessments in individual complaints that Article 27 also protects the material basis of minority culture. As such, land, waters, pastures and other natural resources of importance to traditional Sámi livelihood are protected. In this context, the HRC has stated "that culture manifests itself in many forms, including a particular way of life associated with the use of land resources, especially in the case of Indigenous peoples. That right may include such traditional activities as fishing or hunting" (HRC 1994, para 7). The statement shows that protections against decisions that may affect them are particularly important for Indigenous peoples.

It is therefore not doubtful that a threshold exists beyond which decisions affecting Indigenous peoples violates the Covenant. In *Ángela Poma-Poma vs. Peru* (HRC 2009), the HRC found that this threshold had been crossed when the minority group could no longer receive economic yield from their livelihood due to a state decision, which did not involve them. In the *Sara* case (HR 2017–2428-A), which arose over a dispute concerning the possibilities of the State Reindeer Husbandry Administration to reduce the numbers of reindeer in a *siida* in order to limit presumed overgrazing, the Norwegian Supreme Court confirmed the ruling of *Poma-Poma* a general norm (para. 69). However, the Court found that the *ratio decidendi* of the *Poma Poma* case did not apply considering the facts of the particular case. According to the Supreme Court:

> In the case at hand, it is not the greater society's interference with a minority inter-est that is to be balanced against ICCPR article 27. It concerns a regulation meant to protect the interests of the Sami herders, which raises issues on how the burden of the reindeer cull is to be distributed among them. Legal principles in cases concern-ing encroachment of nature cannot automatically be applied in such a case. (HR 2017–2428-A, para. 71).[2] (see p. 155)

In other words, the Supreme Court found that the measure which imposed for Sara to reduce the number of reindeer in his herd was justified, as the despute concerned an allocation of grazing resources within the *siida* (if Sara was allowed to keep his number of animals, other practitioners had to reduce more than the Reindeer Husbandry Act's requirement for a proportionate reduc-tion, cf. section 60. In other words, HRC's case law was not significant for *internal Sámi issues*.

The decision was, however, opposed by one of the Supreme Court judges (para 109–139).

In accordance with the *Poma-Poma* case, the Supreme Court also confirmed that a repre-sentative body of the minority *generally* must have participated in and given its consent to the decision or law authorising the measure. However, the Supreme Court also found that no such requirement should be applied in the *Sara* case (para. 74–75). As the decision of the court remains controversial, the case has been appealed to the HRC, and the result is awaited with considerable interest.

Norway was the first country to ratify ILO 169, in 1990. As Norway has a dualistic legal system, international law must be incorporated or transformed to take effect as internal law. Although the legislature did not incorporate ILO 169 through the Human Rights Act in 1999, Norway is obliged to by the Convention. ILO 169 is thus an important source of law (NOU 1993:18, 142). Amongst other issues, it demands consultations, that the customs and custom-ary law of Indigenous peoples should be emphasised in the courts and that Indigenous people's property rights or use rights should be recognised and secured. The legislative, executive and judicial powers in Norway have all assumed that the Sámi are covered by this obligation (Doku-ment 8:154 S [2010–2011], 1–5; Nrt 2001, 769). In 2005, the ILO 169 was incorporated in a sectorial manner through the Finnmark Act, also referred to as "sectorial monism", which is explained subsequently. Elsewhere, the ILO 169 is not incorporated but has significance in accordance with the *presumption principle* in Norwegian law, which states that Norwegian law presumes to comply with international law, particularly in cases where several interpretative alternatives to Norwegian law exist.

Article 6 obliges state authorities to consult Indigenous peoples. According to para. 1 (a), the duty of consultation appears "whenever consideration is being given to legislative or administra-tive measures which may affect them directly". This article formed the basis for the agree-ment on consultation procedures between the government and the Sámi Parliament (2005). Held against the HCR's interpretation of ICCPR Article 27 and the UN Declaration on the Rights of Indigenous Peoples Articles 19 and 32, the duty to consult (and right to participate) under certain conditions may include free, prior and informed consent. See also the Norwegian Supreme Court in the *Sara* case (HR-2017–2428-A), which states that conditions of consent for external interventions exist in cases "which completely tore down the livelihood of the complainant and the other members of the minority community she belonged to" (para. 74).

In 2007, the Sámi Rights Committee II proposed a consultation act requiring conditional consent (NOU 2007:13, 824). In 2018, the government presented its bill on the topic without such a requirement (Prop. 116 L [2017–2018]). On 30 April 2019, the bill was heard by the Parliament, which decided to send it back to the government for new hearings (Innst. 253 L [2018–2019]).

Article 8 deals with the customs or customary law of Indigenous peoples. In (1), it is stated that "[i]n applying national laws and regulations to the peoples concerned, due regard shall be had to their customs or customary laws". The provision ensures customary Sámi law as a source of law.

Article 14 deals with Indigenous people's rights of ownership and possession of the lands they have traditionally occupied. The purpose of the provision is the legal recognition of traditional use by Indigenous peoples. The Sámi Rights Committee's International Law Group has described it as the following:

> If a group of people has had a fairly permanent presence in the area, and at the same time have been the only ones who have used the area, the requirements for actual possession will normally have to be seen as fulfilled. If others also have used the area, the use of the group of Indigenous people must have been dominant in relation to the use that has been made by others.
>
> (NOU 1997: 5, 35)

Dominant, and not necessarily exclusive, use is thus sufficient to meet the requirements. This viewpoint has been confirmed by the courts, most recently by the Supreme Court in the *Nesseby* case (HR-2018–456-P, para. 170).

Article 14 (2) states that "[g]overnments shall take steps as necessary to identify the lands which the peoples concerned traditionally occupy, and to guarantee effective protection of their rights of ownership and possession", while (3) states that "[a]dequate procedures shall be established within the national legal system to resolve land claims by the peoples concerned". These two commitments are the direct argument for establishing the Finnmark Commission and the Uncultivated Land Tribunal for Finnmark (Ravna 2019, 451).

The right to property is also protected under the International Convention on the Elimination of All Forms of Racial Discrimination (ICERD) Article 5 (d) (5). This has recently been invoked in the controversial wind power case at Fosen in the South Sámi reindeer husbandry area (Trøndelag County). The Norwegian state has here rejected a request from the Committee on the Elimination of Racial Discrimination (CERD) for a temporary halt in development until a final assessment has been made.

As ILO 169 is incorporated in a sectorial manner by the Finnmark Act, ILO 169 will presumably apply to the process of surveying land rights in Finnmark. The range of incorporation has been evaluated by the Supreme Court in the Stjernøya Case (HR-2016-2030-A). A unanimous Court found that ILO 169 takes precedence over only the Finnmark Act's own provisions – as opposed to over the application of all law applied in the surveying process. The first voting judge referred to the preparatory work when reasoning that the incorporation faced this limitation and then stated, "Even though the Act regulates the procedures for clarifying rights, it does not regulate the substantive rules that the rights shall be clarified on the basis of" (Section 76). This means that ILO 169 does not take precedence over the rules of prescription and immemorial usage, which, according to the Finnmark Act Section 5, paragraph 3, shall be applied as a basis for clarifying rights to land and natural resources in Finnmark. The standpoint has been confirmed in the Nesseby case (HR-2018-456-P), where the Supreme Court found that the Finnmark Estate, not a local Sámi community, had the right to manage and control the rights that had already been recognised by the Finnmark Commission.

On 23 January 2020, the Swedish Supreme Court delivered a judgement in favour of the Girjas Sámi community in a lawsuit against the Swedish state over a long-standing dispute over the right to administer hunting and fishing rights in the Girjas reindeer husbandry area (T 853–18). Although ILO 169 has not been ratified by Sweden, the Court actively uses the Convention to reason for the use of Sámi customs and customary law when applying national law. The Court also states that although the Convention has not been ratified by Sweden, section 8 (on customs or customary laws) can be considered an expression of a general principle of international law. Following this decision, it will be difficult in Norwegian law to interpret ILO 169 as the Norwegian Supreme Court has done in the two aforementioned cases.

Although it is beyond the scope of this chapter to make a comprehensive analysis of the use of the UN Declaration on the Rights of Indigenous Peoples (UNDRIP), it is important to note how the declaration has been put in use in Norway. Despite the Stoltenberg government having noted that it "sets out important guidelines in the further work to determine what rights Indigenous peoples have" (St. meld. nr. 28 [2007–2008], 34), the UNDRIP has not been applied notably in Norwegian courts. In the *Nesseby* case, however, all 17 judges agreed that the UNDRIP "must be regarded as a central document within the Indigenous peoples law, among others because it reflects principles of international law in the area and has received support from very many states" (Section 97). This statement shows that the Court meant that the declaration contains provisions of international customary law. The first voting judge nonetheless stated that this was not of direct importance for the present case.

In addition to the principle of free, prior and informed consent, the UNDRIP incorporates *the right to restitution* in Article 28. In the *Nesseby* case, the Supreme Court agreed that ILO 169 Article 14 (1) ensures such a right (HR-2018–456-P, para. 173). Due to the historical facts of the case, the rule was not applied. In the Karasjok report, where the actual situation is different, the Finnmark Commission exercises the right to restitution (Finnmarkskommisjonen, Rapport Felt 4) when it found the inhabitants, and not the Finnmark Estate, are the landowners.

International criticism

As demonstrated previously, Norway has been criticised by the UN Special Rapporteur of the Rights of Indigenous Peoples. Norway has also been subject to criticism from other UN agencies for how the country has treated its Indigenous obligations in recent years.

In 2015, the CERD noted that despite the Finnmark Act's recognition of the collective and individual rights of the Sámi to land and water, no recognition occurs in practice, which gives the rights limited protection. The CERD also criticised Norway for not following up the Sámi Rights Committee II proposal. In its recommendation, the CERD proposed that Norway "[f]ollow up on the proposals of the Sámi Rights Committee, including by establishing an appropriate mechanism and legal framework, and identifying and recognising Sámi land and resource rights outside Finnmark" (CERD 2015, para 30 [b]).

In the examination of Norway's seventh periodic report on measures taken to implement and safeguard the rights recognised in the ICCPR, the HRC also directed criticism against

Norway for not having followed up the recommendations of the Sámi Rights Committee II from 2007. In its recommendation, the HRC proposed that Norway "[e]nsure effective and speedy follow-up to the proposals of the Sámi Rights Committee of 2007 regarding land and resource rights in Sámi areas outside of Finnmark" (HRC 2018, para. 37 [e]).

In January 2019, the CERD reiterated its criticism of the failure to follow up the proposal of the Sámi Rights Committee II and expressed concern: "That the Government has not yet complied with the Committee's recommendation in its previous observations regarding the legal recognition of rights and resources outside of Finnmark" (CERD 2018, para. 30 [b]).

Sámi law

Sámi customs and legal traditions constitute Sámi law

Sámi customs and legal opinions constitute Sámi law, although it is seldom included in statutory legislation. Such law has significance not only internally amongst the Sámi but also in national law. Furthermore, it is important for the negotiation of bi- or multilateral treaties and conventions between the states that have a Sámi population. These rules may constitute residual remnants of a separate, Sámi legal system, as well as rules developed in a more modern context. They are characterised by having their own legal basis and being complied with due to an opinion of obligation amongst those committed.

Norway's ratification of ILO 169 has strengthened the legal significance of Sámi law and given it a place within Norwegian law, which must be emphasised by both legislators and practitioners. These developments also entail that Sámi legal rules have become more statutory, for example, in the Reindeer Husbandry Act.

The harvesting of outland resources in Sámi areas has, to a relatively small extent, been regulated by law, except the recently adopted Finnmark Act and the general laws on inland fishing and wildlife. Private relations have largely been regulated by customary law. For example, for the snare trapping of ptarmigans, Erik Solem notes that it is mentioned in older literature about the Sámi and "that they often went to remedy themselves concerning such trapping sites" (Solem 1933, 87). Elina Helander further notes that the Sámi participants in her study stated that they possess an internal autonomy for the practice of this hunting tradition and that they are now in danger of losing it. She also notes the belief in a right of use to land and water: "This right of use is considered so strong that it can be equated with a property right, individual or collective, since certain areas are used customarily by different groups during different parts of the year" (Helander 2001, 458).

In the following, I concentrate on customs that have been addressed by the courts. Other Sámi customs can also be elucidated, such as the right to simple housings, as turf huts, in commons to support outland harvesting.

Sámi law in Norwegian courts

It has been difficult for Sámi law to prevail when it contradicts Norwegian statutory legislation. For example, the Supreme Court has concluded that the tradition of letting dogs run freely in outlying areas during the summer (when a period of leash enforcement exists) should not take precedence over the Wildlife Act (Nrt 2001, 1116). Furthermore, spring hunting for ducks in the municipality of Kautokeino is not, according to the Supreme Court, a custom deserving of legal protection (Nrt 1988, 377).

The question of the significance of customary Sámi law has also been tested by the Supreme Court regarding the slaughtering of reindeer (Nrt 2006, 957, 2008, 1789). In no judgements has the claim that customary methods of slaughtering – killing via a small-calibre rifle and a cardiac punch with knife – prevail over the animal welfare act been upheld. However, the legislature has been willing to regulate Sámi customs related to reindeer butchery methods and the spring hunting of ducks by adopting regulations allowing and regulating such activities (Regulation 2008, 2013).

Furthermore, the courts have heard cases of customs related to traditional salmon fishing with nets. Of several judgements regarding fishing in the Tana River, one case stands out, published in (Nrt 2006, 13). In this case, the Supreme Court, unlike the lower courts, found that allowing a person outside of the household to fish on the authority of a relative (who held the fishing rights) violated the Tana Act of 1888. Here, the accused refused to accept a fine for illegal fishing, as he had been fishing on his brother's rights. The Inner Finnmark District Court acquitted the accused, stating that such fishing was in accordance with local custom and legal opinion. On the contrary, the first voting judge of the Supreme Court found that she could "hardly see that there is a basis for ascertaining a customary law that should make exceptions to the system of the law" (para. 15). Recently, a Finnish first-instance court set aside a Finnish act on fishing in the Tana on the grounds that the act violates Sámi law (Paltto 2019). The judgement has been appealed, and it will be interesting to follow this case, which may also impact Norwegian law.

Although the Selbu case directly depends not on Sámi customary law, but rather on an emphasis on the Sámi customary use of nature and pastures, it shows that the Supreme Court has emphasised Sámi customs and the use of nature by ruling on the rights of reindeer pastures (Nrt 2001, 769 [789]). The same is the situation in the Svartskog case, where the Supreme Court emphasized Sámi customary use, Sámi culture and legal opinions of when it concluded that people in a village Troms county had acquired property rights to their common land through immemorial usage. This took place despite the fact that the state had a registered title to the disputed property (Nrt. 2001, 1229).

Nevertheless, Sámi law has encountered difficulties in being recognised and accepted as a source of law in Norwegian courts (Skogvang 2017, 63–76). Although progress has occurred in the acceptance of the duty of the courts to acquire knowledge of customary Sámi law, this knowledge has not become far reaching.

Conclusion

The view of Sámi law and culture has varied over time, as assimilation policies have gradually been replaced by positive attitudes towards Sámi culture. This has led to a legal protection of Sámi language and culture, through national legislation, case law and international obligations undertaken by Norway. Customary Sámi law has simultaneously been recognised as a source of law beyond internal Sámi autonomy.

In Article 108 of the Constitution, Norway has undertaken a duty to safeguard Sámi language, culture and way of life. The ratification of ILO 169 means that customary Sámi law and legal traditions are recognised as Norwegian law, as are the rights of the Sámi to own, use and manage their traditional lands. Despite the seemingly contrary tendency of the Supreme Court, the importance of ILO 169 was strengthened by the incorporation of the Finnmark Act, which itself is a recognition of Sámi rights to land and water (cf. Section 5 of the Act). It can here be referred that the Supreme Court has confirmed that ILO 169 anchored the right to restitution in Article 14, which is emphasized during the survey of land rights in Finnmark.

When the Human Rights Act was adopted in 1999, the most crucial international human rights conventions were given precedence over other Norwegian law. This has also strengthened the protection of Sámi culture, particularly through ICCPR Article 27. Among others, the Covenant sets limits on what interventions can be made in Sámi resource areas.

However, legal developments in recent years have shown that despite international obligations and extensive internal legislation, customary Sámi law has not received that much attention in Norwegian law. In addition, the significance of the international treaties protecting the Sámi language and culture that Norway has endorsed seems to have been reduced by the restrictive interpretations of the Supreme Court. The *Sara* case, the *Stjernøya* case and the *Nesseby* case are examples of this. Despite a solid legal foundation, Sámi law and legal protection still face significant challenges in Norway. In addition to the aforementioned Supreme Court cases, these challenges also manifest through the Finnmark Commission's investigations, the 2014 changes to the Reindeer Husbandry Act and other disputes over reindeer numbers and forced reduction of reindeer. Nevertheless, the latest investigation report of the Finnmark Commission suggests that a change in the direction of the law and outcome can be expected when it comes to the coming survey of land rights in Finnmark.

Questions about self-determination and the self-management of renewable resources have also been pushed into the background of the law. In terms of the Finnmark Act, which, in Section 5, recognises Sámi rights to land and waters, this recognition is currently far from being realised in practice. One issue concerns imprecise regulations regarding the management of outlying lands in Finnmark, which imply that the local populations have not been given the influence their legal basis would indicate. Another issue is that questions raised about the Sámi right to land and water in Finnmark can only be answered after the surveying of the rights by the Finnmark Commission has been completed. This process has been considerably more time consuming than anticipated, as a result of both unnecessary time spent in the start-up phase and time-consuming procedural rules. This means that it will still take many years before ownership and use rights questions are answered; in the meantime, continuous new interventions, through extractive industries, wind power plant constructions and other means, will occur.

In general, it must be said that the social and political situation for the Sámi in Norway is satisfactory. Nevertheless, there is a reason to keep an eye on legal developments. The most precarious issues in recent years can here be linked to three factors: first, the way in which the Norwegian Supreme Court has interpret the incorporation of ILO 169 into the Finnmark Act; second, the way in which the Supreme Court has understood the scope of protection under ICCPR Article 27 in so-called "internal Sámi issues"; and finally, the survey of land rights in Finnmark, where, until December 2019, neither the Finnmark Commission nor the courts had found examples of Sámi collective property rights or usage rights beyond what is already stipulated in the Finnmark Act. The Finnmark Commission's latest report, however, suggests, as mentioned, that other outcomes may be expected in future investigations.

Notes

1 An English (non-updated) translation of the law can be found here: https://app.uio.no/ub/ujur/over satte-lover/data/lov-20050617-085-eng.pdf
2 An English translation of the judgement can be found here: https://www.domstol.no/globalassets/upload/hret/decisions-in-english-translation/hr-2017-2428-a.pdf

References

Literature

Helander, E. 2001. "Samiska rättsuppfatningar i Tana." In NOU [Norwegian public reports] 2001: 34 *Samiske sedvaner og rettsoppfatninger - bakgrunnsmateriale for Samerettsutvalget [Sami Customs and legal opinions - Background Material for the Sami Rights Committee]*, 425–458, Oslo: Statens forvaltnigstjeneste.

Paltto, A.-S. 2019. "Lappi gearretriekti lea hilgon buot áššáskuhttimiid Deanu lobihis bivdoáššis." *YLE Sápmi*, 6 March. Accessed 14 May 2020. https://yle.fi/uutiset/osasto/sapmi/lappi_gearretriekti_lea_hilgon_buot_assaskuhttimiid_deanu_lobihis_bivdoassis/10675636.

Ravna, Ø. 2019. *Same- og reindriftsrett*. Oslo: Gyldendal akademisk forlag.

Ravna, Ø. 2020. "Den tidligere umatrikulerte grunnen i Finnmark: Jordfellesskap framfor statlig eiendom?" *Tidsskrift for Rettsvitenskap* Vol. 133.: 219–263.

Ruud, M., and G. Ulfstein. 2018. *Innføring i folkerett [Introduction to International Law]*. 5th ed. Oslo: Universitetsforlaget.

Skogvang, S. F. 2017. *Samerett [Sámi Rights]*. 3rd ed. Oslo: Universitetsforlaget.

Solem, E. 1933. *Lappiske rettsstudier [Lappish Law Studies]*. Oslo: H. Aschehoug & Co.

Norwegian preparatory works and other official documents

Dokument 8:154 S. 2010–2011. Representantforslag (Bill from representatives of The Parliament) om at Norge trer ut av ILO-konvensjon nr. 169, nedlegger Sametinget, opphever finnmarksloven og avvikler forvaltningsområdet for samiske språk.

Finnmarkskommisjonen, Rapport Felt 1 Stjernøya/Seiland (Finnmark Commission, 20 March 2012).

Finnmarkskommisjonen, Rapport Felt 2 Nesseby (Finnmark Commission, 13 February 2013).

Finnmarkskommisjonen, Rapport Felt 3 Sørøya (Finnmark Commission, 16 October 2013).

Finnmarkskommisjonen, Rapport Felt 5 Varangerhalvøya Øst (Finnmark Commission, 24 June 2014).

Finnmarkskommisjonen, Rapport Felt 6 Varangerhalvøya Vest (Finnmark Commission, 16 October 2015).

Finnmarkskommisjonen, Rapport Felt 4 Karasjok, (Finnmark Commission, 11 December 2019).

Innst. 253 L (2018–2019) om endringer i sameloven mv. (konsultasjoner) (Bill from The Standing Committee to The Parliament).

Innst. O. nr. 80 (2004–2005) om lov om rettsforhold og forvaltning av grunn og naturressurser i Finnmark fylke (finnmarksloven).

Meld. St. 32 (2016–2017) Reindrift – lang tradisjon – unike muligheter (Govermental report).

NOU 1984: 18 *Om sameness redistilling* (Norwegian Public Report – from a law committee).

NOU 1993: 18 *Lovgivning om menneskerettigheter.*

NOU 1993: 34 *Rett til og forvaltning av land og vann i Finnmark.*

NOU 1997a: 4 *Naturgrunnlaget for samisk kultur.*

NOU 1997b: 5 *Urfolks landrettigheter etter folkerett og utenlandsk rett.*

NOU 1999: 19 *Domstolene i samfunnet.*

NOU 1999: 22 *Domstolene i første instans.*

NOU 2001: 34 *Samiske sedvaner og rettsoppfatninger – bakgrunnsmateriale for Samerettsutvalget.*

NOU 2007: 13 *Den nye sameretten.*

NOU 2016:18 *Hjertespråket.*

Ot.prp. nr. 33 (1986–1987) Om lov om Sametinget og andre Samiske rettsforhold (Sameloven). (Bill from the government to the Parliament [until 2009]).

Ot.prp. nr 53 (2002–2003) Om lov om rettsforhold og forvaltning av grunn og naturressurser i Finnmark fylke (Finnmarksloven).

Prop. 89 L (2012–2013) Endringer i reindriftsloven mv. (avvikling av områdestyrene). (Bill from the government to the Parliament [after 2009]).

Prop. 116 L (2017–2018) Endringer i sameloven mv. (konsultasjoner).

Prop. 90 L (2018–2019) Endringer i reindriftsloven (tilgjengeliggjøring av reintall og obligatorisk individmerking).

St. Meld Nr. 28 (2007–2008) Samepolitikken (Governmental report).

Acts and regulations

Forskrift (Regulation) 30. juli 2008 nr. 866 om bruk av krumkniv.

Forskrift (Regulation) 3. juni 2013 om kvoteregulert vårjakt på ender fra og med år 2013 til og med år 2022, Kautokeino kommune, Finnmark.

Lov (Act) 12. juni 1987 nr. 56 om Sametinget og andre Samiske rettsforhold (Sámi Act of 12 June 1987 No 56 Concerning the Sámi Parliament and Other Sámi Legal Matters).

Lov 17. juni 2005 nr. 85 om rettsforhold og forvaltning av grunn og naturressurser i Finnmark fylke (Act of 17 June 2005 No 85 Relating to Legal Relations and Management of Land and Resources in the County of Finnmark (Finnmark Act).

Lov 15. juni 2007 nr. 40 om reindrift (Act of 15 June 2007 No 40 on Reindeer Husbandry (Reindeer Husbandry Act).

Lov 21. mai nr. 30 om styrking av menneskerettighetenes stilling i norsk rett (Act of 21 May 1999 No 30 Relating to the Strengthening of the Status of Human Rights in Norwegian Law).

International treaties, case law and other international documents

Anaya, S. J. "Report of the Special Rapporteur on the rights of Indigenous Peoples. The Situation of the Sami People in the Sápmi Region of Norway, Sweden and Finland." UN General Assembly, 6 June 2011, UN Doc A/HRC/18/35/Add.2.

Committee on the Elimination of Racial Discrimination (CERD). "Concluding Observations on the Combined Twenty-First and Twenty-Second Periodic Reports of Norway." 28 August 2015, UN Doc CERD/C/NOR/CO/21–22.

Committee on the Elimination of Racial Discrimination (CERD). "Concluding Observations on the Combined Twenty Third and Twenty Fourth Periodic Reports of Norway." 14 December 2018, UN Doc CERD/C/NOR/CO/23–24.

Human Rights Committee. "CCPR General Comment No. 23: Article 27 (Rights of Minorities)." 8 April 1994, UN Doc CCPR/C/21/Rev.1/Add.5.

Human Rights Committee. "*Ángela Poma-Poma v. Peru*, Communication No. 1457/2006." 27 March 2009, UN Doc CCPR/C/95/D/1457/2006.

Human Rights Committee. "Concluding Observations on the Seventh Periodic Report of Norway." 25 April 2018, UN Doc CCPR/C/NOR/CO/7.

International Convention on the Elimination of All Forms of Racial Discrimination (adopted 21 December 1965, entered into force 4 January 1969) 660 UNTS 195 (ICERD).

International Covenant on Civil and Political Rights (adopted 16 December 1966, entered into force 23 March 1976) 999 UNTS 171 (ICCPR).

International Labour Organization (ILO). Convention Concerning Indigenous and Tribal Peoples in Independent Countries, C169 (27 June 1989, entered into force 5 September 1991) 28 ILM 1382.

Nordisk samekonvensjon (Draft Nordic Sámi Convention), forslag fra (bill from) Regjeringene i Finland, Norge og Sverige (2017).

Tauli-Corpuz, V. 2016. "Report of the Special Rapporteur on the Rights of Indigenous Peoples on the Human Rights Situation of the Sami People in the Sápmi Region of Norway, Sweden and Finland." United Nations General Assembly, 9 August 2016, UN Doc A/HRC/33/42/Add.3.

United Nations Declaration on the Rights of Indigenous Peoples, 13 September 2007, UN Doc A/RES/61/295 (UNDRIP).

Supreme Court of Norway

Nrt. (Norsk Retstidende [1836–2015], a court archive journal,) 1988, *Andejakt,* 377.

Nrt. 2001, *The Båndtvang Case,* 1116.

Nrt. 2001, *The Selbu Case,* 769.

Nrt. 2001, *The Svartskogen Case,* 1229.

Nrt. 2006, *Fiske i Tana,* 13.

Nrt. 2006, *Customary Sámi law when slaughtering reindeer,* 957.

Nrt. 2008, *Customary Sámi law when slaughtering reindeer,* 1789.

HR-2016–2030-A, *The Stjernøya* case.
HR-2017–2428-A, *The Sara* case.
HR-2018–456-P, *Nesseby* case.

Supreme Court of Sweden

T 853–18, *The Girjas case.*

Uncultivated land tribunal for Finnmark

UTMA-2017-62459, *The Gulgofjord* case.

10

COMPREHENDING THE MANDATE AND INTERACTIONS OF LAND TENURE REFORM IN FINNMARK, NORWAY

Else Grete Broderstad and Eva Josefsen

Abstract

The land management arrangement called the Finnmark Estate (FeFo) established in Finnmark County, the northernmost county of Norway, is built on Indigenous right claims, which implies that particular values, norms and principles are constitutive for and underpin FeFo as an institution. Still, the involved actors – FeFo and the two appointing bodies of the FeFo board, the Sámi Parliament and the Finnmark County Council – have not developed a joint understanding of how to address these principles in order to strengthen the focus of FeFo governance. Based on earlier data compiled from investigations about FeFo and the two appointing bodies of the FeFo board, we focus on the relationship between the three institutions in order to explain the challenges of cooperation between the three parties concerning the management of land and resources by FeFo and how the bodies seek to minimize conflict and contribute to governability. Drawing on the concept of interactive governance, we analyse this governing system in terms of orders of governance by looking into whether the parties have deliberated and developed a set of meta-governance principles that can help in making hard, substantive governance choices easier. We will discuss the challenges that may arise between two political bodies, one with a territorially defined mandate and the other with a mandate to secure Sámi rights, and a governance and management body with a mandate to cooperate in a situation of high-level conflict.

Introduction

When the Norwegian national parliament, the Storting, adopted the Finnmark Act in May/June 2005, a long-lasting period of controversies on land and resource management in Finnmark came to a temporary closure. The Storting adopted the Finnmark Act as a response to an enduring political process where the Sámi, both individuals and organizations, over several

159

decades had challenged the government's claimed ownership of the land. Pursuant to the 2006 Finnmark Act, the ownership of land and resources (95% of the land area in Finnmark) was transferred from the government to the population in Finnmark. Based on Indigenous land claims, a management agency was established, the Finnmark Estate (FeFo), which on the operational level provides Sámi and non-Sámi users the same services. FeFo is a construction framed by a unitary state. The colonial history is different from that of settler states, and Indigenous Sámi and non-Sámi peoples have shared land for centuries. Over several decades, Sámi political empowerment, institutionalization and political integration have taken place, countering impacts of assimilation. Thus, while FeFo, like other land and resource management arrangements in the Circumpolar North, is a result of Indigenous land claims, the previously mentioned contextual aspects also make FeFo unique in terms of implementation of Indigenous rights.

After the adoption of the Finnmark Act in 2005, new challenges soon revealed themselves. One was the role of the executive body – FeFo, responsible for the management of land and resources in Finnmark, the northernmost county of Norway.[1] The board of FeFo consists of six members, three appointed by the Finnmark County Council[2] and three by the Sámi Parliament.[3] The board leader alternates every year between the two groups of appointed members. This construction of the land tenure system can be seen as a way of reconciling the concerns incorporated in the mandate of the Finnmark Act, namely the role of FeFo as a caretaker of a balanced and ecologically sustainable management for the benefit of the residents of the county and particularly as a basis for Sámi culture, reindeer husbandry, use of non-cultivated areas, commercial activity and social life. We pay attention to the concerns of all inhabitants and the protection of Sámi land and resource rights, with a particular focus on the last one.

According to the Finnmark Act, FeFo is a property owner and an independent legal entity, not a public body.[4] Thus, FeFo is not in formal terms subject to the appointing bodies' steering instructions. According to the Finnmark Act, the board members are personally responsible for decisions by the board and shall manage the estate according to the Act and its preamble. According to the Act, the only way the appointing bodies can influence their appointed board members if unsatisfied is by replacing the board member(s).

However, the Finnmark County Council, in the first years after the establishment, conceived itself as an 'owner' of FeFo (Nygaard and Josefsen 2010). Adding to this, the background of FeFo, with the long-lasting struggle for Sámi rights, did not solve the puzzle of the different roles embedded in the mandate (ibid., Broderstad et al. 2015, 2020). The establishment of FeFo implied potentially large changes in the governance of land areas in Finnmark, for example, regarding identification and operationalizing of Sámi land rights (Ravna, this volume). Such possible changes with large symbolic value could create uncertainty, resistance and conflicts (cf. Jacobsen and Thorsvik 1997).

Broderstad et al. (2015, 2020) found that there was general low support for the estate among inhabitants in Finnmark County and explained it as resistance to Sámi rights and Indigenous political initiatives. They concluded that unhandled conflicts could diminish the overall public support for FeFo, which again could result in a failure of the land tenure system to reconcile Indigenous and non-Indigenous land rights. Thus, the purpose here is to illuminate cooperative challenges that may arise when two political bodies – one with a territorially defined mandate, the other with a mandate to secure Sámi rights – are to cooperate with a governance and management body with multiple mandates.

Drawing on the concept of interactive governance (Kooiman et al. 2008; Kooiman and Jentoft 2009; Kooiman and Bavinck 2013), we analyse the governing system in terms of orders of governance by looking into whether the parties have deliberated and developed a set of meta-governance principles that can help making hard, substantive governance choices easier.

Kooiman and Jentoft (2009, 818) argue that such choices are always complicated because the value positions and normative notions contained in them are often in conflict. Our case is framed by the mandate's territorial defined concern versus the concern of securing Sámi rights and FeFo's governing of these considerations. *What are the features of cooperation between the three parties, and how do they seek to minimize the conflict level and contribute to governability?* Kooiman and Jentoft (2009, 819) argue that the choices are made less hard when values, norms and principles are made coherent and explicit. In 2010, Nygaard and Josefsen concluded that FeFo had not sufficiently implemented the Finnmark Act's principle of "management as a basis for Sámi culture," because of strong opposition to Sámi rights, as revealed in a survey on support for FeFo (Broderstad et al. 2015, 2020). This point of departure leads us to focus on the coherence and explicitness of these norms of Sámi rights in the case of FeFo's land governance.

In the next section, we introduce aspects of the conceptual framework of interactive governance and more precisely the enabling and restricting conditions of the concept of orders of governance, that is, how frameworks (e.g. culture and law) limit or widen action potential. Thereafter, in the main part, we empirically account for core elements of how cooperative the stakeholders of FeFo are by addressing the mandate and features of interaction in light of the concepts of meta, second- and first-order governance, and we also include a discussion on limitations and possibilities to strengthen governability before we conclude.

Comprehending interactive governance

While the aim of FeFo is to manage land and resources in accordance with the Finnmark Act, it is too simplistic to see FeFo as a purely management body with a set of tools applied to solve concrete tasks. While attention in management is concentrated on goals and means, in governance, efforts are spent on reflecting and deliberating upon basic values, concerns and principles, implying a process of inclusion, communication and cooperation (Kooiman and Jentoft 2009, 831–832). The estate deals with management of renewable and non-renewable resources, property and businesses. While resource management of fish, wildlife and recreation is specifically defined in the Finnmark Act and a primary task of FeFo, the tasks of managing other resources, issues and land are less regulated and create room for interpreting how and what the responsibility of FeFo should be.

The work of FeFo is underpinned by the Finnmark Act and its strategic plans (2007, 2011, 2015) and is formally attached to the Finnmark County Council and the Sámi Parliament. It is thus about deliberation on goals, including values, norms and principles underpinning them (cf. Jentoft and Chuenpagdee 2009, 555). Hence, an interactive system of governance has elements of deliberative democracy, where a basic requirement is giving reasons for actions (Gutman and Thompson 2004: 3). Governance goes beyond government (Kooiman and Bavinck 2013, 10) and is a complex undertaking, including a range of societal actors with governments in mutual interactions (Bavinck et al. 2015). By applying the conceptual framework of interactive governance,[5] we empirically investigate how interactive the system of FeFo is, and we normatively underline the significance of interactive governance by focusing on the enabling and restricting conditions of interactivity. Interactive governance theory divides societal systems into three parts: a system to be governed, a governing system and the interactions that take place within and between them (Jentoft and Bavinck 2014). The governability of FeFo is then made up of these components: governors, the governed and their interactions contributing to governability (Kooiman and Bavinck 2013, 12). In this chapter, our concerns are the governing interactions at the structural level, understood as "exchanges between actors that contribute to the tackling of social problems and opportunities" (ibid., 11). How has the governability of FeFo developed in response to external conditions? By applying the conceptual framework of interactive governance, we acknowledge the complexity and the dilemmas of the governing interactions and the wickedness[6] of the problems. These dilemmas and wicked problems are no less problematic in Indigenous contexts. Interactive governance suggests that values, principles and goals are not stable and fixed but negotiated and vary according to the relative strength of the participants that come and go (ibid., 12). How do the deliberating actors influence each other's interpretations of the mandate's territorially defined concern versus the concern of securing Sámi rights?

In order to capture the core connection between the elements of the mandate of FeFo, we have chosen, as part of the conceptual framework of interactive governance, to make use of the component of 'orders of governance' that can be divided into first-order, second-order and meta-governance, which allow us to elucidate the question of enabling and restricting conditions of governability in the case of FeFo.

Institutions provide the framework for first-order governance and constitute the meeting ground of the governed and governing. This is where problem-solving and opportunity-creating activities are embedded (Kooiman and Bavinck 2013, 19). First-order governing is about day-to-day affairs, practices and governing tools, that is, concrete management and administrative tasks regarding hunting and fishing; business tasks like sale of gravel, leasing house lots and concrete management of cooperation and disagreements. The operational level is also the level of management, where the distinction between management and governance is found.

The second order of governance focuses on the institutional arrangements within which first-order governing takes place (Jentoft and Bavinck 2014, 74). It is about systems of agreements: rules, rights, laws, norms, roles and procedures, that is, the Finnmark Act's regulations and guidelines, for example, the Sámi Parliament's guidelines for changed use of land and regulations of cooperation with the Sámi Parliament and Finnmark County Council.

Meta-governance includes the values, norms and principles that underline governance and pertains to how ethical principles and other normative notions play a role in actual governance practice. These are ethical in their nature, exist within a particular social field and are about what is perceived as right or wrong (Jentoft and Bavinck 2014). The principles enshrined in international law, like ILO Convention No. 169 concerning Indigenous and Tribal Peoples in Independent Countries (ILO 169) and the United Nations Declaration on the Rights of Indigenous Peoples (UNDRIP) (Ravna, this volume), belong to meta-governance. For the assessment of governability, orders have special importance. Furthermore, are they complementary to one another or at odds (Kooiman et al. 2008, 8)? Have the parties deliberated and developed a set of meta-governance principles that can help make hard, substantive governance choices easier?

The analytical framework we draw upon is summarized in Figure 10.1. The concerns of the mandate of FeFo and features of interaction are investigated through the lenses of meta, second

	Meta-order governance	Second-order governance	First-order governance
Multiple concerns of the mandate	Different understandings of FeFo's mandate	Including the mandate into FeFo regulations	Concrete outcomes reflecting the mandate
Features of stakeholder interaction	Principles of interaction	Function of interaction	Operation of interaction

Figure 10.1 Analytical framework

and first orders of governance. What are the understandings of the mandate, the function of established rules and regulations and the concrete outcomes? What are the principles and the function and operation of interactions?

FeFo as an interactive institution

Background

The history of FeFo is neatly connected with civil disobedience actions against the damming of the Alta-Kautokeino River in Alta, Norway, in the late 1970s, when Sámi land rights were brought into the national public and political agenda, turning into state–Sámi interaction from the early 1980s onwards through a separate commission that investigated these rights. This assessment of Sámi land rights in Finnmark resulted in an Official Norwegian Report in 1997 and the passing of the Finnmark Act by the Norwegian Parliament in 2005 (Ravna, this volume).

FeFo works within a diversity of institutional stakeholders and is, as a governing actor, constrained and enabled by its surroundings, most importantly by the Sámi Parliament and the Finnmark County Council. The Sámi Parliament and the County Council have different roles and tasks. Both are politically elected bodies but with different constituencies (the Sámi of Norway and the people of Finnmark) and different and even conflicting political aims, while FeFo manages natural resources according to the Finnmark Act. The municipalities constitute another significant political structure within which FeFo operates, as they decide land use planning. In size, the municipalities are highly variable, from small with a decreasing and aging population to a few larger municipalities with growth and expansion in both the private and public sectors. Another significant premise provider for FeFo is the state, regulating, for example, the limits for resource extraction through legislation and what should be protected areas and licensing resource exploitation in the county, for example, wind turbines. Also local and regional organizations, be it Sámi or Norwegian, concerned with natural resources and land use (e.g. hunting, fishing, outdoor life and recreation) interact with FeFo, as do traditional and new industries and companies. We will focus on the core governance "triangle" – FeFo, the Finnmark County Council and the Sámi Parliament – but mention other interactions when relevant, such as interactions with resource users (cf. 'Concrete Outcomes – First-Level Governance'). In the next subsections, we will first address the mandate of the Finnmark Act and then the interaction of the stakeholders in light of the three orders of governance in Figure 10.1.

The mandate of FeFo

As mentioned, the preamble of the Finnmark Act points out three main purposes for the management of land and natural resources in the county of Finnmark. The preamble describes the legislators' purpose with the Finnmark Act, and each part of the preamble is equally important for the management of the Act. In this subsection, we discuss the mandate according to how FeFo and the appointing bodies emphasize and interpret the mandate. A guiding question is whether there is a joint understanding of the mandate and how the institutionalized mandate is reflected in FeFo decisions.

Understanding FeFo's mandate – meta-level governance

The policy documents of the appointing bodies reveal opposite approaches to how to understand the preamble and the foundation of Indigenous rights of the Finnmark Act. Sámi politicians fighting to secure Sámi land rights as a new and decisive element in decision-making processes do not necessarily agree with the view of the representatives of the County Council, who see natural resources as a means for economic growth (Nygaard and Josefsen 2010). After the passing of the Act in 2005, the Finnmark County Council quickly drew up a 'policy document' (2005) referring to itself as a co-owner of FeFo, as reflected in the guidelines for how the County Council's appointed members should function. This document has been replaced several times: by a steering document for the appointed members (2008), owner strategies and surplus use (2012), a document on owner strategies (2014) and a document on strategies for FeFo (2018). The Sámi Parliament, on the other hand, had a low profile in the first years regarding the appointed board members and their function. It was not until 2010 that the Sámi Parliament's plenary for the first time commented on FeFo's strategy plan. In 2012, the plenary addressed questions of owner strategies and surplus use; in 2014, it gave a general review of the Finnmark Act and in 2018, it commented on co-governance regarding FeFo. We argue that this handling by the appointing bodies impacts the board members' emphasis and understanding of the preamble, which will reflect on how they interpret and implement the act.

Applying the lens of meta-order governance, we find that the appointing bodies have very different views on international law, including that the Finnmark Act shall apply with the limitations that follow from ILO 169 and be applied in compliance with the provisions of international law concerning Indigenous peoples and minorities. Even if the administration of the County Council, in 2005, had already recommended the need to ensure international law in order for the FeFo board to succeed transferring authority from the state to the region, the County Council's documents on the strategies of FeFo did not refer to Sámi or international law as important frameworks until its strategy document in 2018 (Finnmark County Council 2018). The earlier documents described tasks, challenges and expectations and presented guidelines to the board members appointed by the County Council, but the documents lacked a definition of the content of the Sámi part in the preamble. Explicitly mentioning the concern for Sámi culture could have been anticipated, given the absence of Sámi culture in the earlier land management regime. In 2008, in the County Council Steering document for the board members, the first two parts of the preamble were referred to under the headline "Finnmark County Council's aims for FeFo," leaving out part three on Sámi culture.

The County Council is, however, not alone in having been selective in the content of the preamble. In the 2014 overall review of the Finnmark Act by the Sámi Parliament, the Parliament states that the "main objective of the Finnmark Act is about securing the natural basis for Sámi culture," referring solely to the third section of the preamble (Sámi Parliament 2014). Different from the County Council, this was the first document where the Sámi Parliament addressed the appointed board members. Plenary decisions relating to FeFo prior to this were responses to FeFo's strategic plans and to questions of distribution of surplus from FeFo. While not necessarily surprising that the two appointing bodies prioritize different concerns of the FeFo mandate, it still remains unclear how the core components of the mandate should be secured jointly by the three institutions. Therefore, in search of an overall operationalization of the mandate, we turn our focus to rules and regulations.

Operationalization of the mandate – second-order governance

In this part, we examine how the mandate and specifically section three on Sámi culture of the Finnmark Act have been operationalized. Concerns for Sámi culture were not included in the former management arrangement in Finnmark; thus, parts of the public; some political parties and local, regional and national politicians strongly protested the new act (Olsen 2010, 2011; Eira 2013). At the operational level, the Finnmark Act (§ 4) anchors the guidelines of the Sámi Parliament regarding changes in the use of uncultivated land. In matters concerning changes in the use of uncultivated land, FeFo and state, county and municipal authorities shall assess the significance of such changes on Sámi culture, reindeer husbandry, use of non-cultivated areas, commercial activity and social life. According to the Finnmark Act, the guidelines shall be applied both in public planning of land use and in single issues regarding uncultivated land, and the Sámi Parliament regards them as decisive in securing section three in the mandate (Sámi Parliament 2006, 2007a). When assessing what "changed use" is, the guidelines refer to customary use of land and to whether this use will be able to continue despite the change. Rights holders, users and affected Sámi interests shall also be consulted. By assessing the strategic documents of Finnmark County Council and FeFo in search of whether these documents reflect the guidelines, we can establish to what extent these two parties have institutionalized this concern.

According to FeFo's 2007–2010 Strategic Plan, FeFo shall execute property management in line with the guidelines regarding Sámi interests (FeFo 2007). The Strategic Plan states that these guidelines are the main motivation for FeFo basing its disposal of land on approved municipal plans, because it supposes that municipality planning complies with the guidelines (Nygaard and Josefsen 2010). Nevertheless, a topical issue in the interaction between FeFo and the municipalities has been sale of property, which implies changed use of land and can be contrary to the intentions of the guidelines. While some municipalities want to buy land below market price, FeFo restricts this approach. While justified differently, the Finnmark County Council and the Sámi Parliament seem to indicate a similar scepticism (Sámi Parliament 2006; Finnmark County Council 2018). The Sámi Parliament opposes sale of land in principle while the County Council views sale of land below market price as contrary to EEA regulations. The price has to be market oriented but not price driven. Despite different justifications, a common understanding of the challenges of land sale appears to be shared by the institutions.

In 2014, the Sámi Parliament underlined the importance of an active use of the guidelines and demanded FeFo draw up instructions on how to use the guidelines and, by this, change procedural rules in order to make a concrete and documented assessment of each decision on changed use of land (Sámi Parliament 2014). The 2015 Strategic Plan of FeFo (2015) stated that FeFo will take an active position on all measures that lead to changed use of outlying fields, probably as a response to the Sámi Parliament's demand.

The County Council, on the other hand, does not discuss the guidelines in its "steering documents" with regard to new and non-reversible industry (Finnmark County Council 2008, 2012, 2014, 2018). The County Council's steering documents do, however, specify industry strategies (2014) and existing plans and strategies (2018) as guiding rules for the board representatives (Finnmark County Council 2014, 2018).

We find no indication in FeFo's strategic plans that the content of "Sámi culture" in the preamble has been concretized. According to Nygaard and Josefsen (2010, 28–29), FeFo in 2010 emphasized the second section of the preamble, on the interests of all Finnmark inhabitants, while Sámi culture considerations were toned down. This may have been a response to the public resistance towards Sámi rights that characterized the origin of FeFo. We lack data on whether this has changed after 2010. FeFo's strategic plan (2007–2010) points out that FeFo is responsible for using Sámi language when informing the public, assessing traditional Sámi knowledge and establishing competence on Sámi culture when developing new guidelines for outdoor activities (FeFo 2007). Except for language competence, it remains unclear how competence building on Sámi cultural aspects is institutionalized in FeFo's administration.

Concrete outcomes – first-level governance

The policy documents of the County Council and the Sámi Parliament reflect opposite approaches to the preamble and the foundation of Indigenous rights of the Finnmark Act, a point appearing in documents addressing the board members. How the three elements of the preamble are emphasized by the board reflects the interpretation, follow-up and implementation of the Act. In addition, it reflects how the board members of the two institutions speak about and emphasize the different elements of the Act. Sámi politicians' emphasis on developing and securing Sámi land rights as a new and decisive part in decision-making processes does not necessarily agree with the County Council, who see natural resources as a means for economic growth (Nygaard and Josefsen 2010).

We can also identify a change in how the appointing bodies regard FeFo's leeway. FeFo did start out with a defined aim to be an active landowner. In 2009, seven local energy companies and FeFo joined forces and established Finnmark Kraft (FK) AS as a regionally and locally owned power company in Finnmark. A main push factor for establishing FK was to secure Finnmark a regional share in the value creation of the exploitation of energy resources. In 2005, the County Council Policy Document stated:

> Finnmark has good conditions for wind and hydropower (small power plants). The interest in development is high. As a landowner FeFo can provide a predictable policy and be a facilitator where this is relevant. This can be done by contributing to inter-municipal solutions for wind turbines or the like.
>
> (Finnmark County Council 2005, 6)

In 2008, the County Council emphasized the importance of local anchoring of important companies established on the basis of resources in Finnmark (Finnmark County Council 2008). In 2012, this view had changed. FeFo was solely to engage in companies with a clear foothold in the company's core activity, and FeFo should be cautious engaging in derivative business[7] (Finnmark County Council 2012). The decision does not define the term "core activity" but probably refers to resources specifically mentioned in the Finnmark Act and management tasks taken over from the previous land management regime regarding, for example, hunting, fishing, property leasing contracts and management of gravel and crushed stones. In 2014, this was clarified: FeFo should act "neatly as facilitator for the development of renewable energy, regardless of who is the project owner" (Finnmark County Council 2014, 3).

A similar development can be traced in the Sámi Parliament's documents. In an administrative statement in 2007, the Sámi Parliament referred to windmills and recommended FeFo invoke the value and exploitation potential of an area for electrical production and demand a fair share of this value creation (Sámi Parliament 2007b). In 2010, the Sámi Parliament referred to the Act preamble and FeFo's obligation to do its own assessment based on the guidelines for changed use of uncultivated land. Further, the Sámi Parliament stated that FeFo had to take into account that the role as an active owner may come into conflict with existing land rights (Sámi Parliament 2010). In 2012, the Sámi Parliament recommended FeFo as a landowner to engage in new activity based on natural resources, among others, to "secure local ownership" and "secure the community a greater share of the values created from FeFo's resources" (Sámi Parliament 2012). In 2014, the Sámi Parliament stated that FeFo should only participate in business activities when it, among others, "does not entail changed use of land and the use of property." Further, the Sámi Parliament asked FeFo to assess other alternatives for its ownership in Finnmark Kraft (Sámi Parliament 2014). The FeFo board decided in 2019 to terminate its ownership in Finnmark Kraft by selling its stock in the company (FeFo 2019). By doing this, FeFo adapted to the policy of the appointing bodies.

As illuminated here, board members receive different political signals and expectations from their two appointing institutions, while they are legally obliged to manage all three elements of the preamble. Given the differences of the two appointing institutions in terms of function, different views by these institutions of what FeFo should emphasize do not come as a surprise. The critical question is whether contradicting signals from the County Council and the Sámi Parliament and lack of coordination hamper governability and affect the operationalization and implementation of Sámi land rights by FeFo itself. The lack of translation of the Act's preamble on Sámi rights into concrete management rules suggests this, but the question has not yet been researched.

Besides the guidelines, there are few traces of established rules and regulations concerning the Sámi section of the preamble, and the concern is rarely reflected in concrete outcomes. One of the few examples where FeFo refers to the preamble is the board majority submission/statement on the application from the company Nussir for a license to establish a copper mine. The statement emphasizes that the mining project will provide a basis for positive social development, which in turn is said to provide a basis for safeguarding Sámi interests through increased settlement in a marginalized coastal Sámi area (FeFo 2017). The board members of the Sámi Parliament voted against this statement. The Sámi Parliament found the statement not to be in line with the Finnmark Act's sections regarding changes in the use of uncultivated land (Sámi Parliament 2018). If the appointing institutions fail to unite on core elements of the mandate, at least common ground should be required in terms of well-functioning dialogues (see section 'Features of Interaction').

Summing up points on the mandate

There is a lack of common understanding among the stakeholders on what kind of body FeFo is. Neither the board nor the appointing bodies have focused on overall questions of the mandate. Given the complexity and dilemmas of governing interactions in a multicultural context like the Sámi–Norwegian one, we could have expected the appointing bodies and FeFo to put a more thorough principal emphasis on these questions. As studies of the documents reveal, the appointing bodies emphasize the mandate differently, which in itself can be explained by their respective roles. We turn to the interactions themselves, looking into whether the established arenas of interaction, rules and regulations sustain FeFo governance.

Features of interaction

In this subsection, we look into features of interaction, namely principles linked to the board members' role as appointees and the annual contact meetings between FeFo and the appointing bodies, which we regard as important aspects of interactions at the structural level. The flow of information, assumptions, values, prejudices, communication arenas and channels – among other media and different political policies – creates dynamics as "potentials for change, but can also be disruptive" (Kooiman et al. 2008, 5), as we will see in relation to the contact meetings. We therefore look into how these aspects of interaction contribute to the management and governance of FeFo by looking into how questions of Sámi land rights have been handled. In 2005, the administration of the County Council pointed out the need to ensure cooperation, and in 2008, it established a chief county executive working group to monitor FeFo. What are the principles of interaction, and how does established interaction work? To what degree are the intentions on cooperation reinforced?

Principles of interaction – meta-order governance

Principles of interaction are formally enshrined as well as evolving through practice. In the following, we will take a closer look into how these principles are manifested in appointing bodies' strategic documents. There are few mentions of the working relationship between the Finnmark County Council and the appointed members. The County's strategy documents (2008, 2012, 2014, 2018) state that the board members are to ensure the County Council's interests according to these documents. In 2012, the County Council stated that the appointed board members have to safeguard the interests of County authority in line with the adopted ownership strategy document (Finnmark County Council 2012, 2014). Thus, the County Council relationship to board members seems to have hierarchical features.[8] The County Council emphasis on the role as owner adds weight to this assumption.

In contrast to the County Council, which since 2005 has maintained principles for its board members, the Sámi Parliament addressed its board members for the first time in a plenary document only in 2014. Here it took a "comprehensive review and expresse[d] overall priorities for the further follow-up of Finnmark Act, as well as providing a clear basis for follow-up and support for the work of appointed members in bodies established after" (Sámi Parliament 2014). The decision in 2014 was among others addressed to the appointed board members but did not give any direction for principles for cooperation. In 2018, the Plenary followed up under the heading 'Co-Governance' (Sámi Parliament 2018). It addressed a need for regulating contact with the Sámi Parliament's appointed board members and decided to facilitate, strengthen and develop contact with FeFo and the County Council. The use of the term co-governance (*samstyring*) in the title may be read as an interactive approach. However, the decision contains hierarchical features. The Sámi Parliament directs its appointed board members – for example, the first elected member is candidate to chair the board; appointed members and deputy members are to be summoned to a seminar once every election period and to other meetings when needed; at the beginning of every period, the plenary is to adopt a steering document containing priorities and perceptions on how to fulfil the Act preamble and every year, the plenary should receive an orientation of FeFo's activities in light of the Sámi Parliament's steering document for FeFo.

The appointing bodies' wish to steer "their" board members could potentially collide with the intention of board members' personal responsibility of operation (cf. § 17 of the Finnmark Act) and is a question of how far the appointing bodies perceive their instructive capacity goes. Thus, while the Finnmark County Council took on the role of owner right from the beginning, the Sámi Parliament changed its perception of FeFo from more of an autonomous body to an institution with appointed board members committed to a steering document. This could imply a move towards an understanding of board members with a limited mandate, acting as representatives of the appointing bodies, more than appointees obliged to work as a collegium to fulfil the Finnmark Act's preamble. In the following, we will take a closer look into one concrete aspect of interaction, namely the yearly contact meetings, in search of enabling conditions of interactivity.

Function of interaction – second-order governance

Formal contact meetings are concrete arenas for interaction between the parties. These meetings came into being in 2008. The initial plan was to arrange them twice a year. From 2012, one meeting at the political level has annually been arranged, except for 2017, when the only meeting was arranged at the administrative level. The contact meetings address topics like minerals, windmills, the company Finnmark Kraft, FeFo surplus, ground rent, selling of land and setting of prices, FeFo's strategic plans and the mapping of rights of the Finnmark commission. The contact meetings' duration is between four and six hours.

We have found indications that the contact meetings are not always prioritized by all partners:

> When initiative is taken by the Sámi Parliament, FeFo or the County Council to hold a meeting at the political/administrative level and/or administrative level to discuss concrete and demanding matters, it is assumed that the meetings are actually met by all parties.
>
> (Sámi Parliament SP 037/18)

The topic of cooperation and the question of improving the dialogue between the appointing bodies and FeFo have over the years been on the agenda. At the contact meeting in 2009, communication challenges – both outreach and communication between the partners and FeFo's reputation – were discussed. At the meeting in 2014, the need to improve communication was again raised. The minutes do not, however, describe this topic in detail. Another problem regarding insight is that not all meetings are recorded. We have also been informed that there are no minutes from the 2018 meeting and that the practice of minutes has ended (Sámi Parliament 2019). This practice indicates that contact meetings were used mostly for information exchange. Closed meetings and lack of recording can allow for a more open exchange of positions and standpoints and contribute to trust building. However, minutes contribute to collective memory, while the lack thereof may force the actors to revisit themes.

Operation of interaction – first-order governance

The political processes prior to the establishment of FeFo heightened attention to land and resource management, not seen during the over hundred-year-long state management regime. Due to the central position of Sámi rights in these processes, the estate became highly politicized and faced strong public resistance (Broderstad et al. 2015, 2020). On the other hand, strong support for the specific and concrete management of land and resources was identified (ibid.). As these concrete management tasks are about day-to-day affairs and conflict resolution, they can be read as first-order governing managed according to the Finnmark Act. FeFo directs its concerns towards the different user groups, municipalities and organizations. According to an interview investigation (ibid.) of resource users that actively use nature and local leaders that have experiences with FeFo, small- and big-game hunters were satisfied with FeFo's wildlife management. Those actively using the land, such as recreationists and subsistence harvesters, expressed a relatively high degree of trust towards FeFo. This is the level of management where FeFo provides the same services to Sámi and non-Sámi users, where the interactions with the appointing bodies are more distant but where state legislation and regulations and FeFo's internal guidelines frame the day-to-day management. We cannot assume the appointing bodies, through, for example, their steering documents, want to have more of a say on FeFo's first-order governance. But we will warn against a development where FeFo has to balance different and contradictory management principles of the appointing bodies.

Summing up points on interaction

As noted, enabling and restricting conditions impact the interactivity of institutions. While principles of interaction should enable the governability of FeFo, these principles differ and can restrict governability. The function of contact meetings as an established arena of interaction is evident. A cause of concern is that these meetings do not seem to be prioritized by the appointing institutions. In a situation of institutionalized environments infused with different and even contradicting values and norms regarding the foundation for resource and land management, trust building and dialogue between stakeholders are significant to the quality and efficiency of FeFo's management structures, regulations and concrete management decisions. While first-order governance – daily management – is regulated and thus "manageable," discussions on norms and principles – on meta-governance – which could have "eased" difficult decision-making is lacking. According to Edelenbos (2005, 129), "An ill-considered introduction and incorporation of new (interactive) processes and institutions into existing institutional environments may result in quite a few disappointments for a lot of stakeholders in the process." The two contradicting positions towards FeFo by the appointing bodies in the first years coincided with a period of high negative media and public attention towards FeFo (Eira 2013). Simultaneously, regular contact meetings were not in place from the start. In this situation, with a partly hostile public wanting to uphold the status quo, a partly expectant public demanding significant changes and one active and one more passive appointing institution, FeFo may have tried to reduce conflicts by adapting to the demands of the majority. A weak operationalization of section three on Sámi culture of the Preamble may thus have functioned as a restricting condition of governability. After some years, the plenary of the Sámi Parliament increased its attention towards FeFo by regulating relations with "its" board members and with FeFo as an institution. Whether this changed approach will reinforce partnership or a hierarchy relationship between the parties, and whether interactions of the board itself will be characterized by hierarchical governing of the appointing bodies or by the appointees' work as a collegium, remains to be seen. This last question raises the conflict between each board member being personally and financially responsible for decisions versus a possible development where the appointing bodies act as "owners" instructing "their" board members.

Conclusion

In order to answer the main question of what features of cooperation exist between FeFo and its board composed of members of the Sámi Parliament and the Finnmark County Council and how they seek to minimize conflict and contribute to governability, we have applied the conceptual framework of interactive governance. On the basis of our analytical framework, concerns of the mandate of FeFo and features of interaction have been investigated through the lenses of meta, second and first orders of governance. We have empirically accounted for core elements of how interactive the partners are by addressing different understandings, operationalization and concrete outcomes of the mandate and features of interactions comprehended as principles, functions and operations of interactions.

In practice, we have found that FeFo is subject to different and opposite values of the appointing bodies. This is about hard choices, like when conservation and cultural values are up against development and economic values. While so-called "meta-governance principles" clearly are present, the involved actors have not developed a joint understanding of how to address these principles in order to strengthen the focus of FeFo governance. As clear from our discussion, first- and second-order governance (concrete management and established procedures) contribute to governability, but the parties have not deliberated and developed a set of meta-governance principles that can help in making hard, substantive governance choices easier. Additionally, interaction between parties of conflicting political agendas and implementation expectations needs to be organized into formal structures and routines that can facilitate trust building and reduce tension between interactive processes and different institutional decision-making structures (Edelenbos 2005, 130). A prioritization of regular meetings between the appointing institutions and FeFo is thus expected to strengthen the governability of FeFo, a need which will be emphasized by the new merger of 2020 between the counties of Finnmark and Troms. Still, FeFo has to balance its multiple mandates, and the level of conflict can either be hampered or expanded, depending on the appointing bodies' "steering" policies towards the FeFo board as either hierarchical or co-governing.

Notes

1   The Storting has, however, decided on a new structure of the regional county level in Norway, merging the counties of Finnmark and Troms from January 2020.
2   The county councils are political institutions, elected by the population in each county every fourth year simultaneously with the municipal elections. In a bill to the Storting (Prop. 134L [2018–2019]), the government presents changes to the Finnmark Act as a result of the merger between the counties of Finnmark and Troms and proposes, for example, that the new merged county council appoint members to the FeFo board.
3   The Sámi Parliament is a popularly elected body, with representatives elected by and among Sámi in Norway. The parliament was established according to the Sámi Act and officially opened in 1989. Sámi Parliaments are also established in Finland and Sweden (Mörkenstam et al. 2016).
4   According to the Proposition to the Norwegian Parliament (Innst. O. nr. 80 [2004–2005]), the Finnmark Act gives FeFo status as a regular owner. The law makes certain limitations on the body's ownership interest (distribution of profit, relation to future legislation, others' right to hunt and fish on its grounds, etc.), without changing the legal nature of the Finnmark property.
5   As defined by Kooiman et al. (2008, 2), "The interactive governance approach differs from others by focusing on its applicability and occurrence at different societal scales, from the local to the global and with overlapping, cross-cutting authorities and responsibilities. In addition to horizontal networks, all kinds of vertical governing arrangements between public and private entities are also seen as governance."
6   Problems are wicked in the sense that there are limits to how systematic, effective and rational a governing system can be in solving them (Jentoft and Chuenpagdee 2009, 553).
7   FeFo skal utelukkende engasjere seg i selskaper som har et klart utspring i selskapets kjernevirksomhet. FeFo bør derfor være forsiktig med å engasjere seg i avledende virksomhet", Finnmark County Council, issue 18/12, point 4.
8   In 2008, an administrative working group was established whose task, for example, was to focus on the relationship between the County Council's political leadership and the appointed board members (Finnmark County Council, accessed 30 December 2008).

References

Bavinck, M., S. Jentoft, J. J. Pascual-Fernández, and B. Marciniak. 2015. "Interactive Coastal Governance: The Role of Pre-Modern Fisher Organizations in Improving Governability." *Ocean & Coastal Management* 117: 52–60.

Broderstad, E. G., V. Hausner, E. Josefsen, and S. U. Søreng. 2020. *Local Support among Arcticresidents to a Land Tenure Reform.* Vol 91. Finnmark, Norway: Land Use Policy.

Broderstad, E. G., E. Josefsen, and S. U. Søreng. 2015. *Finnmarkslandskap i endring -Omgivelsenes tillit til FeFo som forvalter, eier og næringsaktør.* Rapport 2015:1. Alta: NORUT Alta.

Edelenbos, J. 2005. "Institutional Implications of Interactive Governance. Insights from Dutch Practice." *Governance: An international Journal of Policy, Administration and Institutions* 18 (1): 111–134.

Eira, S. S. 2013. "Herrer i eget hus; Finnmarksloven i media." *Norsk medietidsskrift* 20(4): 330–346.

Gutman, A., and D. Thompson. 2004. *Why Deliberative Democracy?* Princeton, NJ: Princeton University Press.

Jacobsen, D. I., and J. Thorsvik. 1997. *Hvordan organisasjoner fungerer. Innføring i organisasjon og ledelse.* Bergen: Fagboklaget.

Jentoft, S., and M. Bavinck. 2014. "Interactive Governance for Sustainable Fisheries: Dealing with Legal Pluralism." *Current Opinion in Environmental Sustainability* 11: 71–77.

Jentoft, S., and R. Chuenpagdee. 2009. "Fisheries and Coastal Governance as a Wicked Problem." *Marine Policy* 33 (4):553–560.

Kooiman, J., and M. Bavinck. 2013. "Theorizing Governability – The Interactive Governance Perspective." *Governability of Fisheries and Aquaculture: Theory and Applications, MARE Publications Series* 7: 9–30.

Kooiman, J., M. Bavinck, R. Chuenpagdee, R. Mahon, and R. Pullin. 2008. "Interactive Governance and Governability: An Introduction." *The Journal of Transdisciplinary Environmental Studies* 7 (1): 1–11.

Kooiman, J., and S. Jentoft. 2009. "Meta Governance: Values, Norms and Principles, and the Making of Hard Choices." *Public Administration* 87 (4): 818–836.

Mörkenstam, U., E. Josefsen, and R. Nilsson. 2016. "The Nordic Sámediggis and the Limits of Indigenous Self-Determination." *Gáldu Cála – Journal of Indigenous Peoples Rights* 1: 4–46.

Nygaard, V., and E. Josefsen. 2010. *Finnmarkseiendommen under lupen.* Rapport 2010: 4. Alta: NORUT Alta.

Olsen, K. 2010. "Stat, samer og 'settlere' i Finnmark." *Norsk antropologisk tidsskrift* 21 (2–3): 110–128.

Olsen, K. 2011. "Fefo, reinsdyr og andre vederstyggeligheter; om urfolk, staten og finnmarkingene." *Norsk antropologisk tidsskrift* 22 (2): 116–159.

Ravna, Ø., In this volume. "Sámi Law and Sámi Rights in Norway with a Focus on Recent Development." In *Handbook of Arctic Indigenous Peoples*, edited by T. Koivurova et al., New York: Routledge.

Other documents

International Labour Organization (ILO). Convention Concerning Indigenous and Tribal Peoples in Independent Countries, C169 (adopted 27 June 1989, entered into force 5 September 1991) 28 ILM 1382.

United Nations Declaration on the Rights of Indigenous Peoples (13 September 2007) UN Doc A/RES/61/295 (UNDRIP).

Policy and administrative documents

Finnmark County Council

2005. Policydokument – Det politiske grunnlaget for representantene til styret til Finnmarkseiendommen, 16 November 2005.

2008. Styringsdokument for Finnmark fylkeskommunes styremedlemmer i Finnmarkeiendommen (Fefo).

2012. Finnmark fylkeskommune. FeFo – Eierstrategier og overskuddsanvendelse. Vedtatt i fylkestinget 2012 (sak 12/18).

2014. Finnmark fylkeskommune. Eierstrategier for FeFo. Vedtatt i fylkestinget 2014 (Sak 25/14).

2018. Finnmark fylkeskommune: Strategier for FeFo, (Sak 60/18).

Finnmark Estate (FeFo)

2007. Strategisk plan 2007–2010. Vedtatt av FeFo styret, 20 March 2007.

2011. Strategisk plan. Vedtatt 1. April 2011.

2015. Strategisk plan. Vedtatt i styret 15 December 2015.

2017. Høringsuttalelse på Nussir ASA's søknad om driftskonsesjon etter mineralloven. Board issue 71/2017.

2019. FeFos engasjement i Finnmark Kraft AS. 2019. Board issue 55/2019.

Norwegian government:

Innst. O. nr. 80–2004–2005: Innstilling fra justiskomiteen om lov om rettsforhold og forvaltning avgrunn og naturressurser i Finnmark fylke (finnmarksloven).

Prop. 134L (2018–2019).

Sámi Parliament

2006. Sametingets retningslinjer for endret bruk av meahcci (utmark) i Finnmark fylke, SP 23/06.

2007a. Sametinget retningslinjer etter finnmarkslovens § 4, SP 23/07.

2007b. Letter to FeFo, 21 February 2007: Finnmarkseiendommen – Strategisk plan 2007–2010.

2010. Utforming av Finnmarkseiendommens strategiske plan, SP 45/10.

2012. Finnmarkseiendommen – FeFo – eierstrategier og overskuddsanvendelse, SP 39/12.

2014. Oppfølging av Finnmarksloven, SP 19/14.

2018. Finnmarkseiendommen – Samstyring, SP 52/18.

2019. Innsynsbegjæring. E-mail dated 23 April 2019 from Torvald Falch.

11

THE GIRJAS CASE – COURT PROCEEDINGS AS A STRATEGY TO ENFORCE SÁMI LAND RIGHTS

Malin Brännström

Abstract

Even though Swedish case law clarifies that Sámi land rights are private property rights, the implications for the legal system still remain unclear and debated. Since the political system has failed to clarify Sámi property rights, Girjas sameby, as a representative of the reindeer-herding Sámi, have turned to the judicial system to have these rights enforced. After more than ten years of legal proceedings, at the end of 2019, the Swedish Supreme Court will hand down its decision in the so-called Girjas Case. The heart of the lawsuit is who holds the hunting and fishing rights within a specific high mountain area in the north-western part of Sweden and thus who decides on the granting of licenses to others. The case illustrates the ongoing dispute between the Indigenous Sámi people and the Swedish state over the meaning of property rights with respect to land within Sápmi, the homeland of the Sámi. This chapter describes the historical background of the lawsuit and briefly outlines the legal assessments of the District Court and the Court of Appeal. In addition, the chapter discusses why litigation is being used by the Sámi as a strategy to enforce Sámi land rights within the legal system and the challenges associated with presenting historical perspectives in a court setting.

Introduction

In 1977, the Swedish parliament stated that the Sámi people are to be regarded as an Indigenous people (Prop. 1976/77:80). Ever since then, there has been an ongoing discussion about how Sámi land rights can be understood and implemented within the country's legal system. A central issue has been whether Sweden can ratify the ILO convention no. 169 concerning Indigenous and Tribal Peoples. One of the major obstacles for ratification is that the nature of Sámi fishing and hunting rights is unclear (SOU 1999:25). This is also one of the most disputed issues when it comes to Sámi land rights.

The Sámi lived in the northern part of Sweden long before the nation-state was established. Due to the very northerly location, around and above the Arctic Circle, the winters are long and the vegetation growing period short. Under these harsh conditions, advanced subsistence strategies were developed over thousands of years, and both fishing and hunting became central elements of Sámi livelihood (Bergman et al. 2013).

Through a deliberate colonization process, the Swedish crown gradually gained control over the traditional homelands of the Sámi (Päiviö 2011). As will be described in this chapter, today's complex legal situation is as a result of historical developments associated with various government statements and decisions that have been made over time. This is particularly the case when it comes to Sámi hunting and fishing rights. Ever since the first Reindeer Herding Act was implemented in 1886, the County Administration (Swe: *Länsstyrelsen*) has administered a licensing system to grant people other than Sámi permission to hunt and fish within the high mountain area. For a long period of time, the state seems to have had the view that it was only administering this system. However, since the late 1980s, the state claims hunting and fishing rights as the landowner and thereby also claims the right to decide on licenses. As a consequence of this, the licensing system was reformed in 1993 so that the Sámi had less influence on the decision-making and so that everybody could hunt and fish in the high mountain area. Sámi representatives protested strongly against this claim of land ownership and the changes made in the legal system to open up land for more recreational hunting and fishing (Arnesson-Westerdahl 1994). The Sámi claimed that the hunting and fishing rights belonged to them and that they, therefore, should have considerable influence over the licensing system.

The dispute about hunting and fishing in the high mountain area has been intense ever since 1993. Several public inquiries have been set up to look into Sámi policy and to propose changes in legislation (e.g. SOU 1999:25, 2001:101, 2005:116, 2006:14). However, the proposals submitted have not been adopted, mainly because of strong opposing interests. Forestry, mining and energy production are important and powerful financial interests, and the strengthening of Sámi land rights could adversely affect these interests. Thus, there is an apparent reluctance among politicians to strengthen Sámi land rights (Lantto and Mörkenstam 2008). Sweden has repeatedly received criticism from international institutions in this respect (e.g. Anaya 2011; CERD 2018). After about 25 years of public inquiries, this can only be described as a political failure to properly implement Sámi land rights.

Against this background, Girjas sameby has turned to the court system to clarify Sámi hunting and fishing rights. The lawsuit is supported by the Swedish National Union of Reindeer Herders (Swe: *Svenska Samernas Riksförbund*, SSR). The legal issue at stake is whether Girjas sameby or the Swedish state is the holder of hunting and fishing rights within the high mountain area and who is responsible for deciding whom is issued licenses. The case also addresses the questions of whether the current licensing system discriminates against Sámi and whether Swedish regulations comply with the constitutional protection of property.

In February 2016, Gällivare District Court concluded that Girjas sameby is the holder of exclusive hunting and fishing rights within the area under consideration (Case no. T323-09). The District Court further concluded that the state is not allowed to decide on granting licenses to others and that Girjas sameby can decide on licenses without the state's consent. The Swedish state appealed against the judgment, and, in January 2018, the Court of Appeal for Northern Norrland concluded that Girjas sameby has exclusive rights to hunting and fishing within the relevant area (Case no. T 241–16). Nevertheless, the Court of Appeal found that it is still in accordance with the constitutional protection of property that the state administer the licensing system, and it is not to be considered discriminatory. The judgement raises several questions about the meaning of property rights within the Swedish legal framework (Brännström 2018a). Both Girjas sameby and the state have appealed the judgment. The Swedish Supreme Court has decided to permit a review, and the hearing will take place in September and October 2019 (Case no. T853-18).[1]

The Girjas Case highlights several legal questions concerning Sámi land rights that are still unresolved within the Swedish legal system, which is described in this chapter. The judgment of the Supreme Court is expected to have major implications for the understanding of these rights. This chapter describes the historical background of the Girjas Case and how the political failure to make changes in the legislation regulating Sámi land use has led to the ongoing lawsuit. Furthermore, the judgments of the District Court and the Court of Appeal are briefly described. In addition, the chapter discusses why litigation is used by the Sámi to enforce changes in legislation and the challenges this type of litigation entails given that historical developments are in focus.

It should be noted that the current Reindeer Herding Act (1971:437) states that only the Sámi that are members of a sameby (a geographic area and a legal person that represents reindeer-herding Sámi) are allowed to exercise hunting and fishing rights (Section 1). This means that a large number of Sámi who claim that they are rights holders are not allowed to hunt and fish if they do not have a license. This specific problem will not be dealt with any further in this chapter (see Torp 2015).

The colonization process

Today's complex legal situation can only be understood if historical developments are considered. Therefore, the historical background of the Girjas Case is described in this section.

As already mentioned, Sámi lived in the interior parts of northern Sweden long before the nation-state was established. Historically, Sámi society was divided according to a siida system, meaning allocation of land, water and natural resources between different groups, primarily based on kinships (Bergman et al. 2008). Within the geographic area of a siida, all available resources were utilized, and hunting and fishing were vital elements of Sámi subsistence, along with gathering and – eventually – reindeer herding (Bergman et al. 2013). Movements between different resource areas, such as good fishing lakes, hunting grounds and sites for plant gathering, were key features of historic Sámi land use (Josefsson et al. 2010).

Up until the 1750s, the Sámi lived in these interior areas basically without competition from others (SOU 2006:14). Nevertheless, they were in continual contact with people from the surrounding areas, with a focus on trade; indeed, the Sámi had important commodities that were sought after in Europe, such as game, fur and fish (Hansen and Olsen 2006).

Until the end of the Middle Ages, the primary interest of the Swedish Crown in these northern areas was trade. However, during the 16th century, other interests began to arise, and King Gustav Vasa introduced a new taxation system, with government officials travelling to the north to collect taxes. Taxes consisted mainly of reindeer hides, silver, fish and fur of various kinds. The Crown called the area where the Sámi lived Lappmarker ("the land of the Lapps"), and different Lappmarker divided the area for trade and administrative purposes (Hansen and Olsen 2006).

During the 17th century, the influence of the Crown in the northern areas became stronger (Päiviö 2011). The Crown began to encourage people to settle the interior of northern Sweden. The main reasons for this were tax revenues and to gain control over the northern areas in relation to other nations. Furthermore, mining contributed to the fact that the Crown wanted to generate population growth in these areas. Through a number of royal decrees during the

17th and 18th centuries, the northern inland areas were opened up for colonization. Those who settled there were given special benefits, including exemption from taxation and from military service. The settlers were allowed to hunt and fish in areas around their settlements. Later these settlements were the basis for allocating land ownership through demarcation processes (Swe: *avvittringar*) carried out during the 18th and 19th centuries (Strömgren 2015). During the same period, many Sámi settled down.

In the middle of the 1750s, the inland areas of northern Sweden were exceptionally sparsely populated and virtually exclusively inhabited by Sámi. In 1749, a decree was issued that led to an increase in people moving into the northern inland areas, and eventually the Sámi became a minority here (SOU 2006:14). During the initial period of the colonization process, the Crown was of the opinion that Sámi and settlers could live side by side, as they utilized different resources, the so-called "parallel theory" (Allard 2006). Nevertheless, an important element in the regulations that aimed at regulating the activities of the settlers was to protect the Sámi's interests as far as possible, and this included hunting and fishing. In the regulations, it was emphasized that the activities of the Sámi should not be disadvantaged (SOU 2006:14). However, the assumption of the parallel theory proved to be wrong. Hunting and fishing were important elements of subsistence not only for nomadic Sámi but also for the settlers and the Sámi that settled permanently in one place, and gradually competition for the existing natural resources developed.

In 1751, Lappmarks gränsen ("the Lappland boundary") was established in the counties of Västerbotten and Norrbotten, primarily to prevent coastal peasants from fishing the waters west of the border and thus competing with the Sámi and the settlers (SOU 2006:14; Allard 2006). This boundary, however, did not solve the problem of competing land use. Therefore, the so-called Odlingsgränsen ("the cultivation boundary") was established further west in 1867, closer to the high mountain areas. It was stated that the areas west of the boundary would be reserved for Sámi land use and that no settlements would be allowed there (Swedish Parliament 1867). However, even after this statement, new settlements were established west of the border. It should be noted that the Girjas Case concerns an area west of the cultivation boundary.

As described in this section, the colonization process and the other public decisions taken by the state have clearly contributed to today's complex legal situation with overlapping land rights and an unclear legal situation.

The state's changing approaches to Sámi land rights

Another reason for today's complex legal situation is that the Swedish state's attitude towards Sámi land rights has varied over time. Up until the middle of the 18th century, the land rights of the Sámi were treated as a type of taxpayers' right, comparable with the rights of peasants (Korpijaakko-Labba 1994). Thereafter, up until the end of the 19th century, Sámi land use was considered to be based on customary rights that would be protected in relation to other land use (Päiviö 2011). However, at the beginning of the 20th century, the state began to express the view that Sámi land use was based on what was termed "the Lapp privilege" (Swe: *Lappprivilegiet*), meaning that the law adopted by the state was the foundation of the right to use land and also that the state could regulate Sámi land use through new or amended legislation (Allard 2006). The Sámi opposed the state's position and claimed that they were holders of real property rights.

Against this background, Sámi representatives initiated court proceedings in 1966, in what is known as the Taxed Mountains Case (Swe: *Skattefjällsmålet*), claiming ownership before the state to specific high mountain areas in the county of Jämtland. In 1981, the Swedish Supreme

Court rejected the claim of Sámi ownership (NJA 1981, 1). Nevertheless, the judgment remains central, since the Supreme Court made some important statements to clarify the legal status of Sámi land rights. It was stated that Sámi land use is based on 'immemorial prescription' (Sw: urminnes hävd), that is, their continuous use of the land since 'time immemorial'. In addition, it was stated that these rights are property rights, protected by Chapter 2 Section 15 (formerly Chapter 2 Section 18) of the Constitutional Instrument of Government (hereafter "the Constitution"). In 2011, the legal status of Sámi land rights as property rights was confirmed in the Nordmaling Case, when the Supreme Court stated that winter grazing in the eastern coastal areas is based on the long-term use of land, therefore being customary rights (NJA 2011, 109).

Accordingly, it has long since been clarified within the Swedish legal system that Sámi land rights are private property rights. However, the legislation and administrative orders that were implemented before the statement of the Supreme Court in 1981 were based on the concept of the "Lapp privilege" and did not have Sámi land rights as a starting point. One example is the current Reindeer Herding Act, which regulates Sámi hunting and fishing rights. This legislation was implemented in 1971 (Prop. 1971:51) and has not been amended to properly implement Sámi land rights. Consequently, there are still laws that are based on the view that the Sámi are not holders of real property rights.

Ever since the verdict in the Taxed Mountains Case, changes within the legislation to implement Sámi land rights have been discussed. The Swedish government has established several public inquiries to look into Sámi policy, and all have concluded that Swedish legislation has deficiencies when it comes to the implementation of Sámi land rights (SOU 1999:25, 2001:101, 2005:116, 2006:14). However, the proposals submitted to change the legislation have not been adopted, mainly because of strong opposing interests. Consequently, politicians have not been willing to reform the legislation, since the matter at stake is considered to be of a delicate nature (Bengtsson 2015).

The regulation of hunting and fishing in the high mountain areas

When the first Reindeer Herding Act was enacted in 1886, a system to administer licenses to hunt and fish within the high mountain area was introduced. Because the Crown considered that the Sámi were not capable or organized enough to operate the licensing system, it was decided that the county administrations would run it (Förslag till förordning 1883). Nevertheless, the licensing system was clearly designed to take account of Sámi land use. The Sámi were consulted about licenses, and licenses were not granted if the Sámi would suffer harm from them. Furthermore, revenues were forwarded to the affected Sámi reindeer-herding community and to the Sámi Fund, which was a special fund to support Sámi needs (Reindeer Herding Act of 1886, Section 22).

Initially, licenses to hunt and fish were primarily given to locals and people from the region. However, interest in hunting and fishing for recreation gradually increased and, during the 1980s, some hunting organizations argued for changes in the regulations to open up the licensing system for hunting and fishing to more people (Arnesson-Westerdahl 1994).

The Swedish government did not, before or during the Taxed Mountains Case, explicitly claim any hunting and fishing rights, only the authority to administer the licensing system (Bengtsson 2004). However, a turning point came in a government bill regarding hunting in 1987, wherein the state was described as the landowner with hunting and fishing rights (Prop. 1986/87:58). This statement led to changes within the legal framework a few years later in 1993 to allow everybody to hunt and fish within the high mountain area (Prop. 1992/93:32). It was stated that this would better utilize the hunting resources and provide quality recreation experiences to landless hunters at a low price.

Through the amendment in 1993, the present institutional arrangements of the licensing system were implemented. Today, small-game hunting and fishing provide recreation for local people, other Swedish citizens and individuals from countries within the European Union. Hunting and fishing are also of great importance for tourism. However, the extent of the fishing and hunting causes severe disturbances to reindeer herding, since the reindeer will leave areas that are used. Another disturbance is that Sámi cannot hunt and fish to the same extent as before.

Ever since 1993, the debate about hunting and fishing within the high mountain areas has been intense. Sámi representatives have strongly protested against the state's claim to land ownership and the changes made in the legal framework to open up land for more recreational hunting and fishing (Arnesson-Westerdahl 1994). In 1995, Sámi representatives turned to the European Commission of Human Rights and argued that the Swedish regulation of hunting and fishing infringed their property rights, protected in Article 1 Protocol 1 of the European Convention of Human Rights (hereafter ECHR). The Commission found that the applicants had access to courts in Sweden for the determination of their civil rights. The application was therefore found manifestly ill-founded, and the Sámi were assigned to turn to the Swedish court system with their legal claims (*Könkämä Sámi village and others v Sweden*).

Thereafter a public inquiry was set up to propose a new institutional arrangement to administer the licensing system, and in 2005, a co-management system administering these rights was proposed (SOU 2005:116). However, the proposed system was heavily criticized, and no amendments were implemented. As it became clear that the political system had failed to clarify the legal situation, the interest of having the disputed issues tested in court increased. From 1999 onwards, a lawsuit, commonly referred to as the Nordmaling Case, moved slowly through the Swedish court system (NJA 2011, 109). About 100 landowners in the coastal area had taken legal action against three samebyar in Västerbotten County, claiming that there existed no right to graze reindeer during winter on private land. This was the first time that the Sámi succeeded in persuading the courts that they had established land rights that were to be protected within the Swedish legal system. The outcome of the Nordmaling Case indicated that Sámi could use litigation to advance their interests.

The Girjas Case

Against this background, a lawsuit was prepared by the Swedish National Union of Reindeer Herders to have the content of Sámi hunting and fishing rights clarified. In May 2009, the Union and Girjas sameby took legal action against the Swedish state. Girjas sameby is situated within the county of Norrbotten in the very northernmost part of Sweden. Within the sameby, 18 families live and work with herding reindeer. About 120 people are members of the sameby, and about 400 have some kind of social relationship to it.

Claims, legal arguments and evidence

Girjas sameby claims exclusive hunting and fishing rights in relation to the Swedish state within the territory of the sameby west of the Cultivation border (hereafter "the Area"). As a consequence, Girjas sameby claims the right to decide on licenses issued for small-game hunting and fishing. If these claims are rejected, Girjas sameby claims that it should be established that the sameby and the state both have hunting and fishing rights in the Area and that licenses can only be granted if both right holders agree to them. In short, Girjas sameby argues that the use of land by Sámi over a long period of time has led to exclusive hunting and fishing rights. Girjas sameby presents three alternative legal arguments for exclusive hunting and fishing rights: 1) these follow from the Reindeer Herding Act, 2) they are the result of Sámi land use for reindeer herding as an Indigenous people, and 3) the Sámi have used the Area over a long period of time. Girjas sameby argues that the present regulation of the licensing system is an infringement of the constitutional protection of property in Chapter 2 Section 15 of the Constitution, as well as a case of discrimination, Chapter 2 Section 12 of the Constitution. Furthermore, the regulation is not in compliance with Article 1 of the First Protocol ECHR and Article 14 ECHR.

The Swedish state denies that Girjas sameby is the holder of exclusive hunting and fishing rights in the Area and claims that the state is the holder of these rights as the landowner. The state concedes that the Sámi have established a customary right to land, the so-called reindeer-herding right that is regulated through the Reindeer Herding Act from 1971. The reindeer-herding right includes a right for the Sámi people to hunt and fish to meet their personal needs. However, according to the state, these hunting and fishing rights are subordinate to the rights of the landowner. The state argues that land ownership has been manifested in various ways over time and that the Sámi must have understood that they were not holders of exclusive rights. The licenses granted to others are not based on the rights of the Sámi.

The two parties have presented very diverse and conflicting descriptions of the Area's historical development. Briefly, Girjas sameby has described the historical development like this: the Sámi have lived in the area for a very long time. Land has been divided between *siidas*, and all natural resources available have been important to survive in the harsh environment around and above the Arctic Circle. The Sámi have, over time, developed knowledge to utilize every resource, and hunting and fishing have been central elements of subsistence. For a long period of time, the Sámi lived in the Area without competition from others. The state has not hunted or fished in the Area. The state has acted in the area primarily through administrative measures and as a part of governmental sovereignty. The state has not established hunting and fishing rights through these measures.

The Swedish state, on the other hand, has described the historical development completely differently: the Sámi did not utilize the area to any great extent before the middle of the 18th century. Their use has been fragmented and merely opportunistic in character so that exclusive hunting and fishing rights have not been established. The Sámi have been exposed to competition from others with regard to hunting and fishing. The state has, through public statements, established land ownership, and with this follow the hunting and fishing rights.

Accordingly, a key issue of contention is how Sámi utilized the landscape and resources during historic times. Indeed, assessing evidence about historic land use has become a central part of the court's deliberations, and both parties have referred to an extensive array of documentary material and called expert witnesses from different scholarly disciplines: academics from the fields of history, law and historical ecology have all served as witnesses. Both parties have also referred to many written sources, primarily public material authored by government representatives; these have included public investigations, government bills, court protocols and church records. This material has been interpreted very differently by the two parties.

Since an inspection of the disputed area during the hearing was not allowed by the District Court, Girjas sameby has referred to photographs of the area taken from the late 19th century onwards and also a movie filmed from a helicopter ride over the area with audio commentary from one of the elders. Furthermore, members of Girjas sameby have been heard as witnesses.

A central part of the evidence referred to by Girjas sameby is research about historical land use. An expert in forest history was heard and pointed out that, when historic land use is analysed, the landscape itself should be the analytical starting point, with consideration of basic factors such as topography, seasonal changes and weather conditions. The winters are long and the vegetation growing period is short in Arctic areas. In this testing environment, advanced and sophisticated subsistence strategies have developed, large areas of land are needed for subsistence and resources must be collected at many diverse sites. The expert witness emphasized that findings should be analysed in relation to settlement patterns and prerequisites for mobility. In addition, social, cultural, economic and religious aspects should be considered.

The District Court

Gällivare District Court concluded that Girjas sameby is the holder of exclusive hunting and fishing rights rather than the state within the Area. As a consequence, the court stated that the state is not allowed to decide on licenses to hunt and fish and that Girjas sameby can decide on licenses without the consent of the state. The legal assessment is comprehensive and can only be presented briefly. The starting point of the legal reasoning is that Girjas sameby have proved that Sámi have used the Area for at least one thousand years and that all resources available have been utilized, including hunting and fishing. The Court concludes that Swedish sovereignty was established over the area in the middle of the 16th century and that the state became the owner of the Area through the demarcation process in 1887. Furthermore, the Court concludes that Sámi land rights can be explained through 'immemorial prescription', which should be adjusted to the specific situation of the Sámi and their land use. It is concluded that the Sámi who lived in the area in 1734 were the holders of the hunting and fishing rights, and these rights have not expired subsequently. The Court finds that that the establishment of the state's land ownership has not limited the hunting and fishing rights of the Sámi.

The legal reasoning of the Court reveals that the description of the characteristics of the landscape, such as the harsh environment, and the historic land use are fundamental parts of the legal assessment. Obviously, the court has paid attention to the evidence presented by Girjas sameby about the very special requirements for survival in the Arctic areas.

One element that should be pointed out is that the interpretation of the legal concept of 'immemorial prescription' differs from previous interpretations. This concept has its background in farming societies, where it was used to explain establishment of land ownership. The relevance of it was debated as early as the 18th century, and it has not been used within real estate law for a very long time. The previous understanding was that property rights can be established if land has been utilized intensively enough for approximately 90 years and the usage has not been contested by others (Allard 2015). Based on this approach, the relevant period to determine whether land rights have been established starts from the present day and counts back 90 years (e.g. SOU 2006:14; NJA 2011, 109). However, the District Court applied a completely novel interpretation, since it was concluded that the Sámi had lived in the area for at least a thousand years and for a long period of time without the competition of others. The court concluded that hunting and fishing rights can be established through this type of land use. Thereafter, the court considered the period in which the Swedish state had started to act in the area and concluded that the state had not, through these measures, established any hunting and fishing rights. This type of reasoning, starting with pre-colonial land use, clearly puts historical studies at the forefront of legal assessment.

The Court of Appeal

The state appealed the verdict to the Court of Appeal for Northern Sweden, which concluded that Girjas sameby has exclusive hunting and fishing rights rather than the state within the Area and that the state may not, as the land owner, decide on licenses to hunt and fish in the Area. Nevertheless, the Court of Appeal found that the present licensing system is not in conflict with the constitutional protection of property, nor is it discriminatory. Consequently, the Swedish state can continue to decide on the granting of licenses. The legal assessment of the Court of Appeal is considerably shorter. Initially, the Court of Appeal concludes that the meaning of land ownership has not always been the same. Thereafter, there is a description of the development of the legal situation, particularly during the 19th and 20th centuries. The Court of Appeal concludes that when the high mountain area was discussed in various contexts, reference was made to the hunting and fishing rights of the Sámi. It is also noted that it was stated in the preparatory works that the state was not the holder of such rights. In addition, the Court of Appeal notes various statements made in public inquiries that the Sámi alone should hunt and fish. The Court of Appeal concludes that it was not until 1987 that the state contended that it had hunting and fishing rights. It is concluded that the state cannot establish hunting and fishing rights through this type of statement when it has not claimed any such rights during the last hundred years. The Court of Appeal, therefore, concluded that the state has not established any hunting and fishing rights.

Thereafter, the Court of Appeal examined whether Sámi land use has led to *exclusive* hunting and fishing rights. It is noted that a prerequisite for exclusive rights is that nobody has protested against the exercise of the right. The Court of Appeal states that it is not enough that the Sámi have hunted and fished alone for a long period of time and carried out reindeer herding without competition. In addition, it is required that they have made their claims for exclusive rights visible to others. Since very few people other than the Sámi were in the Area, the Court of Appeal finds that it is not likely that the issue of exclusive rights was brought to light in the interaction with other people in past times. Consequently, the Court concludes that *exclusive* hunting and fishing rights have not been established. Furthermore, it is concluded that there is no evidence for a legal rule at the time of the implementation of the Reindeer Herding Act in 1886 that meant that the use of land could lead to a right to give licenses to hunt and fish. Furthermore, the Court of Appeal concludes that the present institutional arrangement is not in conflict with the constitution, either in terms of protection of property or discrimination. Finally, the Court of Appeal concludes that Girjas sameby has not proven that there is a principle within international law meaning that it has established exclusive usufruct rights to hunting and fishing. The legal reasoning in this part of the judgement is difficult to follow from a property rights perspective, since right holders usually have the right to decide on the use of land (Brännström 2018a).

Both Girjas sameby and the state have appealed to the Swedish Supreme Court, which has decided to give permission for a review. The hearing took place in September and October 2019.

Transformations in the legal system that have opened up opportunities for litigation

The fact that Sámi representatives are turning to the judicial system instead of the political system to force change is part of a more general tendency within the Swedish legal system, where complex legal issues end up in court. This can be explained by transformations in the legal system over decades, which have opened up the possibility of court proceedings to examine the relationship between state regulation and private property. These transformations will be briefly described in this section.

For a long period of time, fundamental rights and freedoms have been rather subordinate within the Swedish legal system (Bernitz 2010/11). Accordingly, the constitutional protection of property has played a modest role. In 1974, fundamental rights and freedoms were enforced in Chapter 2 of the Constitution. However, the presumption was that this regulation was primarily aimed at the Swedish Parliament as the legislator and that enforced statutes were in accordance with the Constitution (Derlén et al. 2016). During recent decades, the function of the fundamental rights and freedoms, including the protection of property, has gradually changed (Åhman 2019). This is partly an effect of Swedish EU membership and the incorporation of the ECHR (Bernitz 2010/11). In addition, amendments were made to the Constitution in 2010 that emphasized that the courts should review whether regulations and public decisions are in accordance with regulations of higher status (Prop. 2009/10:80). Thus, courts now should evaluate whether democratic values within the Constitution, including the protection of property, are maintained in legislation and public decisions (Derlén et al. 2016).

Through these transformations within the constitutional framework, it is now possible for individuals, including Sámi, to challenge the present legal situation through litigation. This means that courts will probably face future judicial proceedings where Sámi representatives argue that legislation or public measures are not in compliance with the requirements of the constitutional protection of property (Brännström 2018b).

Conclusions and discussion

As described previously, Sámi land rights have been examined with reference to the historical context in case law. This puts historical land use at the very centre of any judicial assessment. The historical perspective seems to be reinforced by the Girjas Case, where the courts have concluded that Sámi were using land and resources long before the nation-state was established. In addition, the interpretation of the legal concept of 'immemorial prescription' reveals a distinct historical focus, as the precolonial land use by Sámi is the starting point for the assessment. These are two examples of adjustments in the legal assessments to meet the specific situation of the Sámi.

This type of historical perspective has not been common within Swedish court proceedings. In legal positivism, which has shaped Sweden's legal culture during the 20th century, arguments based on historical developments have been considered of relatively minor importance. However, since the 1990s, Swedish legal culture has gradually changed in this respect: historical perspectives are now increasingly being considered in the legal context (Modéer 2015). Consequently, mature and dispassionate historical deliberation is required as the courts seek to resolve legal problems in other ways than have previously been the norm. This type of legal assessment requires informed references to substantial historical material and a profound understanding of historic land use. However, to handle this type of historical perspective within a civil law process is challenging, both for the courts and for the parties concerned. Some of these challenges are discussed subsequently.

The first challenge is that knowledge about historical Sámi land use has been lacking. There are few written sources from Sámi themselves, since the Sámi language has largely been an oral medium. Consequently, the history of the Sámi has primarily been written by others.

The existing written sources are mainly public material produced by Swedish government representatives of various sorts. The limited knowledge about Sámi lifestyle among public administrators has inevitably affected the quality of the written material that is available (SOU 2006:14).

Another challenge is that courts have primarily focused on written sources. However, the Girjas Case displays a more generous attitude towards other types of evidence about historic land use. It exemplifies how archaeology, history and related disciplines can augment our knowledge of Sámi land use, allowing a deeper understanding and implementation of Sámi land rights. Nevertheless, it should be emphasized that the interpretation of research regarding historical land use requires an in-depth contextual understanding.

A third challenge is that knowledge about Sámi history is limited in the majority of the population as well as among judges. In addition, courts in general have no experts in history. Consequently, courts are dependent on the evidence presented by the parties involved and on analyses of the historical material from specialists within the disciplines of archaeology and history. This is a challenging part of the legal process. A courtroom is an adversarial forum where verbal arguments and other materials have to be presented so that members of the court can comprehend their significance. In the Girjas Case, there was the added difficulty of presenting academic, sometimes highly technical material. Girjas sameby found that it was not sufficient simply to refer to a large number of peer-reviewed academic articles written in English. Instead, an expert in forest history, with a particular focus on the history of ecosystems and human relations, wrote an expert report that described and summarized the conclusions of the academic research. In addition, the expert presented evidence during the hearings.

However, when researchers are heard as witnesses, special considerations are needed, and the judges should be aware of the special conditions of the academic process. One thing to be aware of is the risk that the specialist will interpret the historical material in a way that is questioned by other researchers. This will probably lead to witnesses with contrasting interpretations, which can be difficult to handle within a civil law case with high evidentiary requirements. Another aspect to be aware of is that there is a risk that the court will maintain perceptions that are under debate or could be revaluated in the future.

In addition, there are several formal challenges with this type of litigation. A civil law process follows specific strict rules. Within Swedish civil law proceedings, the court is restricted to following the claims and arguments of the parties. This means that the parties' approaches will constrain the legal assessment of the court. Furthermore, civil law procedure means that only a limited part of a complex problem will be handled within this type of court proceeding.

Despite all of the challenges described previously, the courts still remain reliant on historical material and expert interpretations to develop an understanding of Sámi land rights. Thus, one conclusion is that there is an urgent need for the courts to develop means and methods to handle historical perspectives within court proceedings.

The Girjas Case illustrates how Sámi are now using the court system to force changes in the legal system. For the Indigenous Sámi, this may be a way to persuade the politicians to take action in the future to resolve the legal problems. The final judgment from the Swedish Supreme Court will show whether this is a useful strategy. The political failure to adopt legislation that is in accordance with the status of Sámi land rights as property rights will probably lead to other court proceedings in the future, in which Sámi will argue that the present legislation is not in compliance with the constitutional protection of property. It is not unlikely that the courts will find deficiencies within legislation that need to be evaluated in relation to constitutional requirements.

Note

1 After this chapter was completed and had passed through the peer review process, the Swedish Supreme Court released its verdict concerning the Girjas Case on January 23, 2020: Case No. T 853-18. The Supreme Court based their reasoning on the long-term use of the Sámi that have lived in the relevant area and concluded that they had exclusive hunting and fishing rights by 1750 and that the Swedish state thereafter had not acted in any way to revoke these rights. Hence, the Supreme Court concluded that Girjas sameby has the right to decide on the granting of licenses to others. The Supreme Court also stated that the ILO convention no. 169 is legally binding in Sweden when it comes to Sámi customary rights, although the convention has not been formally ratified. The verdict of the Supreme Court must be regarded as a milestone within the Swedish legal system when it comes to Sámi land rights.

References

Åhman, K. 2019. *Grundläggande rättigheter och juridisk metod: RF 2 kap, Europakonventionen och EU:s stadga och deras tillämpning.* Stockholm: Norstedts Juridik.

Anaya, J. 2011. "Report of the Special Rapporteur on the rights of Indigenous Peoples. The Situation of the Sami People in the Sápmi Region of Norway, Sweden and Finland." United Nations General Assembly. A/HRC/18/35/Add.2.

Arnesson-Westerdahl, A. 1994. *Beslutet om småviltjakten: en studie i myndighetsutövning.* Sweden: Sami Parliament.

Allard, C. 2006. "Two Sides of the Coin, Rights and Duties: The Interface between Environmental Law and Saami Law Based on a Comparison with Aoteoaroa/New Zealand and Canada." PhD diss., Luleå University of Technology.

Allard, C. 2015. *Renskötselrätt i nordisk belysning.* Stockholm: Makadam.

Bengtsson, B. 2004. *Samerätt, en översikt.* Stockholm: Norstedts Juridik.

Bengtsson, B. 2015. "Reforming Swedish Sami Legislation: A Survey of the Arguments." In *Indigenous Rights in Scandinavia: Autonomous Sami Law,* edited by C. Allard and S. F. Skogvang. Surrey: Ashgate.

Bergman, I., L. Liedgren, L. Östlund, and O. Zachrisson. 2008. "Kinship and Settlements: Sami Residence Patterns in the Fennoscandian Alpine Areas around A.D. 1000." *Arctic Anthropology* 45 (1): 97–110.

Bergman, I., O. Zackrisson, and L. Liedgren. 2013. "From Hunting to Herding: Land Use, Ecosystem Processes, and Social Transformation among Sami AD 800–1500." *Arctic Anthropology* 2: 25–39.

Bernitz, U. 2010/11. "Rättighetsskyddets genomslag i svensk rätt – konventionsrättsligt och unionsrättsligt." *Juridisk Tidskrift* 3: 821–845.

Brännström, M. 2018a. "Samiska markrättigheter i förändring? Hovrättens dom i Girjas-målet väcker frågor om innebörden av rättigheter till fast egendom." *Juridisk Publikation* 1: 25–47.

Brännström, Malin. 2018b. "Court Proceedings to Evaluate the Implementation of Sami Land Rights in Sweden." *Retfaerd* 2: 32–45.

Committee on the Elimination of Racial Discrimination. 2018. "Concluding Observations on the Combined Twenty-Second and Twenty-Third Periodic Reports of Sweden." CERD/C/SWE/CO/22–23.

Derlén, M., J. Lindholm, and M. Naarttijärvi. 2016. *Konstitutionell rätt.* Stockholm: Wolters Kluwer.

Hansen, L. I., and B. Olsen. 2006. *Samernas historia fram till 1750.* Stockholm: Liber.

Josefsson, T., I. Bergman, and L. Östlund. 2010. "Quantifying Sami Settlement and Movement Patterns in Northern Sweden 1700–1900." *Arctic* 63 (2): 141–153.

Korpijaakko-Labba, K. 1994. *Om samernas rättsliga ställning i Sverige-Finland.* Helsingfors: Juristförbundets förlag.

Lantto, P., and U. Mörkenstam. 2008. "Sami Rights and Sami Challenges." *Scandinavian Journal of History* 33 (1): 26–51.

Modéer, K. Å. 2015. "Sami Law in Late Modern Legal Contexts." In *Indigenous Rights in Scandinavia: Autonomous Sami Law,* edited by C. Allard and S. F. Skogvang. Surrey: Ashgate.

Päiviö, N.-J. 2011. "Från skattemannarätt till nyttjanderätt. En rättshistorisk studie av utvecklingen av samernas rättigheter från slutet av 1500-talet till 1886 års renbeteslag." PhD diss., Uppsala University.

Strömgren, J. 2015. "The Swedish State's Legacy of Sami Rights Codified in 1886." In *Indigenous Rights in Scandinavia: Autonomous Sami Law,* edited by C. Allard and S. F. Skogvang. Surrey: Ashgate.

Torp, E. 2015. "Sami Hunting and Fishing Rights in Swedish Law." In *Indigenous Rights in Scandinavia: Autonomous Sami Law,* edited by C. Allard and S. F. Skogvang. Surrey: Ashgate.

Other documents

Public commissions of inquiry

SOU 1999:25 Samerna – ett ursprungsfolk i Sverige. Frågan om Sveriges anslutning till ILO:s konvention nr 169.
SOU 2001:101 En ny rennäringspolitik – öppna samebyar och samverkan med andra markanvändare.
SOU 2005:116 Jakt och fiske i samverkan.
SOU 2006:14 Samernas sedvanemarker.

Government bills

Förslag till förordning angående de svenska Lapparne och de bofasta i Sverige samt till förordning angående renmärken afgivna af den dertill utaf Kongl. Maj:ts förordnade komité 1883.
Kungl. Maj:ts proposition 1971:51 med förslag till rennäringslag, m.m.
Prop. 1976/77:80 om insatser för samerna.
Prop. 1986/87:58 om jaktlag.
Prop. 1992/93:32 om samerna och Samisk kultur.
Prop. 2009/10:80 En reformerad grundlag.
Swedish Parliament 1867, Motion FK no. 22.

Court cases

Case no. T323–09 [2016] (Gällivare District Court).
Case no. T214–16 [2018] (Court of Appeal for Northern Norrland).
Könkämä Sámi villages and others v Sweden [1996] (The European Commission of Human Rights) App. No. 27033/95.
Nordmaling Case [2011] (NJA); 109.
Taxed Mountains Case [1981] (NJA); 1.

12

ARCTIC WATERS AS INUIT HOMELAND

Claudio Aporta and Charlie Watt

Abstract

Inuit have used marine areas in a systematic way for as long as they have inhabited the Arctic. They have used the ocean in all seasons and in all states, from open water to solid sea ice. The sea ice, which in most places fastens to the land for most of the year, is a place of significance for Inuit. The sea ice has a well-known topography, where Inuit set their routes, allowing them to travel, reach other shores, harvest, and (in the past) dwell. Open water travel has also been part of Inuit lives through boating in the summer months and also in regions where sea ice does not form as a permanent travelling platform. In times when Arctic waters are gaining interest as a consequence of a significant increase of economic and geopolitical value to multiple stakeholders, understanding Inuit history and mobility patterns in marine regions is paramount.

This chapter will use evidence collected across the Canadian Arctic to propose that Inuit mobility systems reveal systematic and historic use of marine environments, as well as connections among communities through the Inuit circumpolar world. The authors will reflect on the implications of such findings, both for Inuit in Canada and elsewhere.

Introduction

Documenting and mapping Inuit trails have become a mission for both authors. One of us (Aporta) started this work while doing ethnographic research in Igloolik in 2000, as he travelled with hunters by snowmobile and boat, recording their movements, first with GPS and later through participatory mapping sessions. The other, Watt, was actually born in Kuujjuaq and travelled many traditional trails while growing up and as an Inuit hunter. He was involved in the political processes that empowered Inuit communities and promoted Indigenous rights in Canada since the early 1970s. As a senator of Canada and then as president of Makivik, he has supported the mapping of trails across the four Inuit regions in the Canadian Arctic since 2013.

The Inuit trails documented in this project, in partnership with many communities and Inuit regional organizations, are gradually filling an important gap in cartographic representations of Canada. When completed and made available to the public, the maps will show interconnections among all Inuit communities in the Canadian Arctic and extending to other regions in the Inuit circumpolar world. These maps are part of a process that some would call "counter mapping" (Peluso 1995), referring to the ability of marginalized peoples to appropriate what has been traditionally associated with the colonial enterprise (mapping through western cartography) in order to reclaim not only lands and resources (Mazzullo 2018) but also imageries (Hunt and Stevenson 2017).[1] These maps, along with those resulting from multiple community-led or community-sponsored projects,[2] create representations of the Arctic that have not been reflected in public or official maps of Canada since the first colonial encounters with Inuit. The standard maps of Canada are maps without Inuit on them. Even on most maps of Canada today, the Arctic looks as if it were empty, populated only by the scarce and isolated settlements established by the Canadian government for Inuit in the first half of the 20th century. At the most, it is portrayed as a shipping transit area, connecting the Atlantic and Pacific oceans.

For all the efforts to establish Canadian sovereignty in Arctic waters, and more recently to undertake a path of reconciliation with Indigenous peoples, it seems that representing Inuit residence and mobility patterns on Canadian maps would be a step in the right direction in terms of recognizing the historical presence of Inuit on Arctic lands and waters.[3]

While trails are not the only way to represent the social life of the Inuit-inhabited Arctic, they uniquely show the deep connections that Inuit communities have with each other, both presently and historically. They show a sense of community that is connected to the environment, to the animals Inuit depend on, and to places and people at different scales, from local to regional and cross-regional, eventually encompassing the whole Inuit-inhabited Arctic. The trails also show the intrinsic connections between land and ocean, the focus of this chapter.[4]

This chapter will argue that Inuit mobility systems are essential to understanding the Inuit relationship with the Arctic environment and that cartographic representations of trails are unique in illustrating the spatial and, to a lesser degree, temporal dimensions of this relationship. It will also show how trails connect both land and sea, allowing for a sense of *homeland* among Inuit that includes both marine and terrestrial areas. The authors conclude that Arctic marine areas (including those overlapping with important shipping corridors) are an integral part of Inuit homelands.

Inuit and Arctic sovereignty

Arctic states, including Canada, often forget that Inuit are key to claims of Arctic sovereignty, as Inuit have lived in the Arctic for millennia, long before Europeans were even aware of the existence of Arctic geographies. Canada's claims of "effective occupation" of the Arctic is possible only because they have developed partnerships with Inuit, who are actually the only native inhabitants of what today we know as the Canadian Arctic.

Since time immemorial, Inuit have lived on the land and ice-covered water in the Arctic and used the resources of the land and water to grow as a people. Inuit are deeply connected to not just the land but also the Arctic Ocean, as well as to everything in those environments. Inuit are the people who have occupied marine areas and have lived on the ice and hunted and travelled across them before international and territorial boundaries were even conceived.

Canada's claims of sovereignty in the Arctic, both over land and water, depend on its relationship with Inuit. It is clear that an important part of the reconciliation process involves recognition of the Inuit role in Canada's Arctic sovereignty. The preamble to the 1993 Nunavut Land Claims Agreement (NLCA) recognizes "the contributions of Inuit to Canada's history, identity and sovereignty in the Arctic." Article 15 of the NLCA explicitly states that "Canada's sovereignty over the waters of the Arctic Archipelago is supported by Inuit use and occupancy."

The Inuit perspective is shown in the "Circumpolar Inuit Declaration on Sovereignty in the Arctic", adopted by the Inuit Circumpolar Council (ICC), which represents Inuit in Greenland, Canada, Alaska, and Chukotka (Russia) (Simon 2011). The Inuit declaration states that

> sovereignty and sovereign rights must be examined and assessed in the context of the . . . long history of struggle to gain recognition and respect as an Arctic Indigenous people having the right to exercise self-determination over [their] lives, territories, cultures and languages.

Inuit expect Arctic countries to honour their partnerships with them. In Canada, these obligations are created by domestic and international law and are often developed through constitutionally protected Treaties (the comprehensive land claims agreements). The land claims agreements cover not just land but also areas of the ocean. In the Labrador Inuit Land Claims Agreement, Inuit have rights within 48,690 square kilometres of tidal waters, as well as fee-simple rights to an area of the seabed within the treaty area.[5] In every land claim region, Inuit have treaty-protected harvesting rights in marine areas, including rights to fish and harvest marine mammals. Other rights pertain to the management and development of the marine areas. For instance, within the tidal waters that are part of the Treaty area, Labrador Inuit have the right to take part in the creation of a strategy for ocean management and consultation rights concerning any proposed petroleum exploration and extraction.[6] The most recent agreement, the 2006 Nunavik Land Claims Agreement, is focused on the off-shore marine areas and islands. In this agreement, among other rights, Nunavik Inuit are guaranteed a percentage of the royalties from resource production in the Nunavik Marine Region.[7]

However, Canada has not always lived up to its obligations, and the result is that Inuit communities have high rates of poverty and unemployment and face the highest costs of living in Canada.

In 2019, the Senate of Canada issued a report that linked sovereignty to a prosperous north and recommended that Canada should invest in northern communities (Report of the Senate Special Committee on the Arctic 2019). The report connected the prosperity of Inuit communities with Canada's ability to reinforce its sovereignty in the Arctic against other States. In other words, Canada's sovereignty depends not only on the existence but also on the well-being of Inuit communities. Inuit organizations have stated that increased well-being of communities can only happen in a political context of governance approaches that allow for an increase in self-determination. In Nunavik, during the 2018 meeting of the Inuit rights organization (Makivik), participants decided to move forward with negotiations with the government with the ultimate goal of developing Inuit governance arrangements based on Inuit values and identity. Beyond regional differences, Inuit are a circumpolar people with shared ancestry, and many Inuit leaders have shared dreams for an Arctic Nation of Inuit, regardless of international boundaries.

Inuit traditional trails are perhaps the clearest example of 1) connections between Inuit communities across the circumpolar world and 2) Inuit extensive and historical use of both terrestrial and marine areas.[8] Concurrently, the mapping of trails, which is proceeding through the collaboration of the two authors, aims to make visible what conventional maps of the Arctic ignore: 1) that there is a social life in the Arctic embedded in Inuit communities, 2) that all communities across the circumpolar world are interconnected, and 3) that Inuit knowledge and occupancy is tied to both marine and land areas.

Land/sea interface in Inuit homeland

Boundaries of land and sea are interwoven by physical, environmental, and biological processes, such as those that are part of intertidal and coastal and estuary habitats and phenomena. The relationship between fresh and ocean water allows, for instance, for the transit of species such as anadromous fish. In the Arctic, the land/sea distinction is even less clear, as the sea ice becomes attached to the shores, in some areas forming an extension of the land for most of the year. It is not surprising, therefore, that while British explorers in the 17th to the 19th centuries considered the sea ice an obstacle for the navigation of their big ships, Inuit across the Arctic have always conceived the sea ice as part of their homeland. Inuit hunters travel from the land onto the land-fast ice, towards other shores or in search of open water or thin ice where marine mammals can be found. For some species, such as polar bears and caribou, the sea ice equally offers an extension of their habitat or a bridge for movement between two bodies of land.

To most Inuit communities throughout history, the shores were not mere dividers between land and sea but connectors (Aporta et al. 2018; see also Druckenmiller et al. 2013), as both the sea ice and the ice-free ocean are intrinsically intertwined. In the past, Inuit communities would even make dwellings (snow houses) near shore on the sea ice (Aporta 2002), as they would adapt their own residence patterns to seasonal availabilities and conditions, often shifting population gatherings that took advantage of both land and sea resources.

The land/sea divide, however, has permeated most Arctic research and policies since contact. Most of the research, land use studies, and environmental assessments conducted since the 1970s are implicitly organized by underlying assumptions that reflect sea/land distinctions (an example of which is the Nunavut Coastal Resource Inventory (Nunavut Government 2010), which organizes Inuit use of coastal and marine environments mostly by species).

The next sections will propose that Inuit mobility systems (Inuit trails and routes) are essential in understanding the intricate relationship between land and sea environments from the perspective of Inuit communities.

Inuit mobility systems

Conventional maps of Canada show how different provinces, cities, and towns are connected through transportation infrastructures. Railways, roads, highways, and streets show material interconnections across places, reflecting and building senses of community, as well as economic, social, and cultural links. These transportation systems are also connected to supporting infrastructure, from gas stations to hotels, grocery stores, and train stations. All together, roads, rails, and so on are clear indicators of the use, tenure, and residence patterns of the societies that live in those areas. This transportation infrastructure is also an indication of historical use, and it is engrained not only in people's minds but also in cartographic representations of Canada. Nobody would question, for instance, the occupancy of Torontonians over their own city.

Hunter-gatherers have struggled to convey and prove their own occupancy.[9] In the colonial legal system, they have been burdened with the obligation to provide *proof* of historical occupancy, itself a difficult task to peoples who did not have farming practices and land ownership systems recognized by Europeans (Bell and Asch 1997). Nomadic and seminomadic Indigenous groups have had to come up with ways to present "evidence" of their land tenures, often through written documentation of their own oral knowledge and histories and, more recently, with maps.[10] The land claims occupancy studies of the 1970s and 1980s in Canada are clear examples of the burden imposed on Indigenous peoples to document the use of their lands through western cartographic renditions of their knowledge and practices.[11]

Conventional maps of Canada show a Canadian Arctic that is vastly unpopulated, except for the presence of small and isolated Inuit settlements. These cartographic representations are powerful symbols that tell stories of an empty, remote, barren, uninhabited, and harsh Arctic. Those maps not only prevail in the south of Canada but also in the communities where Inuit live, as any visitor of a Canadian Arctic airport would have noticed. More importantly, these representations are huge distortions of a territory that, for Inuit, is experienced as home: a place filled with life, resources, and history and completely embedded with local, regional, and cross-regional social relationships. Those relationships are clearly represented on the maps illustrating and supporting the arguments of this chapter, particularly on those showing networks of routes.

A map of the Canadian Arctic with Inuit routes may be puzzling to non-Inuit observers. The routes are not reflected in permanent physical features. However, as explained elsewhere (Aporta 2009), these routes are as real and historic as roads in the south. They have been used by countless generations for hundreds of years, and they are committed to the memories of Inuit individuals and communities. Crucially, the routes seamlessly connect land and marine spaces.

These routes constitute Inuit mobility systems, in the sense that they connect the Inuit-inhabited Arctic and are well established by historical use. They include sled trails, walking trails (more recently known as ATV trails for their use by all-terrain vehicles), and boat routes.

As shown in Figure 12.1, sled trails become visible as tracks on the snow. Because of their seasonal use (see next section), routes are intrinsically embedded in Inuit social relationships.

Figure 12.1 Photo of sled tracks near the community of Puvurnituq in March 2018

They are, in fact, the arteries through which news, goods, and people have travelled, seasonally, for centuries.

A map of a sample of the Inuit mobility system (Figure 12.2) reflects spatial interconnections across the Canadian Arctic, if not the seasonal subtleties of their use (next section). As will be seen in the last section, this mobility system erases or minimizes the differences between land and marine spaces, with the partial exception of summer travel, where the shores become places of residence as well as anchoring or launching platforms.

Although Inuit occupancy can be shown in many different ways (e.g. maps depicting archaeological sites, harvesting places, etc.), the map in Figure 12.2 is unique in reflecting: 1) the interconnected social dimension of the Arctic, which goes from local to regional and cross-regional; 2) the deep relationship between Inuit communities and the environmental resources available to them (as will be shown subsequently, most trails are connected to the availability of animals); and, indirectly, 3) the historical depth of Inuit residence in their homeland, including in marine areas. The later point is proven by the fact that almost all routes shown on the maps illustrating this chapter have been identified by Inuit communities as "used since time immemorial." Furthermore, a great proportion of them are actually populated by historical remains of use, which could be recent (e.g. a broken sled) or ancient (e.g. tent rings).

Inuit mobility systems show that life, resources, spaces, and histories are experienced as interconnected and that the sea (whether covered by ice or not) is perceived and experienced as an extension of the land. Maps of Inuit routes provide not only more accurate representations of Inuit land and marine use but also a better conceptual framework to understand not only Inuit ontologies regarding their own homeland but also the entire socioecological system and the seasonality of resources in it.

Figure 12.2 A snapshot of the interconnected Arctic, showing trails and place names

Source: Aporta for trails and Inuit Heritage Trust for place names.

A homeland connected to seasonal variations

The sense of Inuit homeland should not be a static conceptualization of use of space or even of communities occupying certain fixed and bounded territories. Inuit are a semi-nomadic culture, and even when living in permanent settlements, their mobility and residence patterns have always been tied to the seasonal presence of animals and environmental phenomena (e.g. the forming of the ice).

While maps of Inuit trails such as the one in Figure 12.2 are good at representing the spatiality of Inuit occupancy, the lines on the map do not accurately represent the subtleties of seasonal variations.

Some of those variations are fairly obvious, such as those lines on the map that represent boat (summer) routes and those that represent sea ice sled trails.[12] Other variations are subtler, as they refer to a number of factors, which will be described in this section, including the intrinsic characteristics of sled trails, the timing with the availability of environmental resources, and the conditions of the sea ice and snow.

Characteristics of sled trails

Sled trails, regardless of whether they are on snow-covered land or on the sea ice, are basically first trodden on the snow by memory at the beginning of each travel season, by which is meant *the beginning of the travel season for a set of trails in a particular area*. The process of trail breaking (described in detail in Aporta 2009) is initiated by a knowledgeable traveller, riding his/her snowmobile or dog sled and leaving the first sled tracks on fresh snow. Other travellers will follow, in some cases allowing for different variations of segments of the trails due to seasonal conditions or personal preferences. The tracks that are followed more extensively become the trails that will be used by all travellers in that season. They, therefore, become physical features in the landscape, their visibility dependent on being trodden by use. As shown in Figure 12.1, some of these trails are so well trodden that become equivalent to highways, allowing for higher speed and smoother travel. As shown by Aporta (2009), these trails are recreated from year to year, and their spatial layouts are transmitted in the contexts of narratives (descriptions of what the trails look like when travelling). The seasonal variations of trails are not trivial because they reflect the fundamental seasonality of Inuit inhabitancy of the Arctic. Inuit trails (as well as other places of significance) are important not only in terms of their discrete spatial locations but as parts of space/time matrixes, defined by the intersecting movements of people and animals, as well as by the dynamics of environmental and atmospheric events (e.g. the presence of snow or the directional clues of winds).

The presence of environmental resources

Inuit have many reasons for travelling, including visiting other communities and places that are emotionally significant,[13] but it is clear that residence and mobility patterns are deeply connected to the presence of animals. Trails are not usually meaningful (or even travelable) if not in the context of hunting or fishing, a factor that was even more important when people depended exclusively on the land and travelled by dog teams. In Figure 12.3, boat routes mapped in Puvirnituq and Salluit (Nunavik) and Cape Dorset (Nunavut) relate to the seasonal presence of walrus on those two islands in the Hudson Strait.

The two islands on the map are frequented by Inuit from several communities in Nunavik, as well as by Inuit from Cape Dorset (north of the map) in Nunavut.

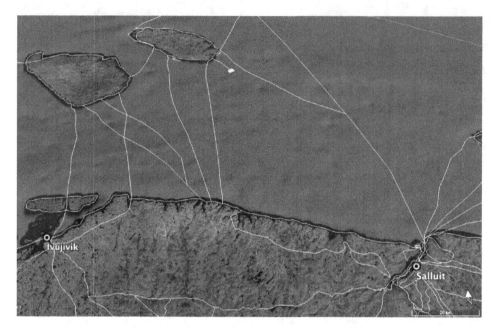

Figure 12.3   Boat trails in the sea, mapped in the Nunavik communities of Salluit and Puvirnituq (south of map) between 2014 and 2018

Given the enormous size of the Inuit-inhabited Arctic and the relatively very small demographic concentration of Inuit communities, seasonality is the main reason for which camps, harvesting sites, and routes can actually be defined as social spaces, as travellers from different communities will converge on similar areas at the same times of the year. In essence, the presence of other people, the availability of resources, and the actual use of the trails are the reasons for which the Canadian Arctic (both its land and marine components) is perceived and experienced as *home* by Inuit. By the same token, the ignorance of such spatial/temporal relationships and social interactions has resulted in non-Inuit experiencing and portraying the Arctic as a harsh and inhospitable space.

It is clear that the seasonal use of every route determines the sociability of the Arctic and is connected to complex understanding and knowledge of environmental phenomena and animal behaviour. Only unexpected environmental events, such as unseasonal temperatures or rain-on-snow, can affect the timing of seasonal movements and create vulnerability in Inuit communities.[14]

The conditions of the sea ice

The temporality of trails is also intrinsically linked to the formation, topographic configuration, and breaking of the sea ice (Aporta 2002). The sea ice is both a changing and recurrent travelling surface, and seasonal variations of sea ice trails are often the consequence of travellers negotiating freezing and melting patterns in the early fall and late spring. The presence of unstable ice at the start and end of the landfast ice season are the only times when travel is kept to a minimum (and eventually stopped), as neither boating nor sled travelling can be undertaken under such conditions. The freezing and melting patterns of the landfast ice will also determine the layout of sea ice crossings, with travellers avoiding unstable ice and often following less direct routes to a destination located on an opposite shore until an efficient trail is consolidated on the safe ice.[15]

Crucially, as explained in the next section, sea ice trails connect Inuit hunters to open water or thin ice areas where sea mammals reside. Sea ice trail systems are, therefore, timed and laid out in direct correlation with the presence of marine mammals in areas of open water or thin ice, including polynyas and tidal cracks. Marine harvesting areas in the sea ice season are therefore reflected on the trails that lead to them, including the very significant trails to and from the floe edge.

The conditions of the snow

Snow is the enabler of sled travel as well as the directional record left by the prevailing winds, which Inuit use for orienting (MacDonald 1998). Trails are, therefore, seasonally aligned to the presence of snow on land and spatially configured so that they avoid unnecessary "bad" travel surfaces, which include lack of snow, areas where the snow is too deep, and areas where the terrain is too rough or too steep.

In practical terms (and as shown in Figure 12.4), this means that the land portion of sled trails follows topographic "opportunities" with good snow and along areas where harvesting can occur. Favoured areas for trails are chains of frozen water, including lakes, rivers, creeks, bays, and fjords.

The connections between land and sea through frozen waterways can be seen even more clearly in Figures 12.5 and 12.6, where watersheds act as arteries across the Labrador Peninsula, providing social spaces (trails, camps, and harvesting areas) for Inuit currently living in Nain

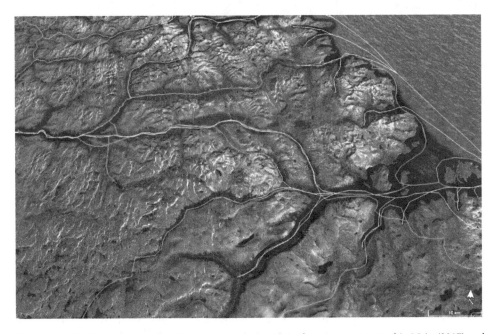

Figure 12.4 Trails in the Labrador Peninsula over fresh and sea frozen water, mapped in Nain (2017) and Kangiqsualujjuaq (2016)

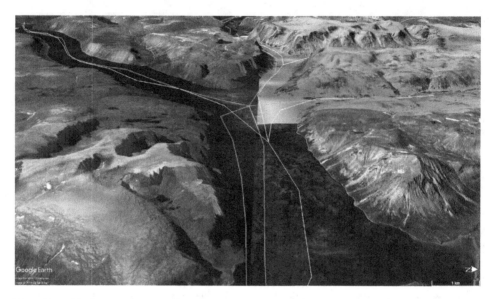

Figure 12.5 Horizon perspective of map in Figure 12.4, seen from the Labrador Sea. Trail data documented in Kangiqsualujjuaq and Nain

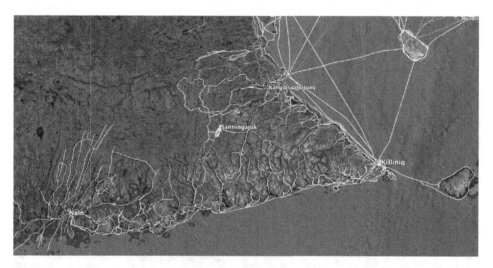

Figure 12.6 At the intersection of Nunavik, Nunatsiavut, and Nunavut: Inuit routes following watershed lines and becoming arteries of the social Arctic

(Nunatsiavut region of Labrador) and Kangiqsualujjuaq (Nunavik region of northern Quebec). The trails in between the two shores are a clearer reflection of Inuit sense of community than the two permanent settlements located in separate jurisdictions.

Between the present-day settlements of Kangiqsualujjuaq and Nain, two significant "in between" places were included in Figure 12.6, as they were recognized as significant in stories documented in both communities. Sanningajuk and Killiniq were and are places of encounters and socialization for Inuit born in the region, regardless of territorial differences. Killiniq is a former Inuit settlement and a former trading and missionary post located within the territory of Nunavut but at the intersection with the borders of Nunatsiavut and Nunavik. From a travel perspective, it is just north of a narrow channel of strong currents connecting the Labrador Sea with Ungava Bay. Sanningajuk is almost exactly in the middle of the peninsula, and it was and is a meeting point for Inuit travelling from both shores.

Land and sea dimensions of Inuit trails

This section will describe in greater detail the different ways through which Inuit mobility systems facilitate and reinforce connections between land and marine areas. It will focus on concrete examples from different parts of the Canadian Arctic, which will show how travellers deal seamlessly with spaces that are not defined by shorelines as hard environmental or spatial dividers.

From land to land, over the ice

Hunters from Igloolik and Hall Beach make the trip every fall to a place they call Majuqtulik, which is roughly 160 km north of Igloolik. From the settlement on Igloolik Island, and weeks after the sea ice has allowed people to travel across the inlet to the mainland (Melville Peninsula), Inuit hunters wait for news of the ice solidifying off the north tip of the mainland, which would allow them to undertake the crossing to Majuqtulik, on the Baffin Island shore. Every year, around November, early travellers to the area report the state of the ice daily and communicate their news to the people in town by CB radio and, in some cases, by satellite phones. When the ice is deemed safe, dozens of hunters initiate their trips, which follow a precise layout of trails. The crossing of the ice off Saglaarjut (Figure 12.7) is often very rough, and, because of the season, the visibility is often poor for lack of light and/or presence of blizzards. During a trip in 2000, as Aporta was snowmobiling with a hunter in the wake of a blizzard, the hunter oriented himself by the feel of the wind and was able to find the cabin in almost complete darkness.

The journey, unremarkable in the sense that it was typical of many other journeys, showed how Inuit communities' sense of homeland is embedded in intricate connections between land, sea, people, animals, and atmospheric events:

- The trip happened at a time of the year that coincided with two expected events: the formation of solid ice in the strait and the presence of caribou around Majuqtulik.
- The trails were well known, and other travellers were expected along the way and in the cabins (note that the same trails and cabins would have been completely empty at most other times of the year).
- The trails followed courses that involved topographic depressions such as valleys and interconnected bodies of frozen water, both on land and sea, highlighting connections between the watershed and the ocean through human mobility.
- The prevailing wind was essential for finding the way in bad visibility, providing a spatial directional framework well known to Inuit across the Arctic.[16]

While Majuqtulik means in Inuktitut *a place for going up* (a reference to fish swimming upriver), the seasonal (late fall) use of this place was not connected to fishing but to caribou hunting. Additionally, Majuqtulik is near well-established traditional outpost camps that were used by Inuit from the Igloolik area, as well as by those from the northern Baffin Island region, around

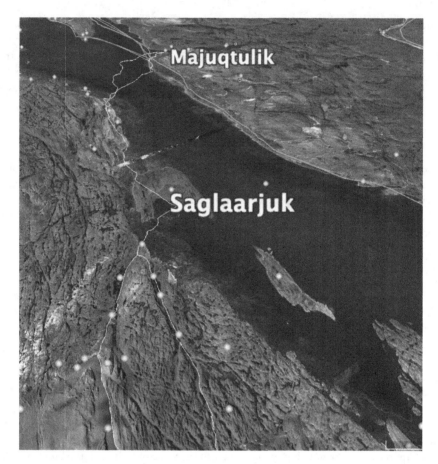

Figure 12.7 GPS tracks of a journey in November 2000. The trails followed frozen fresh water in the mainland, and then crossed to Saglaarjut and to Majuqtulik, on Baffin Island. Points are known place names

the times the permanent settlements were created. The extensive use of these coastal and marine areas was, therefore, associated with different occurrences through the year, reflecting intricate connections between land and sea. In fact, because of seasonal Inuit mobility and residence patterns before permanent settlements were established, the Arctic is filled with historically inhabited residence areas (camp sites) that are outside of their present-day settlements and that acted as meeting points for Inuit from different neighbouring regions. *Home* for Inuit communities was not associated with a single area but a geographic range within which people moved according to well-known seasonal events. Such a sense of home is still observed today, even after generations of living in permanent settlements.

Watershed and sea through the lives of animals and the movement of water

The relationship between watersheds and marine environments through Inuit mobility and residence patterns was discussed previously, but such a relationship is not only limited to the spatial layout of routes and the movement of people. It is also established through the movement of animals whose habitats include both land (or freshwater) and sea. The animals include anadromous fish (Arctic char), polar bears, and even land mammals, such as caribou, that use the sea ice as bridges between two bodies of land. They also include birds that make their nests in the cliffs or on the shores. Inuit observations of and interactions with these animals and processes can be inferred from the meaning and location of traditional Inuit place names.

Near the community of Pond Inlet, most significant coastal features and places are named, including hills, cliffs, glaciers, mouths of bays, and beaches. Sirmiarjuk (Figure 12.8) refers to a small glacier whose "branches" reach out to the sea, connecting it with the watershed. Qarmaarjuit (right in Figure 12.8) is a traditional hunting place for seals, narwhal, and caribou, as well as a place with an abundance of birds. In Qarmaarjuit, these relationships (of people, marine and land mammals, birds, land, and water) are historically rooted, as the place has the remains of very old, small sod houses. Many other names (such as Majuqtulik) reflect the movement of animals through water, and "mouths of bays" are also often named, recognizing marine areas where freshwater mixes with the sea. Inuit have several terms to describe the state of the water in such places.

The extraordinary complexity of the entanglements of people, animals, land, and ocean can be clearly observed in a detail of a map of the region between Arctic Bay and Igloolik (Figure 12.9). Place names (most of them related to harvesting sites, wayfinding markers, and camps) are themselves markers of animal/human encounters, and Inuit routes come and go, connecting freshwater bodies and sea.

Glaciers, sea-ice crossings for caribou migrations, fish migrating between land and sea, nesting sites, presence of marine mammals, tidal and currents occurrences, and wind directions are just some of the many relationships between land and sea that are reflected in place names (themselves a very limited record of the actual knowledge and performances linking land and sea). The very density of place names on the shorelines reflects Inuit residence and mobility patterns in both land and sea, as they relate to coastal camps, launching/landing places for boats and sled trails, landmarks observable from the sea, bays, lakes, and, in general, coastal and marine habitats.

Figure 12.8   Place names west of Pond Inlet

Data source: Inuit Heritage Trust.

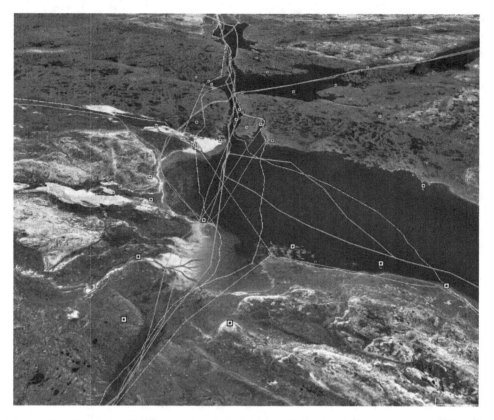

Figure 12.9 Complex sea–land connections through Inuit trails documented in Igloolik, Pond Inlet, and Arctic Bay. Place names are represented as points (Trail data: Aporta; place name data: Aporta and Inuit Heritage Trust)

Floe edge, open water, and thin ice

More significant than the edge of the land are the edges of the ice. These edges mark the marine places during the ice season where marine mammals can be found or can be expected to be. The floe edge, dividing the landfast ice from open water, is a dynamic thing, which in some areas (near polynyas) involves significant "moving ice" environments which, depending on the winds and tides, will attach to or separate from the solid edge of the landfast ice. These areas are dangerous for travel but highly significant because of the abundance of marine mammals.

Inuit elders in Igloolik (Nunavut) and Puvurnituq (Nunavik) reported on the dangers of drifting on thin ice that becomes separated from the solid ice, noting that a deep understanding of tidal shifts and of the effects of winds on moving ice is paramount. Figures 12.10 (Nain, in the Labrador Sea) and 12.11 (Tuktoyaktuk, in the Beaufort Sea) are illustrative of the significance of ice edge, as a number of trails were depicted as reaching a non-visible line on the map (the floe edges).

The significance and variety of harvesting places within marine areas, especially on or near sea ice, along with the crucial travelling surface offered by the sea ice for most of the year, are clearly reflected in the intricacies and locations of Inuit mobility systems, which show that the marine environment is fundamentally connected to the dynamic and recurrent topographies of the sea ice.

Figure 12.10   Trails of Nain (2017)

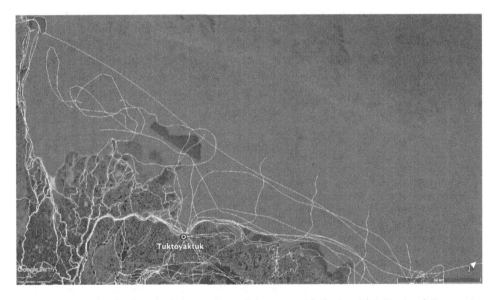

Figure 12.11   Trails of Tuktoyaktuk (mapped in collaboration with the Inuvialuit Regional Corporation in 2018)

Conclusion

Two of the most remarkable underlying assumptions reflected in most maps of Arctic Canada (the perception that the land is empty and the idea that land and sea are clearly delineated) are strongly undermined when Inuit mobility systems are considered. Inuit routes, which can be rendered cartographically on maps, show the intricate connections across the Inuit circumpolar world, linking communities with communities and allowing for the circulation of people, goods, and news. In this sense, the maps of trails illustrating this chapter are strong counter-mapping statements, as they propose alternative visualizations of the Arctic. Since the Inuit mobility system exceeds the geography of the Canadian Arctic to encompass (presumably) all other regions of the Inuit circumpolar world, Inuit routes can also be potentially interpreted as political statements towards a recognition of the existence of an Inuit homeland across (or even beyond) national jurisdictions.

The concept of Inuit homeland articulated in this chapter is not only associated with cartographic representations but with a deeper Inuit ontology of the Inuit Arctic, which is rooted both in space and time and that includes comprehensive relationships between people and the environment where they live. It is actually through consideration of the seasonal use of space that such connections between Inuit and the Arctic environment can be fully understood. In that conceptualization, land and sea environments are intrinsically linked. Inuit routes, even if only represented as static lines on a map, can perhaps open the door to a better and deeper understanding of how the Arctic marine and land environments are intertwined and how they are linked with the very existence of Inuit communities.

In documenting Inuit land use and occupancy, it may be worth wondering how documentation processes could be envisioned that may reflect Inuit ontologies of land and water, species and events, atmospheric and environmental processes, and so on. At the very least, the maps with networks of Inuit routes are helpful reminders of the social dimensions of the Arctic and a clear indication that the sea, in all stages, is an intrinsic part of Inuit homelands. They also implicitly show that Inuit will always be in their homeland, regardless of environmental, economic, and political developments and changes, and in spite of geographic and political boundaries. From a political perspective, there is also a clear corollary to this new Arctic ontology: whatever claims Arctic states may have over Inuit-occupied lands and waters, they must be intrinsically tied to the wellbeing and prosperity of Inuit communities.

Notes

1 For a specific discussion of the impact of participatory mapping among Inuit, see Depaiva 2017. For an Inuit-led documenting project see Gearheard et al. (2011).
2 The list is too long to cite, starting with the land use projects in the 1970s and following with multiple community mapping projects at the present (for contemporary projects in Canada, see, for instance, www.ihti.ca/eng/iht-proj-plac.html; www.ntkp.ca/; https://sikuatlas.ca/index.html).
3 On the debate about how the politics of mapping plays in Arctic sovereignty, see Steinberg and Kristoffersen (2017) and also Bennet et al. (2016). Tobias (2009) produced an exhaustive guide on how to map Indigenous knowledge properly.
4 This work is not, of course, the first to document and map Inuit trails, but it is the first to systematically attempt to map Inuit trails at a pan-Arctic scale. Explorers and anthropologists that mapped Inuit trails include Boas (1888), Hall (1864), Lyon (1824), Parry (1824), Matthiessen (1928), Rasmussen (1929, 1930), Rowley (1996), and Stefánsson (1912), as well as several researchers involved in the land occupancy projects (see, for instance, Brody's work in the ILUOP; Freeman 1976). Several researchers working for Inuit organizations also conducted extensive mapping, including extensive work in Nunavik and Nunavut by Bill Kemp (unpublished). For an atlas of Inuit trails documented in historical texts, see www.paninuittrails.org/ (Aporta et al. 2014).
5 Labrador Inuit Land Claims Agreement, 2005, sections 4.2.3, 4.4.1, 4.4.2, and 4.4.3(a).
6 Labrador Inuit Land Claims Agreement, Part 6.3 and Part 6.6.

7 Nunavik Inuit Land Claims Agreement, Article 15 (pertains to Crown lands, including the seabed, in the Nunavik Marine Region).

8 ICC remarked on the importance of trails in its policy document "The Sea Ice Is Our Highway" (ICC 2008).

9 While *occupancy* is not the right term to refer to Inuit inhabitancy of the Arctic, it is sometimes used in this text, as it is the term of reference in the studies connected to Inuit land claims in Canada. Proving "occupancy," of course, was part of the Inuit effort to obtain rights within the legal system of a colonial state.

10 For an interesting discussion of the processes through which nomadic spaces are produced and known, see Miggelbrink et al. (eds) 2014. For an archaeological study of the significance of mobility systems among other Indigenous peoples, see Borrero et al. 2019.

11 The land claims studies are also wonderful documents, the result of fruitful collaboration between researchers and Inuit communities.

12 Note that the trail data in the maps presented here are aggregated and shown as one type of line. Most of the routes mapped by Aporta were coded as "winter" or "summer" routes, which is also a simplification. Some previous work (e.g. the trails mapped by Bill Kemp in the land occupancy studies in the Nunavik region or the ones mapped by Hugh Brody in the Canadian high Arctic) included four seasonal variations (fall, winter, spring, and summer). The simplification of winter and summer has been intentionally used by Aporta to expedite the participatory mapping process in the communities, mainly due to time constraints but also to avoid participant fatigue.

13 Inuit have historically travelled for many reasons, including some that are not necessarily functional (e.g. curiosity and exploration). In the past, travelling was also associated with trading, even before the establishment of trading posts. Another frequent reason for travel involves finding rocks, bones, antlers, and tusks that are good for carving.

14 These changes can be linked to climate change, but Inuit oral history is rich with references to climatic and environmental changes in the past. For instance, the famine of 1950s around Baker Lake was attributed to caribou changing their migration patterns. It could be argued that life in the settlements may have increased social vulnerability, in part due to disruptions of seasonal encounters with animals (Aporta and Higgs 2005; Aporta 2013; Aporta 2016). In that sense, increasing opportunities for Inuit youth to spend more time on the land may result in an increase in community resilience to change.

15 For a view to the challenges to this process introduced by climatic changes, see Tremblay et al. (2018) and Shake et al. (2018).

16 As documented by MacDonald (1998), prevailing winds are fairly recognizable, but they can shift. Snowdrifts shaped by the prevailing winds are more reliable wayfinding markers.

References

Aporta, C. 2002. "Life on the Ice: Understanding the Codes of a Changing Environment." *Polar Record* 38 (207): 341–354.

Aporta, C. 2009. "The Trail as Home: Inuit and Their Pan-Arctic Network of Routes." *Human Ecology* 37 (2): 131–146.

Aporta, C. 2013. "From Inuit Wayfinding to the Google World: Living within an Ecology of Technologies." In *Nomadic and Indigenous Spaces: Productions and Cognitions*, edited by J. Miggelbrink, J. Habeck, N. Mazzullo, and P. Koch. Farnham: Ashgate.

Aporta, C. 2016. "Markers in Space and Time: Reflections on the Nature of Place Names As Events in the Inuit Approach to the Territory." In *Marking the Land: Hunter-Gatherer Creation of Meaning within Their Surroundings*, edited by R. Whallon and W. Lovis, 67–88. Abingdon, Oxon and New York: Routledge.

Aporta, C., M. Bravo, and F. Taylor. 2014. "Pan Inuit Trails Atlas." Accessed 24 March 2020. http:// paninuittrails.org.

Aporta, C., and E. Higgs. 2005. Satellite Culture. Global Positioning Systems, Inuit Wayfinding, and the Need for a New Account of Technology. *Current Anthropology* 46 (5): 729–753.

Aporta, C., S. Kane, and A. Chircop. 2018. "Shipping Corridors through the Inuit Homeland." *Limn Journal* 10: 50–65.

Bell, C., and M. Asch. 1997. "Challenging Assumptions: The Impact of Precedent in Aboriginal Rights Litigation." In *Aboriginal and Treaty Rights in Canada: Essays on Law, Equity, and Respect for Difference*, edited by M. Asch, 38–74. Vancouver: UBC Press.

Bennett, M., W. Greaves, R. Riedlsperger, and A. Botella. 2016. "Articulating the Arctic: Contrasting State and Inuit maps of the Canadian north." *Polar Record* 52 (6): 630–644. doi:10.1017/S0032247416000164.

Boas, F. 1888. *The Central Eskimo*. Washington: Bureau of Ethnology.

Borrero, L., A. Nuevo Delaunay, and C. Méndez. 2019. "Ethnographical and Historical Accounts for Understanding the Exploration of New Lands: The Case of Central Western Patagonia, Southernmost South America." *Journal of Anthropological Archaeology* 54: 1–16.

Depaiva, A. 2017. "Understanding the Role and Value of Participatory Mapping in an Inuit Knowledge Research Context." MA thesis, Carleton University.

Druckenmiller, M., H. Eicken, J. C. George, and L. Brower. 2013. "Trails to the Whale: Reflections of Change and Choice on an Iñupiat Icescape at Barrow, Alaska." *Polar Geography* 36 (1–2): 5–29.

Gearheard, S., C. Aporta, G. Aipellee, and K. O'Keefe. 2011. "The Igliniit Project: Inuit Hunters Document Life on the Trail to Map and Monitor Arctic Change." *The Canadian Geographer* 55 (1): 42–55.

Hall, C. F. 1864. *Life with the Esquimaux*. London: Sampson Low, Son, and Marston.

Hunt, D., and Shaun A. Stevenson. 2017. "Decolonizing Geographies of Power: Indigenous Digital Counter-Mapping Practices on Turtle Island." *Settler Colonial Studies* 7 (3): 372–392. doi: 10.1080/2201473X.2016.1186311.

Inuit Circumpolar Council (ICC). 2008. *The Sea Ice Is Our Highway: An Inuit Perspective on Transportation in the Arctic*. Ottawa: ICC. Accessed 4 March 2020. www.inuitcircumpolar.com/uploads/3/0/5/4/30542564/20080423_iccamsa_finalpdfprint. pdf.

Lyon, G. F. 1824. *The Private Journal of Captain G. F. Lyon of H.M. S. Hecla during the Recent Voyage of Discovery under Captain Parry*. London: John Murray.

MacDonald, J. 1998. *The Arctic Sky: Inuit Astronomy, Star Lore, and Legend*. Toronto: Royal Ontario Museum & Nunavut Research Institute.

Mazzullo, N. 2018. "Counter-Mapping Commercial Forests and Reclaiming Indigenous Reindeer Herding Pastures in Finnish Upper-Lapland." In *Indigenous Places and Colonial Spaces: The Politics of Intertwined Relations*, edited by N. Gombay and M. Palomino-Schalscha, 127–151. New York: Routledge.

Miggelbrink, Judith, Joachim Otto Habeck, Nuccio Mazzullo and Peter Koch (eds). 2014. *Nomadic and Indigenous Spaces: Productions and Cognitions*. Farnham, Surrey: Ashgate Publishing.

Milton Freeman Research Limited and Canada Department of Indian Affairs Northern Development. 1976. *Inuit Land Use and Occupancy Project: Report*. Ottawa: Department of Indian and Northern Affairs.

Nunavut Coastal Resource Inventory (September 2010). Accessed 1 March 2020. www.gov.nu.ca/environment/information/nunavut-coastal-resource-inventory.

Parry, W. E. 1824. *Journal of a Second Voyage for the Discovery of a Northwest Passage from the Atlantic to the Pacific*. London: John Murray.

Peluso, N. L. 1995. "Whose Woods Are These? Counter-Mapping Forest Territories in Kalimantan, Indonesia." *Antipode* 27 (4): 383–406.

Rasmussen, K. 1929. "Intellectual Culture of the Iglulik Eskimos." In *Report of the Fifth Thule Expedition*. Vol. 7, No. 1. Copenhagen: Gyldendalske Boghandel, Nordisk Forlag.

Rasmussen, K. 1930, 1976. "Iglulik and Caribou Eskimo Texts." In *Report of the Fifth Thule Expedition*. Vol. 7, No. 3. New York: AMS.

Rowley, G. W. 1996. *Cold Comfort: My Love Affair with the Arctic*. Montreal: McGill-Queen's University Press.

Senate Special Committee on the Arctic. 2019. "Northern Lights: A Wake Up Call for the Future of Canada." Accessed 4 March 2020. https://sencanada.ca/content/sen/committee/421/ARCT/reports/ARCTFINALREPORT_E.pdf.

Shake, K., K. Frey, D. Martin, and P. Steinberg. 2018. "(Un)frozen Spaces: Exploring the Role of Sea Ice in the Marine Socio-Legal Spaces of the Bering and Beaufort Seas." *Journal of Borderlands Studies* 33 (2): 239–253.

Simon, M. 2011. Canadian Inuit. *International Journal* 66 (4): 879–891.

Stefánsson, V. 1912. *My Life with the Eskimo*. New York: Harcourt, Brace and Company.

Steinberg, P., and B. Kristoffersen. 2017. "The Ice Edge Is Lost . . . Nature Moved It': Mapping Ice As State Practice in the Canadian and Norwegian North." *Transactions of the Institute of British Geographers* 42 (4): 625–641.

Tremblay, M., J. Ford, and S. Statham. 2018. "Access to the Land and Ice: Travel and Hunting in a Changing Environment." In *From Science to Policy in the Eastern Canadian Arctic: An Integrated Regional Impact*

Study (IRIS) of Climate Change and Modernization, edited by T. Bell and T. M. Brown. Quebec City: ArcticNet.

Tobias, T. N. 2009. *Living Proof: The Essential Data – Collection Guide for Indigenous Use – and – Occupancy Map Surveys*. Vancouver: Ecotrust Canada/Union of BC Indian Chiefs.

Other documents

A Circumpolar Inuit Declaration on Sovereignty in the Arctic (signed April 2009, Inuit Circumpolar Council).

Nunavut Land Claims Agreement Act (S.C. 1993, c. 29).

13

ALASKA NATIVE MARINE MAMMAL HARVESTING

The Marine Mammal Protection Act and the crisis of eligibility

Steve J. Langdon

Abstract

The coastal Indigenous people of Alaska have utilized and depended upon marine mammals from bowhead whales in the Arctic to harbour seals in the Gulf of Alaska down to the present. In 1972, the Marine Mammal Protection Act (MMPA) was passed, outlawing most marine mammal taking in the United States; however, it included an exemption for "coastal Alaska Natives" to make non-wasteful harvests of marine mammals for "subsistence" purposes. A core provision of the MMPA is the definition of "Alaska Native" that was adopted by regulation, incorporating the language found in the Alaska Native Land Claims Settlement Act (ANCSA) of 1971. The definition of Alaska Native in ANCSA stipulates primarily that a person must be a quarter Alaska Native by genealogy to be eligible. Over the past 45 years, demographic changes driven by marriage patterns have resulted in increasing numbers of individuals of Alaska Native identity falling below the requirement of a quarter Alaska Native ancestry, thereby becoming ineligible to hunt marine mammals. Data acquired from the US Bureau of Indian Affairs, who enrol persons of Alaska Native descent for eligibility for Indian Health Services, clearly demonstrate that the current definition and its enforcement pose serious challenges to the future of marine mammal hunting by Alaska Natives in a number of regions. This chapter will address Alaska Native views on the current situation, efforts made by Alaska Natives to address this situation and the current status of the dilemma.

The coastal Indigenous people of Alaska have utilized marine mammals for millennia. From the bowhead whales and walrus taken by the Inupiaq and Sivuqaqmiut in Arctic Alaska to the sea lions and harbour seals used by the Unangan, Alutiiq and Tlingit of the Gulf of Alaska, the connection between humans and marine mammals has been sustained down to the present. In 1972, the Marine Mammal Protection Act (MMPA) was passed by the US Congress, outlawing most marine mammal taking in the United States. The legislation included an exemption for "coastal Alaska Natives" to make non-wasteful harvests of marine mammals for "subsistence" purposes. A core provision of the MMPA is the definition of "Alaska Native" that was adopted by regulation, incorporating the definition found in the Alaska Native Land Claims Settlement

Act (ANCSA) passed into law in 1971. The definition of Alaska Native in ANCSA has two parts. The first provision states that a person must be able to demonstrate via genealogical records that they are of one-fourth Alaska Native descent, and the second states that if a person does not meet the first standard, they are to be deemed Alaska Native if they are so regarded by the community of their membership. Congress incorporated the second provision after Alaska Natives in the Aleut and Alutiiq regions stated that genealogical records were not available for many elders in the region. Both provisions were included in the regulations by the agencies enforcing the MMPA; however, both agencies have chosen not to operationalize the second standard.

Over the past 45 years, demographic changes driven by marriage patterns have resulted in increasing numbers of individuals of Alaska Native identity falling below the requirement of one-fourth Alaska Native ancestry, thereby becoming ineligible to hunt marine mammals. Data acquired from the US Bureau of Indian Affairs on persons of Alaska Native descent registered for eligibility for Indian Health Services clearly demonstrate that the current definition and its enforcement pose serious challenges to the future of marine mammal hunting by Alaska Natives in a number of regions.

This chapter presents information on the history, current status of marine mammal hunting by Indigenous Alaskans, the legal context of that hunting, the implications of that definition based on recently acquired demographic data, Indigenous Alaskan views on the issue, alternative definitions of Alaska Native that might be used and problems associated with redefinition. Information and data in this essay were the result of research supported by the Alaska Federation of Natives presented in the report "Determination of Alaska Native Status under the Marine Mammal Protection Act (MMPA)" (Langdon 2016), which includes additional information.

Historic use

Prior to contact, Alaska Natives in different regions of the state utilized all of the marine mammals present for a variety of purposes. Marine mammals were critical to Alaska Natives' sustained occupation of St. Lawrence Island and coastal Northwest Alaska, where they developed a sophisticated group hunting system to harvest bowhead whales and walrus. South of the Bering Strait, coastal Yup'ik depended on seals, walrus and beluga. In the Gulf of Alaska, Unangan, Alutiiq, Dena'ina, Tlingit and Haida hunted belugas, fur seals, harbour seals, sea otters and sea lions. Marine mammals provide food and materials for clothing and other technologies, products from them were traded and symbology and spirituality associated with marine mammals were present in all societies that engaged with them.

In the late 18th century, coastal Alaska Natives came in sustained contact with people from distant societies not previously encountered. Marine mammals, sea otters initially, were the primary reason that attracted the first wave of outsiders who began coming to coastal Alaska after the explorers. In the southern part of the state, sea otters were the initial reason for Russian penetration of the region, leading to subjugation of the Unangan (Aleuts) and the rapid destruction of sea otter populations. In southeast Alaska, British and American traders engaged in trade with Tlingit and Haida that led to the near disappearance of the species by 1815 (Gibson 1992). Exchange involving marine mammal products with Euroamerican traders, already an ongoing activity among Indigenous Alaskans, soon became important in many coastal areas.

Russians transplanted Unangan (Aleut) to the Pribilof Islands to harvest fur seals used for food by the people who processed the skins, which the Russians then traded. In the Gulf of Alaska, harbour seals became a trade commodity, as the sea otters disappeared by 1815 due to over-harvesting. In the 1850s, Yankee whalers found the bowhead whales in the Arctic Ocean, and a massive slaughter ensued as the whale blubber was rendered into oil for sale in the eastern United States. By the 1870s, bowhead whales had been reduced in numbers to less than 10% of their original size, and walrus populations were likewise decimated, causing starvation and hardship throughout the coastal region (Bockstoce 1986). The population of St. Lawrence Island (Sivuqaq) declined by 90% from 1,500 to less than 150 by 1885 due to disease and starvation (Krupnik et al. 2002).

The United States assumed jurisdiction of Alaska through a transaction with the Russian government in 1867. Sea mammal harvest continued for subsistence uses by Alaska Natives, and soon fur trade companies began exchanges with local groups to acquire skins and other products from the animals. By 1877, excessive harvest of sea otters by crews of outside hunters led to the passage of federal legislation allowing harvest only by Alaska Natives using traditional technology – spears and harpoons. In the 20th century, Alaska Natives continued the harvest and use of all of the marine mammals that provide food, materials and cultural enrichment to their societies. There was relatively little federal regulation during this period. After 1959, state regulation of traditional harvests was minimal until 1971 with the passage of ANCSA.

One of the most controversial and damaging features of ANCSA was the extinguishment of aboriginal hunting and fishing rights inserted in the legislation at the last minute without the knowledge of Alaska Native leaders. The committee report on the legislation stated that it was expected that the State of Alaska and federal agencies would provide for the subsistence needs of Alaska Natives (Case and Voluck 2012). Failure of the state to adequately meet subsistence needs of Alaska Natives led them to seek additional protections for traditional subsistence. Alaska National Interest Lands Conservation Act's (ANILCA) Title VIII, an unsatisfactory patch on this issue for most Alaska Natives, provides a rural priority for federal subsistence but applies to all rural residents and denies subsistence priority to Alaska Natives living in urban or nonsubsistence areas. At the present time, a majority of Alaska Natives reside in nonsubsistence areas and thus are ineligible for the federal subsistence priority.

Current use and cultural significance

While marine mammals are utilized by Indigenous Alaskans throughout coastal Alaska, they are especially important to the coastal Inupiat, who reside around Bering Strait and the Arctic coast and to the Sivuquaqmiut of St. Lawrence Island. Recent subsistence data collected by the Alaska Department of Fish and Game (ADFG) are reported on a census area basis in Table 13.1. In the Nome Census Area, which includes St. Lawrence Island, marine mammals make up the largest category of harvests by a substantial margin and the majority of the resources harvested. In the North Slope Borough, marine mammals are again the largest category of harvest as well as the majority category of harvest. While neither the largest category of harvest nor the majority of the subsistence harvest, marine mammals constitute a very significant component of subsistence in the Northwest Arctic Borough (Kotzebue region). Marine mammals are also important to subsistence harvests in the lower Yukon and Yukon-Kuskokwim delta communities (Kusilvak census area), Yakutat and Valdez (the only

Table 13.1 Estimated harvests of wild resources for home use in Alaska by census area, region and category, 2012

Per capita harvest, kilograms usable weight.

Census area	Salmon	Other fish	Shellfish	Land mammals	Marine mammals	Birds and eggs	Wild plants	All resources
Nome Census Area	29.8	21	1	30.1	**102.7**	9	9.1	202.7
North Slope Borough	3.8	11.8	0.1	54.4	**81.1**	4.3	0.4	155.9
Northwest Arctic Borough	29	78	0.1	80	**47.4**	4.5	6.2	245.2
Arctic Region Subtotal	21.4	34.8	0.5	51.7	**78.6**	6.1	5.5	198.7
Aniak Census Subarea	85	20.3	0	22.7	**0.3**	1.9	6.4	136.6
Denali Borough (portion)	39.5	5.9	1.6	13.5	**0**	1.6	1	63.1
Koyukuk-Middle Yukon	74.8	26.3	0	67.3	**0**	4.2	3	175.6
Southeast Fairbanks (portion)	21.8	12.6	0.2	51.1	**0**	1.6	4.2	91.4
Interior Region Subtotal	61.8	21.4	0.1	55.1	**0.1**	3.1	3.7	145.3
Kodiak Island Borough	25.1	27.3	5.1	10.3	**0.4**	0.4	3.3	72
Cook Inlet (portion)	59.6	20.3	6.5	16.7	**2.1**	0.8	5.2	111.1
Denali Borough (portion)	6.9	2.9	0	33.1	**0**	0.5	2.3	45.7
Chugach Census Area (portion)	35.2	14.5	1.7	22.7	**3.5**	1.1	2.7	81.3
Cooper River Census Subarea	45.1	4.4	0.4	21.3	**0**	0.6	2.8	74.6
South-Central Region Subtotal	43.3	11	2	21.2	**1.7**	0.8	3.2	83.3
Haines Borough	21.9	17.1	5.4	13.9	**0**	0.4	4.6	63.3
Prince of Wales/ Hyder	31.4	26.5	15.4	18	**3.8**	0.6	6.4	102
Sitka Borough	26.2	24.4	12.5	23.1	**3.3**	0.3	3.2	93
Hoonah/Angoon Census Area	31.4	45.6	14.8	27.9	**3.3**	0.5	11.5	135

(*Continued*)

Table 13.1 (Continued)

Per capita harvest, kilograms usable weight.

Census area	Salmon	Other fish	Shellfish	Land mammals	Marine mammals	Birds and eggs	Wild plants	All resources
Petersburg Census Area	26.1	19.3	15.7	11.2	**0.8**	0.3	2.4	75.7
Wrangell Borough	11.6	15.4	27	17.6	**0**	0.6	3.6	75.9
Yakutat Borough	66	39.5	24.6	15.3	**15.7**	1.3	12.4	174.9
Skagway Municipality	8	7	4.1	1.7	**0**	0.2	0.9	21.8
Southeast Region Subtotal	26.3	24.1	14.3	18.4	**2.6**	0.4	4.7	90.8
Aleutians East Borough	64.6	23.1	8.5	9.2	**2.2**	3.6	4.2	115.4
Aleutians West	11.7	32.8	10.2	5.4	**4.1**	0.8	4.7	69.6
Bristol Bay Borough	93.8	5.8	1.9	14.5	**4.2**	2	5.5	127.7
Dillingham Census Area	76.3	15.7	1.3	33	**5.2**	4.7	9.9	146
Lake and Peninsula Borough	121.4	16.9	4.8	38.8	**4.7**	3.3	7.4	197.3
Southwest Region Subtotal	47.4	14.7	3.6	16.1	**3.2**	2.3	5.2	92.5
Lower Kuskokwim	96.3	49.7	0	36.6	**6.2**	10.1	9.6	208.5
Kusilvak	66.9	31.8	0	34.6	**17.1**	5.1	3.9	159.4
Western Region Subtotal	86.8	43.9	0	35.9	**9.7**	8.5	7.8	192.6
Rural State Subtotal	43.4	28.1	4.5	30.4	**18.8**	3.6	5.2	134
Anchorage Municipality	4.1	1.8	0.1	1.6	**0**			7.7
Kenai Peninsula Borough (portion)	8.5	6	0.4	4.2	**0**			19.1
Matanuska-Susitna Borough (portion)	5.1	1.9	0.1	4.8	**0**			11.9
Anchorage Nonsubsistence Area Subtotal	4.9	2.4	0.1	2.6	**0**			10
Fairbanks Nonsubsistence Area	3.2	1.1	0	4.5	**0**			8.9

Per capita harvest, kilograms usable weight.

Census area	Salmon	Other fish	Shellfish	Land mammals	Marine mammals	Birds and eggs	Wild plants	All resources
Juneau Borough	3.6	3	0.3	3	**0**			9.9
Ketchikan Gateway Borough	6	5.2	0.5	3.6	**0.1**			15.4
Valdez	9.8	3.3	0.4	6.4	**0.3**			20.3
Urban State Subtotal	4.6	2.3	0.1	3	**0**			10
STATE TOTAL	11.2	6.7	0.9	7.7	**3.2**	0.6	0.9	31.1

Source: Fall (2016)

nonsubsistence community with significant use of marine mammals). Elsewhere in the state, marine mammal harvests make up relatively smaller portions of the subsistence harvest compared to these regions.

The legal definitions of "subsistence" in state and federal law and the "reasonable opportunity" that is authorized neither recognize nor make provision for the dependence of Indigenous populations on resources for basic nutritional needs and survival. The concept of "food security" has drawn attention to this critical but unaddressed dimension of public policy concerning subsistence harvests and marine mammals in particular, but it has not yet risen to legal standing and attention.

Marine mammals also provide important materials for cultural activities. Skins of bearded seal and walrus are used for the covers of umiaks and qayaqs that continue to be used in Arctic areas. Sealskins are used in the manufacture of clothing such as hats and gloves worn by local populations and sold to others. Walrus ivory taken in subsistence harvests is used for Indigenous artists' carvings, especially on St. Lawrence Island and Nome, but it is illegal to sell raw ivory. Sea otter skins are used to make cloaks, rugs, hats and earmuffs that are sold as products.

Marine mammal products are also significant to a variety of cultural activities. For example, bowhead whale meat and blubber are distributed widely by the harvesters to those who attend ceremonies in North Slope communities and to relatives located in other communities outside the North Slope, including states outside Alaska. Sea otter cloaks are worn at ceremonies such as potlatches by males of certain Tlingit clans, demonstrating the special relationship of the clan to the sea otters.

In the Arctic regions of Alaska, Indigenous societies have persisted in their commitment to their subsistence lifestyles through major changes. What has developed over the past century is a phenomenon referred to as mixed economies, combining cash incomes with subsistence activities and foods. They can in fact be termed subsistence-based societies in which customary and traditional pursuits of fish, animals and birds are the dominant orientation and participation in cash economies supplies money to support these activities and domestic life (Poppel 2010). The bowhead whale hunt conducted by Inupiat in northwest Alaska is iconic as an institution representing and sustaining the communities. The recent Living Conditions in the Arctic (SLiCA) research project used questionnaires and other data to provide perspectives on the status and satisfaction of members of Arctic communities around the globe

(Poppel and Kruse 2009). Table 13.2 compares the number of subsistence activities engaged in by persons from different Arctic societies. The Northern Alaskan sample displays the highest rate comparatively for most activities, and the high rates of participation in these activities indicate their meaningfulness to the respondents and their satisfaction with them (Kruse et al. 2008). Note as a particular example the rates of participation in activities associated with bowhead whaling and seal hunting.

The way of life currently practiced in Arctic Alaska provides great satisfaction to those who engage in it. Table 13.3, also from the SLiCA research project, reports that nearly 90% of northern Alaskan respondents stated that they are very or somewhat satisfied with their way of life. Poppel (2010, 360) offers the following characterization of the mixed economy:

> When the relationship to nature, participation in hunting and fishing and consuming traditional foods are regularly emphasized as significant for Inuit [and Inupiaq] in the Arctic, they . . . indicate . . . relationships and activities which are important for the quality of life of people. *And, seen in this light, market economic activities as parts of the subsistence way of life' offer perhaps a sufficient description of the mixed economy in many of the Arctic communities when observed through local eyes.*
>
> [Emphasis in original]

The Marine Mammal Protection Act and the definition of Alaska Native

Concerns about the status of marine mammals in the United States were growing in the late 1960s. The Marine Mammal Protection Act (1972) is a sweeping law that was passed to address these concerns. The purpose of the MMPA is to protect those animals that were increasingly threatened by the activities of humans and that "they should not be permitted to diminish below their optimum sustainable population" (MMPA 1972, 3). In order to accomplish this purpose, the act imposed a moratorium on all takings, with certain exemptions. The following species were included: cetaceans (whales, dolphins and porpoises), pinnipeds (seals and sea lions), sirenians (manatees and dugongs), sea otters and polar bears within the waters of the United States. Alaska Natives, staggered by the extinguishment of their aboriginal fishing and hunting rights by ANCSA, mounted a massive effort to protect their uses of marine mammals and succeeded in having Senator Ted Stevens include an Alaska Native–specific exemption for them. The exemption, found in Section 101.(b), provides as follows:

> the provisions of this Act shall not apply with respect to the taking of any marine mammal by any Indian, Aleut, or Eskimo who resides in Alaska and who dwells on the coast of the North Pacific Ocean or the Arctic Ocean if such taking – (1) is for subsistence purposes; or (2) is done for purposes of creating and selling authentic native articles of handicrafts and clothing: (3) in each case, is not accomplished in a wasteful manner.

The act authorizes Alaska Natives to capture marine mammals subject to these conditions. Authorities to implement the law and manage marine mammals under its terms are assigned to the United States Fish and Wildlife Service (USFWS) and National Marine Fisheries Service (NMFS).

Table 13.2 Participation in subsistence activities by country

	Canada	Greenland	Chukotka	Northern Alaska	Total
Fish in last 12 month[1]	69%	69%	88%	77%	74%
Pick berries in last 12 months	⋆	71%	73%	70%	71%
Preserve meat or fish in last 12 months	⋆	55%	86%	74%	67%
Prepare or pack for hunting, fishing, camping trip	73%	44%	84%	71%	63%
Make and repair equipment or do household repairs	48%	73%	64%	51%	62%
Maintain a household camp	⋆	40%	92%	46%	56%
Gather greens, roots or other plants in last 12 months	⋆	⋆	45%	53%	48%
Hunt seal or ugruk in last 12 months	⋆	⋆		42%	43%
Hunt waterfowl in last 12 months	59%	40%	26%	44%	43%
Hunt caribou, moose or sheep in last 12 mouths	⋆	35%	21%	53%	34%
Hunt sea mammals	⋆	43%	6%		31%
Help whaling crews by cooking, giving money or supplies, cutting meat in last 12 months	⋆	⋆	29%	33%	30%
Gather eggs in last 12 months	⋆	19%	31%	40%	26%
Make sleds or boats in last 12 months	⋆	17%	43%	23%	25%
Skinned and butchered a caribou in last 12 months	⋆	⋆	44%	53%	25%
Manufacture Native crafts for own use	⋆	20%	26%	37%	24%
Sew skins, make parkas and kamiks in last 12 months	⋆	17%	37%	24%	24%
Member of whaling crew or herded reindeer in last 12 months	⋆	⋆	14%	30%	21%
Make native handicrafts in last 12 months	⋆	12%	15%	36%	17%
Sold meat fish or berries	⋆	10%	23%	7%	13%
Manufacture Native crafts for sale	18%	7%	12%	23%	13%
Trap in last 12 months	11%	4%	15%	11%	9%
Growing crops	⋆	7%	6%	⋆	7%
Estimated Total	22,090	35,240	17,527	10,547	85,404
Mean number of seven subsistence activities in common with Canada:	2.7	2.8	3.2	35	3.0
Mean number of 25 subsistence activities:	⋆	6.5	7.0	8.0	7.3
ANOVA p = 0.000					

⋆No data

Table 13.3 Satisfaction with combination of production activities by country

	Canada	Greenland	Chukotka	Alaska	Total (%)
Very satisfied	a	29%	17%	52%	30
Somewhat satisfied	a	53%	38%	35%	46
Neither satisfied nor dissatisfied	a	14%	a	8%	9
Somewhat dissatisfied	a	2%	33%	3%	11
Very dissatisfied	a	1%	12%	2%	4
	a	100%	100%	100%	100

Source: Poppel and Kruse (2009, 38)

a Data not available

The USFWS is responsible for polar bears, sea otters and walrus. National Oceanic and Atmospheric Administration (NOAA) is responsible for whales (bowhead, beluga, grey), seals (fur, harbour, bearded, ringed, spotted) and sea lions.

The MMPA defines an Alaska Native as: "any Indian, Aleut, or Eskimo who resides in Alaska and who dwells on the coast of the North Pacific Ocean or the Arctic." The statute has no definition of "Indian, Aleut, or Eskimo." Laws passed by Congress are subsequently accompanied by regulations generally developed by the agencies charged with implementing and enforcing the law. The United States Code of Federal Regulations CFR 50 provides these implementing regulations for the MMPA. The definition of "Alaskan Native" in the CFR Title 50 by USFWS and NOAA was modelled on the ANCSA definition in its major provisions. Nevertheless, there are several differences between the ANCSA and MMPA regulatory definitions and especially their implementation that are discussed subsequently.

The US Fish and Wildlife Service current regulation (50 CFR § 18.3) reads as follows:

> Alaskan Native means a person defined in the Alaska Native Claims Settlement Act (43 U.S.C. section 1603(b) (85 Stat. 588)) as a citizen of the United States who is of one-fourth degree or more Alaska Indian (including Tsimshian Indians enrolled or not enrolled in the Metlaktla [sic] Indian Community), Eskimo, or Aleut blood, or combination thereof. The term includes any Native, as so defined, either or both of whose adoptive parents are not Natives. It also includes, in the absence of proof of a minimum blood quantum, any citizen of the United States who is regarded as an Alaska Native by the Native village or town of which he claims to be a member and whose father or mother is (or, if deceased, was) regarded as Native by any Native village or Native town. Any citizen enrolled by the Secretary pursuant to section 5 of the Alaska Native Claims Settlement Act shall be conclusively presumed to be an Alaskan Native for purposes of this part.

The National Oceanic and Atmospheric Administration current regulation (50 CFR § 216.3) reads as follows:

> Alaskan Native means a person defined in the Alaska Native Claims Settlement Act (43 U.S.C. 1602(b)) (85 Stat. 588) as a citizen of the United States who is of one-fourth degree or more Alaska Indian (including Tsimishian [sic] Indians enrolled or not enrolled in the Metlaktla [sic] Indian Community), Eskimo, or Aleut blood or combination thereof. The term includes any Native, as so defined, either or both of whose adoptive parents are not Natives. It also includes, in the absence of proof of

a minimum blood quantum, any citizen of the United States who is regarded as an Alaska Native by the Native village or group, of which he claims to be a member and whose father or mother is (or, if deceased, was) regarded as Native by any Native village or Native group. Any such citizen enrolled by the Secretary of the Interior pursuant to section 5 of the Alaska Native Claims Settlement Act shall be conclusively presumed to be an Alaskan Native for purposes of this part.

The reason for the agencies' separate regulatory definitions is that they have jurisdiction over different marine mammal species, as noted previously, and are required to publish regulations separately. The only difference in language between the two agency regulations is the use of the term "town" twice in the USFWS definition, whereas the word "group" rather than "town" is used twice in the NOAA definition. The term "group" is used in ANCSA, but not "town." Officials for the two agencies reported that in actual practice, the differences have had no impact to date on enforcement.

Research on the agencies' interpretation and implementation of their regulatory definitions revealed a major difference from the manner in which the ANCSA definition was interpreted and implemented by the Bureau of Indian Affairs (BIA) during the period of Alaska Native enrolment in the 1970s (Langdon 2016). Both agencies use only the one-quarter degree of "blood" as the eligibility criteria. Neither has implemented procedures to implement the second portion of the regulatory definition that allows persons of less than one-fourth ancestry to be enrolled based on statements from their Alaska Native "group" or "town" that they are considered Alaska Native. During ANCSA enrolment, these provisions were implemented, and Alaska Native persons from across the state utilized them in order to be enrolled in ANCSA (Langdon 2016).

Some important features of the current law are attractive to Alaska Natives. Alaska Natives are not required to have licenses or permits to hunt marine mammals. With the exception of sea otters, walrus and bowhead whales, Alaska Natives are not required to report their harvests. There are no limits on the harvests of most marine mammals. Unlike state law, marine mammal parts can be sold for cash to other members of the harvester's community. In order to limit Alaska Native harvests of a species, a detailed assessment of the biological status must be conducted that ends in a declaration of "depleted" or "threatened" (MMPA 1972).

There are also provisions that are of substantial concern to Alaska Natives. Any Alaska Native can hunt any marine mammal anywhere in the state if they are a resident of that area. Thus, a Kodiak Alutiiq can take polar bear on the North Slope and a Yup'ik can take sea otter in southeast Alaska if they reside in that area. The requirements for manufacture of "traditional" handicrafts for sale and inability to sell processed raw skins are also considered by Alaska Natives impositions that limit their ability to fully utilize their harvests. However, the most problematic aspect of the MMPA and its regulations is the definition of Alaska Native and the denial of the right of Alaska Natives to make their own determinations about that status and eligibility to hunt marine mammals.

Implications of the Marine Mammal Protection Act regulatory definition of Alaska Native: demographic data

At the time of ANCSA enrolment, Alaska Natives differed substantially in their degree of Alaska Native ancestry. In the Gulf of Alaska region (Aleut, Alutiiq, Dena'ina, Tlingit, Haida and Tsimshian), contact with outside populations occurred in the late 18th century, while in western and northern Alaska, contact with Yup'ik, Sivuqaqmiut and Inupiat populations did not occur until well into the 19th and in some portions of the coastal Yukon-Kuskokwim delta region until well into the 20th. Beginning in the early 1900s, non-native populations sustained contact and established communities associated with various forms of development in the southern areas of the state, and a considerable amount of intermarriage occurred, resulting in declining proportions of Alaska Native ancestry among certain populations. By contrast, the coastal region of Alaska west and north of Bristol Bay has experienced much less development and resulting in-migration of non-Natives. The number of resident non-Natives in the western and northern parts of Alaska is low, and consequently, rates of intermarriage and reproduction are much lower than in the southern regions bordering the Gulf of Alaska. It has been this way for three or four generations. Since ANCSA, 2+ generations have come into existence with these conditions, and an increasing proportion of the descendants of the original shareholding cohort fall below the 25% level each year.

Information on the issue of Alaska Native ancestry characteristics of shareholders of Alaska Native corporations are closely held by the corporations and by the BIA. ANCSA corporations have received original shareholder enrolment data from the BIA and makes its own ancestry determinations of those born after December 1971 if needed for eligibility to become members. The Bureau of Indian Affairs makes determinations of degree of ancestry known as "CDIB" (Certification of Degree of Indian Blood) of applicants by evaluating the birth certificates and other information provided by applicants. Alaska Natives are required to present CDIB cards to be enrolled in many tribes and for obtaining services at Indian Health Service medical institutions in Alaska.

Sealaska Corporation Data. Sealaska, the regional Native corporation for southeast Alaska, provided demographic information on its original shareholders as well as two research reports subsequently prepared for the corporation. In consideration of the future of Sealaska Corporation and the issuance of stock to shareholders born after the enactment of ANCSA, the shareholder committee of Sealaska commissioned two studies on the demographic future of the corporation. The initial study was completed in 2002 and the second in 2005, updating the data and projections of the earlier study. Table 13.4 presents hard data on the blood quantum of the original group of Sealaska shareholders and their descendants (children and grandchildren) born between 1974 and 2001.

Two points of significance from this data need to be highlighted. First, the proportion of the descendant population of less than 25% blood quantum, and therefore ineligible under the MMPA criteria for Alaska Native, had reached 30% by 2001. Second, despite the increase in the ineligible portion of the population, the cohort of the population over 25% had increased the total overall population of eligible MMPA hunters by 13,677, which exceeded the initial shareholder cohort of 13,722 (Table 13.4). Assuming that roughly 10,000 original shareholders

Table 13.4 Blood quantum of original Sealaska population in 1973 and descendants of shareholders and children born 1974–2001

Original shareholders and descendants	Total	100%	75–99%	50–74%	25–49%	<25%
Births to Shareholders and Descendants	19,484	1,126	1,334	4,335	6,882	5,807
Percentage of total		5.80%	6.80%	22.20%	35.30%	29.80%
Original shareholders	13,722	3,843	1,619	4,169	3,973	na
Percentage of total		28.00%	11.80%	30.40%	29%	na

Source: Passel (2002, Table 4)

were still living, this would mean that the total Sealaska original and descendant population eligible under the MMPA had more than doubled.

This information tells us nothing about the actual number of Sealaska shareholders and descendants who participated in MMPA authorized activities with marine mammals.

Table 13.5 presents data taken from the 2005 study that makes projections about blood quantum proportion of cohorts through 2008. The numbers are based on assumptions from the previous hard data on actual blood quantum proportions of descendants through 2001. Data from this study project show that by 2004, there were 21,468 descendants, of which slightly less than half, 9,910 (46%), would be less than a quarter blood quantum (Edmonston 2005, 7). The study further projects that by 2008 there would be 32,424 descendants of whom 19,157, or 59%, will be less than one-fourth blood quantum. Assuming 10,181 original shareholders still living, over 19,000 of new Sealaska shareholders, or 59%, would be ineligible to utilize marine mammals under the current regulatory definition. These figures are projections made nearly a decade ago. When they are compared to CDIB enrolment data from 2011–2016 presented subsequently, the projections appear to overstate the expected increase in number of Sealaska descendants with less than a quarter blood quantum.

Table 13.6 presents data provided by Sealaska Corporation on blood quantum levels of applicants for scholarships in 2014–15 and 2015–2016. While ages of applicants are not included, the age of the college cohort probably falls between 18 and 30 years of age, with possibly a few slightly older. Roughly 40% of the applicants do not reside in Alaska. In 2014–15, 40.4% of 539 applicants were less than 25% blood quantum, and in 2015–16, 39.4% of 426 applicants were less than 25% blood quantum. In both years, those with less than 25% blood quantum made up the most numerous cohorts of applicants by quartile. Thus, the most numerous quartile of Sealaska scholarship applicants in the last two years would be ineligible either to hunt marine mammals or to manufacture handicrafts from marine mammal materials because of being less than a quarter blood quantum.

By contrast, with the projections data presented in Table 13.4, the Sealaska scholarship applicant figures for 2014–2015 and 2015–2016 at about 40% below a quarter blood quantum are comparable to the CDIB enrolment data for Sealaska region presented in Table 13.10 for 2011–2016, which is 42.8%.

Table 13.5 Original Sealaska shareholders and lineal descendants, by blood quantum, at end of year for 1973, 2004 and 2008

Shareholders and descendants	*Original shareholders*	*Total shareholders*	*1/4+*	*1/8 to 1/4*	*Less than 1/8*
1973 original share holders	13,722	13,722	13,604	dkn	dkn
Percentage of total			99.1%	N/A	N/A
2004 shareholder and descendant projections	10,792	21,468	11,558	6,664	3,246
Percentage of total			53.80%	31.00%	15.10%
2008 shareholder and descendant projections	10,181	32,424	13,267	13,111	6,046
Percentage of total			40.90%	40.40%	18.60%

Source: Edmonston (2005)

Table 13.6 Sealaska data on recent scholarship applicants

Blood quantum	Students (out of 539)	Percentage of students
2014–2015		
Under 25%	218 students	40.40
25–49%	214 students	40.00
50–74%	80 students	14.80
75–99%	17 students	3.15
100%	10 students	1.85
2015–2016		
Under 25%	168 students	39.40
25–49%	167 students	39.20
50–74%	65 students	15.25
75–99%	19 students	4.46
100%	7 students	1.64

Source: Sealaska Heritage Institute

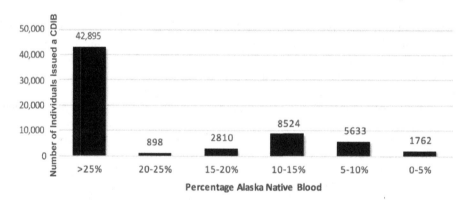

Figure 13.1 Data on all Alaska Native CDIB enrolees and those with less than one-fourth blood quantum by regional affiliation from 7/01/2006–7/08/2016

BIA CDIB Data. Data provided by the BIA enrolment office concerning CDIB enrolments between 7/01/2006 and 7/08/2016 are discussed in this section. Data were provided on the number of Alaska Natives enrolled of less than one-fourth blood quantum by percentage and fraction (for example, 1/8, 1/16 etc.). Data were provided on Alaska Native CDIB enrolees by regional affiliation over the same period, showing total enrolees and number of enrolees with less than one-fourth blood quantum. For data on those of less than one-fourth blood quantum by percentage, only cumulative data are presented. For the data on regional affiliation, cumulative data for 2006–2016 are presented, as well as data for the periods 2006–2011 and 2011–2016 separately.

CDIB enrolees of less than one-fourth blood quantum by percent of blood quantum. Figure 13.1 displays the number of Alaska Natives enrolled between 7/1/06 and 7/08/16 of less than 25%

by blood quantum percentage. Actual data provided by the BIA enrolment office recorded the number of individuals of each blood quantum by percentage and fraction. The data have been grouped into the five ranges under 25% that are displayed. Nearly 20,000 Alaska Natives of less than 25% blood quantum were enrolled during the period. Eight persons were enrolled with 1/128 blood quantum, the lowest of any enrolee recorded.

CDIB enrolees by regional affiliation

During the designated period, 62,522 persons were issued new CDIBs, of which 31.3% were less than one-fourth Alaska Native blood quantum. A portion of the enrolees, 11,642 (18.6%), could not be assigned to a regional affiliation, due primarily to the fact that they were descendants with parentage from more than one regional group. As descendants, 36.2% of the unassigned cohort were less than one-fourth blood quantum, a rate 20% higher than the regionally affiliated cohorts.

The two Alaska Native regions with the highest number of enrolees under one-fourth blood quantum were Sealaska at 11,529 (18.4%) and Calista at 9,234 (14.8). No other region exceeded 5,500 enrolees.

The two Alaska Native regions with the lowest number of enrolees under 25% were Arctic Slope at 228 (14.7%) and Ahtna at 230 (26.5%).

BIA CDIB Enrolees for Coastal Regions. The next three tables display data, but for only the coastal regions where Alaska Natives are eligible to harvest marine mammals. Table 13.8 reports data for the entire period from 2006–2016. For all coastal Natives, the rate of newly enrolled CDIBs under one-fourth-blood quantum is 27.7% over this period. This rate is slightly lower than that for all Alaska Natives, 28.2% over the same period. The difference appears because the unassigned are not included in the coastal cohort. However, it should be noted that some of the unassigned no doubt belong to the coastal Natives cohort.

The data indicate that the coastal Alaska Native regions fall roughly into three groups based on CDIB enrolment blood quantum percentages. Four groups in western Alaska have rates of new CDIBs under one-fourth blood quantum, 20% or less. These are, from lowest to highest, Calista, NANA, Arctic Slope and Bering Straits. The second group of three regional groups displays intermediate rates of CDIB enrolment below one-fourth-blood quantum, between

Table 13.7 CDIBs issued to coastal Alaska Natives less than 1/4 native blood quantum, 7/01/2006–7/08/2016

Alaska regional affiliation	Number issued (less than one-fourth)	All entries for region in same time frame	% of CDIB enrolees with less than one-fourth
Calista	689	9234	7.50
NANA	344	2635	13.10
Arctic Slope	228	1556	14.60
Bering Straits	741	4017	18.40
Sealaska	3684	11529	32.00
Bristol Bay	1116	3415	32.70
Aleut	690	1857	37.10
Chugach	893	1578	56.60
Koniag	1485	2517	59.00
Cook Inlet	3036	5117	59.30
Total records	**12906**	**43455**	**29.70%**

Table 13.8 CDIBs issued to coastal Alaska Natives less than 1/4 Alaska Native blood quantum, 7/2006–7/2011

Alaska regional affiliation	Number issued (less than one-fourth)	All entries for region in same time frame	% of CDIB enrolees with less than one-fourth
Calista	446	6,376	7.00
NANA	197	1,535	12.80
Arctic Slope	153	993	15.40
Bering Straits	431	2,531	17.00
Sealaska	1,720	6,943	24.80
Bristol Bay	706	2,392	29.50
Aleut	481	1,389	34.60
Chugach	651	1,185	54.90
Koniag	1,010	1,830	55.20
Cook Inlet	2,223	3,789	58.70
Total records	**8,018**	**28,963**	**27.70%**

32% and 38%. These are, from lowest to highest, Sealaska, Bristol Bay and Aleut. The third group, consisting of three regional groups with the highest percentage of enrolees of under one-fourth-blood quantum, range from 56% to 59%. From lowest to highest, the groups are Chugach, Koniag and Cook Inlet.

What accounts for the differences between the regions? Two reasons appear most likely. As discussed earlier, the first reason is the history of contact in different parts of Alaska. Regions along the Gulf of Alaska came into direct contact with outsiders much earlier and have been in contact much longer. Some outsiders from the earliest periods settled in the region. One result was intermarriage between the groups that continued over the decades since the late 1700s, especially in the central and western Gulf of Alaska regions. Another wave of outsiders came to the southern regions brought by the onset of the commercial canned salmon industry in the 1880s that also resulted in a substantial amount of intermixing. This means that the southern groups had many more individuals of mixed blood quantum than elsewhere in Alaska. By contrast, groups from western Bristol Bay to the Arctic Slope encountered outsiders much later, and few remained in the region, thus reducing rates of admixture. Central Yup'ik communities of the Calista region are extreme in this regard, with many of them not encountering sustained interaction with outsiders until the first decades of the 20th century. The second reason for the differences is the presence of large concentrations of non-Natives in proximity to Native regional groups at present. In this regard, clearly Cook Inlet but also Sealaska, Chugach and Koniag regional groups are in close and more frequent contact with non-Natives than western and northern Alaska Natives.

Tables 13.9 and 13.10 provide data on two sub-periods, from 2006–2011 and from 2011–2016. It is noteworthy than the total number of CDIBs issued for the coastal Alaska Native regional groups dropped by approximately a third over this period. The primary reason for the steep decline is the substantial increase in the number of unassigned individuals receiving CDIBs who are descendants whose ancestry comes from several regional groups. A second reason is a probable decline in birth rates.

Several important points emerge from comparing the data on CDIB enrolments for coastal Natives during the two periods.

Table 13.9 CDIBs issued to less than 1/4 blood quantum coastal Alaska Natives, 7/1/2011–7/8/2016

Alaska regional affiliation	Number issued (less than one-fourth)	All entries for region in same time frame	% of CDIB enrolees with less than one-fourth
Calista	243	2,858	8.50
NANA	147	1,100	13.40
Arctic Slope	75	563	13.30
Bering Straits	310	1,486	20.90
Bristol Bay	410	1,023	40.10
Sealaska	1,964	4,586	42.80
Aleut	209	468	44.70
Chugach	242	393	61.60
Koniag	475	687	69.10
Cook Inlet	813	1,328	61.20
Total records	**4,888**	**14,492**	**33.70%**

Table 13.10 Changes in percentage of 1/4 blood quantum enrolees of coastal Alaska Native regional groups between 2006–2011 and 2011–2016

Coastal Alaska Native regional groups	Percent increase or decrease
Arctic Slope	− 2.10
Nana	+ 0.30
Calista	+ 1.50
Cook Inlet	+ 1.90
Bering Straits	+ 2.50
Chugach	+ 5.00
Aleut	+ 7.50
Koniag	+ 10.10
Bristol Bay	+ 10.60
Sealaska	+ 10.80

The percentage of CDIB enrolees under one-fourth blood quantum increased from 27.7% to 33.7%, a substantial 22% increase. In keeping with the overall percentage increase in under one-fourth blood quantum CDIB enrolees, all but one of the regional groups showed an increase in the percentage of enrolees under one-fourth blood quantum. The exception is the Arctic Slope region, which declined from a rate of 15.4% to 13.3%, a 15.8% drop.

Table 13.11 identifies the rates of CDIB enrolment increase or decrease for the nine coastal regions from 2006–2011 to 2011–2016.

The rates of percentage increase for the nine groups again fall into three groupings, low, intermediate and high. Four of the regional groups show low rates of percentage increase at +2.5% or less. These are:

Nana	0.3%
Calista	1.5%
Cook Inlet	1.9%
Bering Straits	2.5%

Table 13.11 Primary alternative MMPA regulatory criteria defining Alaska Native

Alternative Criterion	Considerations	How to implement
Reduce blood quantum (for example, 1/8) in this category	Alaska Natives who are less than 1/8 would be ineligible; the number of Alaska Natives will increase in future	Regulatory revision
Use lineal descendancy from original tribal member or original ANCSA shareholder	ANCSA amendments allow non-Native descendants; some Alaska Natives are descended from natives who are neither original shareholders nor tribal enrolees	Statutory amendment or regulatory revision (without blood quantum)
Use CDIB (without blood quantum)	CDIB may not indicate tribal membership or whether person is coastal Alaska Native	Statutory amendment or regulatory revision
Use tribal membership (criteria determined by tribes)	Tribal enrolment criteria vary. Some tribes have non-Alaska Native members; some tribes use blood quantum, while others do not; could result in 200 or more determinations; some Alaska Natives are not tribal members	Statutory or regulatory revision
Use tribal membership plus tribal certification of eligibility (criteria determined by tribes)	Tribes would be required to certify Alaska Native status of members eligible for MMPA; not all Alaska Natives are members of tribes	Statutory amendment or regulatory revision
Implement "regarded as . . . Alaska Native" provision (without blood quantum)	Stipulate that Alaska Native community, village, group or organization considers the person Alaska Native	Statutory amendment or regulatory revision; possibly administrative action

Two of the regional groups show intermediate rates of percentage increase between 5% and 7.5%; these are:

Chugach   5.0%
Aleut   7.5%

Three of the regional groups show high rates of percentage increase of 10% and above.

Koniag   10.1%
Bristol Bay   10.6%
Sealaska   10.8%

Alaska Native views on the Marine Mammal Protection Act and the regulatory definition of Alaska Native

The impact of the regulatory definition on the eligibility of Alaska Natives to hunt marine mammals and manufacture handicrafts from their products gradually increased over the decades as new generations appeared. The earliest formal effort to address the issue by pointing out the severe impacts on communities and families was initiated by the Alaska Native Sea Lion and Sea Otter Commission (TASSC). Under the MMPA, Alaska Natives can create co-management institutions to advise the agencies and the Marine Mammal Commission, a national body created by the MMPA. In 2009, TASSC, expressing the concerns of the Koniag region particularly, submitted a resolution to the Alaska Federation of Natives (AFN) Convention calling for the regulatory definition to be changed to tribal members and lineal descendants of original ANCSA shareholders. A similar but broader resolution was again brought to the AFN in 2010 and delegated this time to the Subsistence Committee, who took no action on the resolution after considerable discussion. Daniel (2011), in a memo to the AFN, discussed the issue and proposed options to change the definition of Alaska Native in the regulations to allow those of less than 1/4 blood quantum to acquire and utilize marine mammals. In 2011, the resolution was again submitted, and the Subsistence Committee, after consultation, requested that the issue be turned over to the Indigenous People's Commission on Marine Mammals (IPCOMM) for study. In 2014, the USFWS and NMFS sent a joint letter to IPCoMM and AFN soliciting Alaska Native views on the definition, noting that changing the criteria would set in motion other administrative adjustments and seeking input on how to manage the new definition. In response to the agencies' letter, in 2015, IPCOMM and the Subsistence Committee commissioned research to provide information on the impact of the regulatory definition and possible alternative definitions of Alaska Native and Alaska Native views on the issue. The results of that research are reported in Langdon (2016). In 2015, the National Congress of American Indians (NCAI) also passed a resolution supporting changing the definition of Alaska Native status to one based on tribal membership. In 2016, the Central Council of Tlingit and Haida Tribes of Alaska (CCTHITA) passed a resolution calling for the definition to be changed to allow members of federally recognized tribes and lineal descendants of original ANCSA shareholders to be eligible to hunt and utilized marine mammals. However, IPCoMM and the AFN Subsistence Committee have never taken a position to support changing the MMPA regulatory definitions of Alaska Native.

Alaska Native views on the determination of Alaska Native status, on whether the MMPA definition should be changed and on what it should be changed to vary substantially. The issue parallels to a significant degree the issue faced by Alaska Natives concerning whether and in what way to allow descendants of Alaska Native original shareholders to be issued stock in the corporation. Regional corporations have varied enormously in the policies they have enacted for the admission of new shareholders born after 1971.

The positions of Alaska Natives on the MMPA regulatory definition can, for purposes of discussion, be seen as supporters of criteria change and supporters of no criteria change. This section summarizes information obtained from documents, interviews and comments at meetings and presentations on the issue to tribal, community and organizational groups.

Supporters of criteria change

Supporters of changes to the MMPA regulatory definition have made their case on multiple occasions through requests to agencies for regulatory changes and through resolutions of their organizations. In those materials, supporters of criteria change assert several arguments. These positions can be categorized as legal, moral and cultural.

Legal

There are three arguments under this category that have been advanced.

First, the regulatory definition in the MMPA is not consistent with language of the law and is more restrictive than the law. They contend that there is nothing in the legislative history of the MMPA that indicates Congress intended to limit the exemption to Alaska Natives who are one-fourth or more Alaska Native ancestry. Another point of inconsistency noted between language in the statute and the regulatory definition is that the statute, in using the term "Indian," does not explicitly identify Alaskan Indians and therefore can be read such that the exemption applies to any Indian or Native American who resides on the coast of Alaska. The regulations, by contrast, stipulate Alaskan Indian, as stated in the ANCSA definition.

Second, the regulatory definition in the MMPA does not promote the purpose of the exemption to make possible the preservation and handing down of cultural traditions and cultural "way of life," as called for in the legislative history of the act. Congress conducted hearings in Bethel and Nome in 1972 on the basis of which the Congressional report notes:

> The findings were that most villages of northern Alaska depend upon marine mammals not only for food, clothing and implements, but utilize products from seals, whales and walruses as the basis for their small cash economy. An Eskimo hunter may bring in a seal, which is cut up for food, pelt and seal oil. Any excess beyond family needs can be sold to neighbors or turned in at a village store for credit on gasoline, fuel oil or ammunition.

Supporters similarly cite Senator Stevens' justification from the legislative history that the exemption was necessary in order for cultural practices to be passed down from generation to generation. Additionally, the testimony of Senator Hollings during floor debate on the MMPA is cited, in which he stated:

> Native Alaskans are proud. They do not ask for special treatment from the Federal Government. But, nonetheless, they, too, have the right to be left alone, to follow their traditional way of life. *It is this way of life I seek to protect in this bill.*
>
> [Emphasis added]

> Alaskan arts and crafts are an artistic and social heritage. This skill, handed down from generation to generation, reveals as much of their history as paintings Rembrandt and other famous European artists reveals of the white man's past history. *Removing the privilege of passing this cultural legacy to future generations will sever children as yet unborn from the past. It will create a cultural Diaspora.*
>
> [Emphasis added]

These statements are presented by those supporting criteria change as evidence of Congressional intent to allow for the transmission of the Alaska Native way of life from generation to generation.

Finally, the application of the regulation is not fully implemented, since there is no operationalization of the section of the regulation that stipulates that "in the absence of proof of blood quantum" those "regarded as" are to be considered Alaska Natives. One commentator observed that the agencies, in their own regulatory definition, have the tools to address the issue through the "regarded as" provision but have chosen not to. It was further observed that Alaska Natives should not have to be proactive on the issue, as it is incumbent on agencies to provide for full implementation of their regulations. A non-Alaska Native commentator who worked for years for a co-management institution stated, "It is as though, for all practical purposes, this clause doesn't exist."

Moral

Four types of positions for changing the criteria that can be classified as moral in nature follow. First, certain Alaska Native communities are dependent on marine mammals for a substantial part of their foods. Others depend heavily on the money obtained from the sale of handicrafts made out of marine mammal materials. If communities are to survive, then younger members, some of whom may be less than one-fourth blood quantum, must be able to provide for their grandparents, parents and themselves. It is morally right to allow people of less than one-fourth blood quantum to hunt to provide for their communities. It is morally wrong to deny such persons the right to hunt and provide for their families and community.

Second, proponents of changing the criteria assert that all families, as a matter of human rights, have the right to transmit their cultural heritage to their descendants, regardless of governmental and legal positions. A number of commentators at presentations made this point in various fashions, some vehemently.

Then, another form of this argument is that it is not fair for some descendants of Alaska Natives to be eligible under the MMPA and others not. Commentators note that the genealogical paperwork necessary to document one-fourth blood quantum is not available uniformly to all Alaska Natives due to a number of factors. The paperwork that is available is also problematic, thus causing unfairness. Several Alaska Native elders noted that they have some grandchildren that are eligible and some that are not eligible. This is very disturbing and painful for them. In their view, it is not fair for some of their descendants to be ineligible to hunt marine mammals, an activity so central to cultural identity and the maintenance of cultural traditions.

Finally, the supporters of changing the criteria note several critical deleterious impacts and implications of MMPA enforcement of the regulatory definition of Alaska Native. A whitepaper on the necessity of changing the Alaska Native eligibility criteria prepared by the Alaska Native Sea Otter and Sea Lion Commission in 2010 stated that:

> In our communities, there are children who do not meet the minimum blood quantum, yet their parents, grandparents and other family members continue their traditional and customary practices including marine mammal hunting and sewing. This is harming our communities. Our traditions are in danger of being lost, or our members are made to feel like criminals and fear prosecution for teaching their kids.

Possible prosecution for teaching children and grandchildren cultural practices and traditions handed down for generations could result from the enforcement of the current regulatory definition of an Alaska Native.

Culture

Two kinds of positions using cultural propositions have been offered for changing the criteria.

First, many Alaska Natives believe that their cultural practices are transmitted by parents and relatives to their descendants in the context of communities. It is a right, an obligation and a practice. The values and practices of the culture are transmitted to all young people in the community. Through exposure to the teachings of elders, the examples of leaders and the observation of appropriate behaviours and practices, young people come to a realization of what they are to do and how they are to behave. Another way of putting this realization is that because of such training and exposure, one comes to see who they are and what they should do. It is not possible, and it is deeply damaging, to deny certain children and grandchildren the right to grow up as full members of their communities.

Second, cultural traditions as a way of life are a totality, and it is fundamentally wrong to separate one part of that tradition from others. Young Alaska Natives of less than one-fourth blood quantum in coastal villages are able to hunt migratory birds and collect their eggs; they are able to hunt sheep, goats, moose, deer and elk and they are able to trap mink, marten, land otter and other furbearers, but they are not allowed to hunt marine mammals or utilize marine mammal parts. This is both injurious and nonsensical.

Supporters of no criteria change

Alaska Natives who are generally in favour of not changing the criteria fall into two camps. The first camp considers that given the contexts of pursuing change, such efforts might lead to loss or substantial modification of the Alaska Native exemption. Therefore, it is unwise and even dangerous to formally pursue change. The second camp who generally favour the current situation have advanced arguments in four areas. These can be classified as imbalance, biology, management and culture.

Imbalance arguments

Imbalance arguments are offered as follows:

> The change in the criteria would dramatically increase the number of eligible hunters. This would likely result in additional harvests and those additional harvests would be at the expense of those already taking the marine mammals who depend on them. This would increase the competition for scarce resources that current hunters are presently in need of.

This comment was made concerning sea otter harvests in southeast Alaska. Those Alaska Natives whose current livelihoods are derived from the creation and sale of traditional handicrafts would be severely damaged or perhaps even lose their livelihoods. One commentator dependent for income on making handicrafts from marine mammal materials stated that markets for the objects created would soon be flooded and expected that income would be dramatically reduced. The danger of changing the criteria to allow those of less than one-fourth blood quantum was so great in terms of the disruption of the present circumstances of those dependent on marine mammals for their economic livelihood that one commentator was comfortable with the fact that several grandchildren of the commentator were ineligible to harvest or use marine mammals.

Biology arguments

The basic claim of the biological justification for maintenance of the status quo is that changing the criteria would increase the number of hunters, resulting in increased harvests of certain marine mammals. This could lead to overharvesting. This would then precipitate limitations on the harvests of present hunters who are providing for their communities and households. It could even lead to closure of hunting or implementation of regulations on seasons and bag limits. These closures would threaten community survival, not to mention the survival of cultural traditions.

Management arguments

The management concern of those opposed to changing the criteria is that it will become significantly more difficult for enforcement agents to identify who is an Alaska Native if the one-fourth blood quantum is reduced. Many ineligible non-Natives will attempt to hunt marine mammals, and Alaska Natives will have to carry their identification papers and respond to more frequent requests from enforcement to prove their status. It would be even more difficult for manufacturers of handicrafts from marine mammal products, as there is much less opportunity for oversight of that activity due to the difficulty of linking an object crafted to its producer.

Cultural arguments

A number of presentations made the argument that what was at stake was the continuity of critical relationships between humans and marine mammals that was at the heart of cultural beliefs and traditions. Put another way, hunting marine mammals, "puts in motion a set of social relations with animals, kin, and the environment that, in their sum define what it is to be a real human being" (Voorhees 2015). In this view, characterized as "relational sustainability" (Langdon 2007), what is crucial to the existence of both humans and marine mammals is respectful actions by humans, the result of which will be the reproduction of the marine mammals and their future return to give themselves to those who have conducted their relations appropriately. Persons that are not raised in the appropriate cultural context with appropriate cultural values threaten everyone by their potentially disrespectful behaviour toward the marine mammals. One inappropriate type of behaviour is reducing the relationship to one of obtaining money and not fully utilizing the animal that has given itself to the hunter. On St. Lawrence Island, the Siberian Yup'ik tribes have codified these requirements for tribal members hunting walrus for full utilization and respectful treatment. The Sitka Marine Mammal Commission similarly requires as part of the permitting process for hunters to affirm that they will hunt in a "culturally appropriate" manner that includes the requirement for full utilization. It is the view of some Alaska Native commentators that those involved in the harvest and use of marine mammals must exhibit the appropriate cultural values and behaviours or the animals will withdraw and be unavailable. This is seen as dependent upon being raised in an environment in which traditional Native values are taught and practiced.

Traditional Native values are most likely to be transmitted when there is an Alaska Native who exhibits traditional values in the daily life of their descendants or other young people on a regular basis. Distance from actual parental and grandparental sources of traditional values and behaviours "dilutes" the cultural understandings due to their lack of available demonstration to the children. To this point, it should be noted that a person of 1/8 blood quantum has only one great-grandparent as the source of their ancestry, while a person of 1/16 blood quantum has only one great-great-grandparent. In such cases, the direct transmission of traditional Native cultural values through persons in contact is substantially reduced. For these commentators, mere biological qualification by blood quantum is regarded as unsatisfactory and virtually irrelevant to the cultural requirements for being Native. As one knowledgeable commentator noted, "There are those who are eligible to be Natives but are not considered Native." Underlying this remark is the cultural premise that Native status and recognition are the result of exposure to traditional cultural values and behaviours of hunting and the demonstration of the appropriate values and behaviours in a community context. Moreover, those in turn are most likely to be performed if they have been inculcated from birth or childhood. The sense of identity and its association with marine mammal hunting is especially strong in northwest Alaska, where utilization of large marine mammals is critical, as data presented previously demonstrated. A recent study paraphrased a long-time Kivalina (an Iñupiat community considered at risk due to vulnerability to increased damage due to climate change) resident as follows:

> When . . . asked why her people don't move – somewhere, anywhere to be safe – she is polite but firm. The land and the water make the Iñupiat who they are. If they moved to Kotzebue, they would be visitors. Moving to Anchorage or Fairbanks, she said, "would be like asking us not to be a people anymore."
>
> (Hamilton et al. 2016, 14)

Voorhees (2015, 5), who worked with the Nanuuq (Polar Bear) Commission and hunters between 2010 and 2014, asserts:

> There is a very different way of understanding how hunting relates to being Alaska Native. This is the view more commonly held by the hunters themselves. In this view, one *becomes* Siberian Yupik, Inupiaq, or Yup'ik . . . through hunting. The view that one acquires social identity [and cultural legitimacy] through hunting is nothing new.

The identity created by the complex of social and cultural practices associated with marine mammal hunting is deeply held, and those who demonstrate commitment to its totality are highly respected in the Indigenous communities of western and north-western Alaska.

Opponents of criteria change nevertheless expressed dissatisfaction with several aspects of the current MMPA regulatory regime. The major concern was the blanket exemption for Alaska Natives that allows a qualifying coastal resident Alaska Native to hunt any marine mammal anywhere in Alaska. One commentator pointed out that the relocation of Alaska Natives from other regions where beluga whales were utilized as an important subsistence species to Anchorage was a significant factor in the reduction of the Cook Inlet beluga population, leading to its placement on the endangered species list in 2008. Another commentator remarked that the recent harvest of a polar bear by an Alaska Native who moved to the North Slope in order to be eligible to hunt them was extremely disturbing to Iñupiat hunters of the region.

A second observation made by commentators at several presentations from three different communities was that tribes should have the power to authorize marine mammal hunting by non-Natives in their customary and traditional territory who are married to Alaska Native women from the community and are providing for their families through their hunting.

The position for local determination of eligibility to hunt marine mammals was also advocated by western Alaska Natives in hearings about the potential transfer of jurisdiction over marine mammals to the State of Alaska held in 1985 (Langdon 1989). It should be noted that tribal constitutions allow for non-Native individuals to be adopted into the tribe if they meet certain cultural and residency requirements. The Sitka Tribal Constitution, for example, stipulates that persons who have demonstrated "social and cultural connections" to the tribe can be enrolled, but that is done on a case-by-case basis. However, federal courts have ruled that such adoption does not change the status of the non-Native for purposes of eligibility for federal programs for Indians or Alaska Natives (Langdon 2016).

While there are substantial differences among Alaska Natives on the issue of who should be deemed Alaska Native for purposes of marine mammal use, there is universal and strong support for the MMPA's exemption allowing coastal Alaska Natives alone the right to take marine mammals for subsistence. With this position paramount, Alaska Natives are deeply sceptical of the process of change, even if they were to come to an agreement on what the definition should be changed to. This position is informed by the profoundly unsatisfying outcome of ANILCA Title VIII (1980) that provides a rural priority for subsistence, thus eliminating non-rural Natives' subsistence priority while allowing rural non-Native participation in customary and traditional Alaska Native subsistence activities. Similarly, regulatory language implementing changes to the Migratory Waterfowl Treaty Act (MWTA) does not recognize an Alaska Native right to harvest in "subsistence areas." In 1998, the USFWS stated "The United States understands that the term 'Indigenous inhabitants' means a permanent resident of a village within the subsistence harvest area, regardless of race" (Case and Voluck 2012, 273). New regulations, 50 CFR Part 92, authorizing harvests, which defined "Indigenous inhabitant" as a "permanent resident of a village within a subsistence harvest area, regardless of race," were subsequently passed. What is apparent in both of these moves is the diminishment and dilution of Alaska Native rights and the assimilation of Alaska Native rights with other non-Natives. The underlying concern of Alaska Native commentators pointing to these outcomes is that there is a substantial danger that efforts to change the MMPA regulatory definition of Alaska Native could result in the loss of the blanket Alaska Native exemption that is presently in place or any kind of Alaska Native preference. At present, all coastally resident Alaska Natives can harvest marine mammals, but the regimes of ANILCA and MWTA both eliminate non-rural Natives while allowing rural non-Native participation in customary and traditional Alaska Native subsistence activities. Either of these results, loss of the exemption or modification of the exemption to a regime similar to ANILCA, is seen as extremely problematic and detrimental to Alaska Natives.

Alaska Natives do not trust the Alaska Congressional delegation, the US Congress nor the agencies to make changes Alaska Natives might recommend. In addition to the examples discussed previously, Alaska Natives are also cognizant of hostility to the Alaska Native exemption among powerful non-Native voices in Alaska (Alaska Outdoor Council) and a state government that cannot recognize the Alaska Native exemption and refuses to work with tribes and Alaska Native organizations on marine mammal issues. All of these contexts and considerations make it extremely difficult for Alaska Natives to see a path forward that will result in an outcome that improves upon and does not diminish the current regime associated with the MMPA.

Alternative possible definitions of Alaska Native in federal law and regulation

The legal status of Alaska Natives in federal law has a complex history as individuals and groups. Do Indigenous Alaskans have the same status in regard to the federal government as "Indians," and are Alaska Natives organized into "tribes"? The short answer to both of those questions is basically yes; however, due to the cessation of treaty-making by the federal government in 1871, Alaska Natives have no treaties that define their relationships with the federal government and identify a territorial base for tribal jurisdiction. The Annette Island Reservation of the Tsimshian Indians in southeast Alaska created by Congressional action in 1891 is the only – Native American community – in Alaska with a territorial base over which to exercise tribal jurisdiction.

A number of alternative criteria for Alaska Native status determination have been proposed since the emergence of the issue in 2009. Alaska Natives have expressed the desire to make their own determinations of that status. ANCSA 1991 amendments provided corporations with wide latitude in deciding about new members and the type of corporate rights they would receive. ANCSA corporations vary enormously in their implementation of these provisions, with all of them using the concept of lineal descendant from original shareholders for eligibility. Virtually all continue to use the one-fourth-blood quantum, but Calista Corporation admits all lineal descendants to equal status with original shareholders regardless of blood quantum. It should be noted that the ANCSA structure is regional in nature, allowing recognition of the diversity among regional views rather than imposing a universal one-size-fits-all definition. Such an approach would allow for regional differences observed in blood quantum figures where southern regions have much higher rates of less than 25% ancestry compared to northern and western regions.

Table 13.14 presents a number of options that have been considered, their implications and the action needed to implement each. The types of action include 1) statutory amendment, 2) regulatory change and 3) administrative action. It should be noted that currently, the Bureau of Indian Affairs and Alaska Native Health Corporations require clients to have a CDIB demonstrating some degree of Alaska Native ancestry but have no blood quantum level for eligibility determination. Most options would likely require much greater administrative bureaucracy and increase the regulatory burden on Alaska Natives to demonstrate their eligibility.

Conclusion

The MMPA (1972) authorizes an exemption for Alaska Natives residing in coastal areas to harvest and utilize marine mammals for subsistence purposes subject to conservation status and wanton waste provisions. Regulations implementing the definition of Alaska Native not provided in the legislation stipulate that to be eligible, a person must have one-fourth Alaska Native ancestry or be recognized as an Alaska Native by the village or tribe of which he/she is a member. In the nearly 50 years since the passage of the act, significant demographic changes have occurred due to Alaska Natives marrying and producing descendants with persons of non-Native ancestry, resulting in a substantial portion of the descendant population having less than one-fourth Alaska Native ancestry and thereby under the only provision of the regulations utilized by the federal agencies being deemed ineligible to harvest marine mammals or manufacture handicrafts. These conditions have been especially apparent in the southern portion of Alaska, where CDIB data obtained from the Bureau of Indian Affairs show that over 40% of the population enrolled between 2006 and 2016 in Cook Inlet, Prince William Sound and Kodiak were less than 25% Alaska Native ancestry (Langdon 2016). These conditions are having enormous social and cultural impacts, as grandmothers are finding that they cannot pass on their crafts to their grandchildren, and grandfathers are finding that some of their grandchildren are eligible to go out and learn to hunt, while others are not (Langdon 2016). The fracturing of family and community resulting from the impacts of the MMPA regulations poses a substantial threat to the future of Alaska Native culture, particularly in the southern part of the state.

Alaska Natives are deeply concerned about the impact of the MMPA regulatory criteria used to define who is an Alaska Native. The issue is sensitive for many reasons, and while many would like to see definitional change, there are key provisions of the MMPA that virtually all Alaska Natives do not wish to see changed. One of these is the Alaska Native–only exemption that allows hunting to continue. A second is that there is no requirement to obtain a license to hunt. The social and political context of the larger Alaskan society and the legislative and administrative processes of change pose grave dangers to both propositions. Alaska Natives have been deliberate in their approach to the issue in order to ensure that rights currently held through the MMPA are not diluted or diminished. Nevertheless, they have expressed a strong desire for the determination of eligibility to be made by Alaska Natives themselves through the tribes or other regional organizations and that the procedures allow for a variety of acceptable outcomes. At this time, there is no process or pathway apparent that would lead to such a resolution.

Acknowledgements

The research through which the data in this chapter were generated was funded by the Alaska Federation of Natives through its Subsistence Committee. I would like to thank Dr. Rosita Worl, Sealaska Heritage Institute, and Mike Miller, Indigenous People's Commission on Marine Mammals, for their support and guidance. Dr. Chuck Smythe, Sealaska Heritage Institute, provided important commentary, for which I thank him. Dr. Meredith Marchioni contributed excellent research assistance, for which I am grateful. Finally, I would like to thank all the Alaska Native leaders and individuals who provided information and perspectives on the topic. If there are any errors or misinterpretations in the report, I am solely responsible for them.

Note

1 Data not available.

References

Bockstoce, J. 1986. *Whales, Ice, & Men: The History of Whaling in the Western Arctic.* Seattle: University of Washington Press.

Case, D., and D. Voluck. 2012. *Alaska Natives and American Laws.* 3rd ed. Chicago: University of Chicago Press.

Daniels, C. 2011. *Marine Mammal Protection Act Regulatory Definition of Alaska Native.* Memorandum submitted to the Alaska Federal of Natives. Manuscript in author's possession.

Edmonston, B. 2005. *Original Shareholders of the Sealaska Corporation and Their Descendants: Estimates and Projections.* Juneau: Sealaska Corporation.

Fall, J. 2016. "Regional Patterns of Fish and Wildlife Harvests in Contemporary Alaska." *Arctic* 69 (1): 47–64.

Gibson, J. 1992. *Otter Skins, Boston Ships and China Goods: The Maritime Fur Trade of the Northwest Coast, 1785–1841.* Montreal: McGill-Queen's University Press.

Hamilton, L., K. Saito, P. Loring, R. Lammers, and H. Huntington. 2016. "Climigration? Population and Climate Change in Arctic Alaska." *Population and Environment*: 1–19.

Krupnik, I., W. Walunga, and V. Metcalf. 2002. "Akuzilleput Igaqullghet [*Our Words Put to Paper*]." *Contributions to Circumpolar Anthropology* 3. Arctic Studies Centre, National Museum of Natural History. Washington, DC: Smithsonian Institution.

Kruse, J., B. Poppel, L. Abryutina, G. Duhaime, S. Martin, M. Poppel, M. Kruse, E. Ward, P. Cochran, and V. Hanna. 2008. "Survey of Living Conditions in the Arctic, SLiCA." In *Barometers of Quality of Life around the Globe*, edited by V. Møller, D. Huschka, and A. C. Michalos. *Barometers of Quality of Life around the Globe*. Springer Social Indicators Research Series. Dordrecht: Springer.

Langdon, S. 1989. "Prospects for Co-Management under the Marine Mammal Protection Act in Alaska." In *Co-Management of Fisheries*, edited by L. Pinkerton, 154–169. Vancouver: University of British Columbia Press.

Langdon, S. 2007. "Sustaining a Relationship: Inquiry into a Logic of Engagement with Salmon among the Southern Tlingits." In *Perspectives on the Ecological Indian: Native Americans and the Environment*, edited by M. Harkin and D. R. Lewis, 233–273. Lincoln, NE: University of Nebraska Press.

Langdon, S. 2016. Determination of Alaska Native Status under the Marine Mammal Protection Act (MMPA). Juneau: Sealaska Heritage Institute.

Marine Mammal Protection Act (MMPA). 1972. 16 U.S.C.: Conservation, 92nd United States Congress 86 Stat. 1027.

Passel, J. 2002. *Original Shareholders of the Sealaska Corporation and Their Descendants: Estimates and Analysis.* Juneau: Sealaska Corporation.

Poppel, B. 2010. "Are Subsistence Activities in the Arctic a Part of the Market Economy, or Is the Market Economy a Part of a Subsistence Based Mixed Economy?" In *Cultural and Social Research in Greenland – Selected Essays 1992-2010.* Nuuk: Ilisimatusarfik/Forlaget/Atuagkat.

Poppel, B., and J. Kruse. 2009. "The Importance of a Mixed Cash- and Harvest Herding Based Economy to Living in the Arctic – An Analysis on the Survey of Living Conditions in the Arctic (SLiCA)." In *Quality of Life and the Millennium Challenge*, edited by V. Møller and D. Huschka, 27–42. Social Indicators Research Series. Vol. 35. Dordrecht: Springer.

Voorhees, H. 2015. "Subsistence Biologies: 'Blood Quantum'-Based Alaska Native Hunting Rights in a Climate of Scarcity." Paper delivered in the session Governing Indigeneity: Bureaucracy, Vulnerability, and the State in the Americans at the annual meeting of the American Anthropological Association held in Denver.

14

REVIEW AND MAPPING OF INDIGENOUS KNOWLEDGE CONCEPTS IN THE ARCTIC

Parnuna Egede Dahl and Pelle Tejsner

Abstract

The importance of using knowledge of Indigenous peoples alongside science in research, management and resource development is increasingly acknowledged. Despite political intentions of including the knowledge of Indigenous peoples, the extent and quality of utilizing their knowledge is uneven in the Arctic. The lack of agreed-upon definitions of various concepts used for the knowledge of Indigenous peoples, and their interchangeable and inconsistent use, creates confusion about their meaning and implications. In this chapter, we review the knowledge concepts and their interrelatedness, developing concept maps to visualize their similarities and differences with a view to clarifying the confusion and aiding a more consistent engagement and utilization of this knowledge. We argue that Indigenous knowledge is the only concept that emphasizes the identity aspect and thus implies the distinct status and collective rights of Indigenous peoples, distinguishing it from other knowledge concepts. Our review suggests that the use of concepts varies significantly in the Arctic, shaped by the colonial and political-economic processes in Greenland, the Canadian Arctic and Alaska. We also observe a transition in use of concepts from traditional knowledge to Indigenous knowledge.

Introduction

Important prerequisites for Indigenous peoples to have survived and thrived in the Arctic for millennia are their abilities to observe, adapt and pass on valuable knowledge (Bravo 2010, 445). Combined with traditions, values and beliefs, these practices have formed holistic knowledge systems in Arctic communities, reflecting the close human–environment relationship (Ibid., 446). These knowledge systems are based on worldviews that reflect contextual frameworks different from scientific knowledge, having their own philosophical, spiritual, cultural and social dimensions (Stevenson 1996), with their own methodologies and validation processes. Knowledge of Indigenous peoples can be considered complementary to scientific knowledge.

Appropriately combined, new knowledge may be generated to inform and improve decision-making and policy development (PPs 2018).

There is an increasing acknowledgement of the importance of respecting and learning from the knowledge of Indigenous peoples in research, management and resource development (Nichols et al. 2004; Nuttall 1998; Riedlinger and Berkes 2001; Wenzel 1999). For example, the International Association of Impact Assessment developed *Principles of Environmental Impact Assessment Best Practices* referring to traditional knowledge and recognizing Indigenous knowledge holders as relevant sources of expert knowledge (IAIA 1999). Another example is the Arctic Council (AC), consisting of the eight Arctic States and six Permanent Participants representing Arctic Indigenous Peoples' Organizations (IPOs). From the beginning, the AC has emphasized the importance of using traditional knowledge and how this is essential to a sustainable future in the Arctic (AC 1996–2017). The legally binding *Agreement on Enhancing International Arctic Scientific Cooperation* was signed at the AC Ministerial Meeting in 2018, committing the Arctic states to promoting utilization of traditional and local knowledge in research activities, to encouraging increased communication between traditional and local knowledge holders and researchers and for holders of traditional and local knowledge to participate in scientific activities (AC 2018). Although traditional knowledge is characterized in a somewhat inconsistent and sometimes restrictive manner in the AC declarations, the use of traditional knowledge continues to be emphasized.

Despite political intentions of including knowledge of Indigenous peoples in research, management and resource development, the extent and quality of engagement of Indigenous peoples and utilization of their knowledge is uneven in the Arctic. The definitions are 'neither clear nor consistent', and there is a lack of guidance on how to implement the knowledge in practice (Usher 2000, 184). Permanent Participants in the AC still call for scientists to include Indigenous knowledge through co-production of knowledge and meaningful engagement of Indigenous peoples in order to fill knowledge gaps (PPs 2018). However, several concepts are used for the knowledge of Indigenous peoples, such as traditional knowledge (e.g. Huntington 2005), traditional ecological knowledge (e.g. Wenzel 1999; Usher 2000), Indigenous knowledge (e.g. Stevenson 1996; ICC 2016) and local knowledge (e.g. Sejersen 1998, 2004). One issue complicating the utilization of knowledge is that there are no single agreed-upon definitions of what each knowledge concept means (Berkes 1999; Huntington 2005; WIPO). In addition, the concepts are often used interchangeably, revealing different understandings of the meaning and implications. The lack of agreed-upon definitions and inconsistent use of concepts creates confusion, which does not aid their utilization and may have significant implications for the Indigenous peoples concerned.

In this chapter, we explore the various knowledge concepts and their interrelations to better understand how they are similar to and different from one another. We first present a literature review of the knowledge concepts, followed by concept maps visualizing their similarities and differences. Throughout the chapter, we use Indigenous knowledge (IK) as the main knowledge concept, as this is the concept mainly being promoted by IPOs in the Arctic (PPs 2018).

Methodology

To explore the different knowledge concepts and their relationships, we use constructivist grounded theory as research method, adding the strategy of informed grounded theory. Grounded theory (GT) is an inductive research methodology aimed at developing grounded theories and conceptual descriptions from data themselves through systematic and iterative strategies for data collection and analysis (Glaser and Strauss 1967; Bryant and Charmaz

2007; Urquhart 2013; Charmaz 2014). Constructivist GT is based on social construction-ism, acknowledging the subjectivity and involvement of the researcher in construction and interpretation of data (Bryant and Charmaz 2007; Charmaz 2014). Informed GT adds litera-ture review strategies that open-mindedly consider pre-existing theories and research findings (Thornberg 2012).

The research was organized into two phases. In the first phase, a literature review of the various knowledge concepts from both academic and grey literature was used to gain an under-standing of the concepts. Academic literature was selected by keyword search in electronic indexes, for example, the Aalborg University search engine Primo, which has access to numer-ous electronic databases such as Scopus, Taylor & Francis Online, SpringerLink and Web of Science. Direct search in the before-mentioned databases was also employed. Search keywords were the various knowledge concepts, for example, traditional knowledge, Indigenous knowl-edge and Inuit knowledge. These were combined with keywords on environmental issues like environmental assessment, research and management of resources. Search hits were filtered according to geographic relevance with keywords such as Arctic and specific countries and ter-ritories in the circumpolar region. Emphasis was put on search hits that also featured keywords related to cultural identity, for example, Indigenous, aboriginal and Inuit. Search results were quality-checked with established scientists within the field. Grey literature was selected from national, regional and Indigenous governments; IPOs and the AC. Selection of grey literature was focused on agreements, declarations, guidelines and position papers referring to knowledge concepts.

To supplement the literature review, a total of 20 qualitative and semi-directed interviews were conducted in 2017–2018 face to face, over the telephone, in video chat, or via email correspondence. Interviewees were chosen among government agencies, IPOs and academics working with environmental issues and Indigenous peoples in Greenland, the Canadian Arctic and Alaska with focus on their perception of knowledge concepts.

In the second phase, the mapping of concepts and visualization of their relationships were conducted using the information on knowledge concepts gathered from the first-phase lit-erature review. The information was used as data for coding and developing categories using constructivist-informed GT. Without conducting semantic analysis as such, the meaning, inter-pretation and implications of words within certain contexts influenced the coding. The cat-egories were then used to develop sets of parameters to visualize the similarities and differences between knowledge concepts.

Results

Phase one: review of concepts

In the following literature review, we first retrace the development of knowledge concepts and their use in different contexts. We then provide an overview of various knowledge concepts, giving examples of definitions and critique while providing our own summarized description of characteristics of each concept. For this review, we have chosen the knowledge concepts of traditional knowledge, traditional ecological knowledge, Indigenous knowledge, Inuit Qauji-majatuqangit/Inuit knowledge and local knowledge. Acknowledging that there are other com-binations (e.g. Indigenous ecological knowledge, Inuit traditional knowledge) and that further concepts exist (e.g. place-based knowledge, user knowledge), the ones chosen are either the most commonly supported by a larger literature or the most frequently used by Arctic govern-ments, IPOs and academics.

Development and use of concepts

Although academic literature on the knowledge of Indigenous peoples dates to at least the 1960s (Lévi-Strauss 1966; Rappaport 1968; Damas 1985), the concepts did not enter mainstream academia until the early 1990s. At that time, a growing interest in the knowledge concepts was primarily expressed through successes such as the documentary-style TV series *Disappearing World* (Granada Television 1970–1993) featuring tribal groups or the *Time* magazine title 'Lost Tribes, Lost Knowledge' describing tribal or Indigenous knowledge (Linden 1991).

The broad-ranging concepts of the knowledge of Indigenous peoples are probably some of the most widely documented, and contested, themes in studies of socioenvironmental relations in the Arctic (Hansen et al. 2016), leading to numerous debates on how to define them, how they may differ from scientific knowledge (Stephens 2000, 10) and which is the most appropriate method of utilizing Indigenous knowledge alongside scientific research (ICC 2010). Whatever concept is used, they are often constructed concepts used by mainly academics and governments to label knowledge for different purposes (Usher 2000). Such labelling is useful, as general reference to the knowledge system and characteristics is often implied. In most cases, the concepts partly reflect a construction created to identify another culture's knowledge, while also reflecting the knowledge that non-Indigenous researchers think Indigenous peoples possess rather than the knowledge itself (McGregor 2009). One unfortunate consequence is that the knowledge 'can be codified in writing and thus taken away, removed, or separated from the cultural context in which it operates' (Usher 2000, 191). Accordingly, such concerns relate to a lack of acknowledgment of other forms of knowledge than scientific knowledge, often at great cost to Indigenous livelihoods (Tejsner 2018). In response to such failures, the utilization of Indigenous knowledge and practice is increasingly part of agendas for improving environmental relations (Berkes 1999; Danielsen et al. 2014).

Concepts can also be used to differentiate between knowledge on a scale from abstract to concrete (L. Kielsen Holm, pers. comm., 13 March 2018). At the abstract scale, concepts like Indigenous knowledge are used on a theoretical, international or national level. Intermediately, concepts like Inuit knowledge or Qitirmiut (Inuit of the Kitikmeot Region in Nunavut) knowledge (Thorpe et al. 2002) can be used to define a subset of Indigenous knowledge on a community-based level. At the concrete scale, knowledge can be further separated into specific topics such as seal hunting or preparation of traditional foods, often used on an individual knowledge holder level.

Traditional knowledge

The concept of traditional knowledge (TK) is perhaps the original and most commonly used concept in addressing the knowledge of Indigenous peoples. The concept of TK is used in international fora when dealing with Indigenous issues – yet its definition is a work in progress and depends on the needs and the context. One example is the Arctic Environmental Protection Strategy, where the term 'traditional knowledge of Indigenous peoples' was used without defining what it meant (AEPS 1991). The Inuit Circumpolar Council (ICC), an international IPO and a Permanent Participant in the AC, used the following definition in 2013:

> Traditional knowledge is a systematic way of thinking applied to phenomena across biological, physical, cultural and spiritual systems. It includes insights based on evidence acquired through direct and long-term experiences and extensive and multigenerational observations, lessons and skills. It has developed over millennia and is still

developing in a living process, including knowledge acquired today and in the future, and it is passed on from generation to generation.

(ICC 2013)

Building further upon this definition, the ICC together with the other Permanent Participants developed the 'Ottawa Traditional Knowledge Principles' with a common definition (not intended to replace individual definitions used by each Permanent Participant), as well as 13 fundamental principles on TK for use in the work of the AC (PPs 2015).

Critique surrounding the concept of TK revolves around two issues. First, despite attempts to define TK in ways that include contemporary knowledge of the present, the word 'traditional' connotes static knowledge of the past and continues to invite misunderstandings and misappropriation (Stevenson 1996; C. Behe, pers. comm., 21 November 2018). Although it has been argued that all knowledge receives meaning from a constantly updated and revised reference frame, Indigenous peoples remain challenged on the view that their knowledge is not relevant to present-day issues, while in fact, culture is dynamic and continuously evolving (see section on IK subsequently). Secondly, in some regions and fora, the use of TK is synonymous with the knowledge of Indigenous peoples, while, in fact, this knowledge concept is not limited to Indigenous peoples and can be used for any non-Indigenous communities with distinct cultures. For example, TK is used for the accumulated knowledge about the environment and interrelation with it that communities living closely with the land and the sea have, for example, Catalan fishermen (Carbonell 2012) or Norwegian non-Indigenous communities (F. J. Theisen, pers. comm., 5 December 2018).

Following this, we summarize the characteristics of TK with this description: A knowledge concept that encompasses social and environmental aspects and which specifically implies a historical continuity in traditions and practices. It does not distinguish on identity and applies to both Indigenous peoples and distinct non-Indigenous cultures associated with a locality, way of life or occupation.

Traditional ecological knowledge

The inclusion of the word 'ecological' in the name of the knowledge concept of traditional ecological knowledge (TEK) places emphasis on parts of knowledge concerned with ecological and environmental issues. For example, Stevenson (1996, 281) elaborates how TEK is 'composed of three interrelated components: 1) specific environmental knowledge . . ., 2) knowledge of ecosystem relationships . . ., and 3) a code of ethics governing appropriate human – environmental relationships'. He notes that humans play a significant role in maintaining key ecosystem and human–nature relationships. Usher (2000, 183) summarized the definition of TEK as 'knowledge about the environment, knowledge about the use of the environment, values about the environment, and the knowledge system itself'.

Regarding critique of TEK, it has been argued that the environmental and ecological aspects are attractive to non-Indigenous people for use in impact assessments and management of living resources, as it seems more accessible due to its resemblance to conventional science topics (Stevenson 1996, 281). For the same reason, understanding of the knowledge by non-Indigenous people may be very narrow and limited to specific aspects that interest particularly social and natural scientists, compromising the holistic nature of the knowledge (Wenzel 2004). When focusing only on environmental and ecological aspects of the knowledge while largely ignoring the social and inter-relational aspects, the risk of losing context and possible misinterpretation is even greater and may not serve the best interest of the Indigenous peoples (Stevenson 1996).

We summarize the characteristics of TEK with this description: A knowledge concept which can be considered a subset of TK, with emphasis on knowledge about, and use of, environment or ecosystem resources; the human–nature relationship and appropriate values, behaviour or governance of resources.

Indigenous knowledge

The concept of Indigenous knowledge emphasizes the Indigenous identity of knowledge holders, thereby building upon the language of Indigenous peoples and their collective rights as set forth in the International Labour Organization Indigenous and Tribal Peoples Convention no. 169 (ILO C196 1989), the United Nations Declaration on the Rights of Indigenous peoples (UN DRIP 2007) and the American Declaration on the Rights of Indigenous peoples (OAS 2016) (Hansen and Bankes 2013). These rights include inherent rights to self-determination, lands, territories and resources, free prior and informed consent (FPIC) and other distinct collective rights. The rights do not form part of the knowledge itself – instead, they necessitate a consideration of the way in which knowledge holders are approached and their rights recognized in the processes. IK has found favour with IPOs and academics when discussing sustainable resource use (Agrawal 1995; Stevenson 1996).

The ICC changed its use of the term TK to IK and otherwise retained the same definition for IK as it used for TK in its most recent policy paper for work done under the auspices of the AC (ICC 2016). The reason behind this transition is to better reflect the Indigenous cosmology and is partly rooted in the critique surrounding the concept of TK (see section on TK previously and section on transition in concepts under 'Discussion'). The emphasis on the identity aspect is considered less controversial and more inclusive and empowering and does not invite misappropriation the way that TK does (Stevenson 1996; C. Behe, pers. comm., 21 November 2018). The word 'Indigenous' is not often used by knowledge holders themselves – still, the label has often proved a useful tool in international negotiations (L. K. Holm, pers. comm., 13 March 2018). While IK in concrete cases can be tied to a specific locality (e.g. 'knowledge that seals feed in this area'), locality is not required as such, since the knowledge is linked to culture (e.g. 'knowledge of how to hunt seals'), which can be widespread over several regions.

We summarize the characteristics of IK with this description: A holistic knowledge concept which encompasses all social and environmental aspects, as well as aspects of Indigenous culture and ways of life. It implies the necessity of acknowledging the collective rights of Indigenous peoples in the approach, thus differentiating the Indigenous from non-Indigenous in a concrete manner.

Inuit Qaujimajatuqangit

Inuit Qaujimajatuqangit (IQ) is an Inuktitut term translated to 'that which Inuit have always known to be true'. It is commonly used in the Canadian Arctic to refer to Inuit knowledge (Thorpe et al. 2002; Wenzel 2004). The concept of IQ first emerged in the context of management and governance following the signing of the Nunavut Land Claims Agreement (Government of Nunavut 1993) and the subsequent process of creating the Territory of Nunavut (Government of Nunavut 1999) (Wenzel 2004; Keenan et al. 2017). The concept was further developed in order to operationalize it in the work of the Government of Nunavut. A report from an IQ workshop by the Nunavut Social Development Council explains IQ as embracing 'all aspects of traditional Inuit culture including values, world-view, language, social organization, knowledge, life skills, perceptions and expectations' (Wenzel 2004) and by the Nunavut

Wildlife Act (2003) as meaning 'traditional Inuit values, knowledge, behaviour, perceptions and expectations'. Accompanying the definition of IQ is a set of guiding principles and concepts that, according to Wenzel (2004), resembles principles of good governance in the way Inuit practice stewardship with respect to their environment and resources, emphasizing how Inuit conceptualize human–animal relations and how this affects their use of wildlife. Wenzel argues that whereas the definition of TEK seems partly covered by the definition of IQ, it does not mean that IQ and TEK can be equated, as IQ is more encompassing of sociocultural content and importance than TEK alone.

We summarize the characteristics of IQ with this description: A holistic knowledge concept which can be considered a subset of IK, as it is specific for the Inuit, with emphasis on both social and environmental aspects and most importantly human–environment interrelations, including values on appropriate behaviour and good governance.

Local knowledge

The concept of local knowledge (LK) has mostly been used about knowledge from people in rural and urban communities using natural resources, including farmers, pastoralists, foresters, hunters, gatherers, fishers, artisans, food processors and traders, as well as local governments and private sectors (Warburton and Martin 1999). LK in this regard relates to 'the whole system of concepts, beliefs and perceptions that people hold about the world around them'. In a report on LK and resource management in the Arctic, it is quoted that 'local knowledge about local resources is limited to local responsibility' (NCM 2015). In much literature, LK is used without being defined as such; at the same time, it is often used interchangeably with other knowledge concepts such as TK, IK and community knowledge (Sejersen 1998, 2004; Danielsen et al. 2014; NCM 2015). An interesting observation is that the knowledge of Inuit hunters and fishers in Greenland is often termed LK, for example, documentation of flora/fauna characteristics or participatory monitoring and co-management of living resources (Sejersen 2004; Danielsen et al. 2014), where in other Arctic regions it would be termed IK, IQ, TK or TEK.

Another example is the Arctic EIA project (2017–2019) in the AC, where the Editorial Committee[1] was unable to find an appropriate definition of LK and created its own definition for its report, stating that it

> refers to knowledge of all Arctic residents, who inhabit a specific geographical area. Local knowledge is adapted to the local culture and environment and is embedded in community practices and institutions. It can include experiences, skills, practices and learning that have been developed, used, sustained and passed on from generation to generation within a community. It can also include knowledge derived from formal schooling.
>
> (AC – Arctic EIA project 2019)

Although this definition contains aspects reminiscent of the historical continuity of traditions and practices embedded in TK, the emphasis is largely on the locality of the knowledge as it is tied to a specific place.

One point of critique regarding the concept of LK revolves around the generality of the word 'residents', as there is no emphasis on identity as such, and the knowledge holders could be Indigenous or non-Indigenous alike. In response, the ICC has stressed that Indigenous peoples should not be conflated with local communities and that 'Indigenous peoples and their knowledge should not be lumped together with non-Indigenous peoples and Local knowledge'

(ICC Alaska 2018). This may relate to the differences in cultural cosmologies as well as the necessity of recognizing the distinct status and rights of Indigenous peoples, including inherent rights to self-determination, lands, territories and resources and other distinct collective rights, as described in the section on IK.

We summarize the characteristics of LK with this description: A knowledge concept with emphasis on the residents inhabiting a specific geographical locality, using natural resources in some way and holding knowledge with local content and context adapted to the specific culture and environment in which they reside.

Phase two: concept mapping

In the following mapping exercise, we use the previous literature review of knowledge concepts and their characteristics to visualize similarities and differences between the knowledge concepts. We first develop categories as a basis for sets of parameters to be used in visual representations. We then use the sets of parameters to develop a Euler diagram for five knowledge concepts and individual concept circles for four knowledge concepts.

The various knowledge concepts and the critique surrounding their definitions, interpretations and connotations of words are used as data for developing codes and categories. The emerging overall categories are suggested as follows (Table 14.1).

In the Identity category, '*Indigenous*' implies the special status and collective rights of Indigenous peoples, including FPIC, which are not applicable to non-Indigenous people. Subsets of Indigenous peoples, and references to people not part of these groups, are also featured. The Social aspects category corresponds in general to aspects considered in Social Impact Assessments of activities related to infrastructure or extractive industries (IAIA 2003). Similarly, the Environmental aspects category corresponds in general to aspects considered in Environmental Impact Assessments of similar activities (IAIA 1999). In the Time frame category, some words can be understood in the sense of historical continuity in traditions and practice, opposed to other words emphasizing the dynamic and evolving nature of knowledge. The Locality category emphasizes knowledge tied to a specific location or distinct to a region. The Format

Table 14.1 Categories for characteristics of knowledge concepts for Indigenous peoples

Categories	Description
Identity	Distinctions of groups based on their cultural identity.
	E.g. *Indigenous, aboriginal, native*. Subsets of specific Indigenous Peoples: *Inuit, Saami*. Including opposites: *non-Indigenous, non-Inuit*
Social aspects	Social and cultural dimensions.
	E.g. *social, cultural, economic, spiritual, values, beliefs, morals, ethics, concerns*
Environmental aspects	Environmental dimensions, including human interrelations.
	E.g. *ecological, environmental, animals, human use of the environment, human/animal/ environment relations*
Locality	Specific geographic locations, knowledge distinct to regions.
	E.g. *local, local community, place-based*
Time frame	Time-related dimensions and the nature of knowledge.
	E.g. *multi-generational, historical, traditional, past*. Including opposites: *present, future, contemporary, dynamic*
Format	How knowledge is obtained, transmitted and approached.
	E.g. *lived experience, practice, way of life, oral, from generation to generation, holistic*

category relates to how the knowledge concepts differentiate themselves from scientific knowledge (Hansen et al. 2016). It is worth noting that in an Indigenous context, these categories are often interrelated, interdependent and even indivisible in accordance with the holistic nature of the knowledge systems. These categories and corresponding sets of parameters (Table 14.2) are merely intended to aid in distinguishing one concept from another by emphasizing their similarities and differences.

Euler diagram

Euler diagrams are useful for illustrating overlapping definitions while only showing relevant relationships between parameters. As the most frequently used knowledge concepts identified in the literature, we chose the concepts of TK, TEK, IK, IQ and LK, with TEK illustrating a subset of TK and IQ illustrating a subset of IK. Table 14.3 features our summarized descriptions of characteristics of each knowledge concepts and relevant parameters from Table 14.2.

In Figure 14.1, the Euler diagram of the knowledge concepts of TK, TEK, IK, IQ and LK shows each concept in rounded rectangles, with parameters shown in ovals within or outside rectangles to indicate whether they are included in the respective knowledge concepts.

The IK and its subset of IQ encompass the Identity parameter, while they share the Time frame parameter with the TK and its subset of TEK. The Locality parameter is encompassed by TK, TEK and LK but not IK and IQ. Likewise, all knowledge concepts encompass the Environmental and Ecological parameters, while only TEK does not encompass the Social, Cultural and Economic parameters.

Concept circles

Concept circles are useful for illustrating the relevant parameters featured in individual knowledge concepts. Using the same knowledge concepts as in the Euler diagram except for IQ, which would be similar to IK, the parameters of Table 14.2 are signified with dots along the circumference of the circle. Parameters within the same categories are shown with similar nuances and patterns. Thick lines denote which parameters are relevant for each knowledge concept, providing distinct visual shapes to ease readings of knowledge concepts.

In Figure 14.2, the concept of TK encompasses all parameters except Identity, since TK can be used about both Indigenous peoples and non-Indigenous people. The Time frame parameter is embedded in the word 'traditional' itself, implying a historical continuity of traditions and

Table 14.2 Categories and corresponding set of parameters used for the Euler diagram in Figure 14.1 and the concept circles in Figures 14.2, 14.3, 14.4 and 14.5

Categories	Set of parameters
Identity	1: Identity
Social aspects	2: Cultural
	3: Social
	4: Economic
Environmental aspects	5: Environmental
	6: Ecological
Locality	7: Locality
Time frame	8: Time frame

Table 14.3 Summarized descriptions of each knowledge concept and specifications of which set of parameters characterize the respective knowledge concepts in Figure 14.1

Knowledge concept	Description	Set of parameters
Traditional knowledge (TK)	A knowledge concept that encompasses social and environmental aspects and which specifically implies a historical continuity in traditions and practices. It does not distinguish identity and applies to both Indigenous Peoples and distinct non-Indigenous cultures associated with a locality, way of life or occupation.	2: Cultural 3: Social 4: Economic 5: Environmental 6: Ecological 7: Locality 8: Time frame
Traditional ecological knowledge (TEK)	A knowledge concept which can be considered a subset of TK, with emphasis on knowledge about, and use of, environment or ecosystem resources; the human–nature relationship and appropriate values, behaviour or governance of resources.	5: Environmental 6: Ecological 7: Locality 8: Time frame
Indigenous knowledge (IK)	A holistic knowledge concept which encompasses all social and environmental aspects, as well as aspects of Indigenous culture and ways of life. It implies the necessity of acknowledging the collective rights of Indigenous peoples in the approach, thus differentiating the Indigenous from non-Indigenous in a concrete manner.	1: Identity 2: Cultural 3: Social 4: Economic 5: Environmental 6: Ecological 8: Time frame
Inuit Qaujimajatuqangit (IQ)	A holistic knowledge concept which can be considered a subset of IK, as it is specific for the Inuit, with emphasis on both social and environmental aspects, and most importantly the human–environment interrelations, including values on appropriate behaviour and good governance.	1: Identity 2: Cultural 3: Social 4: Economic 5: Environmental 6: Ecological 8: Time frame
Local knowledge (LK)	A knowledge concept with emphasis on the residents inhabiting a specific geographical locality, using natural resources in some way and holding knowledge with local content and context adapted to the specific culture and environment in which they reside.	2: Cultural 3: Social 4: Economic 5: Environmental 6: Ecological 7: Locality

practices. Comparing the concept of TK to TEK in Figure 14.3, TEK is considered a subset of TK that only encompasses the Environmental and Ecological parameters and not the Cultural, Social and Economic parameters. In Figure 14.4, the Identity parameter (and the implied Indigenous peoples' rights) is an important feature of the concept of IK, distinguishing it from other concepts. IK include the Time frame parameter with historical continuity of traditions and practices through its intergenerational nature, though Indigenous peoples also place emphasis on the dynamic and developing nature of knowledge, highlighting present and future, thus distancing the concept from TK. The Locality parameter is not required as such for IK, since a culture can be widespread over several regions (e.g. Inuit culture shared in Greenland, Canada, Alaska and Chukotka) and is thus only implied when it comes to concrete cases. In contrast, the Locality parameter is specifically emphasized for the concept of LK in Figure 14.5, whereas the Identity and Time frame parameters are not embedded.

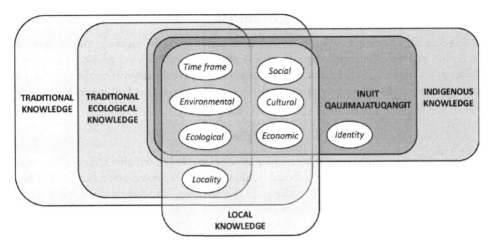

Figure 14.1 Euler diagram of five major knowledge concepts and the sets of parameters that are similar and different between them

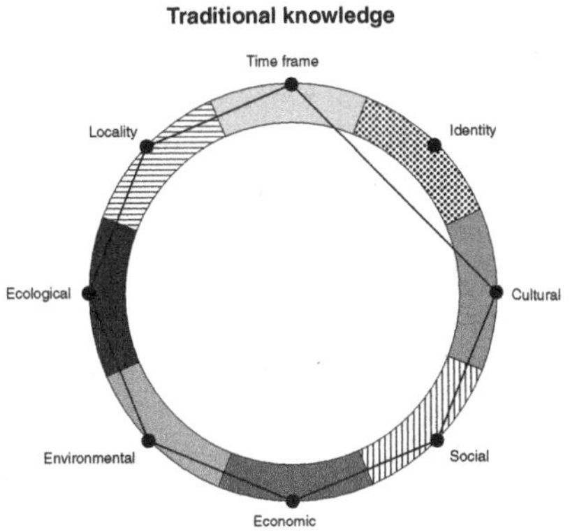

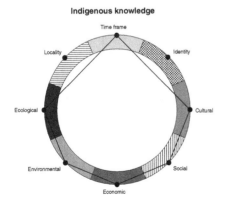

Figure 14.2 Concept circle of TK with relevant parameters connected with thick lines.

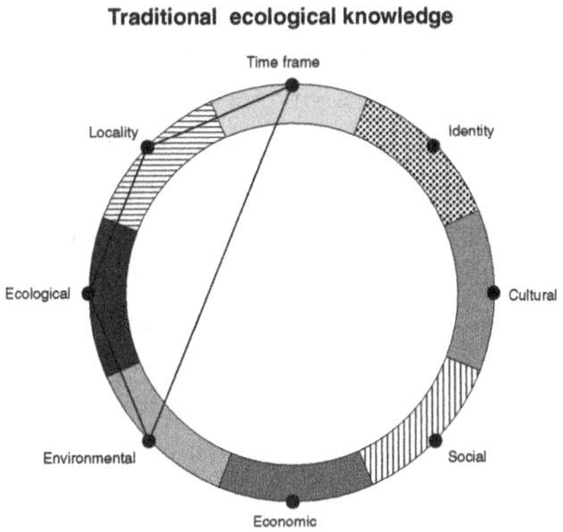

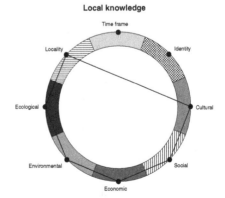

Figure 14.3 Concept circle of TEK with relevant parameters connected with thick lines.

Figure 14.4 Concept circle of IK with relevant parameters connected with thick lines.

Figure 14.5 Concept circle of LK with relevant parameters connected with thick lines.

Discussion

In this chapter, we attempt to clarify the confusion surrounding the meanings and implications of various concepts for knowledge of Indigenous peoples with a view to more consistently utilizing this knowledge in research, management and resource development activities. This can be seen in response to the repeated calls by IPOs for the utilization of Indigenous knowledge.

Use of concepts

Some argue that scientific knowledge and Indigenous knowledge, whether their differences are substantive, methodological, or contextual, cannot really be separated when claims of dichotomy are examined (Agrawal 1995). Others add that labelling of knowledge concepts is a construct serving abstract purposes by others than knowledge holders themselves (e.g. Usher 2000; McGregor 2009; L. K. Holm, pers. comm., 13 March 2018). Some argue that it ultimately comes down to how groups of people wish to define their own knowledge systems (C. Behe, pers. comm., 21 November 2018).

However, in utilizing the knowledge of Indigenous peoples, we argue that it makes a difference whether the distinct status and collective rights of Indigenous peoples are taken into consideration in the process – especially FPIC. These rights are inevitably linked to power relations and governance structures (Nadasdy 1999) and hence possible imbalances between Indigenous peoples and decision-makers. IK is the only knowledge concept that emphasizes identity and thus implies these rights, providing a legal leverage for the empowerment of Indigenous peoples, thus differentiating it from other knowledge concepts. Using other knowledge concepts for Indigenous peoples may seem innocent in following academic traditions in certain research fields or geographical regions, but the implication of ignoring Indigenous peoples' rights can be viewed as facilitating and continuing a colonial dispossession and domination (Cameron 2012).

Our review suggests that use of the concepts for knowledge of Indigenous peoples and their definitions, perceived value and degree of acknowledgment and utilization varies significantly from region to region in the Arctic. Arguably, these relate to the role of colonial and political-economic contexts and how they have shaped the way in which knowledge concepts are expressed and used by governing institutions, stakeholders and Arctic Indigenous peoples (Cameron 2012). For example, the character of Indigenous governance frameworks in the Inuit Homelands of Greenland, the Canadian Arctic and Alaska vary significantly. Greenland has experienced a significant degree of centralization through the semi-autonomous Greenland self-government, influenced by mainly Danish and European frameworks of governance, with Inuit and Saami constituting the only Indigenous peoples in the European Arctic. On the other hand, the Canadian Arctic and Alaskan regions are characterized by historical and ongoing decentralization through Land Claims Agreements and formation of Indigenous governments and territories, influenced by the numerous Indigenous peoples on the North American continent. These differences in governance frameworks seem to be reflected in the choice of knowledge concepts. Researchers and governing institutions in Greenland and Denmark often refer to the knowledge of Greenlandic Inuit as LK, while in the Canadian Arctic and Alaska, the knowledge of Inuit is often referred to as IK, IQ or TK. This is even though the majority of the Greenlandic population is Inuit (Kalaallit) and their status as an Indigenous people has been acknowledged through the Danish and Greenlandic ratification of the ILO C169 and adoption of the UN DRIP (UN HRC 2011). The continued use of the concept of LK for Indigenous peoples constitutes a mismatch linked to the discussion on continued colonial dispossession and domination. While the Greenland self-government has chosen a public, democratic governance framework, it can be argued that the choice of knowledge concept reflects a failure to recognize the need for cultural sensitivity for the Indigenous majority by researchers and the authorities themselves (Stephens 2000). This may be due to a general ignorance in Greenland and Denmark about Indigenous peoples and the application of UN DRIP in the Kingdom of Denmark (A. Lynge, pers. comm., 18 September 2018).

Transition in concepts

Stevenson noted that both Indigenous organizations and academics were beginning to favour the concept of IK instead of TK (1996). As mentioned, the ICC officially changed the use of TK for IK recently. Up until 2015, the ICC was using the concept of TK in its policy paper 'Application of Traditional Knowledge in the work of the Arctic Council' (ICC 2013). As a Permanent Participant in the AC, the ICC contributed to the development of the 'Ottawa Traditional Knowledge Principles' (PPs 2015), offering a combined working definition of the knowledge concept, whereas in 2016, the ICC revised its policy paper for the AC, officially exchanging the term TK to IK but otherwise retaining the same definition of the knowledge concept (ICC 2016). An interview with the ICC (C. Behe, pers. comm., 21 November 2018) revealed that the main reason for this transition was because the choice of language more adequately reflected the Indigenous cosmology of the knowledge. Also, the connotation of the word 'traditional' could give an impression of past knowledge, inviting misappropriation, although the knowledge is continuously revised in relation to the present and future (Stevenson 1996). In addition, even though the concept of TK has been used mainly for Indigenous peoples, it is in no way reserved for these only but can be used for the knowledge of non-Indigenous communities with long-standing and distinct traditions and cultures, for example, Catalan fishermen (Carbonell 2012). The ICC is now promoting the change of concept in the work of the AC. For instance, on October 26, 2018, the Permanent Participants of the AC agreed to change the principles to 'Ottawa Indigenous Knowledge Principles' (PPs 2018), and on May 7, 2019, the Arctic EIA project used the concept of IK for the purpose of its report (AC – Arctic EIA project 2019) – although these discussions within the AC have not yet been concluded, as some Arctic states maintain their emphasis on TK or LK. Supplementary interviews with government agencies, IPOs and academics in Canada and Greenland in 2017–2018 indicate a similar trend in transitioning from the concept of TK to IK concerning Indigenous peoples, although the transition is slow and not always consistent. Further research is needed to understand the details and extent of this discussion.

Finally, using one knowledge concept does not limit the use of other concepts, and while non-Indigenous people cannot hold IK, they and Indigenous peoples may still hold other forms of knowledge. For example, Indigenous peoples may also hold LK tied to their locality or TK tied to occupations not specifically linked to their Indigenous culture, although these forms of knowledge are still filtered through an Indigenous cosmology with cultural interests, values and concerns (Stevenson 1996; C. Behe, pers. comm. 21 November 2018).

Concept maps as visualization tools

Through our review and understanding of the various knowledge concepts, we have developed concept maps using a Euler diagram and individual concept circles. We have thus developed a grounded theory from the literature review data – a model of how the knowledge concepts and their similarities, differences and interrelations can be visualized. These visualization tools will need to be further tested according to the understanding of the knowledge concepts by different groups in the Arctic.

Acknowledgements

We are grateful for the perspectives of the Inuit Circumpolar Council on this topic. We thank Professor Anne Merrild Hansen at Ilisimatusarfik – University of Greenland & Aalborg University – as well as friendly reviews from various fellows for valuable feedback on the concept maps. We also thank Rasmus Egede Dahl for helping with graphical design of concept circles.

Note

1 Disclaimer: Parnuna Egede Dahl was part of the editorial group of the Arctic EIA project and had significant influence on the co-created definition of LK.

References

Arctic Council (AC). 1996–2017. "All Arctic Council Declarations." Accessed 2 August 2019. https://oaarchive.arctic-council.org/handle/11374/94.

AC. 2018. "Agreement on Enhancing International Arctic Scientific Cooperation." Accessed 26 September 2018. https://oaarchive.arctic-council.org/handle/11374/1916.

AC – Arctic EIA project. 2019. "Good Practices for Environmental Impact Assessment and Meaningful Engagement in the Arctic – Including Good Practice Recommendations." Accessed 28 May 2019. www.sdwg.org/wp-content/uploads/2019/05/EIA_Report_Screen_Lores_Spreads.pdf.

Arctic Environment Protection Strategy (AEPS). 1991. *Guidelines for Environmental Impact Assessment (EIA) in the Arctic*. Forssa: Finnish Ministry of the Environment.

Agrawal, A. 1995. "Dismantling the Divide between Indigenous and Scientific Knowledge." *Development and Change* 26 (3): 413–439.

Berkes, F. 1999. *Sacred Ecology: Traditional Ecological Knowledge and Resource Management*. Philadelphia: Taylor & Francis.

Bravo, M. 2010. "Epilogue: The Humanism of Sea Ice." In *SIKU: Knowing Our Ice*, edited by I. Krupnik, C. Aporta, S. Gearheard, G. J. Laidler, and L. Kielsen Holm. London: Springer.

Bryant, A., and K. Charmaz. 2007. *The SAGE Handbook of Grounded Theory*. Los Angeles: SAGE Publications.

Cameron, E. S. 2012. "Securing Indigenous Politics: A Critique of the Vulnerability and Adaptation Approach to the Human Dimensions of Climate Change in the Canadian Arctic." *Global Environmental Change* 22 (1): 103–114.

Carbonell, E. 2012. "The Catalan Fishermen's Traditional Knowledge Climate and the Weather: A Distinctive Way of Relating to Nature." *International Journal of Intangible Heritage* 7: 62–75.

Charmaz, K. 2014. *Constructing Grounded Theory*. 2nd ed. London: SAGE Publications.

Damas, D. 1985. *Handbook of North American Indians. 5. Arctic*. Washington: Smithsonia Books.

Danielsen, F., E. Topp-Jørgensen, N. Levermann, P. Løvstrøm, M. Schiøtz, M. Enghoff, and P. Jakobsen. 2014. "Counting What Counts: Using Local Knowledge to Improve Arctic Resource Management." *Polar Geography* 37 (1): 69–91.

Glaser, B., and A. L. Strauss. 1967. *The Discovery of Grounded Theory – Strategies for Qualitative Research*. Chicago: Aldine.

Government of Nunavut. 1993. "Nunavut Land Claims Agreement." Accessed 17 August 2020. www.gov.nu.ca/sites/default/files/Nunavut_Land_Claims_Agreement.pdf.

Government of Nunavut. 1999. "Creation of Nunavut." Accessed 17 August 2020. https://www.gov.nu.ca/sites/default/files/gn_info_package_-_creation_of_nunavut.pdf.

Government of Nunavut. "Consolidation of Wildlife Act," S.Nu. 2003, c. 26. Current to: 6 November 2012. Accessed 17 August 2020. www.gov.nu.ca/sites/default/files/consSNu2003c26.pdf.

Granada Television. 1970–1993. "Disappearing World." Accessed 17 August 2020. https://www.imdb.com/title/tt0835398/.

Hansen, A. M., P. Tejsner, and P. P. Egede. 2016. "Traditional Knowledge and Industrial Development." In *Perspectives on Skills: An Anthology on Informally Acquired Skills in Greenland*, edited by R. Knudsen. Copenhagen: Copenhagen University Press.

Hansen, K. F., and N. Bankes. 2013. "Human Rights and Indigenous Peoples in the Arctic." In *Arctic Oil and Gas: Sustainability at Risk?*, edited by A. Mikkelsen and O. Langhelle, 291–317. New York: Routledge.

Huntington, H. P. 2005. "'We Dance around in a Ring and Suppose': Academic Engagement with Traditional Knowledge." *Arctic Anthropology* 42 (1): 29–32.

IAIA/Vanclay, F. 2003. "International Principles for Social Impact Assessment." *Impact Assessment and Project Appraisal* 21 (1): 5–11.

International Association of Impact Assessment (IAIA). 1999. *Principles of Environmental Impact Assessment Best Practice*. Fargo, North Dakota: IAIA International Headquarters & Institute of Environmental Assessment.

Inuit Circumpolar Council (ICC). 2010. "Inuit Arctic Policy." Accessed 14 October 2019. https://icca laska.org/wp-icc/wp-content/uploads/2016/01/Inuit-Arctic-Policy-June02_FINAL.pdf.

ICC. 2013. "Application of Traditional Knowledge in the Arctic Council." Accessed 1 June 2017. http:// iccalaska.org/wp-icc/wp-content/uploads/2016/03/Application-of-Traditional-Knowledge-in-the-Arctic-Council.pdf.

ICC. 2016. "Application of Indigenous Knowledge in the Arctic Council." Accessed 6 June 2018. http:// iccalaska.org/wp-icc/wp-content/uploads/2016/03/Application-of-IK-in-the-Arctic-Council.pdf.

ICC Alaska. 2018. "Press Release: ICC Chair Addresses Arctic Biodiversity Congress, Emphasizes the Interconnectedness of Climate Change, Biodiversity, and Indigenous Rights." Accessed 14 October 2018. https://iccalaska.org/wp-icc/wp-content/uploads/2018/10/Press-Release-ICC-Chair-Addresses-ABC-.pdf.

International Labour Organisation (ILO), Convention Concerning Indigenous and Tribal Peoples in Independent Countries (adopted 27 June 1989, entered into force September 5, 1991) 72 ILO Official Bull. 59 (ILO C169 1989).

Keenan, E., L. M. Fanning, and C. Milley 2017. "Mobilizing Inuit Qaujimajatuqangit in Narwhal Management through Community Empowerment: A Case Study in Naujaat, Nunavut." *Arctic* 71 (1): 27–39.

Lévi-Strauss, C. 1966. "Science of the Concrete." In *The Savage Mind*, Ch. 1. London: Weidenfeld and Nicholson.

Linden, E. 1991. "Lost Tribes, Lost Knowledge." *Time*, 23 September: 46–54.

McGregor, D. 2009. "Linking Traditional Knowledge and Environmental Practice in Ontario." *Journal of Canadian Studies* 43 (3): 69–100.

Nadasdy, P. 1999. "The Politics of TEK: Power and the 'Integration' of Knowledge." *Arctic Anthropology* 36 (1–2): 1–18.

Nichols, T., F. Berkes, D. Jolly, N. B. Snow, and the Community of Sachs Harbour. 2004. "Climate Change and Sea Ice: Local Observations from the Canadian Western Arctic." *Arctic* 57 (1): 68–79.

Nordic Council of Ministers (NCM). 2015. *Local Knowledge and Resource Management – On the Use of Indigenous and Local Knowledge to Document and Manage Natural Resources in the Arctic*. Denmark: TemaNord 2015:506.

Nuttall, M. 1998. "An Environment at Risk: Arctic Indigenous Peoples, Local Livelihoods and Climate Change." In *Arctic Alpine Ecosystems and People in a Changing Environment*, edited by J. B. Ørbaek, R. Kallenborn, I. Tombre, E. N. Hegseth, S. Falk-Petersen, and A. H. Hoel, 19–35. Berlin: Springer.

Organisation of American States (OAS), American Declaration on the Rights of Indigenous Peoples, AG/ RES.2888 XLVI-O/16 (adopted 15 June 2016).

Permanent Participants (PPs). 2015. "Ottawa Traditional Knowledge Principles." Accessed 24 October 2018. www.saamicouncil.net/fileadmin/user_upload/Documents/Eara_dokumeanttat/Ottawa_TK_Principles.pdf.

PPs. 2018. "Press Release: Arctic Indigenous Peoples at the Second Arctic Science Ministerial (including the updated Ottawa Indigenous Knowledge Principles)." Accessed 30 November 2018. https://icca laska.org/wp-icc/wp-content/uploads/2018/10/ICC_PP-press-release_1026.pdf.

Rappaport, R. A. 1968. *Pigs for the Ancestors: Ritual in the Ecology of a New Guinea People*. Long Grove, IL: Waveland Press.

Riedlinger, D., and F. Berkes. 2001. "Contributions of Traditional Knowledge to Understanding Climate Change in the Canadian Arctic." *Polar Record* 37 (203): 315–328.

Sejersen, F. 1998. "Hunting in Greenland and the Integration of Local Users' Knowledge in Management Strategies." In *Aboriginal Environmental Knowledge in the North*, edited by L.-J. Dorais, M. Nagy, and L. Müller-Wille. Quebec: Gétic.

Sejersen, F. 2004. "Local Knowledge in Greenland – Arctic Perspectives and Contextual Differences." In *Cultivating Arctic Landscapes: Knowing and Managing Animals in the Circumpolar North*, edited by D. G. Anderson, and M. Nuttall, 33–56. Oxford: Berghahn Books.

Stephens, S. 2000. *Handbook for Culturally Responsive Science Curriculum*. 2nd ed. Fairbanks: Alaska Native Knowledge Network.

Stevenson, M. G. 1996. "Indigenous Knowledge in Environmental Assessment." *Arctic* 49 (3): 278–291.

Tejsner, P. 2018. "Indigenous Modes of Ownership: Reopening the Case for Communal Rights in Greenland." In *Governance of Arctic Offshore Oil and Gas*, edited by E. M. Basse and C. Pelaudeix. New York: Routledge.

Thornberg, R. 2012. "Informed Grounded Theory." *Scandinavian Journal of Educational Research* 56 (3): 243–259.

Thorpe, N., N. Hakongak, S. Eyegetok, and Kitikmeot Elders. 2002. "Reviewed Work(s): Thunder on the Tundra: Inuit Qaujimajatuqangit of the Bathurst Caribou." *Arctic* 55 (4): 395–396.

United Nations (UN) General Assembly, Human Rights Council (HRC). "National Report Submitted in Accordance with Paragraph 15 (a) of the Annex to Human Rights Council Resolution 5/1 – Denmark." (17 February 2011) UN Doc A/HRC/WG.6/11/DNK/1.

United Nations (UN) General Assembly, United Nations Declaration on the Rights of Indigenous Peoples (13 September 2007) UN Doc A/RES/61/295 (UN DRIP 2007).

Urquhart, C. 2013. *Grounded Theory for Qualitative Research – A Practical Guide*. Los Angeles, CA: Sage.

Usher, P. J. 2000. "Traditional Ecological Knowledge in Environmental Assessment and Management." *Arctic* 53 (2): 183–193.

Warburton, H., and A. Martin. 1999. *Local People's Knowledge in Natural Resource Development. Socio-Economic Methodologies for Natural Resources Research*. Chatham: Natural Resources Institute.

Wenzel, G. W. 1999. "Traditional Ecological Knowledge and Inuit: Reflections on TEK Research and Ethics." *Arctic* 52 (2): 113–124.

Wenzel, G. W. 2004. "From TEK to IQ: Inuit Qaujimajatuqangit and Inuit Cultural Ecology." *Arctic Anthropology* 41 (2): 238–250.

World Intellectual Property Organization (WIPO) website. "Traditional Knowledge." Accessed 12 December 2018. www.wipo.int/tk/en/tk/.

SECTION 3

Indigenous peoples and self-determination in the Arctic

Dorothée Cambou and Timo Koivurova

With the adoption of the United Nations Declaration on the Rights of Indigenous Peoples (UNDRIP) in 2007, the right of Indigenous peoples to self-determination has been authoritatively recognized in international law. According to the Declaration, in exercising this right, Indigenous peoples have the right to autonomy or self-government in matters relating to their internal and local affairs. They also have the have the right to maintain and strengthen their distinct political, legal, economic, social and cultural institutions while retaining their right to participate fully, if they so choose, in the political, economic, social and cultural life of the State.

> The affirmation of these dual aspects reflects the widely-shared understanding that Indigenous peoples are not to be considered unconnected from larger social and political structures. Rather, they are appropriately viewed as simultaneously distinct from, yet joined to, larger units of social and political interaction, units that may include Indigenous federations, the states within which they live, and the global community itself.
>
> (Anaya 2009, 193)

Yet the adoption of the UNDRIP and recognition of the right of Indigenous peoples to self-determination have not solved all difficulties, and the implementation of the right raises many controversies in practice.

In the Arctic, the efforts of Indigenous peoples to achieve greater self-determination have been fundamental to empower Indigenous peoples and ensure their representation and participation in Arctic decision-making processes at all scales. At the same time, a number of challenges still need to be solved in order to ensure that Indigenous peoples can determine their future across all territories in which they live. Those challenges are multiple. They range from demands for better political autonomy, representation and participation in decision-making processes to a protection of their cultural values and the ability to maintain and develop those values from the local to the regional levels. In addition, these demands are closely intertwined with the right of Indigenous peoples to govern and manage their traditional land and natural resources, which is a fundamental basis of the exercise of their right to self-determination. Achieving sustainable self-determination therefore requires changes in several domains and necessitates the adoption of a holistic approach that is "culturally viable and inextricably linked to Indigenous relationships to the natural world" (Corntassel 2008, 108) in order to ensure durable changes at all levels of the Arctic governance system.

In this book section, we have collected a number of chapters that examine the multiple challenges raised by the recognition of the right of Indigenous peoples to self-determination in the Arctic. Starting from a general overview of self-government arrangements in the Arctic and ending with a chapter concerning the principle of Free Prior and Informed Consent, the purpose of this section is to provide a general understanding of the legal and political issues raised by self-determination across the Arctic and at different scales. From the outset, even though Indigenous peoples are jointly entitled to the right of self-determination, it is important to acknowledge that there is in practice not a one-size-fits-all model of self-determination. Indigenous peoples in the Arctic are diverse. They share common but different histories of colonial domination, forced assimilation and continuous encroachment on their traditional land and resources. Their demands for self-determination are also shaped by different political, geographical and societal contexts. As a result, the implementation of self-determination necessarily calls for separate arrangements at the state and regional levels.

The variety of models that can advance the right of Indigenous peoples to self-determination in the Arctic is first described by Kuokkanen in Chapter 15. Her chapter focuses on Indigenous self-government in the Arctic and assesses the scope and legitimacy of several arrangements in Nunavut, Greenland and Sápmi territories. As she describes and examines, the exercise of self-determination in the Arctic is various. It can take the form of extensive self-government arrangements that can eventually give rise to independence – as recognized in the Self-Government Act of Greenland – but can also materialize in more or less constrained autonomous arrangements, as is the case for the Inuit people living in Nunavut or for the Sámi people living in Finland, Norway and Sweden. In this regard, important political and legislative changes have occurred in all those territories to accommodate Indigenous self-determination. However, if significant progress has been made in recent years, Kuokkanen also demonstrates in her chapter that important challenges remain to fully operationalize the right in practice.

Furthermore, the challenges raised by self-determination cannot be solved all at once. In fact, it is now well established that self-determination is a process rather than a one-time right. As well argued by Argetsinger in Chapter 16 concerning Inuit self-determination in Alaska, achieving self-determination is a process, "an evolving story that cannot be confined to any one agreement or piece of legislation". In effect, the success of any development supporting Indigenous self-determination can therefore only be measured against the prosperity and wellbeing it provides for its peoples over time. In this regard, there is no perfect model of self-determination, and each arrangement must be adequately designed to support the needs and aspirations of all peoples. This being said, Indigenous peoples can nonetheless learn from one another in order to achieve greater self-determination. In this regard, Argetsinger argues that Alaskan Inuit could gain inspiration from Canadian Inuit land claims organizations whose arrangements appear to better align with the priorities of their peoples. To the same extent, he also demonstrates that reforms are still needed to ensure Inuit self-determination in Alaska.

Ultimately, one of the most crucial issues concerning self-determination lies in the assessment of what it truly means for Indigenous peoples and how it can adequately benefit their diverse needs and aspirations. However, Morkenstäm et al. are correct to indicate that only a few systematic studies have analyzed how persons belonging to Indigenous peoples perceive the right to self-determination in practice. To fill this gap, Chapter 17 is specifically devoted to the analysis of Sámi self-determination from the perspective of the Sámi electorate in Sweden. Based on the study of data from the second Swedish Sámi election study in 2017, the authors of Chapter 17 examine what matters for Sámi voters and what they mean when claiming self-determination. While they demonstrate that voters have different perspectives on the issues that need to be addressed in order to ensure Sámi self-determination, their conclusion also demonstrates that an overwhelming majority support the view that Sámi self-determination needs to be increased through the Sámediggi (Sámi Parliament in Northern Sámi) as their representative body. Ultimately, these conclusions support the general view that Sámi self-determination is not adequately implemented in Sweden, nor in other Nordic States, and that further reforms need to be enacted in order to ensure the realization of this right at the state level.

Similarly, the realization of the right of self-determination of Indigenous peoples in the Russian Federation is far from being achieved. In their critical analysis of the land rights of Indigenous peoples in the Russian context, Liubov and Rodion Sulyandziga describe how the development of the legal framework in Russia has disempowered Indigenous peoples over the years to the benefit of government and corporate interests. In line with Chapter 8, this analysis underscores how the Russian legal framework fails to provide the right for Indigenous peoples to control their territories or to fully participate in decision-making processes affecting their homelands in accordance with international legal standards (see also Kryazhkov 2013, 155). This chapter also signals the importance of the right of land and natural resources for the achievement of Indigenous self-determination. As well described by the authors, "Without land rights and rights over natural resources, the right of self-determination and other rights would be meaningless or merely become '*paper*' rights, as it has happened in the case of Russia (Corntassel 2008, 108)." Hence, the chapter concludes that

> Russia's declarative laws do not translate into progress in its domestic Indigenous policy, a situation that greatly contrasts with the four decades of what was labelled "the most progressive stage in the history of development of Indigenous peoples' rights and freedoms" at the international level.

Together with the preceding chapters, this analysis therefore demonstrates that much remains to be achieved to make the human right of Indigenous peoples to self-determination a reality in all Arctic states in which they live.

While Indigenous self-determination necessarily requires self-government for Indigenous peoples in matters concerning their local affairs within the state, it also requires their right to participate in decision-making processes affecting them outside state borders. In a world where governance structures are increasingly regionalized, internationalized and transnationalized, it becomes evident that limiting the exercise of self-determination to the realm of state borders is to a great extent obsolete. In order to determine their future, the representation and participation of Indigenous peoples in governance structures that affect them at the regional and international levels are therefore also required (Cambou 2018). In this regard, in Chapter 19, Cambou and Koivurova examine the participation of Indigenous peoples in the governance of the Arctic region. The purpose of this chapter is to provide an overview of the importance and significance of the participation of Arctic Indigenous Organizations such as the Inuit Circumpolar Council and the Sámi Council in the governance of the Arctic, notably by highlighting their unique status as permanent participants in the Arctic Council. Although it is difficult to appraise to what extent Arctic Indigenous Organizations have managed to infiltrate Arctic policy in practice, the chapter demonstrates that their unique position at the Arctic Council has strengthened their representation as active participants in the governance of the Arctic, which also tends to consolidate their self-determination at the regional level.

Thus, the right of Indigenous peoples to self-determination calls for various adjustments and the creation of novel arrangements in various domains and at all scales. It is a process that requires genuine developments for guaranteeing that all actors, including states and corporations, continuously uphold this right in practice. In this context, a core principle that forms the basis of self-determination is the principle of Free, Prior and Informed Consent (FPIC). FPIC is a principle recognized by international human rights standards and which has found application in all decision-making processes affecting Indigenous peoples, including in the Arctic. Despite its status and significance, the principle remains nevertheless contested. In this regard, Heinämäki in Chapter 20 examines the concept of FPIC and its normative content and defines key elements of its implementation process in order to clarify the meaning of the principle. With this analysis, her goal is also to encourage states and corporate actors to recognize and implement this right in practice and to dissipate the expectations of some Indigenous peoples who problematically interpret FPIC as a right to unilateral veto. As argued in her chapter, "FPIC should be understood as Indigenous peoples' right to a qualified consultation and assessment process, which is guided by and honours Indigenous peoples' right to self-determination". However, if FPIC does not provide a unilateral right to veto, it remains crucial for ensuring Indigenous peoples' self-determination because it functions as a counterweight to the exercise of state sovereignty by requiring the duty of government to ensure that any decision affecting Indigenous peoples will not threaten their survival or significantly impact their traditional livelihoods (Anaya and Puig 2017).

On the basis of these chapters, this section therefore highlights multifarious aspects of the right to self-determination and sets out a broad range of perspectives on the topic as it connects to the specific and diverse demands of Indigenous peoples in the Arctic. Although the chapters are limited in scope, the vast arrays of self-determination practices and challenges examined in this section also demonstrate that self-determination is not a topic that we can do justice to in just one book section. Nevertheless, it is our hope that, together, these chapters will provide a solid basis for understanding this topic, spark debates on the issue and generate adequate and effective reforms to ensure greater self-determination for all Indigenous peoples across the Arctic region.

References

Anaya, J. 2009. "The Right of Indigenous Peoples to Self-Determination in the Post-Declaration Era." In *Making the Declaration Work: The United Nations Declaration on the Rights of Indigenous Peoples*, 184–199. Copenhagen: IWGIA.

Anaya, J., and S. Puig. 2017. "Mitigating State Sovereignty: The Duty to Consult with Indigenous Peoples." *University of Toronto Law Journal* 4 (67): 435–464.

Cambou, D. 2018. "Enhancing the Participation of Indigenous Peoples at the Intergovernmental Level to Strengthen Self-Determination: Lessons from the Arctic." *Nordic Journal of International Law* 1 (87): 26–55.

Corntassel, J. 2008. "Toward Sustainable Self-Determination: Rethinking the Contemporary Indigenous-Rights Discourse." *Alternatives* 33: 105–132.

Kryazhkov, V. A. 2013. "Development of Russian Legislation on Northern Indigenous Peoples." *Arctic Review* 2 (4): 140–155.

15

INDIGENOUS SELF-GOVERNMENT IN THE ARCTIC

Assessing the scope and legitimacy in Nunavut, Greenland and Sápmi[1]

Rauna Kuokkanen

Abstract

This chapter considers Indigenous self-determination in the Arctic from a comparative perspective with a focus on Canada, Greenland and Sápmi. Drawing on fieldwork with Indigenous research participants from the three regions, this chapter examines the meaning, scope and status of self-determination in the Arctic. The main results demonstrate that while the objectives and challenges regarding implementing Indigenous self-determination are similar across the Arctic, the circumstances and thus the prospects of self-determination vary considerably from region to region. Specifically, the author considers and compares the self-government institutions in Nunavut, Greenland and Sápmi.

Introduction

Arctic Indigenous peoples, particularly the Inuit and Sámi, have been at the vanguard advancing self-determination since the early 1970s. The Arctic Peoples Conference held in Copenhagen in 1973 marked the beginning of close cooperation of circumpolar Indigenous peoples from Canada, Greenland and Scandinavia. The conference resolution called for Arctic Indigenous peoples' full and equal participation in negotiations dealing with land claims and other rights (Kleivan 1992). Four years later, the Inuit Circumpolar Conference (today Council, ICC) was established to represent the Inuit across the Arctic with regional offices in Greenland, Canada, Alaska and Chukotka, Russia. Since its inception in 1977, ICC has been at the forefront of the international advocacy for Indigenous self-determination. Inuit rights and political autonomy, together with the protection of Arctic environment, have been the organization's key policy areas. In Sápmi, the Nordic Sámi Council was established in 1956 to advance Sámi rights in Norway, Sweden and Finland. The Russian Sámi joined formally the Sámi Council in 1996.

The Inuit Circumpolar Council and the Sámi Council were among the first Indigenous non-governmental organizations in the world and have been among the most active Indigenous peoples globally in the work advancing self-determination in multilateral institutions such as the United Nations and Arctic Council.

This chapter considers Indigenous self-government in the Arctic. Indigenous governance arrangements in the Arctic consist of diverse models, from public governments, such as local boroughs in Alaska and the governments of Nunavut and Greenland, to Indigenous elected assemblies, corporations and resource management regimes. While some Arctic Indigenous peoples have been relatively successful in regaining and implementing political autonomy (e.g., Greenland and Nunavut), others are in the process of negotiating self-government agreements with states. In this chapter, my specific focus is on the scope and structures of the existing self-government arrangements in three regions: Nunavut, Greenland and Sápmi. I examine two interrelated issues: the degree of political autonomy and jurisdiction of each self-government model and the extent to which they are regarded as a form of Indigenous governance.

Indigenous governance is a term that recognizes that Indigenous peoples have had and, in many cases, continue to have their own forms and institutions of governance and law, ranging from local, often fairly informal, deliberative and decision-making processes to complex, formal and centralized structures. Self-government is a political theory and arrangement that enables a group to govern themselves according to their own will and through their own institutions. Indigenous self-government commonly refers to contemporary arrangements with the state in which an Indigenous people have been delegated certain administrative, representational or consultative authority and tasks. Given the limited space, I first provide a brief overview of the background and existing self-government structures and then discuss the two main questions posed for this chapter.[2]

Nunavut: creating Inuit government

Since the 1980s, there have been a number of studies and initiatives in Canada, most notably the Special Committee on Indian Self-Government and its Penner Report (1983) and, in the 1990s, the Royal Commission on Aboriginal Peoples (RCAP) that have recommended the development of a process and framework for the implementation of Aboriginal self-government. The RCAP. recommended a number of ways to restructure the relationship with the state, including the nation model, the public government model and the community interest model. The public government model was implemented with the establishment of the territory of Nunavut in the eastern Arctic in 1999. While the Inuit form the numerical majority (85%), the public government guarantees the participation of all residents of the territory.

As the newest, largest, northernmost and least populous territory of Canada, Nunavut was formally separated from the Northwest Territories on April 1, 1999, via the Nunavut Act and the Nunavut Land Claims Agreement Act (NLCA). The discovery of oil in the Arctic during the 1960s and 1970s stimulated Indigenous groups to bring several land claims against the Alaskan and Canadian governments and led to land claims negotiations with federal and provincial governments. In 1971, the Inuit Tapirisat of Canada (ITC) initiated a study of Inuit land use and occupancy, demonstrating the extent of Inuit aboriginal title in the Arctic and forming the geographic basis of the Nunavut Territory. A plebiscite held in the Northwest Territories in 1982 resulted in 54% of the voting public favouring the division of the territory. A separate land claims agreement was being simultaneously negotiated, leading to an Agreement-in-Principle in 1990. The Final Agreement was signed in 1993, following Inuit ratification vote. A separate political accord, the Nunavut Act of 1993, was also negotiated to deal with division of powers, financing and timing for the political and legal framework of a new territory.

The Nunavut Land Claims Agreement provides Aboriginal title to the Inuit of approximately 20% of the territory of Nunavut (the rest remaining as the Crown land) and establishes harvesting rights and a range of wildlife, resource and environmental management boards, among others. The agreement further created regional five-year economic development programs, supported Native development corporations and provided for the training and development of an administration and staff to implement the settlement. The implementation, however, has been deficient, and the organization in charge of overseeing the land claim, Nunavut Tunngavik Inc., filed a lawsuit against the federal government for failing to live up to the terms of the NLCA in 2006. A settlement agreement was reached in 2015, providing funding particularly for training for Inuit employment.

The government of Nunavut was created in stages over 16 years. It consists of a unicameral legislative assembly of 22 members who are elected individually. There are no parties, and the legislature is consensus based. The head of the government, the premier and the cabinet are elected by and from the members of the legislative assembly. One of unique aspects of the government of Nunavut has been its explicit objective to create a public government structured and operating according to Inuit values and ways known as Inuit Qaujimajatuqangit (IQ). Since the 1970s, the Inuit vision of the Eastern Arctic has been not only strong political autonomy and jurisdiction over their territories but a form of self-government informed by the Inuit worldview and culture. This has, however, proven a monumental if not an impossible task, given the twin challenge of capacity-building (such as filling the vacancies in the administration) and seeking to imbue a Westminster-style governing structure by values that in many ways stand in opposition to standard bureaucratic practices and policy-making (Timpson 2006; White 2009). Another major difficulty has been different understandings of the substance and details of Inuit Qaujimajatuqangit within communities and among generations and gender groups (Wachowich et al. 1999).

Another central component in the endeavour of creating Inuit government has been decentralization. The experience from the Northwest Territories had been a high level of concentration of political and administrative power in the capital of Yellowknife, geographically distant from the Eastern Arctic. With Nunavut, there was a desire to establish government closer to the people and communities through a complex system of more localized administration and dividing the territory into three main regions, Kitikmeot, Kivalliq and Qikiqtaaluk/Baffin, which would also assist spreading "the economic benefits of government employment [and associated economic benefits] beyond the capital into communities where private-sector jobs are scarce" (Hicks and White 2015, 7).

A third unique component recommended by the Nunavut Implementation Commission in 1994 was the gender parity proposal. The idea behind the proposal was to create a political system that would restore and guarantee equal voice and equitable representation of women and men, something which was seen by proponents as a central part of traditional Inuit values (Minor 2002). In practice, it would have meant that "one male and one female MLA would represent each electoral district. Voters would elect two candidates: they would cast one ballot for their preferred male candidate and the other for their preferred female candidate" (Minor 2002, 83). There was, however, considerable confusion among the inhabitants of Nunavut about the gender parity proposal, and it was defeated in a plebiscite in 1997. Some Inuit women even feared that increased political participation of women would increase violence and social problems (Dahl 1997; Altamirano-Jimenez 2008).

In spite of the distinct features put in place to create a uniquely Inuit public government, particularly the incorporation of IQ and decentralization, the Government of Nunavut continued on the legislative and political path set by its predecessor, the Government of Northwest Territories, including the application of the NWT legislation until amended or repealed and the non-partisan 'consensus' parliamentary system (Hicks and White 2015, 8).

Twenty years since the establishment of Nunavut, some serious challenges remain, including "the lack of infrastructure, distances from markets, and sparse population," which seriously impede economic development of the territory (Hicks and White 2015, 6). Limited revenue creation cripples the government's ability to deal with persistent social problems related to housing shortage, domestic violence, suicide and others. Capacity building remains a major issue in building bureaucracy with Inuit representation numerically, culturally and linguistically (White 2009).

Greenland: self-government as a road to independence

Before the Home Rule Act was introduced in 1979, Greenlanders were involved in running their own country and affairs in very limited terms. As a result of growing dissatisfaction with the Danish rule, an internal Home Rule Committee was created in 1973 with the intention of considering Greenland's increased political autonomy within the Danish realm. The Committee submitted a proposal for negotiation to the Danish government in 1975, and the joint Greenland-Danish Home Rule Commission was set up that same year. During this period, the political party Siumut ("Forward") was established, which gained widespread and long-lasting support in most towns and settlements among hunters, fishers and workers. Siumut became the leading voice in formulating the political propositions of the Home Rule Commission. The final report of the commission suggested a Home Rule Act, which was approved first by the Danish Parliament in 1978 and in a referendum in 1979. The same year, Home Rule was initiated in Greenland.

The overarching principle of the Greenland Home Rule Act was the devolution and delegation of legislative and executive authority from Danish to Greenlandic authorities, within certain areas of jurisdiction. These included domestic affairs, taxation, fisheries, planning, trade, church affairs, social welfare, labour market, education, cultural affairs, health, housing, supply of goods, transportation and environmental protection. One of the priorities of the Home Rule government was to reverse colonial urbanization and forced settlement policies and to improve living conditions in the settlements. Later, however, Home Rule enacted economic reforms that revoked economic benefits previously enjoyed by the settlements since the colonial era, such as the one-price system. Regardless, the settlements owe their continued existence to Home Rule (Dahl 2010). Though it was sometimes hailed as a leading example of Indigenous governance, Greenland Home Rule was a delegated authority with sovereignty vested firmly in the Danish crown. It did not recognize the *sui generis* self-determination of the Inuit Greenlanders. Home Rule was a public government focused on the building of the Greenlandic nation structured around key concepts and institutions of Western nation-state such as democracy and parliamentarianism.

After two decades of Home Rule, the Greenland Landstyre had assumed the responsibility of practically all areas of jurisdiction stipulated in the Home Rule Act. Recognizing the need for a reform of Greenland's political and legal status within the Danish realm, the Landstyre established the Greenland Commission on Self-Governance: "we knew we had to take another step from Home Rule. . . . The *Home-Rule Act*, and the framework that it represented was becoming too tight. So, the leading politicians were saying that the 'Anorak' was [getting] too tight."[3] The Commission recommended expanding Greenland's autonomy in its 2003 report. The following year, a Danish-Greenlandic Commission was set up to develop a framework for a greater self-governance in Greenland. The joint Commission submitted its final report to the Danish and Home Rule governments in 2008. Later in the same year, a referendum was held in Greenland on expanded self-rule and 75% voted in favour. As the result, the Greenland Self-Government replaced the Home Rule arrangement on 21 June 2009.

The Act on Greenland Self-Government (commonly referred to as the Self Rule Act) establishes new political and legal opportunities for Greenland to gain extensive self-governance and ultimately independence if the population of Greenland so chooses in the future. The Act contains 33 areas of jurisdiction for the self-rule government (Naalakkersu-isuit or the Government of Greenland) to exercise legislative and executive authority over. The most important of these is the mineral resources. Two other issues of major significance include the recognition of the Greenlanders as a people in international law and adoption of Greenlandic as the official language. Within the framework of the Self-Government Act, Denmark retains the control of the constitution, citizenship, Supreme Court, foreign affairs, defence and currency. Denmark is, however, expected to involve Greenland on foreign affairs and security matters that affect or are in the interests of Greenland. Moreover, since Home Rule, Greenland has been permitted to have missions in countries of special interest to Greenland.

Through the Home Rule Act and the Self-Government Act, Greenland has the right to elect its own parliament and government, the latter with executive authority over the areas of jurisdiction included in the Acts. The elected assembly or the Parliament of Greenland (Inatsi-sartut) consists of 31 members, who are elected by the population of Greenland for a four-year period. The elected assembly approves the government, which is responsible for the central administration, headed by a premier with a cabinet. The Parliament also appoints the premier, who nominates the ministers for the cabinet. There are currently eight ministers, all of whom are Inuit Greenlanders.

Greenland is a parliamentary democracy in which the party system has played a vital role. Political parties emerged in the mid-1970s before the establishment of Home Rule, as the well-educated Greenlandic elite was "eager to use the party system to take power from the Danes" (Loukacheva 2007, 56). The dominant social democratic party Siumut is supported by hunters, fishermen and workers, especially in settlements and smaller towns. Siumut has had a leading role in the process leading to Home Rule, and it has formed the government since, except the first term of the Self-Rule era (2009–2013). The first elections held after Self-Rule was won by the socialist Inuit Ataqatigiit (IA, Inuit Community) with its support base mostly among urban educated Greenlanders.[4]

Since Home Rule, Greenland's governance structure has been a Nordic-style unicameral cabinet-parliamentary system. The Home Rule administration was adopted from Denmark "element for element and law for law" (Nielsen 2001, 232). There were no changes to the governance structure in the Self-Government Act except a cosmetic name change: the Danish terms for the parliament (Landsting) and the government (Landstyre) were replaced with Greenlandic ones (Inatsisartut and Naalakkersuisuit, respectively). The "institutional inertia" and the lack of interest in developing specifically Greenlandic institutions have been explained by Greenland's "organizational dependency" on Denmark (Jonsson 1997).

During the self-government agreement negotiations in the early 2000s, there was no discussion of Inuit values or governance. According to a civil servant attending the negotiations, considering 'the Inuit' separately from 'the people of Greenland' would have been regarded as undemocratic. The implicit assumption during the negotiations was that because the Inuit constitute an overwhelming majority of Greenland's population, the Inuit character of the new self-government regime is guaranteed and thus self-government would represent *de facto* Inuit governance. The focus of the negotiations was on voting rights and eligibility, resulting in the creation of a new category, 'the people of Greenland,' rather than considering Inuit principles of governance. While the term 'Greenlander' is commonly understood to refer only to the Inuit in Greenland, the term 'the people of Greenland' encompasses the entire population: the Inuit, Danes and everyone else who has lived in Greenland more than six months.[5] The Act does not recognize the Inuit as a people and consequently does not deal with the question of the self-determination of the Inuit.

Sápmi: limited cultural autonomy

Historically, Sámi nationhood was recognized in the 1751 Lapp Codicil, a largely forgotten and little-known legal addendum to the Peace Treaty of Strömstad that demarcated the border between Norway (then a Danish territory) and Sweden. In spite of its historical significance, the Lapp Codicil has not featured in legal or political discourses of Sámi self-government or considerations with regard to the source, legitimacy or authority of Sámi self-determination.[6] An exception is the Draft Nordic Sámi Convention. Considered a development and renewal of the existing Sámi rights codified in the Lapp Codicil, the draft Nordic Sámi Convention is an international human rights instrument aiming to confirm and strengthen the rights of the Sámi people. The draft Convention was submitted in November 2005, and the negotiations commenced in 2011. The final draft was released in January 2017, with a number of compromises pertaining to the central rights of the Sámi as an Indigenous people.[7]

In the Nordic countries, there has been a general tendency to consider Indigenous rights as little as possible by the state and its institutions (Hannikainen 1996; Alfredsson 1999). While Indigenous political mobilization in Canada and Greenland has resulted in land claims and self-government agreements, Sámi land rights are only partially recognized in the northernmost county of Norway (through the Finnmark Act). Of the three Nordic countries, Norway (with the majority of the Sámi population) has had the most progressive Sámi policy since the 1980s. It has been of great strategic value for other Sámi, and it has served as a model to influence Sámi policy in Finland and Sweden.

Sámi policy in Norway took a radical shift as a result of the Alta River conflict in the early 1980s. In the late 1970s, the Norwegian government had decided to dam the Alta-Kautokeino River at the heart of the Sámi region in Northern Norway. In its original form, the hydroelectric dam was going to submerge the Sámi village of Máze (Masi) and a considerable portion of important reindeer grazing and calving areas in a significant reindeer-herding region. The government plans were met with unexpected resistance by a coalition of the Sámi, environmentalists and fishermen, the latter of whom were concerned about the destruction of a significant salmon river. The conflict culminated in a major demonstration at the construction site by the river, a hunger strike at the front of the Norwegian Parliament building in Oslo in 1979 and an occupation of the office of the Norwegian prime minister Gro Harlem Brundtland by 14 Sámi women in 1981. Viewed as a "political earthquake," the conflict "shook the political establishment in Norway [and] turned the traditional views on the legitimacy of policymaking towards the Norwegian Sámi upside down" (Josefsen et al. 2015, 44). Broad national and international attention to the conflict pressured the government to promptly address Sámi rights, while Norway did not want its status of a leading international human and Indigenous rights advocate be tarnished (Semb 2001).

One of the outcomes of the Alta conflict was the appointment of the Sámi Rights Commission in 1980 to examine Sámi rights and draft a new Sámi policy in Norway. The report of the Commission, released in 1984, recognized the Sámi as a distinct people with the right to enjoy their culture according to standards established in international law, notably in Article 27 of the UN Covenant on Civil and Political Rights. The Sámi Parliament in Norway was established by the Sámi Act in 1987, and in 1988, the Constitution was amended to include Article 110a (as a result of 2014 constitutional revision, now Section 108) that obligates the state "to create the conditions necessary for the Sámi to protect and develop their language, their culture and their society" (quoted in Henriksen 1999, 37). The Sámi Parliament is considered the main means of implementing Section 108 of the Constitution. The Sámi Parliament is structured as a Western parliamentary system of a legislative assembly and the executive council, consisting of the President and four or more appointed members. The assembly consists of 39 representatives elected every four years by Sámi registered in a specific electoral roll. In order to be eligible to register, certain criteria must be met.[8] There are seven electoral districts that cover the entire country, and representatives are elected from lists created by Sámi national organizations, Norwegian political parties and local coalitions.

Since the first reindeer-herding acts in the late nineteenth century, the Swedish Sámi policy has been characterized by a two-pronged approach. The policy of segregation was specifically issued for the reindeer-herding Sámi whose way of life was nomadic and to be shielded from outside influence.[9] The rest of the Sámi were deemed for assimilation. In the 1950s and 1960s, Sámi policy in Sweden was integrated into national, interventionist welfare state policies aiming at modernization and rationalization of reindeer herding, now considered solely an economic venture among other industries. The Swedish Sámi Association, established in 1950 with a specific focus on the reindeer-herding rights, played a critical role in opposing Swedish Sámi policy, particularly with regard to Sámi land and resource rights. The government view of Sámi land rights as a privilege granted by the state was challenged by the organization, which posited that the Sámi use of their territories preceded the state. Since the 1960s, the association began to increasingly employ the discourse of Sámi as an Indigenous people as the foundation of their rights to land (Lantto and Mörkenstam 2008). It also launched a suit against the government known as the Taxed Mountain case, in which the organization sought the recognition of its resource rights (Beach 1995).

In 1981, after 15 years of litigation, the Supreme Court of Sweden decided in the Taxed Mountain case that the Sámi only possess a strong usufruct right, not an ownership right to land or water in the disputed reindeer-herding region. Formed in 1982, the Sámi Rights Commission released its final report in 1989, recommending the Swedish government revise legislation to conform to international law and the Taxed Mountain decision in a way that recognizes the rights of the Sámi as an Indigenous people, including their right to self-determination. Following the example from Norway, the recommendations also included the establishment of the Sámi Parliament. In Sweden, the Act on the Sámi Parliament came into effect in 1993. The legal status of the Sámi Parliament in Sweden is limited to a state administrative authority to which the Sámi can elect representatives. As part of the state bureaucracy, the capacity of the Sámi Parliament to independently represent the Sámi in Sweden is repeatedly questioned by Sámi and others (Kuokkanen 2019). The two roles are seen as in conflict with one another, placing the institution in a bind. Scholars have long maintained that the double bind that characterizes the establishment of the Sámi Parliament in Sweden represents unworkable circumstances. Johan Eriksson called it "undoubtedly an impossible situation," and argued: "if a Parliament representing an ethnic minority [sic] is to enjoy any legitimacy and to show any kind of power, it is obvious that it cannot at the same time represent its major opponent, in this case the state" (Eriksson 1997, 162). Others have similarly pointed out how the dual function is problematic when considering the Sámi right to self-determination and stands as a major obstacle in further developing the power and legitimacy of the Sámi Parliament (Vars 2009, 397; Lawrence and Mörkenstam 2012, 207).

In Finland, the establishment of the Sámi Parliament followed quite a different trajectory than in Norway and Sweden. In the early 1970s at the height of the global civil rights movement, the desire to establish an independent national Sámi organization in Finland emerged, especially among the generation of young, educated Sámi. The idea did not, however, take off, mostly due to a split between the younger, more radical Sámi and the older generation of politically active Sámi, who preferred an elected, representative body within the state. The growing demands of the Sámi movement and the pressure it created propelled the state into action by solving the "Sámi question" by containing it. The Sámi Committee was established in 1971 to examine the economic, social, educational and legal status of the Sámi and recommended holding "experimental elections" for an elected representative body for the Sámi in Finland (Lehtola 2005).

The first elections of the "experimental" assembly called the Sámi Delegation were held in late 1972. Besides the establishment of a representative body for the Sámi, one of the key recommendations of the 1973 report was the enactment of the Sámi Act – a proposal put forward by the first Committee of Sámi Affairs report already in 1952. Sámi land rights had been studied since the 1970s, followed by drafting legislation to recognize Sámi land rights. The proposal for the Sámi Act was submitted to the government in 1990, which coincided with the international adoption of the ILO Convention 169 the year earlier.[10]

The proposed Sámi Act would have recognized the Sámi rights to their territories and transferred the ownership and management of them to Sámi siidas[11] from the state. It failed, however, due to vehement opposition by non-Sámi population in Northern Finland and several state bodies refuting Sámi right claims to their territories. As a result of the broad-based antagonism, the question of Sámi land rights was separated from what became termed "Sámi cultural autonomy." This led to the passing of the Act on the Sámi Parliament and the constitutional recognition of the right of the Sámi, as an Indigenous people in Finland, to "cultural autonomy" in 1995. Section 121 of the Constitution stipulates the scope of Sámi cultural autonomy: "In their native region, the Sámi have linguistic and cultural self-government, as provided by an Act" (Ministry of Justice Finland 1999). The administration and implementation of "linguistic and cultural self-government" was delegated to the newly established Sámi Parliament, which replaced the Sámi Delegation. While the Act on the Sámi Parliament somewhat strengthened the weak position of the Sámi Delegation, the Sámi Parliament in Finland has no power or decision-making authority except in a limited number of internal matters such as hiring its own staff and allocating funding to projects related to the Sámi language, education and culture.

The Constitutions of Norway and Finland have recognized a form of Sámi cultural, non-territorial autonomy exercised through elected, representative bodies of the Sámi Parliaments. The three Sámi Parliaments are also government agencies in charge of administering Sámi-related affairs, specifically Sámi cultural policy. All three Sámi Parliaments have somewhat ambivalent mandates, but all have been established as mainly consultative or advisory bodies rather than self-governing institutions. In Sweden, legislation explicitly states that the Sámi Parliament is not a self-government institution but rather a special state authority responsible for Sámi cultural affairs. The Sámi Parliaments exercise limited decision-making authority over their own affairs, mainly through the administration and dissemination of state funding in areas of education, language, health and social services. In addition, the Sámi Parliament in Norway has been delegated the sole authority over Sámi cultural heritage, including responsibility for sacred sites.

The emphasis of the two somewhat incompatible functions – elected representative bodies and administrative authorities – between the three Sámi Parliaments varies significantly. In the past few years, the Sámi Parliament in Norway has been able to increase its authority and political influence on the state administration due to its open-ended mandate and a range of gradual political, structural and legal changes (Josefsen et al. 2015). Whereas the Sámi Parliament of Norway has unquestionably the most clout as a representative assembly with some decision-making authority, the function of the Sámi Parliament in Sweden is constrained mainly to a government agency, creating conflicts with regard to the decision-making authority and, more fundamentally, constituting a structural obstacle to Sámi self-determination (Lawrence and Mörkenstam 2012).

The scope of Indigenous political autonomy in the Arctic

The three regions considered in this chapter are shaped by specific geopolitical contexts and informed by their specific histories, cultures and the various political contexts. Indigenous political discourse in Canada is very different from that of Scandinavia or Greenland, and these differences inform the way in which self-determination is conceptualized and understood in different regions. The Inuit in both Nunavut and Greenland have achieved extensive self-government powers beyond most Indigenous peoples in the world. Yet, ironically, legislation in neither jurisdiction affirms the right of the Inuit as an *Indigenous people* to self-determination or self-government. The Governments of Greenland and Nunavut are public governments that guarantee the participation of all residents of the territory. Both are, however, de facto Inuit governments thanks to the numerical majority of the Inuit in both territories. Both public governments are also Westminster-style parliamentary systems. In other ways, the two governments are very different from one another. The legislature in Nunavut is a consensus-based system with no political parties, while in Greenland, the party system has driven self-government institutions and public life since the early days of Home Rule. In Nunavut, there was a strong and explicit desire to establish an explicitly Inuit government through the incorporation of Inuit values and knowledge in the governance structures, a work force representative of the Inuit population and decentralizing the administration into a number of communities. In Greenland, on the other hand, there has been no broader public interest in establishing a government based on Inuit values and governance principles.

The Inuit in Nunavut and Greenland have achieved extensive political autonomy and jurisdiction over their own affairs. Greenland has achieved the most extensive political autonomy of all Indigenous peoples in the world. In a way, Inuit Greenlanders are only a referendum away from a full political independence. This would obviously come with its own challenges, by far the biggest being the required economic self-sufficiency. This would unavoidably mean aggressively exploiting Greenland's considerable mineral resources, including offshore oil and gas. This in turn would require foreign investment and a large-scale entry of multinational corporations, which would create new dependencies.

As the third and newest territory in Canada, Nunavut has a wide-ranging decision-making authority its own affairs, including housing, health, education, social services and language. The federal government, through Indigenous and Northern Affairs, continues to maintain jurisdiction over land and resource management and development on Crown lands (which cover 80% of Nunavut's territory). This would change, however, with a devolution agreement that would allow the government of Nunavut to receive royalties from economic development. The devolution negotiations are currently ongoing with the federal government, but there are some concerns that Nunavut is not ready to take additional governing responsibilities, considering, for example, that a quarter of government positions remain vacant (Murray 2018).

The scope of Sámi self-government is, by and large, confined to cultural affairs, allocation of state funding and consultation with the governments. In Norway, where the scope of autonomy is considered most extensive, the Sámi Parliament is increasingly burdened by its growing administrative responsibilities with regard to financial and grant management, leaving less and less capacity for political and policy development (Falch et al. 2016). In Sweden, the work of the Sámi Parliament largely revolves around culture and language, even though it is also tasked with serving as governmental expert agency on reindeer herding. As a government administrative agency, the Swedish Sámi Parliament has not been

> granted any actual political influence or real power, such as a right of participation in decision-making, veto-rights concerning administrative decisions, or a legal status as a body to which proposed legislative measures on Sámi issues ought to be referred for consideration by other administrative authorities.
>
> (Lantto and Mörkenstam 2008, 39)

In Finland, although the Sámi Parliament has not been entrenched in "the state machinery" quite extensively as in Sweden (Nyyssönen 2011, 86), it does not exercise self-government authority beyond the areas of culture and language.

All the self-government institutions considered in this chapter have turned out to be quite different from their early visions in the 1970s. The Inuit Land Use and Occupancy Research Project by the Inuit Tapirisat of Canada in the early 1970s outlined Inuit self-determination as its overall prospect, in which Inuit political development and land claim/rights form an indivisible core. The negotiated land claim agreement (NLCA), however, does not "refer to self-government or self-determination but to the specific cultural traits of a particular group and that programs and services should reflect this difference" (Henderson 2008, 231). Thus, Inuit authority pertains more to cultural policies than political self-determination, suggesting that "the NLCA has more in common with multiculturalism policy than the accommodation of desires for self-government" (ibid). This, together with other factors discussed previously, has resulted in a situation where the initial vision for Nunavut remains a distant dream but has led Inuit leadership to think of new governing structures "that has Inuit involved in the design of programs and prioritizes Inuit education, culture and language" (Frizzell 2019).

In Greenland, not everybody agrees that having a government and a parliament consisting of Greenlandic representatives amounts to Inuit decision-making authority. During Home Rule, Henriette Rasmussen, who served as the minister for Social Affairs (1990–95) and the Minister for Culture, Education and Research (2003–2005), observed:

> Our society is not acceptable as it is. I keep searching for our cultural heritage in every step we take politically. The standards and values we are applying in our society have

failed. Yet, nobody considers the values of our culture heritage – they have no immediate monetary value.

(cited in Thomsen 1990, 252)

In the self-government era, Rasmussen maintained that the Inuit

are still not in charge. Our whole administration is done by outsiders. . . . If you look at the structure of the law making, for example, the laws are *not* borne out from the parliament, they are borne out from the administration.

(Interview, March 21, 2013)

Another participant claimed that self-government reflected Inuit values only on paper and posited:

The way it works, the government is still an old model of Denmark and . . . most of the employees of the home-rule government who tried to implement all the things are still Danish people, with a Danish view. . . . So, there is a big gap between what politicians really say and what is really being done, in my opinion.

(Interview with educator, April 10, 2013)

The three Sámi Parliaments have long been criticized for the same reasons: many view them as centralized, bureaucratic institutions lacking community engagement and serving the needs of state bureaucracy rather than those of the Sámi people (Kuokkanen 2011, 2019). As an example, an older woman who has long been involved in Sámi politics (but never a member of the Sámi Parliament) lamented the fact that the Sámi Parliament turned out to be something else she had envisioned: "I thought we would be capable of building something that would be ours, that we wouldn't have to think how the Norwegian Parliament works, how Norwegian political parties work" (cited in Kuokkanen 2011, 52).

The early vision of many Arctic Indigenous people of creating their own culturally meaningful institutions of political autonomy and decision-making has not disappeared but remains a challenge that is not unique to Indigenous peoples in the Arctic. Self-government arrangements may seem empowering and creating greater autonomy but can, in reality, have a reverse, negative effect of increased state control and influence, entrenching and incorporating Indigenous peoples more extensively and intricately into the state structures. By definition, self-government refers to an arrangement with the state of delegated administrative, representative or consultative authority and tasks. A more fundamental, yet largely unrecognized, challenge is the incompatibility of Indigenous social and political orders of direct democracy with structures of centralized parliamentary democracy and its institutions that pressure Indigenous people to accept and adopt a range of foreign premises and practices of power and governance (Nadasdy 2012).[12] This has an alienating effect to many Indigenous individuals and communities who envision real political authority and more localized and culturally meaningful avenues and platforms for collective deliberation and decision-making.

Notes

1 This chapter partly draws on my book, *Restructuring Relations: Indigenous Self-Determination, Governance and Gender*. New York: Oxford University Press, 2019.

2 For more extensive considerations of the self-government models in Greenland and Sápmi, see, for example, Kuokkanen, R. (2019); and in Nunavut, see, for example, Hicks, J. and White, G. (2015); Loukacheva, N. (2007).

3 Interview with civil servant, 3 April 2013.

4 The other main parties include the liberal conservative Atassut (Solidarity) emphasizing interdependence with Denmark and centre-right Democrats. Recently Greenland's political party system has seen the phenomenon of splinter groups forming new parties; the left-wing separatist Partii Inuit, split from Inuit Ataqatigiit in 2013, and centrist Partii Naleraq, split from Siumut in 2014.

5 Interview with civil servant, 3 April 2013.

6 For Lapp Codicil, see Pedersen (1987).

7 Several legal scholars considered the previous draft a groundbreaking instrument in international law as well as a global example of good practice of Indigenous rights (see Scheinin, M. 2006; Koivurova, T. 2008; Fitzmaurice, M. 2009; Bankes, N. 2013).

8 This is the same with all three Sámi Parliaments. Although there are some considerable differences, all the criteria focus, on varying degrees, on the Sámi language either as a mother tongue or language spoken at home.

9 Part of the Swedish Sámi policy was the system known as the Lapp Administration headed by a Lapp bailiff and later assisted by regional supervisors. The Lapp Administration operated from 1885 to 1971 in the three northernmost counties with the purpose of supervising that the Sámi observed the reindeer-herding legislation and informing the Sámi and the settlers of their respective rights and obligations. See, for example, Lantto, P. (2005).

10 Of the three Nordic countries, only Norway has ratified the convention.

11 Historically, the Sámi territory was organized into siidas, Sámi villages which held political and legal authority over their members and territory.

12 For traditional forms of Indigenous direct democracy and governance in the Arctic, see Fondahl and Irlbacher-Fox (2009).

References

Alfredsson, G. 1999. "The Rights of Indigenous Peoples with a Focus on the National Performance and Foreign Policies of the Nordic Countries." *Heidelberg Journal of International Law* 59: 529–542.

Altamirano-Jimenez, I. 2008. "Nunavut. Whose Homeland, Whose Voices?" *Canadian Woman Studies* 26: 128–134.

Bankes, N. 2013. "The Forms of Recognition of Indigenous Property Rights in Settler States: Modern Land Claim Agreements in Canada." In *The Proposed Nordic Saami Convention: National and International Dimensions of Indigenous Property Rights.* Nigel Bankes and Timo Koivurova, 351–378. Oxford and Portland, OR: Hart Publishing.

Beach, H. 1995. "The New Swedish Sámi Policy – A Dismal Failure: Concerning the Swedish Government's Proposition 1992/93:32, Samerna och samisk kultur m.m. (Bill)." In *Indigenous and Tribal Peoples' Rights – 1993 and after,* edited by E. Gayim and K. Myntti. Rovaniemi: Northern Institute for Environment and Minority Law, University of Lapland.

Dahl, J. 1997. "Gender Parity in Nunavut." *Indigenous Affairs* 3/4: 42–47.

Dahl, J. 2010. "Identity, Urbanization and Political Demography in Greenland." *Acta Borealia* 27: 125–140.

Eriksson, J. 1997. "Partition and Redemption: A Machiavellian Analysis of Sami and Basque Patriotism." PhD diss., Umeå University.

Falch, T., P. Selle, and K. Strømsnes. 2016. "The Sámi: 25 Years of Indigenous Authority in Norway." *Ethnopolitics* 15: 125–143.

Fitzmaurice, M. 2009. "The New Developments Regarding the Saami Peoples of the North." *International Journal on Minority and Group Rights* 16: 67–156.

Fondahl, G., and S. Irlbacher-Fox. 2009. "Indigenous Governance in the Arctic." A Report for the Arctic Governance Project. Toronto, Walter and Duncan Gordon Foundation.

Frizzell, S. 2019. "'We've Done Good': Nunavummiut See Progress, More Work as Territory Turns 20." CBC *News*, 1 April. Accessed 13 April 2019. www.cbc.ca/news/canada/north/nunavut-20th-anniversary-self-government-1.5064412.

Hannikainen, L. 1996. "The Status of Minorities, Indigenous Peoples and Immigrant and Refugee Groups in Four Nordic States." *Nordic Journal of International Law* 1: 1–71.

Henderson, A. 2008. "Self-Government in Nunavut." In *Aboriginal Self-Government in Canada: Current Trends and Issues*, edited by Y. Belanger. Saskatoon: Purich.

Henriksen, J. B. 1999. *Saami Parliamentary Co-Operation: An Analysis*. Guovdageaidnu, Norway: Nordic Sámi Institute.

Hicks, J., and G. White. 2015. *Made in Nunavut. An Experiment in Decentralized Government*. Vancouver: UBC Press.

Jonsson, I. 1997. "Greenland – From Home Rule to Independence: New Opportunities for a New Generation in Greenland." *In NARF-Symposium: Dependency, Autonomy and Conditions for Sustainability in the Arctic*. Nuuk: Institute of Economics and Management.

Josefsen, E., U. Mörkenstam, and J. Saglie. 2015. "Different Institutions within Similar States: The Norwegian and Swedish Sámediggis." *Ethnopolitics* 14: 32–51.

Kleivan, I. 1992. "The Arctic Peoples' Conference in Copenhagen, November 22–25, 1973." *Études/Inuit/Studies* 16: 227–236.

Koivurova, T. 2008. "The Draft Nordic Saami Convention: Nations Working Together." *International Community Law Review* 10: 279–293.

Kuokkanen, R. 2011. "Self-Determination and Indigenous Women – 'Whose Voice Is It We Hear in the Sámi Parliament?'." *International Journal of Minority and Group Rights* 18: 39–62.

Kuokkanen, R. 2019. *Restructuring Relations: Indigenous Self-Determination, Governance and Gender*. New York: Oxford University Press.

Lantto, P. 2005. "Raising Their Voices: The Sami Movement in Sweden and the Swedish Sami Policy, 1900–1962." In *The Northern Peoples and States: Changing Relationships*, edited by A. Lette. Tartu: Tartu University Press.

Lantto, P., and U. Mörkenstam. 2008. "Sami Rights and Sami Challenges: The Modernization Process and the Swedish Sami Movement, 1886–2006." *Scandinavian Journal of History* 33 (1): 26–51.

Lawrence, R., and U. Mörkenstam. 2012. "Självbestämmande genom myndighetsutövning?" *Statsvetenskaplig tidskrift* 2: 207–239.

Lehtola, V. P. 2005. *Saamelaisten parlamentti. Suomen saamelaisvaltuuskunta 1973–1995 ja Saamelaiskäräjät*. Inari: Sámediggi.

Loukacheva, N. 2007. *The Arctic Promise: Legal and Political Autonomy of Greenland and Nunavut*. Toronto: University of Toronto Press.

Ministry of Justice Finland. 1999. *The Constitution of Finland*. Unofficial translation. 731/1999.

Minor, T. 2002. "Political Participation of Inuit Women in the Government of Nunavut." *Wicazo Sa Review* 17: 65–90.

Murray, N. 2018. "Nunavut Likely to Miss Its Target on Reaching Agreement-in-Principle on Devolution." *CBC News*, 8 March. Accessed 12 April 2019. www.cbc.ca/news/canada/north/nunavut-devolution-john-main-paul-quassa-1.4567030.

Nadasdy, P. 2012. "Boundaries among Kin: Sovereignty, the Modern Treaty Process, and the Rise of Ethno-Territorial Nationalism among Yukon First Nations." *Comparative Studies in Society and History* 54: 499–532.

Nielsen, J. K. 2001. "Government, Culture and Sustainability in Greenland: A Microstate with a Hinterland." *Public Organization Review* 1: 229–243.

Nyyssönen, J. 2011. "Principles and Practice in Finnish National Policies toward the Sámi People." In *First World, First Nations. Internal Colonialism and Indigenous Self-Determination in Northern Europe and Australia*, edited by G. Minnerup and P. Solberg. Brighton: Sussex Academic Press.

Pedersen, S. 1987. "The Lappcodicill of 1751 – Magna Charta of the Sami? Mennesker og Rettiheter." *Nordic Journal on Human Rights* 3: 31–38.

Scheinin, M. 2006. "Ihmisen ja kansan oikeudet – kohti Pohjoismaista saamelaissopimusta." *Lakimies* 1: 27–41.

Semb, A. J. 2001. "How Norms Affect Policy – The Case of Sami Policy in Norway." *International Journal on Minority and Group Rights* 8: 177–222.

Thomsen, M. L. 1990. "Women and Politics – A Talk with Three Greenlandic Women." In *Gossip: A Spoken History of Women in the North*, edited by M. Crnkovich. Ottawa: Canadian Arctic Resources Committee.

Timpson, A. M. 2006. "Stretching the Concept of Representative Bureaucracy: The Case of Nunavut." *International Review of Administrative Sciences* 72: 517–530.

Vars, L. S. 2009. "The Sámi People's Right to Self-Determination." PhD diss., University of Tromsø.

Wachowich, N., A. A. Awa, R. K. Katsak, and S. P. Katsak. 1999. *Saqiyuk: Stories from the Lives of Three Inuit Women*. Montreal and Kingston: McGill-Queen's University Press.

White, G. 2009. "Governance in Nunavut: Capacity vs. Culture?" *Journal of Canadian Studies* 43: 57–81.

16

ADVANCING INUIT SELF-DETERMINATION AND GOVERNANCE IN ALASKA AND CANADA AMIDST RENEWED GLOBAL FOCUS ON THE ARCTIC

Tim Aqukkasuk Argetsinger

Abstract

The 1971 Alaska Native Claims Settlement Act (ANCSA) constitutes the largest land claims settlement with Indigenous peoples in U.S. history. The terms of the settlement included the termination of Aboriginal title in exchange for land and cash and the creation of for-profit Alaska Native corporations to manage landed assets and generate monetary wealth for their Alaska Native shareholders. ANCSA was novel for its time and served as an early reference point for Inuit in Canada, who settled four land claims agreements between 1975 and 2005. However, unlike ANCSA, Inuit land claims agreements are comprehensive, affirming specific rights that enable Inuit representative organizations to play a more dynamic and efficient role in creating prosperity for their beneficiaries. In this chapter, I draw on specific examples to argue that Inuit land claims agreements serve as useful reference points for Alaskan Inuit seeking to remedy the lasting challenges created by ANCSA. I argue that Inuit self-determination is an evolving story that cannot be confined to any one agreement or piece of legislation and that bold actions and ideas are required in Alaska and other regions of Inuit Nunaat to enact governance models that create the best possible outcomes for our people.

Introduction

This chapter evaluates the capacity of Inuit-led institutions in Alaska and Canada to respond to and navigate risks and opportunities associated with the renewed global focus on the Arctic. The Arctic is viewed by many nation states as a globally strategic region for advancing their economic and political priorities. The United States and Canada control nearly half of the Arctic coastline, yet have largely ignored the Arctic until recently. Climate change is likely to continue driving warming trends that could result in ice-free summers across the Inuit homeland as early as 2036 (Peng et al. 2018), further opening the region to shipping, resource extraction, and military activity. In Alaska and Canada, Inuit are rights holders and are the majority population throughout the region, yet our rights, representative institutions, and priorities are often overlooked by nation states, despite the disproportionate burden Inuit-led institutions face in navigating risks and opportunities associated with any activity carried out in our homeland. Strong Inuit-led institutions in Alaska and Canada are therefore needed for securing Inuit-state sovereignty; supporting marine and coastal management; protecting biological diversity; and navigating the environmental, social, and economic challenges and opportunities associated with the renewed global focus on the Arctic. This chapter considers the leadership role our people must play in shaping the Arctic policy of the United States, Canada, and other nation states and the advancements in Inuit self-determination and governance that are likely required in order to do so effectively.

Inuit are one people living throughout Inuit Nunaat, the Inuit homeland, encompassing parts of Chukotka (Russian Federation), northern Alaska and Canada, and Kalaallit Nunaat (Greenland). Inuit Nunaat encompasses the entire Arctic coastline for the United States, Canada, and the Kingdom of Denmark and includes significant offshore areas. Our people form the majority population throughout this distinct geographic, cultural, and political region; co-manage its lands and waters with governments; and are, collectively, the largest private landowners in the world. In addition, Inuit remain central to the international growth of the Arctic as a definable global region governed by peaceful principles rather than military force.

Inuit in Canada have made positive strides in establishing cohesive, Inuit-led representative institutions between 1971 and 2005 that have been recognized as global models for advancing Indigenous self-determination (Wilson and Selle 2019). These institutions continue to evolve through democratic governance structures at the regional and national levels that help facilitate accountability among Inuit leadership as well as unified national Inuit policy positions and initiatives. By contrast, Inuit-led institutions in Alaska face challenges stemming from overlapping and sometimes conflicting mandates that can sow political division and undermine the prosperity of Inuit communities. Despite the absence of cohesive Inuit governance and representation at the regional or state-wide levels, Inuit-led institutions in Alaska possess significant experience and capacity in the areas of governance, economic development, and service delivery that could be instructive for Inuit-led institutions in Canada.

Inuit in Alaska began to organize politically in the 1950s in order to resist the theft and exploitation of our territory by the U.S. government, creating Inuit-led institutions for advancing Inuit self-determination in areas such as news media, sea mammal harvesting and management regimes, and land claims. Inuit-led institutions were established in an ad-hoc manner during the past six decades to serve practical needs such as land and resource management, economic development, and service delivery. However, this approach has led to multiple Inuit-led institutions with overlapping mandates serving the same constituency. Cohesive and unified representative institutions, supported by democratic governance structures that consolidate governance, resources, decision-making, and power, are needed to more effectively address the health and wellness of our people and navigate potential challenges and opportunities associated with the renewed global focus on the Arctic.

Inuit-led institutions shaping Arctic policy

Increasing shipping traffic in parts of the Northeast Passage and Northwest Passage pose risks for Inuit communities that include potential for restricting and endangering travel over ice and water, disturbances to biodiversity, local wildlife, and the contamination of traditional food sources, among others (Arctic Corridors Research 2019; Porta et al. 2017). Longitudinal data on sea vessel traffic throughout Inuit Nunaat is scarce, but overall vessel traffic in the Canadian Arctic has increased in the past decade, with total shipping distances tripling between 1990 and 2015, including among bulk carriers, passenger ships, government vessels, and pleasure craft (Dawson 2018). In Nunavut in 2018 alone, dozens of cruise ships visited 13 of the territory's 23 communities over the span of three months, discharging thousands of tourists that in some cases resulted in significant revenue for local governments (Government of Nunavut 2018). Tourism and other marine shipping activities could have a variety of consequential impacts for Inuit, including in the areas of Inuit harvesting and food security, economic development, infrastructure and employment, and public safety and emergency response.

Cargo ships could begin regularly traversing the Northwest Passage during ice-free summers in the future in order to reduce transoceanic shipping times, further contributing to risks that are compounded by profound infrastructure gaps in Alaska and Canada that include the absence of deep-water ports, harbours, and seaways, as well as other critical infrastructure needed to support shipping traffic. The global climate emergency could also prove to be economically lucrative for those seeking to extract and ship natural resources from the Arctic, possibly generating indirect economic benefits to Inuit communities through impact benefit agreements. At the same time, resource exploration and extraction could lead to more shipping and disturbances to communities, wildlife, and the environment. It is estimated that as much 90 billion barrels of oil, 1,669 trillion cubic feet of natural gas, and 44 billion barrels of natural gas liquids may remain to be found in the Arctic, motivating future exploration and extractive activities throughout Inuit Nunaat and other parts of the Arctic for decades to come (USGS 2008).

Inuit from Alaska and Canada are the majority population in Inuit Nunaat and have used, occupied, and travelled throughout it for millennia. Most Inuit communities throughout the region are located on the coast and rely on fish and wildlife harvesting for cultural continuity, food security, and to help offset high living costs. Inuit therefore face the greatest impacts from increased shipping and activities that rely on shipping throughout this region, where, despite being under-resourced, we already play pivotal roles in the areas of sovereignty and defence (Government of Canada 2017) and marine search and rescue (Senate of Canada 2018). Inuit representative organizations, such as the Inuit Circumpolar Council, an international NGO, have urged Canada to partner with Inuit to advance solutions to complex, emerging challenges in relation to Arctic shipping and tourism that are effective for Inuit communities and for Canada, as well as to include Inuit in diplomatic dialogue with nation states (Inuit Circumpolar Council Canada 2019).

However, Inuit are too often absent from or on the peripheries of discussions, treaties, and polices dealing with Arctic shipping and activity throughout the waters and lands encompassed by Inuit Nunaat, despite the probable impacts on our communities and territory from these and other activities. Neither national nor international legislation governing Arctic shipping, for example, acknowledges the use of sea ice by Inuit, a form of natural infrastructure that we rely on for travel and to feed our families (Porta et al. 2017). Inuit are permanent participants to the Arctic Council through the Inuit Circumpolar Council, an international NGO, yet this forum has proven ill-equipped to meaningfully advance Inuit self-determination and governance in all aspects of Arctic diplomacy. Moreover, Inuit are not directly engaged in activities of the International Maritime Organization, the U.N. agency with responsibility for the safety and security of shipping and the prevention of marine and atmospheric pollution by ships, whose International Code for Ships Operating in Polar Waters came into force on January 1, 2017.

Chatter about the Arctic and its strategic significance among nation states, investors, and researchers has grown to a roar in the past two decades. The Arctic Circle Conference assemblies that began convening annually in Reykjavik in 2013 are perhaps the most high-profile reminder that many states view the Arctic as little more than an investment opportunity – a strategic geographic region within the public domain, free for those with the largest GDPs and militaries to plunder. The 2018 assembly was attended by 60 countries and featured plenary speeches from state dignitaries such as Japan's foreign affairs minister, who opined on what an "ideal Arctic" should look like (Kono 2018). In the Arctic policies of states like Japan, China, and Russia, Inuit and other Indigenous peoples tend to be characterized as passive stakeholders rather than rights holders that must be included as partners in guiding any actions that could impact our people and territory.

The Government of Canada's Arctic and Northern Policy Framework released in September 2019 is a welcome exception to this disturbing trend (Government of Canada 2019). Inuit in Canada partnered with the federal government in co-developing the Framework and developed a separate, complementary Inuit Nunangat Chapter intended to guide how the Framework is implemented in each of the four Inuit regions of Canada (Inuit Tapiriit Kanatami 2019). The Framework's goals and objectives align with Inuit priorities, and its introductory text specifically cites the importance of respecting and supporting Inuit self-determination in the implementation of those goals and objectives. Inuit in Canada managed to leave their imprint on the Framework in large part due to the cohesive and unified nature of Inuit representation and governance in Canada, which extends from the regional to international levels.

Inuit have also acted to centre our voices and priorities in relation to Arctic sovereignty and shipping in and around the Northwest Passage. In 2009, the Inuit Circumpolar Council released *A Circumpolar Inuit Declaration on Sovereignty in the Arctic*, urging Arctic states to include Inuit as partners in charting the future of the region (Inuit Circumpolar Council 2009). Inuit Tapiriit Kanatami, the national Inuit representative organization in Canada, has documented Inuit perspectives on the Northwest Passage, shipping, and marine issues, including those of former Inuit Circumpolar Council-Canada President Nancy Karetak-Lindell. In the words of Karetak-Lindell:

> A sense of security is strong when we know who is coming, from where, and their purpose. But our sense of security and sense of control is eroding because not only are we facing increased Canadian traffic, but also foreign traffic coming through. We can no longer have confidence that we know who is traveling in our waters and lands or why. . . . Inuit have to foremost be at the discussions and consulted before any decisions are made and any research is done. We have to be part of the planning and be at the table to decide what framework will be used to come to decisions and what research will be done and how. This will ensure solid decisions are made that will mitigate changes that we know are coming in the future. We cannot turn back the tide, especially on climate change, but we can be better prepared to successfully adapt.
> (Inuit Tapiriit Kanatami 2017)

The rhetorical war over the Arctic is well underway, with nation states such as the United States and China disputing Canada's claims to sovereignty over the portion of the Northwest Passage threading through its territory. China, an Arctic Council observer since 2013, has been particularly aggressive on the Arctic, releasing its own Arctic policy in January 2018 in which it absurdly characterizes itself as a "near-Arctic state" (China, January 2018). China's Arctic policy includes a commitment to invest a portion of its $1 trillion Belt and Road Initiative infrastructure fund into developing Arctic shipping lanes, a "Polar Silk Road" open to all states. Russia has similarly outlined plans for major infrastructure investments, including ports and icebreakers, throughout its Arctic coastline (Isachenkov and Titova 2019). In May 2019, U.S. Secretary of State Mike Pompeo upended the 1988 Arctic Co-operation agreement between the United States and Canada by declaring in a speech to Arctic Council ministers that Canada's sovereignty claims over the Northwest Passage are illegitimate (Blanchfield 2019).

Unified and cohesive Inuit-led institutions that can authoritatively represent their respective Inuit regions and constituencies are needed to facilitate Inuit participation in U.S. and Canadian Arctic policy, diplomacy, and decision-making at all levels of government. At the same time, such institutions are needed within our respective jurisdictions to effectively reduce social and economic inequities throughout Inuit Nunaat and secure the infrastructure investments required to sustain Inuit communities. The four Inuit land claims organizations in Canada, though motivated by a variety of factors, are already evolving toward self-government: in addition to the Nunatsiavut Government, an Inuit self-government established in 2005, one Inuit land claims organization is exploring self-government (Nunavut Tunngavik Inc 2018), another has created an Inuit constitutional task force and is actively seeking to commence negotiations for self-government (Makivik Corporation 2019a), and yet another is actively engaged in negotiating a final Inuit self-government agreement (Inuvialuit Regional Corporation 2019).

Alaskan Inuit political organization

There are 27 majority Inuit (Iñupiat) communities in Alaska; however, this analysis focuses only on the approximately 11,000 Alaskan Inuit living in 19 communities within the boundaries of the Northwest Arctic Borough and North Slope Borough, two public regional governments governed by elected assemblies. These two jurisdictions combined encompass a land area of more than 337,000 square kilometres. Alaskan Inuit make up approximately 63 percent of the total combined population of the Northwest Arctic Borough and North Slope Borough (U.S. Census 2018). Alaskan Inuit must navigate complex layers of government and representation that can be grouped into three main categories: Inuit-owned, for-profit corporations; public regional governments; and federally recognized tribes and regional tribal health and social service providers. Despite the fractured and sometimes conflicting mandates and priorities of these institutions, they each possess significant experience and capacity that could be more effectively leveraged through a more cohesive and unified model of representation and governance.

Inuit political organization in Alaska is largely rooted in the Indigenous land claims movement of the 1960s, evolving in an ad-hoc fashion to meet community and regional priorities such as education, service delivery, and economic development. In the late 1960s and early 1970s, Alaskan Inuit played a pivotal role in advancing Inuit self-determination by spearheading land claims and establishing the North Slope Borough as a home rule government, the first majority Inuit public government in Inuit Nunaat. The Indigenous land claims movement in Alaska culminated in 1971 with the passage of the Alaska Native Claims Settlement Act (ANCSA) by the U.S. Congress. This legislation purports to terminate the aboriginal title of Indigenous peoples in Alaska and transferred the land title to approximately 180,000 square kilometres of territory as private property to Indigenous peoples. The federal government transferred the title to these lands to 12 regional corporations and more than 200 community-based corporations that are owned by their Indigenous shareholders.

North Slope Inuit initially opposed the settlement, which provides a total land allocation for all Indigenous peoples in Alaska that is less than the total land area claimed for their region alone. ANCSA was also opposed by many other Indigenous peoples throughout the state because

Congress failed to secure their consent during negotiations. Instead, the federal government opted to accept the consent of a handful of unelected Indigenous representatives, unilaterally imposing the legislation on Indigenous peoples in order to expedite the construction of the trans-Alaska pipeline (Berger 1985). Unlike comprehensive Inuit land claims in Canada, whose beneficiaries consented by referendum, the legislative content that became ANCSA was never consented to outside of a small circle of Indigenous leadership. Furthermore, ANCSA's purported termination of aboriginal title and its silence on Indigenous harvesting rights mean that to this day, Indigenous peoples must adhere to most of the same state and federal harvesting regulations that apply to all other U.S. citizens carrying out harvesting activities on state and federal lands.

North Slope Inuit established the North Slope Borough under state law in 1972 in order to collect revenue through taxes from the Prudhoe Bay oil fields and exercise jurisdiction over land zoning, as well as to advance local self-determination for the region's eight communities, including through the administration of primary and secondary education. Inuit in Northwest Alaska eventually followed, establishing the Northwest Arctic Borough in 1986 following the discovery of lead and zinc deposits that today remain the primary source of revenue for this jurisdiction. Like other public governments, the Northwest Arctic Borough and North Slope Borough governments are led by elected mayors and assemblies and oversee the delivery of key services such as education, public safety, and wildlife monitoring and species management.

Inuit in Alaska have achieved a measure of economic self-determination through the Inuit-owned economic development corporations established under ANCSA. The legislation directed that 12 for-profit regional corporations be created under Alaska state law, which were to be the vehicles for distributing the settlement's land and monetary benefits to their Indigenous shareholders (U.S. Government Accountability Office 2012). ANCSA transferred the title to approximately 178,000 square kilometres of land area, or approximately 12 percent of Alaska's landmass, and $962 million, to these corporations as start-up capital to begin generating revenue for their shareholders. However, ANCSA does not affirm specific rights for Inuit and other Indigenous peoples in Alaska, instead transmuting collective Indigenous land ownership based on aboriginal title into corporate stock held by Indigenous peoples in their regional and/ or community corporations. Although the basic corporate framework established by ANCSA has not changed in the last half century, the legislation has been amended, including amendments that authorize the issuance of stock to Indigenous peoples born after the land claim was settled and extend the prohibition on the sale of stock, as well as amendments that exempt the corporations from regulation by the U.S. Securities and Exchange Commission (SEC) (U.S. Government Accountability Office 2012, 2). NANA Regional Corporation (NANA) and the Arctic Slope Regional Corporation (ASRC) are the two main Inuit-led ANCSA corporations for the Northwest Arctic and North Slope regions, respectively, in addition to one community-based ANCSA corporation in the Northwest Arctic and eight community-based ANCSA corporations on the North Slope. Although the mission statements of NANA and ASRC are broad, both focus almost exclusively on economic development and do not directly administer programs and services.

ANCSA corporations are peculiar creatures. Though Indigenous peoples own ANCSA corporations and are therefore de facto landowners, ANCSA corporations are not rights holders, and they do not represent Indigenous peoples in our relationships with governments. This distinction has become increasingly blurred as Inuit-led and other ANCSA corporations have transformed economic prosperity into political influence. NANA and ASRC have in recent years declared billions of dollars in gross revenue and paid their shareholders millions in dividends. In 2017 alone, for example, ASRC spent nearly $590,000 in lobbying fees in support of opening up a section of the Arctic National Wildlife Refuge for oil exploration and development (Fountain and Eder 2018). NANA and ASRC have spent millions on super PACs and public relations campaigns to protect and further their economic development interests and activities.

State-wide, ANCSA corporations are among the largest employers in Alaska and make significant contributions to their regional economies as well as the state and national economy. According to one recent ranking, ANCSA corporations made up nine of the top ten corporations in Alaska ranked by gross revenue, two of which were NANA and ASRC (Alaska Business 2018). Inuit-led and other ANCSA corporations are politically active, and, despite their branding as purveyors of Inuit culture and values, their diverse subsidiaries located throughout the U.S., investment strategies, and mainly non-Inuit employees tend to resemble those of other sprawling U.S. corporations (U.S. Senate Committee on Homeland Security and Governmental Affairs 2009). NANA's investments in renewable energy projects throughout the Northwest Arctic Borough to help curb high energy costs are a noteworthy exception, among other positive local investments and activities spearheaded by the corporation (Herz 2018).

Alaskan Inuit tribal governments

Unlike ANSCA corporations, U.S. federal Indian law identifies federally recognized tribes as the legal representatives of Indigenous peoples in our fiduciary or "trust" relationship with the federal government. There are 229 federally recognized tribes in Alaska, representing more than one-third of all federally recognized tribes in the United States. Despite being one people in Alaska and throughout Inuit Nunaat, Inuit in Alaska are enrolled in many federally recognized tribes whose citizens and territory roughly cohere to the lands used and occupied by the ancestors of Inuit in and around present-day communities. In the Northwest Arctic Borough, for example, there are 11 federally recognized tribes – 1 representing each of the region's communities and a further 8 within the North Slope Borough.

Federally recognized tribes in Alaska are distinct from most other U.S. federally recognized tribes in that they do not possess a land base over which to exercise their jurisdiction. However, tribes do exercise jurisdiction in critical areas, including child welfare, education, and the delivery of health services. In an effort to share resources and achieve economies of scale, Alaska tribes have developed tribal consortia in order to deliver services within their respective regions. In the Northwest Arctic Borough, for example, Maniilaq Association, the regional tribal health association and largest employer, is mandated by the region's 11 federally recognized tribes to administer health and wellness services through a compacting arrangement with the federal government under the 1975 Indian Self-Determination and Education Assistance Act. Through this arrangement, Maniilaq administers the region's hospital, nursing home, and community clinics; Maniilaq also administers a range of social services, including a children's home, senior and disability services, and behavioural health services. The Arctic Slope Native Association, the North Slope equivalent of Maniilaq, administers similar services throughout the North Slope Borough.

This fragmentation of power, resources, mandates, and jurisdiction among Inuit in Alaska can create conflict between Inuit-owned, for-profit corporations; public regional governments; and federally recognized tribal governments, even though they all serve the same population. In 2011, for example, the Native Village of Point Hope, a federally recognized tribe on the North Slope, opposed plans by Shell Oil to drill exploratory wells in the community's hunting waters in the Chukchi Sea, clashing with the community's Inuit-led ANCSA corporation as well as with ASRC (Yardley and Olsen 2011). In 2010, the Native Village of Kivalina, a federally recognized tribe within the Northwest Arctic Borough, along with the Native Village of Point Hope and two environmental groups, challenged a federal water pollution discharge permit issued by the Environmental Protection Agency to Teck Alaska, Inc., the operator of an open pit lead and zinc mine located on lands it leases from NANA. The EPA's Environmental Appeals Board eventually rejected the challenge, leading to a lawsuit, *Native Village of Kivalina IRA Council v. United States Environmental Protection Agency*, that the tribe lost (Hansen 2013). Similar litigation is ongoing: as recently as March 2019, the Native Village of Nuiqsut, a North Slope tribe, along with five environmental groups, filed a lawsuit against the federal Bureau of Land Management, alleging that the agency did not conduct a thorough enough environmental review of drilling activity in the National Petroleum Reserve-Alaska, which borders the community and which, ostensibly, ASRC and the North Slope Borough stand to benefit from (Oliver 2019).

Such conflict is difficult to avoid between institutions with varying capacity and resources, sometimes conflicting or overlapping mandates, and overlapping jurisdiction. However, such conflict only serves to undermine Inuit self-determination and representation among our small population and is politically and economically unsustainable. Moreover, such conflict between Inuit-led institutions is likely to occur more frequently if renewed global focus on the Arctic leads to greater investment in the Arctic, shipping, and resource extraction throughout the region that will inevitably be perceived as business and economic development opportunities for some and as existential threats to others.

Canadian Inuit political organization

Canadian Inuit political organization is characterized by Inuit-led rights-holding institutions with streamlined and unified representation and governance structures that extend from the regional to international levels. Canadian Inuit are one of three Indigenous peoples recognized in section 35 of Canada's constitution, the majority of whom live in Inuit Nunangat, the Inuit homeland encompassing the Inuvialuit Settlement Region in the Northwest Territories, Nunavut, Nunavik in northern Quebec, and Nunatsiavut in northern Labrador. Inuit Nunangat encompasses nearly one-third of Canada's landmass, its entire Arctic coastline, and significant offshore areas. Approximately 73 percent of the 65,000 Inuit in Canada live in the 51 Inuit communities throughout this distinct geographic, political, and cultural region (Inuit Tapiriit Kanatami 2018). Most Inuit are registered beneficiaries of one of four comprehensive Inuit-Crown land claims agreements completed between 1975 and 2005. Although Inuit have been recognized as being within core federal legislative jurisdiction since 1939, Inuit are not included in the Indian Act, the draconian federal statute underpinning the system of programs, services, and First Nation membership and administration.

Subsequent to the negotiation of the James Bay and Norther Quebec Agreement in 1975 by Nunavik Inuit, Inuit in Canada developed regional organizations to advance land claims that were supported by a national organization. Four Inuit land claims organizations negotiated their respective land claims agreements with the Crown as well as with provincial and territorial governments between 1975 and 2005 and today advocate for the implementation of those agreements by governments. The four Inuit land claims organizations are: Inuvialuit Regional Corporation, Nunavut Tunngavik Inc., Makivik Corporation, and Nunatsiavut Government. The leaders of each Inuit land claims organization are elected by the enrolled beneficiaries of each Inuit land claims organization and serve out set terms. In addition to being beneficiaries of their respective land claims agreements, Inuit who live in Inuit Nunangat are residents of the territories and provinces whose borders crisscross their homeland. At the national level, Inuit

Tapiriit Kanatami (ITK) is the national Inuit representative organization and works to ensure federal programs, policies, and initiatives are aligned with Inuit priorities. ITK is directed by the presidents of the four Inuit land claims organizations.

Inuit-Crown land claims agreements vary widely in their scope and content, yet all affirm specific collective rights in exchange for the purported termination of Aboriginal title. They do not preclude Inuit from exercising our right to self-determination and self-government. This basic characteristic of Inuit land claims agreements, coupled with the fact that each was democratically ratified through referenda by the prospective Inuit beneficiaries of each agreement, sets them apart from ANCSA, which does not affirm specific rights for Indigenous peoples in Alaska, nor does it establish a legal framework or basis upon which to do so. The specific rights affirmed by Inuit land claims agreements include rights in relation to Inuit land, resource, and wildlife harvesting and management, as well as social and cultural rights.

Moreover, three of the Inuit-Crown land claims agreements include provisions that led to the creation of subnational or regional governments: the 1975 James Bay and Northern Quebec Agreement created the Kativik Regional Government in northern Quebec, a public regional government similar to Borough governments in Alaska; the 1993 Nunavut Agreement led to the creation of a new territory in 1999 and the Government of Nunavut (GN); and the 2005 Labrador Inuit Land Claims Agreement established the Nunatsiavut Government (NG), the only Inuit self-government in Canada. Inuit land claims organizations are generally not involved in service delivery. Nunavut Tunngavik Inc. may administer Inuit-specific programs and initiatives for its beneficiaries in Nunavut, for example, but it does not administer government services, most of which fall within the jurisdiction of the GN, a public territorial government. The GN and its 21-member legislative assembly have a de facto Inuit-focused mandate by virtue of the territory's majority Inuit population, though the territorial government does have legal obligations to Inuit under the Nunavut Agreement, including in the areas of Inuit employment and Inuit consultation in the development of social and cultural policy.

The NG is a unique institution within Alaskan and Canadian Inuit political organization. It is unique among Inuit land claims organizations in Canada because in addition to being the rights holder for Nunatsiavut Inuit beneficiaries, NG also oversees the administration of many services throughout Nunatsiavut's five communities, thereby consolidating institutions of government with the functions of an Inuit land claims organizations. This model differs from Nunavut and Nunavik, where the mandates of Inuit land claims organizations and public governments may differ, contributing to political conflict between Inuit rights holders and public governments that can undermine Inuit rights in areas such as education, employment, and housing (Pigott 2016).

Canadian Inuit have partnered with the government of Canada to proactively protect and manage waters in and around their communities, including waters within the Northwest Passage. In 2017, the NG signed a statement of intent with the federal department of Environment and Climate Change Canada to launch the Imappivut Initiative to manage and protect Nunatsiavut Inuit interests in the 48,690 square kilometres of coastal and marine area encompassed by the Labrador Inuit Land Claims Agreement. Though still in the data collection phase at the time of writing, the initiative will include a marine co-management plan between the NG and federal government developed to protect a stretch of coastline longer than California's (Imappivut 2019). The NG's decision to pursue the development of a co-management plan was in part motivated by its desire to manage the impacts and opportunities of increased shipping traffic throughout the region (Bissett 2017).

The Qikiqtani Inuit Association, an economic development–focused member organization of Nunavut Tunngavik Inc., similarly partnered with the federal government to establish Tallurutiup Imanga in August 2019, the largest national marine protected area in Canada. Tallurutiup Imanga encompasses approximately 108,000 square kilometres of the Lancaster Sound in the north-eastern region of Nunavut, within the Northwest Passage. In addition to being an ecologically rich harvesting area for Inuit, the creation of the national marine conversation area enables Inuit and the federal government to monitor the area and co-manage fisheries and marine transportation activities in a more ecologically holistic manner under the terms of the Inuit Impact Benefit Agreement (IIBA) negotiated by the two parties as part of the establishment of the protected area (Parks Canada 2018). The IIBA includes significant federal investments in infrastructure, research, local employment, and economic development that are intended to benefit the five communities located in and around Tallurutiup Imanga (Qikiqtani Inuit Association 2019).

At the national level, Inuit in Canada are represented by Inuit Tapiriit Kanatami, whose president is elected every three years by the leaders of Inuit land claims organizations plus delegates, and whose work is directed by the leaders of the four Inuit land claims organizations. ITK has spearheaded several national initiatives, including the release of a *National Inuit Climate Change Strategy* (Inuit Tapiriit Kanatami 2019), *National Inuit Suicide Prevention Strategy* (Inuit Tapiriit Kanatami 2016), *National Inuit Strategy on Research* (Inuit Tapiriit Kanatami 2018a), and an *Inuit Tuberculosis Elimination Framework* (Inuit Tapiriit Kanatami 2018c). In 2017, ITK brokered the formation of the Inuit-Crown Partnership Committee with the federal government in order to advance work on eight shared priority areas. Co-chaired by the prime minister and ITK president, the Committee includes Inuit leaders and federal cabinet ministers, whose tri-annual meetings are guided by workplans under each priority area.

Inuit self-government in Canada

Canadian Inuit-led institutions are evolving in response to the barriers faced by Inuit land claims organizations in advancing their priorities within the framework of public governments. Canadian Inuit political organization is evolving toward Inuit self-government in order to bring decision-making closer to their communities through streamlined, Inuit-focused governance structures that consolidate representation and power in ethnic Inuit governments. Inuit land claims organizations, despite being rights holders under the Canadian constitution, continue to be marginalized by provincial and territorial governments in the development and implementation of programs, policies, and initiatives that impact Inuit, as well as in federal budget allocations that are intended to benefit Inuit yet are too often absorbed by public governments (Inuit Tapiriit Kanatami 2018b). With the exception of the NG, Inuit land claims organizations are either actively negotiating self-government agreements or have signalled intentions to do so. Self-government could be a positive and powerful step toward strengthening Inuit self-determination and will likely enable Inuit to play a more assertive and active role in charting the region's future.

Taking the NG as a basic model, Canadian Inuit self-government from an Alaskan Inuit perspective can be understood as borough governments and tribal governments rolled into one, with broader jurisdiction, such as law-making authority. The government of Canada recognized the right to self-government as an inherent right in 1995 under section 35 of the Constitution Act, 1982. This inherent right policy recognizes that Indigenous peoples "have the right to govern themselves in relation to matters that are internal to their communities, integral to their unique cultures, identities, traditions, languages and institutions, and with respect to their special relationship to their land and their resources" (Government of Canada 1995).

The Inuvialuit Regional Corporation has been engaged in self-government negotiations since 2006 with the Crown and government of the Northwest Territories. The parties reached an agreement-in-principle in 2015 and have now begun negotiation of a final self-government agreement, financial agreements, and an implementation plan (Inuvialuit Regional Corporation). Conclusion of a final self-government agreement would lead to the development of an Inuvialuit constitution and establishment of a council of elected representatives from each of the region's six communities; Inuvialuit beneficiaries would elect the leader of their new government.

Although linked to the Inuvialuit Final Agreement, the Inuvialuit self-government agreement will not conflict with or replace this agreement, which deals with land and harvesting rights, participation in the management of land and wildlife, and financial compensation. The self-government agreement will include law-making authorities of the Inuvialuit Government, the Government of the Northwest Territories and federal government; it would enable the Inuvialuit Government to pass laws and deliver programs and services established by those laws (Inuvialuit Regional Corporation). Once concluded, an Inuvialuit self-government agreement is expected to outline Inuvialuit jurisdiction over matters such as language and culture, health and social services, social assistance, education, economic development, and justice. The agreement will provide the power to make and enforce Inuvialuit laws, design policies and programs, and deliver services to Inuvialuit beneficiaries throughout the region.

Makivik Corporation has entered into self-government negotiations with Canada, and Nunavut Tunngavik Inc. is exploring self-government. A declaration adopted during a meeting of representatives from all Nunavik organizations and communities in May 2018 provides Makivik with a mandate to negotiate with the governments of Canada and Quebec in order to establish "a form of Indigenous government based on Inuit values, identity, culture and language, with the contribution of the concerned Nunavik organizations who are at the front line" (Makivik Corporation 2018). The Government of Canada began negotiations with Makivik in October 2018 on a draft Memorandum of Understanding that will guide the negotiations (Makivik Corporation 2019). In June 2019, the completed MOU, which outlines Inuit objectives, a timeline for discussions, a work plan, dispute resolution, and funding, was signed by the president of Makivik and the federal Minister of Crown-Indigenous Relations (Rogers 2019).

In October 2018, the NTI board of directors passed a resolution during its annual general meeting directing the organization to commission a study of Inuit self-determination and government in Nunavut. The rationale for the action is premised, in part, on the failure of the federal and territorial governments to fully implement their Nunavut Agreement obligations. The resolution suggests that Inuit self-government may be a mechanism for Nunavut Inuit to promote their rights and self-determination and to ensure Inuit participation in decision-making (Nunavut Tunngavik Inc.). Throughout the past 20 years, the GN has often failed to meet key legal obligations to Inuit under the Nunavut Agreement, including in the areas of Inuit employment in government and Inuit participation in the development of social and cultural policies and in the design of social and cultural programs and services (Nunavut Tunngavik Inc 2016, 2017). For example, as of December 2018, Inuit make up approximately 27 percent of the GN public service when considering the total number of filled and vacant[1] positions in the territory, falling far short of the GN's proportional Inuit employment obligation under the Nunavut Agreement.

Social and economic inequities and the need for
Inuit self-government in Alaska

In contrast to Canadian Inuit political developments, Alaskan Inuit-led institutions have remained largely unchanged for the past three decades. Despite our small population, Alaskan Inuit-led institutions have developed significant experience and capacity in the areas of economic development and health service delivery. However, a partial outcome of public regional governments, Inuit-led corporations, and federally recognized tribes focusing on their respective mandates is that power and representation are dispersed among institutions throughout these two regions. This fragmented model of representation is likely to continue to undermine the potential effectiveness and impact of representation and governance in these two regions as they navigate challenges created by shipping, climate change, extractive activities, Inuit harvesting and food security, militarization, and myriad other issues.

Healthy people and communities are foundational to the sustainability of our communities and to the sustainability of Inuit Nunaat as a distinct geographic, cultural, and political region. However, many Alaskan Inuit experience profound social and economic inequities that should call into question the effectiveness of the existing model of fragmented Inuit-led institutions for creating prosperity for our people. Although health and wellness data for these two regions are limited, what we do know paints a distressing picture despite the economic success of NANA, ASRC, and some community-based corporations. The life expectancy of Alaskan Inuit was fully 10 years less than U.S. whites between 2009 and 2013 (Alaska Native Tribal Health Consortium, August 2017). The elevated rate of suicide among Alaskan Inuit is the most distressing indicator of poor health and wellness experienced by too many families. The suicide rate among Alaskan Inuit between 2012 and 2015 in the Northwest Arctic Borough was 65.5 per 100,000 and 28.3 per 100,000 for the Northwest Arctic Borough and North Slope Borough, respectively, compared to 14.3 per 100,000 for U.S. whites (Alaska Native Tribal Health Consortium, February 1, 2017). Many Alaskan Inuit continue to live below the poverty line even as the corporations we own have ballooned into economic giants in the state: 27 percent and 16 percent of Northwest Arctic Borough and North Slope Borough residents were living in poverty between 2011 and 2015 (Alaska Native Tribal Health Consortium 2017).

Alaskan Inuit, like Canadian Inuit, experience high household crowding. High household crowding is a risk factor for a range of negative health and wellness outcomes, including stress, violence, and poor mental and physical health (World Health Organization 2018). Access to housing is a fundamental human right, and limited access to housing in communities that can only be accessed by air magnifies the negative impacts of other social and economic inequities on families, communities, and society as a whole. Although disaggregated, Inuit-specific data are not available, high household crowding in Northwest Arctic Borough and North Slope Borough communities is indicative of high crowding among Inuit households in these two majority Inuit regions. Nearly 40 percent of homes in the Northwest Arctic Borough and 27 percent in the North Slope Borough are crowded or severely overcrowded, nearly 12 and 8 times the national average, respectively (Alaska Housing Finance Corporation 2017; Alaska Housing Finance Corporation 2017a).

Equally alarming is the moribund status of Iñupiatun, the dialect of the Inuit language spoken in Alaska. According to the results of one survey carried out in 2005, only 14 percent of Northwest Arctic Borough residents self-identify as fluent speakers of the language, the majority of them over the age of 65 (Northwest Arctic Leadership Team 2006). Indigenous languages are a vital element of cultural continuity, and their status is a key indicator of cultural wellness and vitality. Alaskan Inuit-led institutions have overseen the decline of Iñupiatun, even as those institutions have experienced economic success and made important strides toward greater self-determination. Today, English dominates nearly every sector of Alaskan Inuit society, a trend that reflects the limited prioritization of the language among fragmented institutions and which is likely to result in language death within a generation unless significant economic, political, and human capital is invested in language revitalization, maintenance, and promotion.

These and other health and wellness challenges are not unique to Alaskan Inuit; Canadian Inuit health and wellness indicators are, with few exceptions, comparable and in many cases worse than those found among Alaskan Inuit (Anctil 2008; Egeland 2010a, 2010b, 2010c). One such exception is in the area of language: 58 percent of Canadian Inuit report being able to speak the Inuit language well enough to conduct a conversation, and 40 percent report that it is the language used most often at home (Inuit Tapiriit Kanatami 2018d).

Re-imagining Inuit representation and governance

The strides made by Inuit in Canada in the area of self-determination and self-government are promising, but the larger significance of these developments for Inuit health and wellness should be interpreted with cautious optimism. After all, the NG, the only Inuit self-government in Canada, has struggled for nearly 15 years to curb poverty, household crowding, elevated rates of suicide, tuberculosis, and other poor health and wellness indicators among its beneficiaries. Kalaallit Nunaat (Greenland), an autonomous majority-Inuit country within the Kingdom of Denmark, similarly struggles to create prosperity for its people despite enjoying sweeping powers of self-government that most Indigenous peoples, and indeed most Inuit, can only dream of. Self-government, in and of itself, is not a panacea to the complex social and economic challenges facing too many Inuit.

Similarly, the strides made by ANSCA corporations in terms of business and economic development have not directly translated into improved health and social conditions for most Inuit. Despite the robust amount of economic activity generated by ANSCA corporations, it is clear that the generation of revenue, periodically distributed in a trickle-down fashion to Alaskan Inuit shareholders through dividends and scholarships, is insufficient to create prosperous communities.

However, self-government can be a viable path toward streamlined and cohesive Inuit-led institutions that more effectively and efficiently serve our people and withstand changes in government and bureaucracy that too often lead to differential treatment. The consolidation of representation and governance among Inuit land claims organizations in Canada has strengthened Inuit self-determination at the regional and national levels and enables closer working relationships with governments that benefit Inuit in their day to day lives. However imperfect, Inuit land claims organizations enable Inuit to speak with one voice and act decisively through democratically elected leadership.

The challenges and potential opportunities associated with increased tourism, shipping, pollution, and natural resource extraction throughout Inuit Nunaat can only be navigated through streamlined, rights-based Inuit representation and governance. By contrast, the fractured and sometimes conflicting mandates and priorities of public governments, Inuit-led ANCSA corporations, and tribes in Alaska have contributed to political divisions that undermine the sustainability and prosperity of our communities. Alaskan Inuit-led institutions are bursting with potential; they possess significant experience and capacity in the areas of governance, economic development, and service delivery that could be consolidated and leveraged to replace the existing fragmented model of representation and governance with a new governance model.

Such a model should consolidate the 19 federally recognized Inuit tribes into two regional Inuit self-governments that oversee Northwest Alaska and the North Slope. Tribes already possess significant experience and capacity in health and wellness service and program delivery, and many of the programs and services currently administered by the Northwest Arctic Borough and North Slope Borough could be devolved to and subsumed within regional Inuit self-governments. This would likely mean dissolving the Northwest Arctic Borough and North Slope Borough governments. The jurisdiction and mandates of Maniilaq Association and the Arctic Slope Native Association, as tribal health service delivery agents, could simply be expanded to encompass additional areas of service delivery, with these two entities evolving into Inuit self-governments. Within such a hypothetical model, the bylaws, ownership, governance structures, and mandates of Inuit-led ANCSA corporations could be modified and aligned with the priorities of Inuit self-governments, serving as the business arm of Inuit self-governments. Doing so would align economic development with the priorities of democratically elected regional Inuit tribal self-governments that, similar to governments in other jurisdictions, measure success not in revenue and dividends but in the health and wellbeing of our people over time.

Achieving such an outcome would likely require addressing significant legal barriers that could include amending ANCSA, as well as the state laws governing the development and administration of borough governments. Equally challenging would be convincing those who have a vested interest in ANCSA corporations, including their shareholders, that ceding power and repurposing the current model is in the best interests of Inuit in Alaska. Although challenging, the development of such a new governance model is achievable and would likely result in more effective and impactful Inuit representation and governance at a time of renewed global focus on the Arctic.

Conclusion

Canadian Inuit political organization is characterized by relatively cohesive and streamlined Inuit-led institutions that are better positioned than their Alaskan counterparts to manage the risks and opportunities associated with increased shipping traffic and activity throughout Inuit Nunaat. Strides made by Canadian Inuit land claims organizations in the areas of self-determination and self-government may further align the rights and obligations of Inuit land claims organizations with the priorities of their beneficiaries. By further consolidating power and resources in their respective regions, they seek to build on the strengths of Inuit land claims organizations. In contrast, Alaskan Inuit oversee fractured Inuit-led institutions whose mandates are often in conflict, diminishing their effectiveness in creating prosperity for our people. The Northwest Arctic Borough and North Slope Borough, Inuit-led ANCSA corporations, and tribes possess significant experience and capacity in the areas of governance, economic development, and service delivery that could be more effectively leveraged by regional Alaskan Inuit tribal self-governments. Nothing prevents our people from undertaking this political exercise. Doing so would strengthen Alaskan Inuit self-determination and governance and help ensure that we are able to more effectively create social and economic equity for our people while navigating and managing social, economic, and environmental risks and opportunities associated with the renewed global focus on the Arctic.

Note

1 The number is 50 percent when accounting for filled positions only. There are 1,475 vacant positions in Nunavut, 37 percent of which are in the health sector alone.

References

Alaska Business. 2018. "2018 top 49ers." October. Accessed 5 June 2019. www.akbizmag.com/lists/top-49ers/.

Alaska Housing Finance Corporation. 2017a. "2017 Alaska Housing Assessment: Northwest Arctic Borough." Accessed 5 June 2019. www.ahfc.us/application/files/5615/1510/4576/Final_-_Northwest_Arctic_Borough_Summary.pdf.

Alaska Housing Finance Corporation. 2017b. "2017 Alaska Housing Assessment: Arctic Slope Regional Corporation ANCSA Region." Accessed 5 June 2019. www.ahfc.us/application/files/4815/1510/4491/Final_-_Arctic_Slope_Regional_Corporation_Summary.pdf.

Alaska Native Tribal Health Consortium. 2017. "Alaska Native Health Status Report: Second Edition." 26 August. Accessed 5 June 2019. www.anthctoday.org/epicenter/publications/HealthStatusReport/AN_HealthStatusReport_FINAL2017.pdf.

Anctil, M. 2008. *Survey Highlights – Nunavik Inuit Health Survey 2004, Qanuippitaa? How Are We?* Nunavik Regional Board of Health and Social Services. Accessed 5 June 2019. www.inspq.qc.ca/pdf/publica tions/774_ESISurveyHighlights.pdf.

Arctic Corridors Research. 2019. Accessed 5 June 2019. www.arcticcorridors.ca/reports/.

Berger, T. R. 1985. *Village Journey: The Report of the Alaska Native Review Commission.* New York: Hill & Wang.

Bissett, K. 2017. "Ottawa Signs Deal with Labrador Inuit to Co-Manage Large Marine Protected Area." *National Post*, 29 September. Accessed 5 June 2019. https://nationalpost.com/pmn/news-pmn/canada-news-pmn/ottawa-signs-deal-with-labrador-inuit-to-co-manage-large-marine-protected-area.

Blanchfield, M. 2019. "U.S. Secretary of State Mike Pompeo Says Canada's Claim to the Northwest Passage Is 'illegitimate'." *National Post*, 6 May. Accessed 5 June 2019. https://nationalpost.com/news/canada/pompeo-says-canadian-claim-to-northwest-passage-is-illegitimate.

China. 2018. "China's Arctic Policy." *Xinhuanet*, 26 January. Accessed 5 June 2019. www.xinhuanet.com/english/2018-01/26/c_136926498.htm.

Dawson, J., L. Pizzolato, S. Howell, L. Copland, and M. E. Johnston. 2018. "Temporal and Spatial Patterns of Ship Traffic in the Canadian Arctic from 1990 to 2015." *Arctic* 71 (1): 15–26. Accessed 4 June 2019. https://journalhosting.ucalgary.ca/index.php/arctic/article/view/67736/51632.

Egeland, G. M. 2010a. "Inuit Health Survey 2007–2008: Inuvialuit Settlement Region." In *International Polar Year Inuit Health Survey: Health in Transition and Resiliency.* Accessed 5 June 2019. www.mcgill.ca/cine/files/cine/adult_report_-_inuvialuit.pdf.

Egeland, G. M. 2010b. "Inuit Health Survey 2007–2008: Nunatsiavut." In *International Polar Year Inuit Health Survey: Health in Transition and Resiliency.* Accessed 5 June 2019. www.mcgill.ca/cine/files/cine/adult_report_-_nunatsiavut.pdf.

Egeland, G. M. 2010c. "Inuit Health Survey 2007–2008: Nunavut." In *International Polar Year Inuit Health Survey: Health in Transition and Resiliency.* Accessed 5 June 2019. www.mcgill.ca/cine/files/cine/adult_report_nunavut.pdf.

Fountain, F., and S. Eder. 2018. "In the Blind of an Eye, the Hunt for Oil Threatens Pristine Alaska." *New York Times*, 3 December. Accessed 5 June 2019. www.nytimes.com/2018/12/03/us/oil-drilling-arctic-national-wildlife-refuge.html.

Government of Canada. 1995. "The Government of Canada's Approach to Implementation of the Inherent Right and the Negotiation of Aboriginal Self-Government." Accessed 18 August 2020. https://www.rcaanc-cirnac.gc.ca/eng/1100100031843/1539869205136.

Government of Canada. 2017. "Canadian Rangers: A Systemic Investigation of the Factors that Impact Health Care Entitlements and Related Benefits of the Rangers." In *Office of the Ombudsman for the Department of National Defence and the Canadian Forces*, September. Accessed 4 June 2019. www.ombudsman.forces.gc.ca/assets/OMBUDSMAN_Internet/docs/en/rangers/rangersreport_nov-30-2017_reducedsize.pdf.

Government of Canada. 2019. "Canada's Arctic and Northern Policy Framework, September." Accessed 14 September 2019. www.rcaanc-cirnac.gc.ca/eng/1560523306861/1560523330587.

Government of Nunavut. 2018. "Nunavut Cruise Itinerary 2018." In *Department of Economic Development and Transportation.* Accessed 15 January 2020. www.gov.nu.ca/master-itinerary/nunavut-cruise-itinerary-2018.

Hansen, C. 2013. "Native Village of Kivalina IRA Council v. United States Environmental Protection Agency." *Public Land and Resources Law Review.* Article 10.

Herz, N. 2018. "A Solar Project in Rural Alaska Takes Aim at Sky-High Electric Bills." *KTOO Public Media*, 24 October. Accessed 5 June 2019. www.ktoo.org/2018/10/24/a-solar-project-in-rural-alaska-takes-aim-at-sky-high-electric-bills/.

Imappivut. 2019. "About – What Is Imappivut?" Accessed 5 June 2019. https://imappivut.com/about/.

Inuit Circumpolar Council. 2009. *A Circumpolar Inuit Declaration on Sovereignty in the Arctic*. Accessed 18 August 2020. https://www.iccalaska.org/wp-icc/wp-content/uploads/2016/01/Signed-Inuit-Sovereignty-Declaration-11x17.pdf.

Inuit Circumpolar Council Canada. 2019. "Submission of the Inuit Circumpolar Council Canada to the Special Senate Committee on the Arctic Regarding the Arctic Policy Framework and International Priorities." March. Accessed 4 June 2019. https://sencanada.ca/content/sen/committee/421/ARCT/Briefs/InuitCircumpolarCouncilCanada_e.pdf.

Inuit Tapiriit Kanatami. 2019. "National Inuit Climate Change Strategy." Accessed September 2019. www.itk.ca/wp-content/uploads/2019/06/ITK_Climate-Change-Strategy_English.pdf.

Inuit Tapiriit Kanatami. 2017. "Nilliajut 2: Inuit Perspectives on the Northwest Passage – Shipping and Marine Issues." Accessed 5 June 2019. www.itk.ca/wp-content/uploads/2018/01/NilliajutTextPages_Draftv4_english_web.pdf.

Inuit Tapiriit Kanatami. 2016. "National Inuit Suicide Prevention Strategy." Accessed 5 June 2019. www.itk.ca/wp-content/uploads/2016/07/ITK-National-Inuit-Suicide-Prevention-Strategy-2016.pdf.

Inuit Tapiriit Kanatami. 2018a. "National Inuit Strategy on Research." Accessed 5 June 2019. www.itk.ca/wp-content/uploads/2018/03/National-Inuit-Strategy-on-Research.pdf.

Inuit Tapiriit Kanatami. 2018b. "Position Paper: Development and Implementation of the Arctic Policy Framework." Accessed 5 June 2019. www.itk.ca/wp-content/uploads/2018/11/2018-APFPolicyPositionPaper-FINAL.pdf.

Inuit Tapiriit Kanatami. 2018c. "Inuit Tuberculosis Elimination Framework." Accessed 5 June 2019. www.itk.ca/wp-content/uploads/2018/12/FINAL-ElectronicEN-Inuit-TB-Elimination-Framework.pdf.

Inuit Tapiriit Kanatami. 2018d. "Inuit Statistical Profile 2018." Accessed 5 June 2019. www.itk.ca/wp-content/uploads/2018/08/Inuit-Statistical-Profile.pdf.

Inuvialuit Regional Corporation. "Frequently Asked Questions." Accessed 5 June 2019. http://irc.inuvialuit.com/frequently-asked-questions.

Inuvialuit Regional Corporation. 2019. "Self-Government." Accessed 4 June 2019. https://irc.inuvialuit.com/self-government.

Isachenkov, V., and I. Titova. 2019. "Putin Outlines Ambitious Arctic Expansion Program." *AP News*, 9 April. Accessed 5 June 2019. www.apnews.com/d0c2eb39a3b44b40ac8ddb1749ebe143.

Kono, T. 2018. "Speech of the Minister for Foreign Affairs of Japan at the Arctic Circle 2018 Opening Session." 19 October. Reykjavik, Iceland. Accessed 4 June 2019. www.mofa.go.jp/files/000410409.pdf.

Makivik Corporation. 2018. *2018 Nunavik Inuit Declaration*. Kuujjuaq, Nunavik, 23–25 May. Accessed 5 June 2019. www.makivik.org/wp-content/uploads/2018/12/1-All-Orgs-Declaration.pdf.

Makivik Corporation. 2019a. "Makivik Corporation Announces Members of Nunavik Inuit Constitutional Task Force." 28 May. Accessed 5 June 2019. www.makivik.org/makivik-corporation-announces-members-of-nunavik-inuit-constitutional-task-force/.

Makivik Corporation. 2019b. "Nunavik Inuit Pass Significant Self-Determination Resolution." 21 March. Accessed 5 June 2019. www.makivik.org/nunavik-inuit-pass-significant-self-determination-resolution/.

Northwest Arctic Leadership Team. 2006. "Quad 2006: Growing Strong Together." Accessed 5 June 2019. www.nwarcticleadershipteam.com/files/6612/8593/7973/2006nwaltquadbooklet.pdf.

Nunavut Tunngavik Inc. 2016. "2014/15 Annual Report on the State of Inuit Culture and Society." Accessed 5 June 2019. www.tunngavik.com/files/2016/09/SICS-Report-2015-ENG.pdf.

Nunavut Tunngavik Inc. 2017. *The Cost of Not Successfully Implementing Article 23: Representative Employment for Inuit within the Government*. PriceWaterhouseCoopers. Accessed 5 June 2019. www.tunngavik.com/files/2017/09/2017-09-12-PWC-Report-Eng-1.pdf.

Nunavut Tunngavik Inc. 2018. "Resolution RSA#: 18–10–11 on Inuit Self-Government." Annual General Meeting, 23–25 October. Accessed 5 June 2019. www.tunngavik.com/files/2018/10/RSA-18-10-11-Self-Government-eng.pdf.

Oliver, S. G. 2019. "Nuiqsut Tribe Sues BLM Over Drilling." *The Arctic Sounder*, 7 March. Accessed 5 June 2019. www.thearcticsounder.com/article/1910nuiqsut_tribe_sues_blm_over_drilling.

Parks Canada. 2018. "Important Milestone Reached for Tallurutiup Imanga National Marine Conservation Area." *Government of Canada*, 4 December. Accessed 5 June 2019. www.canada.ca/en/parks-canada/

news/2018/12/important-milestone-reached-for-tallurutiup-imanga-national-marine-conservation-area.html.

Peng, G., J. L. Matthews, and J. T. Yu. 2018. "Sensitivity Analysis of Arctic Sea Ice Extent Trends and Statistical Projections Using Satellite Data." *Remote Sensing* 10 (230).

Pigott, C. 2016. "Vote Down Education Act Changes, Nunavut Tunngavik tells MLAs." *CBC News*, 21 October. Accessed 5 June 2019. www.cbc.ca/news/canada/north/nunavut-education-act-inuit-schools-1.3815799.

Porta, L., E. Abou-Absi, J. Dawson, and O. Mussels. 2017. "Shipping Corridors As a Framework for Advancing Marine Law and Policy in the Canadian Arctic." *Ocean and Coastal Law Journal* 22 (1): 63–84. Accessed 11 July 2019. https://digitalcommons.mainelaw.maine.edu/cgi/viewcontent.cgi?article=1341&context=oclj.

Qikiqtani Inuit Association. 2019. "Tallurutiup Imanga and Tuvaijuittuq Agreements." 6 August. Accessed 8 September 2019. www.qia.ca/tallurutiup-imanga-and-tuvaijuittuq-agreements/.

Rogers, S. 2019. "Nunavik Inuit Sign Deal with Ottawa to Launch Self-Government Negotiations." *Nunatsiaq News*, 20 June. Accessed 11 July 2019. https://nunatsiaq.com/stories/article/nunavik-inuit-sign-deal-with-ottawa-to-launch-self-government-negotiations/.

Senate of Canada. 2018. *When Every Minute Counts: Maritime Search and Rescue.* Report of the Standing Committee on Fisheries and Oceans. Accessed 4 June 2019. https://sencanada.ca/content/sen/committee/421/POFO/reports/MaritimeSARReport_e(forweb)_e.pdf.

U.S. Census Bureau. 2018. "QuickFacts Northwest Arctic Borough, Alaska." Accessed 5 June 2019. www.census.gov/quickfacts/fact/table/northwestarcticboroughalaska/PST045218.

U.S. Government Accountability Office. 2012. *Regional Alaska Native Corporations: Status 40 Years after Establishment, and Future Considerations.* Washington, DC: U.S. Government Accountability Office.

USGS. 2008. "Circum-Arctic Resource Appraisal: Estimates of Undiscovered Oil and Gas North of the Arctic Circle." *Fact Sheet 2008–3049.* Accessed 18 April 2019. https://pubs.usgs.gov/fs/2008/3049/fs2008-3049.pdf.

U.S. Senate Committee on Homeland Security and Governmental Affairs. 2009. *New Information about Contracting Preferences for Alaska Native Corporations (Part I).* Subcommittee on Contracting Oversight. Accessed 5 June 2019. www.hsgac.senate.gov/imo/media/doc/SubcommitteMajorityStaffAnalysisofPubliclyAvailableANCData62309.pdf?attempt=2.

Wilson, G. N., and P. Selle. 2019. "Indigenous Self-Determination in Northern Canada and Norway." *Institute for Research on Public Policy*, 5 February. Accessed 9 August 2019. https://irpp.org/research-studies/Indigenous-self-determination-in-northern-canada-and-norway/.

World Health Organization. 2018. *WHO Housing and Health Guidelines.* Geneva. Accessed 5 June 2019. www.ncbi.nlm.nih.gov/books/NBK535293/pdf/Bookshelf_NBK535293.pdf.

Yardley, W., and E. Olsen. 2011. "Arctic Village Is Torn by Plan for Oil Drilling." *New York Times*, 25 October. Accessed 5 June 2019. www.nytimes.com/2011/10/26/us/arctic-village-split-by-oil-drilling-plan.html.

Other documents

A Circumpolar Inuit Declaration on Sovereignty in the Arctic (signed April 2009, Inuit Circumpolar Council)

17

INDIGENOUS PEOPLES' RIGHT TO SELF-DETERMINATION

Perceptions of self-determination among the Sámi electorate in Sweden

Ulf Mörkenstam, Ragnhild Nilsson and Stefan Dahlberg

Abstract

On an international level, we have the last decades witnessed a remarkable development of Indigenous rights, mainly as a result of Indigenous peoples' political struggle and mobilisation on a local, national and international level. Paramount in this context is the third article of the UN Declaration on the Rights of Indigenous Peoples (UNDRIP) recognising Indigenous peoples as peoples with a right to self-determination. The Nordic countries had a common response to rights-claims from their Indigenous people, the Sámi, in establishing representative institutions – Sámediggis (Sámi Parliaments in Northern Sámi) – consisting of popularly elected Sámi representatives. Today, the Sámediggis are considered the main vehicles to safeguard Sámi self-determination. What the right to self-determination implies in political practice – that is, for domestic constitutional, legal and institutional reforms – is, however, still most controversial. How Sámi self-determination ought to be implemented in the Nordic countries has also been recurrently debated, especially the role of the Sámediggis. There are, however, few systematic studies analysing how persons belonging to Indigenous peoples perceive the right to self-determination: On what matters are self-determination of importance to Indigenous persons? Are there differences between persons identifying with the same Indigenous group? And how are Indigenous persons' understanding of self-determination related to international law and contemporary national policies? The aim of this chapter is to analyse Sámi self-determination from the perspective of the Sámi electorate in Sweden based on data from the second Swedish Sámi Election Study in 2017.

Introduction

On an international level, we have the last decades witnessed a remarkable development of Indigenous rights, mainly as a result of Indigenous peoples' political struggle and mobilisation on a local, national and international level (see, e.g., Anaya and Rodriguez-Piñero 2018; Lightfoot 2016). In contemporary politics, there are two instruments of specific importance for Indigenous peoples: the 1989 International Labour Convention (ILO) Convention

Concerning Indigenous and Tribal Peoples in Independent Countries (No. 169) and the UN Declaration on the Rights of Indigenous Peoples (UNDRIP), adopted by the UN General Assembly in 2007. Only 23 countries have ratified the binding ILO convention (C169) during its first 30 years of existence – Denmark and Norway among the Nordic countries – while, in contrast, 144 countries voted for the UNDRIP, including all the Nordic countries. Although not legally binding, the UNDRIP reflects in many ways customary international law on Indigenous peoples' rights and, as such, it serves as a point of reference in the interpretation of other human rights instruments (see, e.g., Diergarten 2019; Lightfoot 2016; Koivurova 2011).

Paramount in this context is the third article of the UNDRIP recognising Indigenous peoples as peoples with a right to self-determination – paraphrasing the first articles of the two 1966 UN Covenants on Civil and Political Rights (ICCPR) and on Economic, Social and Cultural Rights (ICESCR) on all peoples' right to self-determination – that "Indigenous peoples have the right to self-determination. By virtue of that right they freely determine their political status and freely pursue their economic, social and cultural development". Indigenous peoples' self-determination challenges a traditional nation-state centred understanding of territorial jurisdiction, democracy and political rights. What this right ought to mean in political practice in terms of constitutional, legal and institutional reforms is thus – not surprisingly – still most controversial, and national responses (if any) to demands for Indigenous self-determination vary significantly over the world. International law provides no or little guidance here, since it lacks a "more precise scope and content of Indigenous peoples' self-government and autonomy rights" (Scheinin and Åhrén 2018, 74).

The global Indigenous mobilisation, starting in the 1960s and 70s, inspired the political mobilisation of the Sámi in the Nordic countries, and early on, the Sámi became active in the international arena (see, e.g., Minde 2003). With Norway as blueprint, the Nordic countries have since the 1980s developed a common response to rights-claims from their Indigenous people, the Sámi, in establishing representative institutions – Sámediggis (Sámi Parliaments in Northern Sámi) – consisting of popularly elected Sámi representatives (in Norway in 1989, in Sweden in 1993 and in Finland 1995).[1] From an international perspective, this has often been described as a radical approach to safeguarding the right to self-determination of an Indigenous people, and the Sámediggis have been described as important models "for Indigenous self-governance and participation in decision-making that could inspire the development of similar institutions elsewhere in the world" (Anaya 2011, para. 37). The role of the Sámediggis in the implementation of Sámi self-determination has, however, been recurrently debated, and they have been criticised for their formal mandate as government agencies and for having limited decision-making power on the most topical issues within the Sámi societies in respective state as well as for not having any independent financial resources (see, e.g., Falch and Selle 2018; Heinämäki et al. 2017; Josefsen 2014; Lawrence and Mörkenstam 2016). Especially the Sámediggis in Finland and Sweden have been criticised for lacking both the autonomy and influence to actually safeguard the Sámi right to self-determination (Mörkenstam et al. 2016). As Else-Grete Broderstad (2015, 77–78) points out: "Without institutionally anchored rights and established procedures securing Indigenous participation in state decision-making processes, the situation will only become more critical".

In contrast both to the vivid debate on Indigenous self-determination in legal and political theory and to the ongoing debate on the role of the Sámediggis in the Nordic countries, there are few extensive and systematic empirical studies on how persons belonging to Indigenous peoples perceive the right to self-determination: On what matters are self-determination of importance to Indigenous persons? Are there differences between persons identifying with the same Indigenous group? And how are Indigenous persons' understanding of self-determination related to international law and contemporary national policies? The aim of this chapter is to analyse Sámi self-determination from the perspective of the Sámi electorate in Sweden based on data from the second Swedish Sámi Election Study in 2017.

In the following part, we briefly present our conceptual framework based on the contemporary understanding of self-determination in legal and political theory. Thereafter, we introduce the Swedish Sámediggi and the Sámi electorate. Then, we present our data, and in the following part our results. The concluding part relates our findings to the contemporary debate on Indigenous peoples' right to self-determination and Swedish Sámi politics.

Self-determination in legal and political theory: a conceptual framework

The adoption of the UNDRIP in 2007 signalled "an international consensus that individual and minority rights had fallen short of fully protecting the legitimate interests of Indigenous peoples" (Quane 2012, 79). It was preceded by more than two decades of negotiations between Indigenous representatives, states, NGOs and UN officials, and the participation of Indigenous peoples was indispensable for finalising the document (see, e.g., Anaya and Rodriguez-Piñero 2018; Daes 2011; Lightfoot 2016). "The Declaration", as Julian Burger argues, "owes its existence and its content to the advocacy of Indigenous peoples" but still constitutes "a compromise between the text solicited by Indigenous peoples on the basis of the draft prepared by experts and that [version] finally accepted by States" (Burger 2016, 42). The process leading to the adoption of the UNDRIP is important for its legitimacy, and it established that Indigenous peoples are equal to other peoples within international law – with a right to self-determination – and that Indigenous rights are human rights (Allen and Xanthaki 2011). Political rights, "or sovereign rights if one wants" (Scheinin and Åhrén 2018, 63), have always been in the forefront of Indigenous peoples' political struggle. Today, there exists a consensus among both political theorists and legal scholars that it is the will of the Indigenous peoples that ought to determine their political status and their economic, cultural and social development in order to protect, preserve and develop their own culture (Hohmann and Weller 2018; Mörkenstam 2015; Xanthaki 2007).

In international law, however, Indigenous self-determination is most often interpreted as a qualified right, since it is limited to self-determination within the borders of already existing nation-states or with respect to existing borders in the case of transnational Indigenous peoples (like the Sámi) for whom the right to exercise their self-determination across separating national borders is crucial. The statements made by Indigenous representatives "that the right to self-determination does not necessarily imply a right of a separate sovereign existence" was also crucial in the preparatory work of the UNDRIP (Anaya 2009, 60), and this limited understanding of Indigenous self-determination was decisive in the process leading to its adoption (Weller 2018; Wiessner 2008). Two articles in the UNDRIP emphasise this limited version: first, article four limits the scope of self-determination: "Indigenous peoples, in exercising their right to self-determination, have the right to autonomy or self-government in matters relating to their internal and local affairs", and, second, article 46 explicitly states that "[n]othing in this Declaration may be interpreted as . . . authorizing or encouraging any action which would dismember or impair, totally, or in part, the territorial integrity or political unity of sovereign and independent States". The right to self-determination thus "applies to Indigenous peoples through domestic recognition and action" (Åhrén 2016, 188). This limited understanding of the right to self-determination has, of course, been criticised by Indigenous representatives for being discriminatory, since "a special Indigenous version of the right to self-determination" makes a distinction between categories of peoples (Baer 2005, 229–230). To condition Indigenous self-determination is to deny Indigenous peoples a right already accorded to other peoples and thus to perpetuate a hierarchical societal order in which one people may continue to dominate the other(s) (see, e.g., Buchanan 2004; Tully 1995; Weller 2018; Xanthaki 2007).

The traditional understanding of self-determination in international law is that it has two aspects: an external and an internal. The former aspect refers to a people's right to define their own political status as an independent state and to be free from interference from the outside and the latter to a people's right to democratic governance and to political participation (see, e.g., Cassese 1995, Ch 4–5; Xanthaki 2007, 159–169). With a limited understanding of Indigenous peoples' right to self-determination, the external aspect must obviously be reinterpreted, since the core of this aspect – the right for a people to define their own political status as an independent state – is denied Indigenous peoples. In the contemporary debate, the external aspect has therefore often been interpreted as a right for Indigenous peoples "to represent themselves on the international level, including the United Nations (UN) and other international institutions" (Åhrén 2016, 131).

With Indigenous self-determination to be exercised within the territorial jurisdiction of already existing states, the internal aspect of self-determination has to be reinterpreted as well. Democratic participation on equal terms as other members of society could never be sufficient to realise Indigenous self-determination, since Indigenous peoples in most parts of the world constitute permanent minorities in the states in which they live (in Sweden, the Sámi are, for instance, outnumbered 450 to 1).[2] If there are differences on fundamental issues, for instance, concerning the use of land and natural resources, an Indigenous people will constantly run the risk of being outvoted by the majority population (Buchanan 2004, 360–362). Indigenous self-determination through democratic participation on an individual level in the majority society would thus render the right to self-determination void and meaningless.

Instead, the right to internal self-determination must be understood as a right to preserve the autonomous functions that the Indigenous society needs in order to survive. This includes the right to develop their own culture, language, religion and industries and to establish their own social, economic, cultural and political institutions. From this perspective, the internal aspect "requires a governing order under which individuals and groups are able to make meaningful choices in matters touching upon all spheres of life on a continuous basis", something James Anaya (2004, 106) calls "on-going self-determination". In short, Indigenous peoples must have political power in the capacity of being self-determining peoples, not "merely power delegated by a higher political unit and subject to being overridden or revoked by the latter" (Buchanan 2004, 333). There must thus be a handover of power from national political bodies to Indigenous peoples' own representative institutions (Scheinin and Åhrén 2018). Within an already-existing state, this requires, for instance, language and cultural rights, social rights, rights to their own legal system and self-government rights (see, e.g., Hohmann and Weller 2018; Xanthaki 2007, 196–279; Åhrén 2016, 149–199).

Of specific importance in this context are property rights to land and natural resources: "Indigenous cultures are intrinsically rooted in their traditional territories and dispossession of their traditional lands throughout colonisation has had, and continues to have, 'disastrous effects' on Indigenous peoples", depriving them of their political sovereignty, as well as having "contributed to dislocation and loss of cultural integrity, language and cultural connection" (Lawrence and Åhrén 2017, 149). Hence, the cultural, social and economic life of Indigenous communities is dependent on and conditioned by the natural resources available, especially since

Indigenous peoples' traditional livelihoods most often are linked to land and water (see, e.g., Anaya 2015; Scheinin 2008; Åhrén 2016).

In the following, we will analyse how the perceptions of self-determination within the Sámi electorate in Sweden relate to international law. Does, for instance, the Sámi electorate share the understanding of self-determination in international law as a right limited to internal self-governance within the territorial jurisdiction of the Swedish state? Is the electorate in favour of political rights in terms of increasing the scope of the Sámediggi's decision-making power? And what are the electorate's opinions on cultural, social and economic rights, as well as on control of natural resources? First, however, we will introduce the Swedish Sámediggi and its electorate.

The Swedish Sámediggi, the Sámi electorate and self-determination

Already in the late 19th century, the Swedish Sámi claimed the right to local self-government, property rights to land within the traditional Sámi settlement area (today called Sápmi), as well as the right to political representation on a national level through Sámi representatives in the Swedish national parliament, the *Riksdag*. These arguments were based on resistance to what was perceived as discriminatory national legislation (see, e.g., Laula 1904). It is well known that this early Sámi political mobilisation did not have any major impact on Swedish politics; Sámi self-determination was basically a non-issue in Swedish politics until the mid-1980s and the work of the first Sámi Rights Commission, and the question of Sámi property rights is still most controversial, as seen in the many legal processes since the 1990s (see, e.g., Mörkenstam 2019).

Sweden recognised the Sámi as an Indigenous people in accordance with international law during the 1980s, and in 1993, a popularly elected Sámediggi was established to function as a representative body of the Sámi people in order to grant the Sámi cultural autonomy. The Sámediggi is an institution for non-territorial autonomy, where the right to vote and to run for office is based on ethnicity and not geography. It shall thus represent all Sámi within the country, those who live outside Sápmi as well. The Sámediggi (just like its counterparts in Norway and Finland) lacks legislative power and has no independent financial resources. In this respect, the translation of Sámediggi into the English label Sámi Parliament is misleading. The Sámediggi's actual political power has been delegated from the Swedish state, and it is executed as part of the responsibilities it has as a government agency (Lawrence and Mörkenstam 2016).

The general mission of the Swedish Sámediggi – as a government agency – is to promote a living Sámi culture which, for instance, includes allocating funding to cultural activities and Sámi organisations, appointing the board of directors for the Sámi School and guiding and directing work regarding Sámi languages. It also has administrative duties related to the traditional Sámi livelihood reindeer herding (Fjellström et al. 2016). The formal status as a government agency constitutes a major problem, since it is both an administrative authority under the Swedish government and a popularly elected body representing the Sámi people in Sweden. Obviously, this construction with dual roles may cause severe tensions within the Sámediggi, as well as between the Sámediggi and the Swedish government, when there are conflicts of interests between the policy of the government and the interests of the Sámi electorate. This is often the case, for instance, in issues on the exploitation of natural resources within Sápmi (see, e.g., Lawrence and Mörkenstam 2016). This construction has recurrently received critique from UN monitoring bodies, recently, for instance, in a report from the UN special rapporteur on the rights of Indigenous peoples encouraging "Sweden to introduce reforms to ensure that the Sami Parliament has greater independence from State institutions and authorities" (Tauli-Corpuz 2016, para. 81).

Following the development in international law, Sweden recognised the Sámi as a people having rights "in the international law sense . . . including the right to self-determination" in 2006 (Prop. 2005/06:86, 38), voted for an adoption of the UNDRIP in 2007 and recognised the Sámi special status constitutionally for the first time in 2010. However, the actual content of this right has not been defined further in negotiations or discussions between the Sámi and the state, and no new Sámi Rights Commission has, for instance, been appointed (Mörkenstam 2019). Instead, Sámi self-determination has so far been made equal to extending the administrative duties of the Sámediggi in its capacity as government agency.

This defence of political status quo stands in sharp contrast to the rights–claims made by the Sámediggi, which are based on the fact that the Sámi are a people with their right to self-determination recognised both nationally and internationally (Lantto and Mörkenstam 2015). The Sámediggi has also presented two reports on the right to self-determination. The first, from 2004, was a thorough analysis of the contemporary understanding of Indigenous peoples' right to self-determination in international law, what self-determination could contain from a Sámi perspective and, to some extent, how it could be implemented on the Swedish side of Sápmi (Sametinget 2004). The starting point in the report was the common first article of the ICCPR and the ICESCR, mentioned previously, stating that "[a]ll peoples have the right to self-determination. By virtue of that right they freely determine their political status and freely pursue their economic, social and cultural development". According to the report, the content of Sámi self-determination must be based on the extent to which the Sámi have decision-making power over land issues, including water and natural resources. The economic, cultural and social aspects of self-determination were seen as dependent on the availability of natural resources. Land, water and natural resources were thus conceived of as prerequisites for the Sámi society to be able to become self-sufficient, that is, to develop schools, child and elderly care, industries, culture and so on (Sametinget 2004, 93). Moreover, the formal position of the Sámediggi as a government agency was severely criticised from a self-determination perspective.

Four years later, in 2008, a second report was presented by the Sámediggi focusing on the legislative changes needed to implement Sámi self-determination with the newly adopted UNDRIP as backdrop (Sametinget 2008). Much of the main core of the first report was now absent. Self-determination was, for instance, discussed in terms of distinct spheres – economic, social and cultural – that could be developed independently from each other. Regarding the role of the Sámediggi in the implementation of self-determination, the previous critique of its formal position was not repeated with the same intensity; instead, it was considered decisive to what extent the Swedish government was ready to transfer additional authority from other government agencies to the Sámediggi (Sametinget 2008, 67).

In parallel to this development on a national level, it is important to mention the ongoing work on an international treaty between Sweden, Finland and Norway, the Nordic Sámi Convention (see, e.g., Bankes and Koivurova 2013; Heinämäki and Cambou 2018; Åhrén et al. 2007). Stemming from a Sámi proposal already made in the late 1980s, an expert group appointed by the three Nordic states presented a first draft in 2005, and a revised version was presented in 2016. The Draft Convention takes the development in international law as its starting point, and, like the UNDRIP, the Sámi Convention's third article concerns the right to self-determination:

> As a people, the Saami has the right of self-determination in accordance with the rules and provisions of international law and of this Convention. In so far as it follows

from these rules and provisions, the Saami people has the right to determine its own economic, social and cultural development and to dispose, to their own benefit, over its own natural resources.

In the draft conventions, the Sámi are recognised as an Indigenous people with interests differing from the three states, and the Sámi right to self-determination within the borders of respective state is recognised in parallel with the three states' right to self-determination. The Plenary of the Swedish Sámi Parliament voted for an adoption of the first draft convention presented by the expert group (accepting all articles). However, the revised version of the convention, presented after negotiations between representatives of the three governments in cooperation with the three Sámi parliaments, has met severe critique from the Sámi for being a weaker version not fulfilling international norms (see, e.g., Lantto and Mörkenstam 2015; Samefolket 2017). This critical view seems also to be widespread among the Sámi electorate – as shown in the second Swedish Sámi Election Study in 2017 (see subsequently) – only 37 percent of the voters thought it was a very good or quite good proposal to ratify the latest draft of the Nordic Sámi Convention.

If we finally turn to the electorate's opinions on self-determination, a previous study on political cleavages showed that there existed a clear conflict between the Sámi and the Swedish state, as analysed through the electorate's opinions on the influence of the Sámediggi on certain political issues (Nilsson et al. 2016). No explicit questions were, however, asked about self-determination. The cleavage found showed clear signs of being a territorial–cultural conflict following the nation-building processes of states involving – in the words of Seymour M. Lipset and Stein Rokkan (1967, 10) – "the typical reactions of peripheral regions, linguistic minorities, and culturally threatened populations to the pressures of the centralizing, standardizing and 'rationalizing' machinery of the nation-state". The profundity of this Sámi-state cleavage was accentuated by the fact that the voters of all parties represented in the parliament on a general level were positive towards an increase in the Sámediggi's influence. Still, however, there were differences discernible between the voters of different parties.

Self-determination has also been analysed in a study explaining voter turnout among the Swedish Sámi electorate in Sámi, national and European elections (Dahlberg and Mörkenstam 2019). The analysis showed that the main factor explaining voter behaviour was social integration in the Sámi society: it had a positive effect on voter turnout in Sámediggi elections, clear negative effects on turnout in elections to the *Riksdag* and almost no effect on European Parliament elections. The different effects of social integration on turnout in different elections were partly explained by the strong relation between social integration in the Sámi society and opinions about Sámi self-determination: for every unit increase in social integration in the Sámi society, the stronger the claims for self-determination (something we return to subsequently). "These claims for self-determination might explain why voting for the Sámi Parliament is more important for a part of the Sámi electorate – it is a potential vehicle for self-determination – but also why the same part of the electorate tend to abstain from voting for the *Riksdag*" (Dahlberg and Mörkenstam 2019, 104).

In the following, we will analyse the attitudes of the Sámi electorate on self-determination in relation to the formal position of the Sámediggi and the Sámediggi's own reports on self-determination. First, however, we will describe our data.

Survey data

Our analysis is based on the second-ever Swedish Sámi Election Study in 2017; the first study was carried out during the election of 2013 (Nilsson et al. 2016). The 2017 study was based on a random sample drawn from the Sámi electoral roll (2000 persons out of 8788) after permission granted from the Sámediggi and after approval of the study by the Ethical Review Board in Stockholm. With this random sample, all our respondents are self-identifying as Sámi, since self-identification is the first criteria for registering in the Sámi electoral roll.

In these kinds of studies, the design of the surveys is decisive, and it was designed from within a Sámi social and political context (with the Electoral Board of the Sámediggi as reference group). The data collection started immediately after the election in May of the election year, and it was conducted through a combination of postal and web questionnaires in three languages (Northern and Southern Sámi and Swedish), with four reminders, two postcards and two telephone messages. The response rate was 43 percent.

The dataset has an overrepresentation of (self-reported) voters, 83.9 percent in the sample compared to 57.7 percent in the population. This may affect the results, since less engaged and motivated people usually have a lower response rate in surveys (see, e.g., Dahlberg and Persson 2014). However, it is hard to know how this would affect the results: if the voters, for instance, are less engaged or motivated by a dissatisfaction with the work or representativeness of the Sámediggi and the politicians/parties, we could perhaps expect lower support for Sámi self-determination realised through the Sámediggi; if the voters are less engaged and motivated due to dissatisfaction with Swedish Sámi politics and the formal position of the Sámediggi as a government agency, that would probably increase the support for self-determination.

All research that involves Indigenous peoples raises some ethical considerations. On the one hand, we know that statistics have been used through history to characterise Indigenous peoples from an outside perspective. Therefore, the results must not recreate or help to sustain prejudices and stereotypes that the majority society has produced throughout history. On the other hand, it is important for Indigenous communities themselves to get statistical knowledge about their own society and that the knowledge produced be of relevance for, in this case, the Sámi society (Drugge 2016; Pettersen 2016). Ethical considerations being raised in the work with the Sámi Election Study have been about the relevance of the questions in the survey, language difficulties and if we can apply established theories, with the constraints these may have for Indigenous peoples (Josefsen et al. 2017; Kukutai and Walter 2015; Walter and Andersen 2013). Our approach in this study is twofold: first, and as mentioned previously, the survey has been designed with the Electoral Board of the Swedish Sámediggi as reference group, and the results have been presented for the members of the Sámediggi at several occasions; second, we have been particularly careful with the theoretical assumptions and categorisations we make and have kept in mind the effects and consequences of previous Swedish state policy towards the Sámi society in the analysis.

Results: the Sámi electorate and the right to self-determination

In the election study in 2017, one survey item was explicitly related to international law, namely the recurrent proposal since 1989 that Sweden ought to ratify C169; a vast majority of the voters – 70 percent – thought it was a very good or quite good proposal, and only 8 percent found it a bad or very bad proposal. Two survey items specifically targeted the electorate's opinions on self-determination. One question was formulated in general terms, "Do you believe that the Sámediggi should have increased self-determination?", and one specifically asked on what issues the Sámediggi ought to have self-determination, followed by 18 different issues. The results from these two items are summarised in Table 17.1.

It is obvious that a vast majority of the voters, three out of four, want to increase the Sámediggi's self-determination, which is not surprising giving the limited self-determination the

Table 17.1 Attitudes within the Sámi electorate on self-determination (percent)

	Yes	No	N:
The Sámi Parliament should have increased self-determination	74	26	659
Childcare	62	38	648
Duodji/Sámi handicrafts	84	16	591
Defence	28	72	605
Mining	75	25	720
Healthcare	58	42	684
Culture	89	11	735
Land and water management	76	24	723
Reindeer herding	79	21	743
Carnivore management	77	23	736
Taxation	35	65	614
Forestry	64	36	693
Education	68	32	713
Environment protection (nature reserves etc.)	74	26	712
Small game hunting	68	32	714
Language	87	13	738
Windpower	70	30	727
Elderly care	68	32	708
Moose hunting	56	44	692

Note: Yes consists of the responses "to a large extent" and "to a fairly large extent", and No of the responses "to a fairly small extent" and "not at all".

Sámediggi has today. These findings confirm the results found in the 2013 survey that there is a conflict between the Sámi and the Swedish state revolving around self-determination (Nilsson et al. 2016; Nilsson and Möller 2016), although this time, the survey questions explicitly targeted self-determination.

Moreover, the results show that self-determination is more urgent on certain matters than on others. In terms of the right to "freely pursue their economic, social and cultural development", as stated in the UNDRIP, there are clear differences between these spheres, where self-determination on cultural issues (culture, language and duodji) are the most important within the electorate. Almost nine out of ten voters want to increase the Sámediggi's self-determination on these matters.

Considering the importance of property rights and rights to use the land as a prerequisite for economic development and to maintain Sámi culture and language, it comes as no surprise that between 56 and 79 percent of the voters want to increase Sámi self-determination on reindeer herding and the use of natural resources within Sápmi (mining, forestry, small game hunting, land and water management, environment protection, carnivore management, moose hunting and wind power).

On matters related to the social sphere (childcare, healthcare, elderly care and education), the need for self-determination seems to be a little less, even if the numbers are still high: between 58 to 68 percent want to increase Sámi self-determination on these matters. One explanation for the slightly lower support for increasing Sámi self-determination on social issues can be the high general level of social welfare in Sweden, where increased self-determination might raise some fear of not getting enough resources to obtain the same level of social services.

The two issues related to the external aspect of self-determination, defence and taxation, did not receive support from a majority of voters. Still, however, 28 and 35 percent of the electorate want to have self-determination concerning defence and taxation, respectively. The latter issue has previously been on the Sámediggi's agenda in the form of a motion from one political party, with a proposal that "Sámi persons should be able to direct a part of their taxation to the Sámediggi" (Min Geaidnu 2009). It did not get support from a majority within the Sámediggi plenary.

A majority of the Sámi voters are thus in favour of increasing the Sámediggi's self-determination – mainly manifested through internal self-determination – although the electorate opinion varies between different political issues and different spheres. There are, however, differences in opinions within the electorate, something we analyse both on a party level and on an individual level in the following.

Opinions on self-determination and party choice

Since the first election to the parliament in 1993 up until 2009, the parties representing the reindeer herders' interests were in the majority in the Sámediggi (in Sámi parliamentary elections, only Sámi parties participate). Dominant among these parties – especially during the first three parliamentary terms – has been Sámiid Riikkabellodat, a party that has developed out of Sámiid Riikkasearvi, the first national Sámi organisation with its base in the reindeer-herding communities.[3] The party has never had a majority of seats, however, and has relied on coalitions with other parties to form a majority, for instance, Guovssonásti, with its origin in the reindeer owners' organisation (Renägarförbundet), and Vuovdega-Skogssamerna, representing the forest reindeer-herding Sámi. However, the largest party in the Sámediggi during the most recent terms has been Jakt- och Fiskesamerna, primarily representing the interests of the Sámi not engaged in reindeer herding but in hunting and fishing. They are working closely with Landspartiet Svenska Samer, originating from the largest organisation representing Sámi outside reindeer herding, and Álbmut/Folket. Between these two party blocs, there are two parties that have been in coalition with both sides over the years to form a majority: Min Geaidnu and Samerna/Sámit (Fjellström et al. 2016; Dahlberg and Mörkenstam 2016). After the 2017 election, five parties formed an alliance to constitute the Board of the Sámediggi – Sámiid Riikkabellodat, Guovssonásti, Vuovdega-Skogssamerna, Samerna/Sámit and the newly elected Samiska folkomröstningspartiet – with the Chairman of the Board representing Sámiid Riikkabellodat.

The existing party structure with two blocs can, at least partly, be explained by the Swedish state's Sámi policy developed and established already in the late 19th century, when a system of Sámi rights was constructed around reindeer husbandry excluding all Sámi occupied in other Sámi livelihoods, like fishing, hunting and handicraft. Thus, the early legislation created two categories of Sámi – members of reindeer-herding communities (with specific rights attached to reindeer herding, like the right to hunt and fish on 'Crown land') and non-members (without rights) – still maintained in contemporary legislation. This "category-split", as it was called by the Sámi leader Israel Ruong (1982, 187–188), was further enforced by a dual state policy of both segregation and assimilation of the Sámi: the nomadic reindeer herders were to be segregated from the Swedish society and "civilisation"; Sámi in other livelihoods were to be assimilated in the Swedish society (see, e.g., Lantto and Mörkenstam 2015). As a consequence of the state's legislation and dual policy, the opportunity to upheld traditional livelihoods, culture and language became unequally distributed within Sámi society, reinforcing this "category-split".

Index on Self-Determination

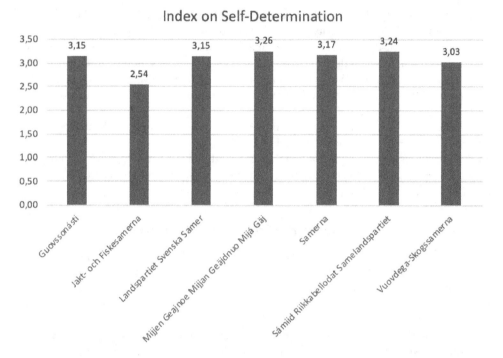

Figure 17.1 Voters' opinions on self-determination and party choice (mean values). Too few of the survey respondents had voted for Álbmut/Folket and Samiska folkomröstningspartiet to include the two parties in the analysis.

Figure 17.1 shows the differences in opinions between voters of different parties represented in the Sámediggi. It is worth noticing that a majority of the voters of all parties favour an increase in self-determination (the mean values differ between 2.54 and 3.26 on a scale from 0–4). In addition, the differences between six of the seven parties included in the analysis are very small: the voters of Min Geaidnu are most positive towards increasing the Sámediggi's self-determination, closely followed by the voters of Sámiid Riikkabellodat and the four other parties. Only the voters of the party Jakt- och Fiskesamerna differ in any significant way, although they also favour increased self-determination. However, these noticeable differences indicate that opinions on self-determination are politicised within the Sámi electorate (see also Saglie et al. 2020).

These results also indicate that opinions on self-determination and on a Swedish ratification of C169 are closely related. In the first Swedish Sámi Election Study in 2013, the most politicised political issue within the Sámi electorate (among 16 different issues) was if Sweden ought to ratify C169 (Dahlberg and Mörkenstam 2016). Fifty percent of the voters of Jakt- och Fiskesamerna supported ratification, in comparison to between 82 and 98 percent of the voters of the other parties.[4]

Individual factors explaining opinions on self-determination

We have also analysed the individual factors explaining opinions on self-determination. Our starting point here is in the results from previous research showing that social integration in

Sámi society is one of the main factors explaining voter behaviour and attitudes (Bergh et al. 2018; Dahlberg and Mörkenstam 2019). Social integration is measured through an index of eight survey items (Cronbach's alpha .89). Three questions ask about language skills: whether a respondent can speak, read and write in any of the Sámi languages. Two items revolve around affinity: with the local Sámi environment and with Sápmi, the traditional Sámi settlement area. The three remaining questions in the index are about how many friends are registered in the election roll, if the respondent has grown up in a Sámi community and how many friends they have in everyday life that are Sámi. Our index thus defines social integration through knowledge of the vernacular language, group affinity and the day-to-day interaction with other group members.

In Table 17.2, we explain opinions on self-determination on an individual level by conducting a simple linear regression analysis. Model 1 shows the bivariate relationship between social integration in Sámi society and opinions about self-determination, and the model serves as a benchmark against the results for models 2 and 3 in order to judge how much of the variation is being shared by the different sets of control variables. The initial correlation in model 1 is substantial in the sense that one step in increase on the social integration index yields a predicted change in .56 on self-determination measured on a scale from 1 to 4.

In model 2, we control for politically related variables such as interest in politics, political trust, media consumption and party choice. When introducing these variables, the differences presented in the simple figure previously are accentuated; self-determination seems to be a highly politicised issue. The initial bivariate correlation from social integration decreases by 30 percent at the same time as the amount of explained variance (R^2) increases by twice the size (from 16 percent in the bivariate model to 33 percent in model 2). Model 2 thus explains twice as much of the variance in the dependent variable compared to the bivariate case in model 1, in this case about one-third. What we also find in this model is a positive covariation with interest in Sámi politics and trust in the Sámediggi, while a strong and negative correlation is found in voting for Jakt- och Fiskesamerna. We can also see that party choice does not explain opinions on self-determination among voters of the other parties: Guovssonásti is used as baseline for party choice (0), and the figures of the other parties are showing differences in comparison to baseline.

In model 3, we also introduce a set of social-demographic control variables, but none of these variables affect the results established in model 2 to any large extent. We find a small but significant negative impact from income and residence in Sápmi, but the strong positive effect from social integration and the small but significant positive effects of interest in Sámi politics and trust in the Sámediggi remain unaffected, just like the negative effect from voting for Jakt- och Fiskesamerna. Also, the amount of explained variance in opinions on self-determination stays the same.

In Figure 17.2, we have plotted the predicted linear relationship from social integration on opinions on self-determination from model 3 in Table 17.2. Clearly, the two variables are closely related even after controlling for residency and party sympathy.

Our analysis thus confirms the findings of previous studies showing that social integration within the Sámi community has a strong explanatory value for voter behaviour and opinions on an individual level and also to explain the Swedish Sámi electorate's opinions on Sámi self-determination.

Table 17.2 The controlled relationship between social integration and opinions about self-determination (OLS)

	Model 1	Model 2	Model 3
Social Integration in the Sámi Society	0.561★★★	0.369★★★	0.379★★★
	(0.067)	(0.086)	(0.087)
General Political Interest		0.030	0.026
		(0.059)	(0.060)
Interest in Sámi Politics		0.165★	0.146★
		(0.065)	(0.065)
Media Consumption		−0.012	−0.010
		(0.029)	(0.029)
Institutional Trust		−0.070	−0.099★
		(0.046)	(0.047)
Trust in Sámi Parliament		0.200★★★	0.195★★★
		(0.036)	(0.036)
Jakt- och Fiskesamerna		−0.580★★★	−0.558★★★
		(0.132)	(0.134)
Landspartiet Svenska Samer		−0.093	−0.137
		(0.153)	(0.155)
Min Geaidnu		0.098	0.101
		(0.175)	(0.174)
Samerna/Sámit		0.003	−0.033
		(0.157)	(0.159)
Sámiid Riikkabellodat		−0.029	−0.038
		(0.123)	(0.121)
Vuovdega-Skogssamerna		−0.136	−0.128
		(0.147)	(0.151)
Member in Reindeer-Herding Community		−0.093	−0.036
		(0.095)	(0.095)
Age			0.243★
			(0.121)
Age2			−0.023
			(0.014)
Sex			0.064
			(0.072)
Edu. Low			0.000
			(.)
Edu. Medium			0.068
			(0.112)
Edu. High			0.216
			(0.116)
Employment			0.003
			(0.088)
Cohabitation			0.107
			(0.092)
Income			−0.046★★
			(0.016)
Residence in Sápmi			−0.191★
			(0.093)

	Model 1	*Model 2*	*Model 3*
Constant	1.776***	1.536***	1.247**
	(0.142)	(0.287)	(0.382)
R-squared	0.162	0.330	0.355
N	359	359	359

* $p < 0.05$, ** $p < 0.01$, *** $p < 0.001$
Comment: See Appendix 1 for variable coding. Guovssonásti is used as baseline for party choice. Too few of the survey respondents had voted for Álbmut/Folket and Samiska folkomröstningspartiet to include them in the analysis.

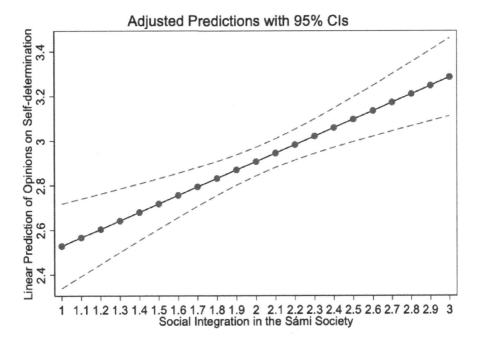

Figure 17.2   Linear prediction of social integration in Sámi society on opinions about self-determination

Conclusion

Self-determination has been one of the most crucial issues discussed and debated within the Sámediggi since its establishment, and our study clearly shows that self-determination is also a topical issue within the Sámi electorate. In this concluding part, we want to highlight three of our findings in relation to international law and Swedish Sámi policy. First, an overwhelming majority of the voters, 74 percent, want to increase the Sámediggi's self-determination. The limited responsibilities that the Sámediggi has today are obviously not enough for the Sámi electorate: the Sámi are a people with a right to self-determination not yet implemented by the state. The analysis also shows that slightly more than one out of four voters want to have increased self-determination over defence and taxation – issues related to external self-determination – although issues related to internal self-determination have the strongest support.

Most important to the electorate is, however, self-determination on cultural issues and on language. In this context, it may be worth noticing that these are matters already defined as part of the Sámediggi's administrative duties. Moreover, cultural rights and language rights are often rather non-controversial within the majority society, since these kinds of rights do not affect the majority population to any higher degree or run the risk of changing the balance of power between the minority and the dominant society (see, e.g., Åhrén 2016). Self-determination is, however, also of great importance to the electorate on more controversial issues, such as property rights and rights to use and manage the national resources within Sápmi, as well as on welfare issues.

Second, self-determination is not only highly politicised in relation to the Swedish state, it is also politicised within the Sámi electorate. Our study shows that the voters of all parties are positive towards an increase in the Sámediggi's self-determination, although the voters of one party, Jakt- och Fiskesamerna, are less positive. A similar difference on a party level is also to be found on opinions on ratification of C169, indicating that international law is closely related to the development of Indigenous self-determination on a nation-state level.

Third, social integration in Sámi society is an important factor explaining opinions on self-determination on an individual level. One explanation of this finding can be found in theories showing that differences in voter behaviour and attitudes between group members self-identifying with the same group – as in Sweden illustrated by the first subjective criteria to register in the Sámi electoral roll – may be explained by distinguishing between group identity and group consciousness. Group identity is commonly described as "an individual's awareness of belonging to a certain group and having a psychological attachment to that group", while group consciousness is an individual's "group identification *politicized* by a set of ideological beliefs about one's group's social standing, as well as a view that collective action is the best means by which the group can improve its status and realize its interests" (McClain et al. 2009, 474, 476). Why some persons develop group consciousness and others do not is often related to an "individual's social interaction with, exposure to, and integration into the community in which they live and act" (Anderson 1996, 111). Opinions on Sámi self-determination could thus be interpreted as a result of a process of politicisation within parts of the electorate well integrated within the Sámi society.

The strong opinions within the Sámi electorate for an increase in self-determination through their representative body, the Sámediggi, are clearly a rejection of the prevailing Swedish politics. It may thus be the time to finally initiate negotiations between the Sámi and the state on what the Sámi right to self-determination ought to mean in a Swedish context, as well as between the states and the Sámi on a Nordic level, taking the opinions of the electorate(s) as starting point.

Notes

1 The Sámi are the only people living within European borders that are recognised as an Indigenous people, and they have through the course of history been divided between four nation-states: Finland, Norway, Sweden and Russia. The estimates of the number of Sámi differ depending on the sources used, but the figures most often seen vary between 80,000 and 100,000, of which 50–65,000 reside in Norway, 20–40,000 in Sweden, around 8,000 in Finland and 2,000 in Russia (Sápmi, Antalet samer i Sápmi. Accessed 16 April 2019 www.samer.se/1536.

2 They have, however, the same rights to participate democratically – both as peoples and as individuals – as other members of the society in which they live, that is, rights to non-discrimination (see, e.g., Gover 2018). This is, for instance, stated in the second article of the UNDRIP: "Indigenous peoples and individuals are free and equal to all other peoples and individuals and have the right to be free from

any kind of discrimination, in the exercise of their rights, in particular that based on their Indigenous origin or identity".

3 The reindeer-herding communities are administrative units created by the Swedish state as a part of the regulation of reindeer herding through the first Reindeer Grazing Act in 1886.

4 Voters of Álbmut/Folket were not included in the analysis, due to having too few survey respondents.

References

Åhrén, M. 2016. *Indigenous Peoples' Status in the International Legal System.* Oxford: Oxford University Press.

Åhrén, M., M. Scheinin, and J. B. Henriksen. 2007. "The Nordic Sami Convention: International Human Rights, Self-Determination and Other Central Provisions." *Gáldu Čála – Journal of Indigenous Peoples Rights* 3.

Allen, S., and A. Xanthaki. 2011. *Reflections on the UN Declaration on the Rights of Indigenous Peoples.* Oxford: Hart Publishing.

Anaya, S. J. 2004. *Indigenous Peoples in International Law.* 2nd ed. Oxford: Oxford University Press.

Anaya, S. J. 2009. "Why There Should Not Have to Be a Declaration on the Rights of Indigenous Peoples." In *International Human Rights and Indigenous Peoples*, edited by S. J. Anaya, 58–63. Chicago: Wolters Kluwer Law & Business.

Anaya, S. J. 2011. "Report of the Special Rapporteur on the Situation of Human Rights and Fundamental Freedoms of Indigenous People." 12 January. A/HRC/18/XX/Add.Y.

Anaya, S. J. 2015. "Report of the Special Rapporteur on the Rights of Indigenous Peoples on Extractive Industries and Indigenous Peoples." *Arizona Journal of International & Comparative Law* 32 (1): 109–142.

Anaya, S. J., and L. Rodriguez-Piñero. 2018. "The Making of the UNDRIP." In *The UN Declaration on the Rights of Indigenous Peoples: A Commentary*, edited by J. Hohmann and M. Weller, 38–62. New York: Oxford University Press.

Anderson, C. J. 1996. "Political Action and Social Integration." *American Politics Quarterly* 24 (1): 105–124.

Baer, L-A. 2005. "The Right of Self-Determination and the Case of the Sámi." In *Operationalizing the Right of Indigenous Peoples to Self-Determination*, edited by P. Aikio and M. Scheinin, 223–231. Åbo: Institute for Human Rights.

Bankes, N., and T. Koivurova. 2013. *The Proposed Nordic Saami Convention. National and International Dimensions of Indigenous Property Rights.* Oxford: Hart Publishing.

Bergh, J., S. Dahlberg, U. Mörkenstam, and J. Saglie. 2018. "Participation in Indigenous Democracy: Voter Turnout in Sámi Parliamentary Elections in Norway and Sweden." *Scandinavian Political Studies* 41 (4): 263–287.

Broderstad, E. G. 2015. "Implementing Indigenous Self-Determination: The Case of the Sámi in Norway." In *Restoring Indigenous Self-Determination. Theoretical and Practical Approaches*, edited by M. Woons, 72–79. Bristol: E-International Relations Publishing.

Buchanan, A. 2004. *Justice, Legitimacy, and Self-Determination: Moral Foundations for International Law.* Oxford: Oxford University Press.

Burger, J. 2016. "From Outsiders to Centre Stage: Three Decades of Indigenous Peoples' Presence at the United Nations." In *Handbook of Indigenous Peoples' Rights*, edited by C. Lennox and D. Short, 325–330. London: Routledge.

Cassese, A. 1995. *Self-Determination of Peoples. A Legal Reappraisal.* Cambridge: Cambridge University Press.

Daes, E-I. 2011. "The UN Declaration on the Rights of Indigenous Peoples: Background and Appraisal." In *Reflections on the UN Declaration on the Rights of Indigenous Peoples*, edited by S. Allen and A. Xanthaki, 11–40. Oxford: Hart Publishing.

Dahlberg, S., and U. Mörkenstam. 2016. "Partiernas väljare och väljarnas val av parti." In *Sametingsval: väljare, partier och media*, edited by R. Nilsson, S. Dahlberg, and U. Mörkenstam, 155–180. Stockholm: Santérus förlag.

Dahlberg, S., and U. Mörkenstam. 2019. "Social Identification, In-Group Integration and Voter Turnout in Three Parliamentary Elections: An Analysis of the Swedish Sámi Electorate." *Electoral Studies* 59: 99–108.

Dahlberg, S., and M. Persson. 2014. "Different Surveys, Different Results? A Comparison of Two Surveys on the 2009 European Parliamentary Election." *West European Politics* 37 (1): 204–221.

Diergarten, Y. 2019. "Indigenous or Out of Scope? Large-Scale Land Acquisitions in Developing Countries, International Human Rights Law and the Current Deficiencies in Land Rights Protection." *Human Rights Law Review* 19 (1): 37–52.

Drugge, A-L. 2016. *Ethics in Indigenous Research: Past Experiences – Future Challenges.* Umeå: Vaartoe – Centre for Sámi Research.

Falch, T., and P. Selle. 2018. *Sametinget: Institusjonalisering av en ny samepolitik.* Oslo: Gyldendal.

Fjellström, A-M., U. Mörkenstam, R. Nilsson, and M. Knobloch. 2016. "Sametingets formella ställning, valsystem och partier." In *Sametingsval: valdeltagande, representation och media,* edited by R. Nilsson, S. Dahlberg, and U. Mörkenstam, 77–102. Stockholm: Santérus förlag.

Gover, K. 2018. "Equality and Non-Discrimination in the UNDRIP." In *The UN Declaration on the Rights of Indigenous Peoples: A Commentary,* edited by J. Hohmann and M. Weller, 179–212. New York: Oxford University Press.

Heinämäki, L., and D. C. Cambou. 2018. "New Proposal for the Nordic Sámi Convention: An Appraisal of the Sámi People's Right to Self-Determination." *Retfærd: nordisk juridisk tidsskrift* 41: 3–18.

Heinämäki, L., et al. 2017. *Actualizing Sámi Rights: International Comparative Research.* Helsinki: Prime Minister's Office.

Hohmann, J., and M. Weller. 2018. *The UN Declaration on the Rights of Indigenous Peoples: A Commentary.* New York: Oxford University Press.

Josefsen, E. 2014. *Selvbestemmelse og samstyrning. En studie av Sametingets plass i politiske processer i Norge.* Tromsö: Universitetet i Tromsö.

Josefsen, E., U. Mörkenstam, R. Nilsson, and J. Saglie. 2017. *Ett folk, ulike valg. Sametingsvalg i Norge og Sverige.* Oslo: Gyldendal Akademisk.

Koivurova, T. 2011. "Jurisprudence of the European Court of Human Rights Regarding Indigenous Peoples." *International Journal on Minority and Group Rights* 18: 1–37.

Kukutai, T., and M. Walter. 2015. "Recognition and Indigenizing Official Statistics: Reflections from Aotearoa New Zealand and Australia." *Statistical Journal of the IAOS* 31 (2): 317–326.

Lantto, P., and U. Mörkenstam. 2015. "Action, Organisation and Confrontation: Strategies of the Sámi Movement in Sweden during the Twentieth Century." In *Indigenous Politics: Institutions, Representation, Mobilisation,* edited by M. Berg-Nordlie, J. Saglie, and A. Sullivan, 135–163. Colchester: ECPR Press.

Laula, E. 1904. *Inför lif eller död? Sanningsord i de lappska förhållandena.* Stockholm: Wilhelmssons Boktryckeri AB.

Lawrence, R., and M. Åhrén. 2017. "Mining as Colonization: The Need for Restorative Justice and Restitution of Traditional Sámi Lands." In *Nature, Temporality, and Environmental Management. Scandinavian and Australian Perspectives on Peoples and Landscapes,* edited by L. Head et al., 149–166. London: Routledge.

Lawrence, R., and U. Mörkenstam. 2016. "Indigenous Self-Determination through a Government Agency? The Impossible Task of the Swedish *Sámediggi.*" *International Journal of Minority and Group Rights* 23 (1): 105–127.

Lightfoot, S. R. 2016. *Global Indigenous Politics. A Subtle Revolution.* London: Routledge.

Lipset, S. M., and S. Rokkan. 1967. "Cleavage Structure, Party Systems, and Voter Alignments: An Introduction." In *Party Systems and Voter Alignments,* edited by S. M. Lipset and S. Rokkan. New York: The Free Press.

McClain, P. D., J. D. Johnson Carew, E. Walton Jr., and C. S. Watts. 2009. "Group Membership, Group Identity, and Group Consciousness: Measures of Racial Identity in American Politics?" *Annual Review of Political Science* 11: 457–478.

Minde, H. 2003. "Urfolksoffensiv, folkerettsfokus og styringskrise: kampen for en ny samepolitik 1960–1990." In *Samer, makt og demokrati. Sametinget og den nye samiske offentligheten,* edited by B. Bjerkli, and P. Selle, 87–123. Oslo: Gyldendal Norsk Forlag.

Min Geaidnu. 2009. M369, dnr 2009–1500 [motion in the Sámediggi]. Available on file with author.

Mörkenstam, U. 2015. "Recognition *As if* Sovereigns: A Procedural Understanding of Indigenous Peoples' Right to Self-Determination." *Citizenship Studies* 19 (6–7): 634–648.

Mörkenstam, U. 2019. "Organised Hypocrisy? The Influence of International Norms on the Indigenous Rights Regime in Sweden." *The International Journal of Human Rights* 23 (10): 1718–1741.

Mörkenstam, U., E. Josefsen, and R. Nilsson. 2016. "The Nordic Sámediggis and the Limits of Indigenous Self-Determination." *Gáldu Cála – Journal of Indigenous Peoples Rights* 1: 6–46.

Nilsson, R., S. Dahlberg, and U. Mörkenstam. 2016. *Sametingsval: väljare, partier och media.* Stockholm: Santérus förlag.

Nilsson, R., and T. Möller. 2016. "Social tillit." In *Sametingsval: väljare, partier och media*, edited by R. Nilsson, S. Dahlberg, and U. Mörkenstam, 231–246. Stockholm: Santérus förlag.

Nilsson, R., U. Mörkenstam, and R. Svensson. 2016. "Politiska skiljelinjer vid sametingsval." In *Sametingsval: väljare, partier och media*, edited by R. Nilsson, S. Dahlberg, and U. Mörkenstam, 181–200. Stockholm: Santérus förlag.

Pettersen, T. 2016. *Sámi Ethnicity as a Variable. Premises and Implications for Populations-Based Studies on Health and Living Conditions in Norwa*. Tromsö: Universitetet i Tromsö.

Quane, H. 2012. "A Further Dimension to the Interdependence and Indivisibility of Human Rights? Recent Developments Concerning the Rights of Indigenous Peoples." *Harvard Human Rights Journal* 25: 49–84.

Ruong, I. 1982. *Samerna i historien och nutiden*. Stockholm: Bonnier fakta.

Saglie, J., U. Mörkenstam and J. Bergh. 2020. "Political Cleavages in Indigenous Representation: The Case of the Norwegian and Swedish Sámediggis." *Nationalism and Ethnic Politics* 26. doi:10.1080/135 37113.2020.1754555.

Samefolket. 2017. "Nordisk Samekonvention – en katastrof för samerna som urfolk." Accessed 27 September 2018. https://samefolket.se/nordisk-samekonvention-en-katastrof-for-samerna-som-urfolk/.

Sametinget. 2004. "Betänkande av det svenska Sametingets kommitté med uppgift att ta fram ett förslag till strategi för en implementering av det samiska folkets rätt till självbestämmande på den svenska sidan av Sápmi." Accessed 15 April 2019. www.sametinget.se/1328.

Sametinget. 2008. "Sveriges grundlagsanpassning till gällande rätt. Ett delbetänkande av Iesjmierredimjuogos." SáOU 2008: 1. Accessed 15 March 2019. www.sametinget.se/80653.

Scheinin, M. 2008. "The Right of a People to Enjoy Its Culture: Towards a Nordic Saami Rights Convention." In *Cultural Human Rights*, edited by F. Francioni and M. Scheinin. Leiden: Martinus Nijhoff Publishers.

Scheinin, M., and M. Åhrén. 2018. "Relationship to Human Rights, and Related International Instruments." In *The UN Declaration on the Rights of Indigenous Peoples: A Commentary*, edited by J. Hohmann and M. Weller, 63–86. New York: Oxford University Press.

Tauli-Corpuz, V. 2016. "Report of the Special Rapporteur on the Rights of Indigenous Peoples on the Human Rights Situation of the Sami People in the Sápmi Region of Norway, Sweden and Finland." 9 August. A/HCR/33/42/Add.3.

Tully, J. 1995. *Strange Multiplicity. Constitutionalism in an Age of Diversity*. Cambridge: Cambridge University Press.

Walter, M., and C. Andersen. 2013. *Indigenous Statistics. A Quantitative Research Methodology*. Walnut Creek: Left Coast Press.

Weller, M. 2018. "Self-Determination of Indigenous Peoples: Articles 3, 4, 5, 18, 23, and 46(1)." In *The UN Declaration on the Rights of Indigenous Peoples: A Commentary*, edited by J. Hohmann and M. Weller, 115–149. New York: Oxford University Press.

Wiessner, S. 2008. "Indigenous Sovereignty: A Reassessment in the Light of the UN Declaration on the Rights of Indigenous Peoples." *Vanderbilt Journal of Transnational Law* 41: 1141–1176.

Xanthaki, A. 2007. *Indigenous Rights and the United Nations Standards. Self-Determination, Culture and Land*. Cambridge: Cambridge University Press.

Other documents

Convention Concerning Indigenous and Tribal Peoples in Independent Countries (adopted 27 June 1989, entered into force September 5, 1991) 72 ILO Official Bull. 59 (ILO Convention No. 169).

Draft Nordic Sámi Convention. 2005. Accessed 19 July 2019. www.sametinget.se/105173.

Draft Nordic Sámi Convention. 2016. Accessed 19 August 2019. www.sametinget.se/111445.

Prop. 2005/06:86, *Ett ökat samiskt inflytande* [Prop. is an abbreviation of *Proposition*, Government Bill].

United Nations Declaration on the Rights of Indigenous Peoples (13 September 2007) UN Doc A/RES/61/295 (UNDRIP).

Appendix 1
COMMENTS ON TABLE 17.1

Social integration in the Sámi society is an index consisting of eight survey items (Cronbach's alpha .89): three questions about language skills, whether a respondent can speak, read and write in any of the Sámi languages; two about affinity with a) the local Sámi environment and b) Sápmi; three about a) how many friends that are registered in the election roll, b) if the respondent has grown up in a Sámi community and c) how many friends in everyday life that are Sámi. The response options were rescaled into three-point scales (min, medium, max).

General interest in politics is measured on a four-point scale stretching from: "Not at all interested", "not very interested", "somewhat interested" and "very interested".

Interest in Sámi politics is measured on the same scale as general interest.

Media consumption measures how frequently the respondent has taken part in the Sámi election campaign in 2013, starting with "never", "seldom", "1–2 days a week", "3–4 days a week", "5–6 days a week" and "daily".

General institutional trust is measured on a five-point scale, where 1 equals "no trust at all", 2 "not much trust", 3 "neither trusting nor distrusting, 4 "somewhat trusting" and 5 "trusting to a large extent". The variable is an index based on trust in the Swedish parliament, the Swedish government, the municipality council, courts, the police, the European Union (Cronbach's alpha .81).

Trust in the Sámi parliament is measured in the same battery of trust questions.

Jakt- och Fiskesamerna, Landspartiet Svenska Samer, Min Geaidnu, Samerna/Sámit, Sámiid Riikkabellodat, Vuovdega – Skogssamerna are measured by voting for one of these parties. *Guovssonásti* is used as baseline for party choice.

Member in a reindeer-herding community is dichotomous, where 1 = member and 0 = non-member.

Age is based on eight categories: 18–21, 22–30, 31–40, 41–50, 51–60, 61–70, 71–80, 80–.

Sex is a dichotomous variable for men and women, where 1 = women and 0 = men.

Education is measured in terms of low (elementary school), medium (upper secondary school) and high education (university).

Employment is entered as a dichotomous variable, where 1 corresponds to employed in the public or private sector or self-employed, while 0 corresponds to unemployed, retired, students, being on sick leave, being in labour market policy programs or other.

Cohabitation is a dichotomous variable, where 1 = cohabitation with a partner and 0 = living alone.

Income is measured by 12 categories divided by 99,000 SEK, starting with 100,000 or less as 1 to 12, more than 1,001,000.

Residence in Sápmi is dichotomous, where 1 corresponds to living in the counties of Norrbotten, Västerbotten, Jämtland/Härjedalen, Västernorrland and Gävleborg (the traditional Sámi settlement areas) and 0 to all other Swedish counties.

Opinions on self-determination is an additive index (Cronbach's alpha .95) constructed from 18 survey items asking about self-determination in different areas: a) childcare, b) duodji, c) defence, d) mining, e) healthcare, f) culture, g) soil and water, h) reindeer herding, i) predators, j) taxes, k) forestry, l) school/education, m) protected areas (nature reservations etc.), n) small game hunting, o) language, p) wind power, q) elderly care, r) moose hunting. The response options are: To a large extent, To a fairly large extent, To a fairly small extent, Not at all. The index is reversed so that higher numbers mean more self-determination.

18

INDIGENOUS SELF-DETERMINATION AND DISEMPOWERMENT IN THE RUSSIAN NORTH

Liubov Sulyandziga and Rodion Sulyandziga

Abstract

By the end of the second decade of the 21st century, Russian Indigenous communities have found themselves outcast in their own lands. Inhabiting the country's richest area, the Arctic, they are confronted with the unmatched severity of extractive industries' activities, facilitated by a government policy that emphasizes the Arctic as a resource treasure house. This chapter examines the topic of Indigenous self-determination as a complexity of pathways to Russia's Indigenous disempowerment. We show how the emphasis on supporting extractive industrial development that feeds the Russian budget continuously weakens the modest Indigenous peoples' rights protection that was introduced in the immediate aftermath of the Soviet Union. Our research findings show that instead of devising a comprehensive policy designed to address Indigenous accommodation, the Russian state is far more inclined to remove the "last obstacle" standing in the way of its "Arctic dream", according to which the world's largest country's budget as well as its energy and military security is guaranteed by the Arctic.

Introduction: the significance of the Arctic for Russia

Russian Indigenous peoples inhabit lands rich in natural resources, including oil, gas, and minerals, and are thus heavily affected by large industrial projects. In particular, the Arctic region has been depicted as a resource repository, capable of feeding the whole country. This chapter is an attempt to raise fundamental questions about the nature of contemporary Russian policy towards its Indigenous population in the Arctic and shed light upon the various characteristics that have come to define Russia's response to Indigenous as problematic.

Russia holds approximately 40% of the Arctic territory (Geopolitical Futures 2017). The Russian Arctic zone constitutes approximately 25% of the country's landmass and holds a disproportionately large part of the natural resources (Breyfogle and Dunifon 2012). In terms of demography, the Russian Arctic territory is the most populous and stretches along eight territorial districts. Out of the total Arctic population of 4 million people, approximately half live in the Russian Federation (Fert-Malka 2018; Heleniak and Napper this volume).

For centuries, the Arctic has served Russia as a resource periphery, from which source materials such as furs, silver, gold, diamonds, wood, coal, oil, gas, and other goods were to be extracted for state benefit. To date, the Arctic region continues to be the driver of the Russian economy: about 20% of Russia's GDP and about one-fifth of its exports are generated in the region; 60%–80% of country's natural resources are produced in the North, including 93% of natural gas, 76% of oil, 100% of diamonds and platinum, 90% of nickel, and 63% of gold. This shows the critical significance of the region to the Kremlin (Stamatopoulou 2017). As a result, analysts have described the Arctic as a "global energy corridor" and a "floating pipeline" of Russian oil and gas (Breyfogle and Dunifon 2012). As such, the approach of the Russian federal government to Arctic development constructs the region as a vital resource base and emphasizes its crucial importance to energy security and state development.

Historically, the Soviet state invested far more in the Arctic than any other country, showed much more interest in the integration of its North, and claimed that "no other country has made so many explorations in the Arctic as the Soviet Union" (Zenzinov 1944, 65). Although like in other countries, the target of the exploration of the Arctic lands was primarily resource extraction through industrial colonization, the Soviet Union developed its Arctic regions very differently from Canada or Alaska. While other Arctic states conceived the development of territories through shift work, the Soviet Union concentrated on large-scale development of permanent settlements and full-scale industrial facilities and infrastructure as incentives for the population to stay permanently in the Arctic (Bolotova and Stammler 2010; Laruelle 2015; Stuart 1980). Development of the North was accompanied by a mass northward migration of workers. As a result, from the second half of the 20th century, the Russian North has experienced one of "the most extensive and spectacular resource development in the Arctic" (Crate and Nuttall 2003, 89).

Since its exploration, the discourse of *osvoenie severa* ("acquiring the North") fed the national fascination (Breyfogle and Dunifon 2012; Rowe and Blakkisrud 2014) and quickly captured the Russian imagination. Based on top officials' words, the Far North occupies a particular place in the national imaginary and plays a role far more important than just a driving force of the Russian economy and strategic energy battleground (Hille et al. 2016). It is considered a fundamental and inseparable part of Russian identity, a "reservoir of national authenticity" and an essential component of state patriotism (Honneland 2017, 401). The region has been referred to by various authors as a "Russian dream", "heart of most Russian people", "one of the foundation blocks of statehood", "the shrine of Russia's national idea, a new political and spiritual continent, a promised land, Russian destiny" (Harvey and Walker 2013; Honneland 2017).

In terms of administrative organization, the Russian Arctic is characterized by a top-down approach, with an enormous share of the wealth diverted from the region. (Fert-Malka 2018). The Kremlin's dependence on the resource-rich Arctic, with more than 55% of federal revenues derived from the use and export of natural materials, has resulted in a peculiar situation with Russian central regions feeding off a periphery that itself remains largely underdeveloped, although it hosts the majority of the resources that generate this revenue. The Russian government not only centralized the share and distribution of taxes[1] but registered headquarters of strategic corporations operating in the Arctic in Moscow or St. Petersburg, where they pay regional taxes. As a result, these central parts of Russia generate an unequally large regional gross product, while the Arctic regions are characterized by massive underinvestment.

The growing emphasis on Arctic economic development on Russia's political agenda has led to a duality in the authorities' relations to other actors with an interest in Arctic natural resources. While the Kremlin believes it to be vital to directly influence everything that is going on in the region, state officials consistently present the Arctic as the region of shared concern, harmony of interests and collaboration in the international arena (Rowe and Blakkisrud 2014). Under the circumstances of economic decline and a limited federal budget, Russia cannot sustain the development and potential use of its Arctic zone without the other Arctic states as partners. The strategy is therefore to balance its expansionist aspirations with constructive regional engagement in order to maximize the potential for economic growth and ensure the support of Nordic countries for investment, technology, know-how, and markets to enhance its own energy production and regional development (Andreassen 2016).

Indigenous rights, from international progress to Russian realities

Within the course of the past decades, many achievements have been made with reference to Indigenous rights standards, primarily through Indigenous engagement and dedication within global society. After 50 years of active participation in the global arena, Indigenous rights movements continue to gain momentum, transforming into one of "the most visible civil society grouping across the UN" (Morgan 2011, 2). As a result of adoption of international standards and guidelines in addition to the establishment of institutions that specifically target the concerns of Indigenous people, today Indigenous peoples are more mobilized than any other time. The Russian Federation is a notable exception from this trend, as despite a promising beginning of professional Indigenous activism in the early 2000s, Russian Indigenous groups saw further division into yet more separate paths in contrast to international Indigenous development (Eckert 2012). While the protection of Indigenous peoples' rights and interests is becoming an important global goal and an essential sphere of international cooperation, domestically there are still some fundamental imbalances in power, rights, and inclusion of Indigenous peoples in decision-making processes.

Among Indigenous claims, one of the most significant presuppositions held by Indigenous peoples is that their inalienable rights to lands and resources override the subsequent claims by dominant societies (Rogers 2000). In fact, land issues have always been fundamental in Indigenous struggles, with the restitution of Indigenous lands seen as an act of overcoming historical injustice. This assertion is grounded in the fact that Indigenous livelihoods are inseparable from the lands and resources, which form a basis for traditional activities such as hunting, fishing, gathering, and nomadism, as well as religious, spiritual, and ceremonial practices (Minde 2008).

As James Anaya (2004, 396) states, "They are Indigenous because their ancestral roots are embedded in the lands much more deeply than others. They are peoples because they represent distinct communities and have culture and identity that link them with their nations of the ancestral past".

In other words, many Indigenous communities see themselves as part of the land they have resided on for centuries. Natural resources, in turn, are not only the sources of livelihoods for many Indigenous peoples but also a source of their identity and a means to preserve their traditions and customs. The loss of land would thus mean a threat to their entire culture. Henceforth, securing access to these territories and natural resources and legal recognition of land tenure rights are an essential foundation to empower Indigenous peoples with civil, social, cultural, political, and economic rights (Alcorn 2013).

The Indigenous peoples' strong attachment to the environment and surrounding ecosystems have resulted in complex and distinct tenurial arrangements that are often at odds with the formal legal management regimes of the state. Whereas Indigenous peoples have not operated under the concept of private land ownership (Berg-Nordlie 2015), which means that

Indigenous land was instead governed by customary tenure based on the principles of long-term and uninterrupted land use, inheritance, and oral agreements with neighbours (Sirina 2005; Stammler 2005, 225–227), governments viewed Indigenous lands as terra nullius ("nobody's land") or previously ownerless and therefore open for utilization by newcomers. Particularly, although Indigenous peoples constitute one of the most vulnerable populations on earth as a result of centuries of marginalization and discrimination, their territories often contain abundant natural resources. As a result, Indigenous territories become objects for land acquisition for agriculture, biodiversity conservation, appropriation by outside interests, and other development initiatives, both private and governmental (Alcorn 2013). From the perspective of the industries in particular, these lands are frequently regarded as a source of income generation "rather than as heritage to be cherished" (Glennie 2014). Indigenous peoples, in turn, have to live adjacent to extractive facilities that generate enormous wealth for their owners and do not stand to gain economically or socially from the projects either collectively or as individuals (O'Faircheallaigh 2013). The compensation that is sometimes provided by companies cannot cover the deterioration inflicted on the land, which frequently becomes unfit for Indigenous practices (Stamatopoulou 2017).

Indigenous struggles on their land with incomers: concepts and examples from Russia

In 2011, the Russian mining company Yuzhnaya started its activities near Kazas settlement in the Kemerovo region in Southwest Siberia – one of the major coal districts of the Russian Federation. Kazas, the territory of traditional residence of Russia's Indigenous peoples – the Shors – has been subject to decades of environmental destruction and fatal effects of the coal industry (IWGIA et al. 2017). At the end of 2012, Yuzhnaya started buying households in Kazas to expand its industrial activities. By 2013, only five families refused to sell their houses and leave the ancestral lands. On 2 November 2013, during a meeting with the villagers, the CEO of the company threatened to set on fire all the remaining houses if the families refused to sell them to the company. The first house was burnt a week later. At the end of December, the second one was set on fire. In January 2014, two houses burnt down. The last one was struck in March 2014 (Sulyandziga 2016).

In 2012, Sergei Nikiforov, the leader of the Amur Evenki people, was sentenced to four years in prison for allegedly extorting money from the Petropavlovsk gold mining company after he led a protest movement against the company's attempts to take over native reindeer pastures and hunting grounds (IWGIA et al. 2017).

In 2013, 1 million tons of oil was discovered on the bottom of Lake Imlor in Khanty-Mansi Autonomous Okrug, Russia's leading oil-producing region. The same year, Surgutneftegaz company obtained a license to explore oil and gas deposits under the lake, which is sacred to the Indigenous Khanty people. With their land under threat and alternative job prospects, the majority of Khanty people have left the ancestral land. In 2015, Sergei Kechimov, a Khanty shaman, the only person left living near the lake, was accused of uttering death threats to a worker of Surgutneftegas oil company and sentenced to imprisonment (Stamatopoulou 2017; Lerner et al. 2017; CESCR 2017b).

Just a couple of months before the launch of criminal investigation against Kechimov, the 113th Session of the UN Human Rights Committee was attended by an unprecedented number of representatives of the Russian Federation, "presenting their shadow reports denouncing a wide range of human rights violations" (IWGIA 2015). A couple of months after Kechimov's hearings in the court, the Russian Federation also attended the Third Committee of UN General Assembly, where it was stated that the

> Russian Federation has always supported and continue to support Indigenous peoples in full and effective implementation of their rights. . . . We are confident that the main instrument for the practical implementation of the UNDRIP provisions and the outcome document of the World Conference on Indigenous Peoples should be the goodwill of states, coupled with the daily hard work to support the Indigenous population and protect their rights and freedoms, as it is done in Russia.
>
> (Statement by the representative of the Russian Federation/ Agenda Item 70 "Indigenous peoples rights" of the Third Committee of UN General Assembly 2015)

However, Russia abstained during the vote for the adoption of the UNDRIP in 2007. A closer look at Russian Indigenous legislation, particularly that on land rights, helps to contextualize the case.

Legal disempowerment

Since the beginning of the 2000s, with the increasing presence of resource extraction activities on Indigenous homelands in Russia, discussions of management of nature use, industrial development of Indigenous lands in the context of ethnic and environmental problems, and the legacy of state development policies, Indigenous participation in the management of their lands and resources has been on the rise (e.g. Xanthaki 2004; Fondahl and Sirina 2006; Stammler and Peskov 2008; O'Faircheallaigh 2013; Wilson and Stammler 2016; Tulaeva and Tysiachniouk 2017).

Historically, the question of Indigenous land ownership has been complex. Indigenous territory has never been regarded as a form of private property by the aboriginal population; instead, Indigenous land was used and managed collectively (Sirina 2005; Stammler 2005, 225–227). With regard to Russia, the approach to land has been developed differently from other Arctic states, such as Canada or the United States, where legally binding contractual evidence supporting Indigenous peoples' rights to land exists. On the contrary, Russian Indigenous peoples have not been involved in legal relationships with the state on the matter of land ownership; they have neither sold their lands nor received any compensation or delegated the right to supervise their lands to a third party.

After the Russian revolution, all land was considered state property. The Soviet Union, therefore, simply declared Indigenous territories state lands and managed them at its own discretion. Since there were never any treaties signed between Indigenous peoples and the state, the best outcome Indigenous groups can hope for is a long-term lease; that is, "the title to land is not even on the table" (Eckert et al. 2012, 45). Henceforth, Russia's Indigenous groups' claims are much more modest than those of Indigenous communities in the West, focusing on the right to preserve a traditional lifestyle and some type of limited property rights to land and resources (ibid.).

Codified federal law on Indigenous rights in the Russian Federation

Many experts note that Russian legislation includes rather strong state obligations to protect Indigenous peoples' rights. The Russian Federation adopted three national framework laws establishing the framework of cultural, territorial, and political rights of Indigenous peoples and community obshchinas (see Fondahl et al. this volume) (Federal Law on the Guarantees of the Rights of the Indigenous Small-Numbered Peoples of the Russian Federation adopted in 1999; Federal Law on General Principles of Organization of Obshchina of Small-Numbered Indigenous Peoples of the North, Siberia and the Far East of the Russian Federation adopted in 2000; Federal Law on Territories of Traditional Nature Use of the Small-Numbered Indigenous Peoples of the North, Siberia and the Far East of the Russian Federation adopted in 2001).

Originally, these laws guaranteed Indigenous peoples' rights to use the land; to participate in the implementation of control over land use and in decisions about protecting their traditional lands and way of life, economy, and activities through conducting ecological and ethnological expertise; and to be compensated for damages to their traditional lands resulting from industrial and economic activity (On Guarantees, Article 8). Although these laws have offered the basis for the Indigenous population to make claims to lands and to establish self-government, recent years have been marked by intense efforts to legally disempower and exclude Indigenous peoples from the management of their ancestral territories. Recent amendments to all these laws have made full implementation of Indigenous peoples' collective rights to land and resources virtually impossible (Zaikov et al. 2017; Kryazhkov 2013). Even already modest provisions that were included in this legislation have today lost their power.

Land rights legislation and problems of its implementation

Attempts to create a legal framework for Indigenous peoples' land rights date back to the early 1990s, when several Russian regions elaborated their own Indigenous land rights regimes. The earliest attempt was the introduction of "patrimonial lands" adopted in 1992 in Khanty-Mansi Autonomous Okrug (On the Statutes of Patrimonial Lands of Khanty-Mansi Autonomous Okrug 1992). Later, in 2001, the state initiated the creation of the so-called "Territories of Traditional Nature Use" (TTPs hereafter; see Fondahl et al. this volume) designed to protect Indigenous land from industrial encroachment, exclude these lands from the real estate trade, and provide the Indigenous population with secure plots of land "in perpetuity" assigned to traditional economic sectors – reindeer herding, fishing, marine animal hunting, harvesting, and so on – that provide the main employment and main source of income for Indigenous communities (Turaev 1998; Colchester n.d.; Fondahl et al. this volume). Under the legislation, companies which pursue industrial activity within the officially designated TTP should reach an agreement with the Indigenous population about land use and are obliged to compensate for damaging traditional lands. The law also provides Indigenous peoples the right to participate in assessments of sociocultural impact on the Indigenous communities by extractive companies (Article 6.8).

In 2001, at the same time the Law on TTP was adopted, the Russian Federation enacted the Land Code, which ruled out any form of land tenure other than rented and private property:

> Citizens cannot be granted permanent (indefinite) use [rights] over plots of land. Judicial persons, except those named under item 1 of this provision are obliged to have their right to permanent (indefinite) use of land plots transferred into the right to rent the given plots or to obtain the plots as property.
>
> (Article 20)

This effectively means that Indigenous lands can become the private or long-term leasehold property of industrial companies (Vinding 2002). Given that nomadic Indigenous communities typically migrate with their herd throughout the year in search of pastures following the cycle of reindeer herding and hence use substantial areas, up to several thousand hectares (300 hectares for one reindeer), neither purchase nor rent are financially viable options for Indigenous groups (Basov 2018; Fondahl et al. this volume).

The hierarchy of Russian legislation means that the Land Code – which does not recognize Indigenous traditional resource or land rights – will override the Indigenous rights legislation. Thus, in practice, if a traditional resource use area is threatened by an oil, gas, or mining project, no real protection is offered by regulations (Murashko 2008; Wilson and Swiderska 2009). Furthermore, in 2007, the phrase "in perpetuity" disappeared from the TTP Law (Yakovleva 2014; Fondahl et al. 2019).[2] In 2014, the Land Code stipulated that lands free of charge can be granted to Indigenous peoples only for the construction of buildings or other facilities needed for development and conservation of the Indigenous traditional lifestyle for a period of no more than ten years (Article 39.10). The provision in the Land Code that had explicitly stated that in places of Indigenous traditional residence, authorities decide on location of industrial objects (i.e., infrastructure, extraction facilities, etc.) based on the results of information gathered from Indigenous communities was entirely removed (CESCR 2017a).

Another problem is that almost all lands that might be candidates for TTP status are either partly or wholly situated on federal land (70% of Russia territory is categorized as forest fund, which is also federal property); therefore, local and regional organs do not have the authority to transfer control over such lands to Indigenous peoples. Only the federal government has the authority to do so (Eckert et al. 2012). As a result, since the adoption of the law on TTP by the State Duma in 2001, no TTP has been designated on the federal level at all. And while regional authorities have, however, created over 500 TTPs, none of them has been confirmed by the federal government. The existing TTPs, therefore, have "no guaranteed legal status and no effective protection from being dissolved or downsized, as often happens" (CESCR 2017a, 5). In effect, due to the government's failure to confirm existing TTPs, their status is open to changes at any time.[3]

The situation was aggravated further in 2013, when the law amending Federal Law No. 33-FZ on protected areas (Articles 5 and 6) was approved without public discussion, despite petitions from lawyers and ecologists. One of the most significant pitfalls was the downgrading of TTP status from "Specially Protected Conservation Areas" to "Specially Protected Areas" (CESCR 2017a).[4] As a result, the word "conservation" (alluded to as "nature") was removed from the TTP definition. While "specially protected conservation areas" is a term stipulated in environmental legislation of the Russian Federation which creates specific safeguards for Indigenous participation and consultation rights, the designation "specially protected areas" does not exist in Russian law and, as such, is not identified in state legislation. As a result, now, allocation of land and projects for economic activity (construction of roads, pipelines, and industrial facilities) are no longer subject to ecological assessment, and evaluation of negative impacts on Indigenous lives by industrial projects is no longer required (Stammler and Ivanova 2016, 1234).

Markedly, these territories have been eliminated from the real estate trade as well. The amendment also changed the rules for the removal of land plots from TTPs. Originally, in the event of Indigenous peoples' removal from their ancestral land, the state was obligated to provide Indigenous communities with an equivalent plot of territory and natural objects in exchange. After the revision, the expression "Compensation for losses in case of alienation of plots of land for state or municipal needs" disappeared from the land legislation as a whole.

When the law on obshchinas was introduced in 2001, many Indigenous peoples organized into communes to pursue their traditional activities (Colchester n.d.). The original intent behind the introduction of the obshchina concept was multifaceted: for one thing, obshchinas were supposed to carry out functions of local self-administration, participate in decision-making processes in the interests of Indigenous peoples (Sirina 2005), provide services in the domain of culture and education, and, at the same time, function as economic cooperatives through which Indigenous peoples could pursue their traditional economic activities in a viable and sustainable manner (Stammler 2005, 254–260). The obshchina was seen as a rightful unit of property management. Initially, Indigenous peoples had the right to use obshchina lands in perpetuity and without charge (Fondahl et al. this volume). In 2004, the law was changed; the notion "in perpetuity and without charge" was revoked, and rent was introduced. Since then, many communities have lost their rights to the lands granted to them for traditional subsistence practices (Evengard et al. 2015; Stamatopoulou 2017). In many regions, Indigenous obshchinas are now regarded as competitors by private businesses, especially in the fishing industry, some of which are affiliated with the local administration and spare no effort to push Indigenous communities out of business. Another troublesome aspect of the law is its restriction to pursue "traditional" types of activity. They can be terminated if they stop engaging in traditional economic activities (Eckert et al. 2012). In contrast to the initial idea, obshchina lands do not provide a comprehensive solution to either Indigenous land rights or environmental protection of Indigenous homelands. More importantly, they cannot become self-governing bodies without being given an authority over a territory, natural resources, and economic independence (Turaev 2018).

Accordingly, the provisions on preferential allocation and free use of various categories of land by Indigenous peoples, originally stipulated in the Land, Forest, and Water Codes of the Russian Federation, have been withdrawn. Originally, some provisions of sectoral legislation (e.g., land, forest, and water codes, as well as acts on subsoil) stipulated the rights of Indigenous peoples for preferential use of resources in areas of their traditional residence. With regard to one of the main economic activities of many Indigenous communities, fishing, already in Soviet times, economically profitable fishing attracted the attention of business. As a result, Indigenous communities have been gradually pushed out of the activity. The initial provision that gave permission to Indigenous peoples for preferential use of lands for fishing without competition was recognized as invalid (Article 39 FZ-166 2004). Now fishing grounds belong to people or businesses who won quotas to pursue commercial activities (Mamontova 2012).

In fact, since 2008, all Indigenous territories for hunting or fishing have to be distributed through auctions only, and there are no exceptions for the Indigenous communities inhabiting those territories. Indigenous peoples are obliged to compete in commercial tenders for hunting and fishing grounds with usually more competitive private businesses who lease these lands for long-term tenure (up to 49 years). As a result, traditional fishing and reindeer herding and hunting grounds can now be shared with other users, and many Indigenous communities have lost their traditional lands since that time. By clearing a way for businesses opportunities, these provisions substantially endanger Indigenous access to their sources of subsistence, food, and income and have been identified as one of the principal obstacles preventing Indigenous peoples from enjoying their fundamental rights (CESCR 2017b). In realities where economic incentives outweigh the importance of Indigenous interests, Indigenous rights have been entirely ignored.

The path to federal legal stagnation?

All in all, the period from the 2000s onwards has been referred to as "legal stagnation for Indigenous rights" in Russia (Kryazhkov 2012, 29; Fondahl 2014). Major institutions dealing directly with Indigenous peoples in Russia have been liquidated as well. During the 1990s, responsibility for Indigenous minority policy shifted rapidly between different State Committees and Ministries, leaving Indigenous policy field institutionally "homeless" in the period 2000–2004. In 2001, the Ministry of Federal Affairs, National and Migration Policy was disbanded. In 2004, Indigenous policy was handed to the Regional Development Ministry, which was responsible for elaboration of state policy on Indigenous peoples and normative relations of socioeconomic development of Indigenous groups in regions with Indigenous population and also managed ethnic interrelations that for security reasons were much higher on the political agenda (Chyebotaryev and Gladun 2015). The Russian Association of the Indigenous Peoples of the North (RAIPON) and this ministry established relatively good working relations. In 2014, however, the Ministry was dissolved and Indigenous policy transferred to the Ministry of Culture (Berg-Nordlie 2015). Indigenous representatives argued that this change placed limitations on Indigenous affairs, which were now framed within constraints of sponsorship for "singing and dancing", "whereas rights, land and development would be off the table" (IWGIA 2014a).

Martin (2001, 13) described this approach as a strategy of depoliticizing ethnicity "through the aggressive promotion of symbolic markers of national identity: folklore, museums, dress, food, costumes". While specific programmes are actively supported by both local and central governments, the measures are limited to cultural events without any rights granted. In other words, Indigenous customs and traditions are treated as valuable, yet they are not identified as sources of rights. As such, the Russian state came to promote exclusionary categories of its ethnic diversity (Etkind 2014) and to narrowly frame Indigenous rights by focusing on state support on traditional cultures while taking the focus away from more substantive discussions regarding the reclamation of Indigenous territories, livelihoods, natural resources, and self-government (Corntassel 2008). In this context, Indigenous policy remains highly restrictive and limited to cultural rights, while Indigenous demands for special representation and political rights have little room to manoeuvre. Legal stagnation created an organizational void and institutional capture of Indigenous agency incapable of developing self-defence mechanisms. As of today, on the federal level, Indigenous policy remains poorly institutionalized. Indigenous issues lost the ministerial level, and the Federal Agency for Ethnic Affairs is responsible for all Indigenous issues at the national level.

This agency's most recent project is the National Registry of Indigenous peoples, which came into force on May 7, 2020. Its purpose is that the new registration mechanism would simplify access to natural resources, benefits, preferences, and privileges for Indigenous people, because a person would need only to go through the process of proving his/her Indigenous identity once, by submitting a number of documents. The Federal Agency for Ethnic Affairs is then in charge of creating the Registry according to these submissions. However, the regulations for inclusion to the registry suggest that most likely only candidates leading a traditional way of life will be considered. This can create considerable tensions within Indigenous groups, as it presumes that some Indigenous peoples are more Indigenous than others and that some groups can be excluded from the indigeneity discourses, disadvantaging them economically and politically in the process. In the worst case, the registry, instead of providing an opportunity for responsive change, could become an instrument for control.

Another tendency in federal Russian policy that also contributes to legally disempowering Indigenous peoples is the merging of regions with Indigenous autonomies into bigger units. From 2003 to 2008, the Taymyr Dolgano-Nenets, Evenk, and Koryak Autonomous Okrugs were liquidated by incorporating them into bigger "parent" units, the Krasnoyarskiy and Kamchatka provinces. Now the same may happen with the Nenets Autonomous Okrug in North

West Russia: the governor of the region agreed with the governor of Arkhangelsk Oblast for a road map of incorporation by the end of 2020.[5]

Conclusion

According to numerous scholars, clear land tenure is a prerequisite for the effective implementation of Indigenous rights. Without land rights and rights over natural resources, the right of self-determination and other rights would be meaningless or merely become "paper" rights, as happened in the case of Russia (Corntassel 2008, 108). Clear tenure helps to ensure and secure property rights, as well as the right to access natural resources. Land rights are also a basis for claiming benefits. Clear tenure facilitates their allocation and lowers the potential for conflicts over benefits linked to resources. Unclear or insecure tenure in turn has long been known as a factor that impedes proper natural resource management, whereas the conflicts over land are recognized as a barrier to Indigenous empowerment.

While the Constitution of the Russian Federation allows for varied forms of land and natural resource ownership (private, state, municipal, and otherwise), most of the land and subsoil resources in Russia remain under state control.[6] In comparison to Canadian and US contexts, there is not the same sort of legally binding contractual evidence supporting Indigenous peoples' rights to land. There were never any treaties signed, and the question of native title to land "is not even on the table" (Eckert et al. 2012, 45). In other words, whereas Indigenous peoples are accorded rights to use the land and its resources, title ownership remains with the state. At most, Indigenous peoples participate in guarding the territories: they may use their lands, but they are not allowed to be in full control of the territory.

Federal laws do not grant any special rights that let Indigenous peoples participate in the decision-making process concerning the lands and resources. Similarly, there is no regulated system ensuring consultation, cooperation, agreements, and other forms of Indigenous participation. Only a few regions, such as the Republic of Sakha (Yakutia), Yamal-Nenets, and Khanty-Mansi Autonomous Okrugs, developed legislation or regional programs regulating Indigenous participation, thus acting even before the federal state on Indigenous needs (see Stammler and Ivanova 2016, 1236; Fondahl et al. this volume for more information).

TTPs served as a guarantee for the future solid self-development of Indigenous territories (Turaev 1998). The original idea behind their creation was that these lands would be mostly off limits to industrial development (Evengard et al. 2015). These lands were meant to be managed, or at a minimum co-managed, by Indigenous communities. Importantly, TTPs and obshchinas were not only created to fulfil economic rights of Indigenous groups by giving them the possibility to ensure their traditional economic activities. Their creation reflected the existing link of Indigenous culture and the traditional economy; as such, allocation of lands to Indigenous groups was crucial to preservation of their unique traditions. In this regard, the TTP was seen as an "indivisible foundation" of Indigenous community aimed at preservation of the environment in which that community has been formed. In the same vein, established obshchinas were seen as a sole subject of use (ownership) in TTP management as an institution of economic autonomy and environmental management (Turaev 1998). In practice, however, obshchinas have been formalized not as decision-making or land-owning bodies but something more akin to civil society formations instead of Indigenous self-governing bodies (Øverland and Blakkisrud 2006; Berg-Nordlie 2015).

Neither the creation of TTPs nor obshchinas was supported by a set of measures for the development of the traditional economy or mechanisms for the socioeconomic development of territories. As a result, the formation of TTPs and obshchinas is seen mainly as a political action, turning out to be merely a product of the era of the democratic "romanticism" of the 1990s.

Some regional regulations provide considerably more opportunities for Indigenous participation. However, because of jurisdictional uncertainty and weak regional power vis-à-vis the federal government, the federal government usually overrides regional law in areas of shared jurisdiction – land use, natural resources, and Indigenous peoples (Newman et al. 2014). Thus, insufficient regulatory potential, lack of mechanisms to implement the declared rights, jurisdictional vagueness and non-concreteness, and authoritative federal power represent the biggest obstacles for Indigenous communities seeking adequate protection (Newman et al. 2014; Gladun and Ivanova 2017). Due to the lack of normative and legal mechanisms that provide for Indigenous rights' realization, the existing system of Russian domestic legal regulation is full of gaps, inconsistencies, and contradictions and has yet to be redeveloped according to current international standards. Legislative decrees and presidential edicts are often left ignored by most regional jurisdictions. In other cases, authorities implement federal laws in a very selective way, especially with respect to natural resources and lands issues. In particular, even at times when Indigenous peoples were seemingly backed up by already modest, yet existing, legislation, the state moved the finish line by withdrawing and changing the few laws designed for Indigenous protection.

These exemptions to legal norms can be seen in companies' ignorance of obligations to assess possible negative impacts of projects on the traditional way of life of Indigenous peoples or the permission to define, downsize, and resize TTPs. Often, these exemptions are claimed to act upon federal approval. All in all, as Kryazhkov (2012, 35) stated, "Russian legislation concerning Indigenous minority peoples could be characterized as unstable, contradictive, often imitational, only initially developed, and not enough adjusted with international law". By looking at the legislative system and existing institutions dealing with Indigenous issues in the country, the major role of law in Indigenous disempowerment becomes apparent. As a result, fragmented governmental systems, instead of providing an opportunity for responsive change, become an instrument for control (Cunneen 2011). On its steady way to becoming the largest oil and gas producer and increasing its production capacity of oil and gas pipelines (located primarily on Indigenous lands), coupled with a powerful lobby of extractive industry and business representation in political structures, Russia's authorities have been largely unsuccessful in protecting Indigenous rights (Nikolaeva 2017).

It has been frequently observed that Indigenous peoples have captured the world's attention and conscience (Watt-Cloutier 2019). How does Russia's approach to its Indigenous peoples fit in the four decades of what was labelled "the most progressive stage in the history of development of Indigenous peoples' rights and freedoms"? Unfortunately, Russia's declarative laws do not translate into progress in its domestic Indigenous policy. State and industrial actors do not hesitate to use a variety of instruments to disempower Indigenous communities legally, economically, and politically. Indigenous peoples are left without powerful counterbalance to dominant players, trapped between misplaced responsibilities of thirsty-for-profit companies and unactionable authorities (Petrov and Francis 2015). With more companies circling closer and closer around Indigenous territories and becoming richer, and government siding with business, Indigenous peoples have become outcast on their own lands.

Indigenous empowerment is not a matter of course but a political achievement. The perpetual crisis that Indigenous peoples found themselves trapped in does not refer to the crisis of Indigenous rights, struggles, and determination per se but points to a perpetual failure of policy reforms. A brief revival of rights-based conversations in the 1990s has not stood the test of time and is coming to terms with decades of failure. Promising laws sooner or later were made ineffective, changed, or withdrawn. At the same time, regional variations of Indigenous rights are developing across the Russian North; certain regions are known for more progressive legislative foundation for Indigenous rights' protection and more stringent rights protection mechanisms than those guaranteed by the federal government. As an illustration, the Republic of Sakha (Yakutia), where Yakut people enjoy the status of the so-called titular nation, adopted its constitution even before the Russian Federation, having enacted proactive Indigenous policies, including Russia's first and only law on anthropological expert review (or an ethnocultural impact assessment), designed to assess the sociocultural and economic impacts on Indigenous communities affected by extractive projects (Ivanova and Stammler 2017; see also Ivanova et al. this volume).

All in all, 28 years after the disintegration of the Soviet Union, it is evident that the Russian Federation does not prioritize a search for a more inclusive Indigenous policy, and what is more, it is neither an aspiration nor a political ideal of the federal government. Under circumstances where the Russian state not only turns a blind eye to Indigenous peoples but nullifies any progress, Indigenous groups are not in the position to lead decision-making processes in the country.

Notes

1 Oil companies in Russia pay taxes to the federal government (income, profit, mineral production, and value-added tax), to the regional government (property taxes, transport tax), and to local governments (land tax). Most of the taxes paid by oil companies go to the federal budget, leaving a much smaller share for the regions. Until 2002, 60% of taxes from mining operations levied to the budgets of resources-producing regions, while 40% accrued to the federal budget. With the increase in oil and gas prices in recent years, the federal government decided to change the tax sharing ratio between the central authority and the regions in its favour. As the result, the regional share of mineral tax on oil and natural gas declined significantly. In 2002, the share of taxes on oil production falling on regional budgets dropped from 60% to 20%; in 2003, the share declined to 15% and, in 2005, to 5%. Since 2004, tax revenues from natural gas production are going exclusively to the federal budget. The continental shelf is solely owned by the Russian Federation; therefore, regions are not entitled to taxes and revenue from oil mining on the shelf.
2 All these contradictions in laws make it hard to reveal whether Indigenous peoples pay for the use of land (IP representatives contend that they do pay such fees, and even if these are small, they nonetheless impose an economic burden on Indigenous communities). Notable exceptions include the Sakha Republic, who managed to "enable its residents to more fully benefit from the land allocation and to simultaneously remove lands from allocation to outsiders" (see Fondahl et al. 2019 for more details).
3 On 15 January 2015, the Court of Appeals rejected an appeal by the administration of Oleneksky district of the Sakha Republic challenging the legality of a license issued by the regional resource authority, Yakutnedra, for the exploration and extraction of mineral resources in TTP that had been established by the local authorities in Oleneksky district. The court rejected the appeal because the boundaries of the specified TTP had not been determined by the federal government. As noted previously, this is true for all currently existing TTP, such that they are all unprotected from similar encroachments.
4 Two acts passed in 2014 significantly weakened the law on TTP, these being Federal Law 171-FZ dated 23.06.2014 and 499-FZ, dated 31.12.2014.
5 http://gazetazp.ru/news/ekonomika/melochi-ukrupneniya-arhangelsk-i-nenetskiy-okrug-planiruyut-obyedinenie-po-obraztsu-taymyira-i-evenkii-.html, accessed 19 May 2020.
6 Ninety-two percent of Russian land is publicly owned, either at the federal, regional, or municipal level (the rest is held by legal entities and individuals). Agricultural, forest, pasture, and other land parcels utilized by private entities are primarily leased from the government.

References

Literature

Alcorn, J. 2013. "Tenure and Indigenous Peoples: The Importance of Self-Determination, Territory, and Rights to Land and Other Natural Resources." *USAID Issue Brief* [online]. U.S. Agency for International Development. Accessed 28 April 2019. https://www.land-links.org/wp-content/uploads/2016/09/Tenure-and-Indigenous-Peoples.pdf.

Anaya, J. 2004. *Indigenous Peoples in International Law*. 1st ed. Oxford: Oxford University Press.

Andreassen, N. 2016. "Arctic Energy Development in Russia – How 'Sustainability' Can Fit?" *Energy Research & Social Science* 16: 78–88.

Basov, A. 2018. "Dialogue and Bureaucratic Procedures: 'etnologicheskaya ekspertiza' (ethnological impact assessment) in the Republic of Sakha (Yakutia)." *Sibirskie istoricheskie issledovaniya* 2: 91–123.

Berg-Nordlie, M. 2015. "Two Centuries of Russian Sámi Policy: Arrangements for Autonomy and Participation Seen in Light of Imperial, Soviet and Federal Indigenous Minority Policy 1822–2014." *Acta Borealia* 32 (1): 40–67.

Bolotova, A., and F. Stammler. 2010. "How the North Became Home. Attachment to Place among Industrial Migrants in Murmansk Region." In *Migration in the Circumpolar North: Issues and Contexts*, edited by L. Huskeyand and C. Southcott, 193–220. Edmonton: CCI Press.

Breyfogle, N., and J. Dunifon. 2012. "Russia and the Race for the Arctic." *Origins: Current Events in Historical Perspective* 5 (11).

Chyebotaryev, G., and E. Gladun. 2015. "Co-Management by Indigenous Minorities of the North over Arctic Territories in the Period of Their Commercial Development." *Journal of Russian Law* 3 (5).

Colchester, M. n.d. "Indigenous Peoples and Communal Tenures in Asia." *Fao.org*. Accessed 28 April 2019. www.fao.org/3/y5407t/y5407t07.htm.

Corntassel, J. 2008. "Toward Sustainable Self-Determination: Rethinking the Contemporary Indigenous-Rights Discourse." *Alternatives: Global, Local, Political* 33 (1): 105–132.

Crate, S., and M. Nuttall. 2003. "The Russian North in Circumpolar Context." *Polar Geography* 27 (2): 85–96.

Cunneen, C. 2011. "Postcolonial Perspectives for Criminology." In *What Is Criminology*, edited by M. Bosworth and C. Hoyle. Oxford: Oxford University Press.

Eckert, J. 2012. *Law against the State*. Cambridge: Cambridge University Press.

Eckert, J., B. Donahoe, C. Strümpell, and Z. Biner. 2012. *Law against the State: Ethnographic Forays into Law's Transformations*. Cambridge: Cambridge University Press.

Etkind, A. 2014. "Post-Soviet Russia: The Land of the Oil Curse, Pussy Riot, and Magical Historicism." *Boundary* 2 41 (1): 153–170.

Evengard, B., J. Nymand Larsen, and Ø. Paasche. 2015. *The New Arctic*. Cham, Switzerland: Springer.

Fert-Malka, M. 2018. "Hierarchy and Development in the Russian Arctic." *World Policy*, 21 June. Accessed 29 April 2019. https://worldpolicy.org/2018/06/21/hierarchy-and-development-in-the-russian-arctic/.

Fondahl, G. 2014. "Where Is Indigenous? Legal Productions of Indigenous Space in the Russian North." In *Nomadic and Indigenous Spaces: Productions and Cognitions*, edited by J. Miggelbrink, J. Habeck, and P. Koch, 77–89. Farham: Ashgate.

Fondahl, G., and A. Sirina. 2006. "Rights and Risks: Evenki Concerns Regarding the Proposed Eastern Siberia-Pacific Ocean Pipeline." *Sibirica* 5 (2): 115–138.

Fondahl, G., et al. 2019. "Niches of Agency: Managing State-Region Relations through Law in Russia." *Space and Polity* 23 (1): 49–66.

Fondahl, G., V. Filippova, A. Savvinova, and V. Shadrin. This volume. "Changing Indigenous Territorial Rights in the Russian North." Chapter 8.

Geopolitical Futures. 2017. "The Arctic: A Russian Vulnerability – Geopolitical Futures." 4 August. Accessed 30 April 2019. https://geopoliticalfutures.com/arctic-russian-vulnerability/#.

Gladun, E., and K. Ivanova. 2017. "Preservation of Territories and Traditional Activities of the Northern Indigenous Peoples in the Period of the Arctic Industrial Development." In *The Interconnected Arctic – UArctic Congress 2016*, edited by K. Latola and H. Savela. Springer Polar Sciences. Cham, Switzerland: Springer.

Glennie, J. 2014. "Why Are Indigenous People Left Out of the Sustainable Development Goals?" *The Guardian* [online]. Accessed 27 April 2018. https://www.theguardian.com/global-development/poverty-matters/2014/aug/14/Indigenous-people-sustainable-development-goals.

Harvey, F., and S. Walker. 2013. "Arctic Oil Spill Is Certain if Drilling Goes Ahead, Says Top Scientist." *The Guardian*, 19 November. Accessed 29 April 2019. www.theguardian.com/world/2013/nov/19/arctic-oil-drilling-russia.

Heleniak, T., and O. Napper. This volume. "The Role of Statistics in Relation to Arctic Indigenous Realities." Chapter 1.

Hille, K., V. Kortekaas, and S. Ager. 2016. "Frozen Dreams: Russia's Arctic Obsession." Accessed 9 May 2019. www.ft.com/video/7b097901-a609-300f-a841-58d544986e9d.

Honneland, G. 2017. *International Politics in the Arctic. Contested Borders, Natural Resources and Russian Foreign Policy*. London: I.B. Tauris & Co. Ltd.

Ivanova, A. A., M. P. Okorokova, F. Stammler, and E. Wilson. This volume. "What Makes a Good Political Leader? Young People's Perceptions from the Republic of Sakha (Yakutia)." Chapter 3.

Ivanova, A. A., and F. M. Stammler. 2017. "Mnogoobraziye upravlyayemosti prirodnymi resursami v Rossiyskoy Arktike [The Diversity Natural Resource Governance in the Russian Arctic]." *Sibirskiye istoricheskiye issledovaniya* 5 (4): 210–225.

IWGIA. 2015. "Russia: Denial of Indigenous Peoples' Rights Concerns UN Human Rights Committee." IWGIA [online]. Accessed 25 May 2019. https://www.iwgia.org/en/russia/2245-russia-denial-of-Indigenous-peoples-rights-concern.

Kryazhkov, V. 2012. "Russian Legislation about Northern Peoples and Law Enforcement Practice: Current Situation and Perspective." *State and Law Magazine* 5.

Kryazhkov, V. 2013. "Development of Russian Legislation on Northern Indigenous Peoples." *Arctic Review on Law and Politics* 4 (2): 140–155.

Laruelle, M. 2015. *Russia's Arctic Strategies and the Future of the Far North*. Armonk, NY: Routledge.

Lerner, M. A., V. Koshurina, O. Chistanova, and A. Wheeler. 2017. "Mitigating the Risks of Resource Extraction for Industrial Actors and Northern Indigenous Peoples." *Arctic Review on Law and Politics* 8.

Mamontova, N. 2012. "Traditional Economy of the Sakhalin Nivkhs: Between Sustainability and Development." *Ethnographic Review* 1: 133–150.

Martin, T. 2001. *The Affirmative Action Empire*. Ithaca: Cornell University Press.

Minde, H. 2008. *Indigenous Peoples: Self-Determination, Knowledge and Indigeneity*. Delft: Eburon Publishers.

Morgan, R. 2011. *Non-State Actors in International Law, Politics and Governance Series: Transforming Law and Institution: Indigenous Peoples, the United Nations and Human Rights*. Farnham: Ashgate.

Murashko, O. A. 2008. "Protecting Indigenous Peoples' Rights to their Natural Resources – The Case of Russia." *Indigenous Affairs* 3–4 (8): 48–59.

Newman, D., M. Biddulph, and L. Binnion. 2014. "Arctic Energy Development and Best Practices on Consultation with Indigenous Peoples." *Boston University International Journal* 32 (2): 101–160.

Nikolaeva, S. 2017. "Post-Soviet Melancholia and Impossibility of Indigenous Politics in the Russian North." *UMASA Journal* 34: 116–125.

O'Faircheallaigh, C. 2013. "Extractive Industries and Indigenous Peoples: A Changing Dynamic?" *Journal of Rural Studies* 30: 20–30.

Øverland, I., and H. Blakkisrud. 2006. "The Evolution of Federal Indigenous Policy in the PostSoviet North." In *Tackling Space. Federal Politics and the Russian North*, edited by H. Blakkisrud and G. Hønneland, 163–192. Lanham: University Press of America.

Petrov, A., and E. Francis. 2015. "Resources and Sustainable Development in the Arctic: Can, Should and Will Resource-Based Development Be Sustainable?" In *NSF Arctic- FROST Annual Network Meeting*. St. Petersburg: Arctic-FROST. Accessed 28 April 2019. https://arctic-frost.uni.edu/wp-content/uploads/2017/04/ReportFROST_2015.pdf.

Rogers, N. 2000. *On the Ground Research: The Impact of Large-Scale Mining on Local Communities*. Ottawa, ON: MiningWatch Canada.

Rowe, E., and H. Blakkisrud. 2014. "A New Kind of Arctic Power? Russia's Policy Discourses and Diplomatic Practices in the Circumpolar North." *Geopolitics* 19.

Sirina, A. 2005. "Clan Communities among the Northern Indigenous Peoples of the Sakha (Yakutia) Republic: A Step to Self-determination?" In *Rebuilding Identities: Pathways to Reform in Postsoviet Siberia*, edited by E. Kasten, 197–216. Berlin: Reimer (Siberian Studies).

Stamatopoulou, E. 2017. *Indigenous Peoples' Rights and Unreported Struggles: Conflict and Peace.* New York: Columbia University.

Stammler, F. 2005. *Reindeer Nomads Meet the Market: Culture, Property and Globalisation at the End of the Land*, edited by C. M. Hann et al. Münster: Lit publishers (Halle Studies in the Anthropology of Eurasia).

Stammler, F., and A. Ivanova. 2016. "Resources, Rights and Communities: Extractive Mega-Projects and Local People in the Russian Arctic." *Europe-Asia Studies* 68 (5): 1220–1244.

Stammler, F., and V. Peskov. 2008. "Building a 'Culture of Dialogue' among Stakeholders in North-West Russian Oil Extraction." *Europe-Asia Studies* 60 (5): 831–849.

Stuart, S. 1980. *Human Settlements in the Arctic.* Oxford: Pergamon Press.

Sulyandziga, P. 2016. "Government and Businesses Interests Silence Russia's Indigenous Peoples." Business & Human Rights Resource Centre. Accessed 1 June 2019. www.business-humanrights.org/en/russia-govt-businesses-destroying-Indigenous-peoples-communities-says-member-of-un-working-group-on-business-human-rights.

Tulaeva, S., and M. Tysiachniouk. 2017. "Benefit-Sharing Arrangements between Oil Companies and Indigenous People in Russian Northern Regions." *Sustainability* 9 (8).

Turaev, V. 1998. "Territorial approach to ethnic problems at the Russian Far East" [Территориальный подход к решению этнических проблем на российском Дальнем Востоке.] Slavic Research Center [online]. Accessed 28 April 2019. http://src-h.slav.hokudai.ac.jp/sympo/97summer/vadim.html.

Turaev, V. 2018. "Self-Determination and Self-Government of Indigenous of Few Peoples of the Far East: From Hope to Reality. Bulletin of the Irkutsk State University." *Geoarchaeology, Ethnology, and Anthropology Series* 23: 164–185.

Vinding, D. 2002. *Indigenous World. 2001–2002 Series.* Copenhagen: International Work Group for Indigenous Affairs (IWGIA).

Watt-Cloutier, S. 2019. *Testimony of Sheila Watt-Cloutier at the Inuit Circumpolar Conference.* Accessed 29 April 2019. www.ciel.org/Publications/McCainHearingSpeech15Sept04.pdf.

Wilson, E., and F. Stammler. 2016. "Beyond Extractivism and Alternative Cosmologies: Arctic Communities and Extractive Industries in Uncertain Times." *The Extractive Industries and Society* 3 (1): 1–8.

Wilson, E., and K. Swiderska. 2009. *Extractive Industries and Indigenous Peoples in Russia: Regulation, Participation and the Role of Anthropologists.* London: International Institute for Environment and Development.

Xanthaki, A. 2004. "Indigenous Rights in the Russian Federation: The Rights Case of Numerically Small Peoples of the Russian North, Siberia, and Far East." *Human Rights Quarterly* 26 (1): 74–105.

Yakovleva, N. 2014. "Land, Oil and Indigenous People in the Russian North: A Case of Oil Pipeline and Evenki in Aldan." In *Natural Resources Extraction and Indigenous Livelihoods: Development Challenges in an Era of Globalization*, edited by E. Gilberthorpe and G. Hilson, 147–178. Aldershot: Ashgate.

Zaikov, K., A. Tamitskiy, and M. Zadorin. 2017. "Legal and Political Framework of the Federal and Regional Legislation on National Ethnic Policy in the Russian Arctic." *The Polar Journal* 7 (1): 125–142.

Zenzinov, V. 1944. "The Soviet Arctic." *Russian Review* 3 (2).

Russian legal documents

Land code of the Russian Federation. 2001. N 136-FZ. Accessed 20 May 2020. http://docs.cntd.ru/document/542638716.

On Amendments to the Land Code of the Russian Federation and Certain Legislative Acts of the Russian Federation. 2014. N 171-FZ. Accessed 20 May 2020. http://docs.cntd.ru/document/420202723.

On Amendments to the Russian Federation Land Code and Certain Legal Acts of the Russian Federation. 2014. N 499-FZ. Accessed 20 May 2020. http://docs.cntd.ru/document/420242951.

On General Principles of Organization of Obshchina of Indigenous Peoples of the North, Siberia and the Far East of the Russian Federation. 2000. N 104-FZ. Accessed 20 May 2020. http://docs.cntd.ru/document/901765288.

On Guarantees of the Rights of the Indigenous Minorities of the Russian Federation. 1999. N 82-FZ. Accessed 20 May 2020. http://docs.cntd.ru/document/901732262.

On Territories of Traditional Nature Use of Indigenous Peoples of the North, Siberia and the Russian Far East. 2001. N 49-FZ. Accessed 20 May 2020. http://docs.cntd.ru/document/901786770.

On the Statutes of Patrimonial Lands of Khanty-Mansi Autonomous Okrug. 1992. N 69. Accessed 20 May 2020. http://docs.cntd.ru/document/991004572.

On Amendments to the Federal Law on Protected Areas. 2013. N 406-FZ. Accessed 20 May 2020. http://docs.cntd.ru/document/499067417.

International documents

CESCR. 2017a. "Sixth Periodic Report of the Russian Federation on the Implementation of the Provisions of the International Covenant on Economic, Social and Cultural Rights." 16 September 2016, UN Doc E/C.12/RUS/6.

CESCR. 2017b. "Economic, Social and Cultural Rights of Indigenous Minority Peoples of the North, Siberia and the Far East of the Russian Federation." CESCR 60th Session, Pre-Sessional Working Group.

IWGIA. 2014a. "Russia: Ministry in Charge of Indigenous Affairs to Be Dissolved." Accessed 19 April 2019. www.iwgia.org/en/russia/2120-russia-ministry-in-charge-of- Indigenous-affairs-to.

IWGIA, INFOE, Myski local civic organisation "Revival of Kazas and the Shor People" and Greenpeace Russia. 2017. "Discrimination against Indigenous Minority Peoples of the North, Siberia and the Far East of the Russian Federation." In CERD 93rd Session. UN Doc CERD/C/RUS/23–24.

19

THE PARTICIPATION OF ARCTIC INDIGENOUS PEOPLES' ORGANIZATIONS IN THE ARCTIC COUNCIL AND BEYOND

Dorothée Cambou and Timo Koivurova

Abstract

The purpose of this chapter is to examine the participation of Arctic Indigenous peoples' organizations (IPOs) in the policy and law making of Arctic governance. By focusing on the importance of the status of Permanent Participants in the Arctic Council, the chapter describes how Arctic IPOs have cemented their unique status as non-state actors in the Arctic Council and also emerged as important players in the development of Arctic policy outside this forum. Although there remain important challenges that still limit the influence of Arctic IPOs, their contribution in shaping Arctic policy has helped over the years to integrate and acknowledge the values and rights of Indigenous peoples in the governance of the Arctic. In this chapter, the Sámi Council and Inuit Circumpolar Council are also singled out to illustrate their specific role in this context.

Introduction

Indigenous peoples in the Arctic have been world leaders in establishing and developing cooperation arrangements at the transnational and international levels in order to foster their interests in regional and international policymaking. From the beginning of the 1980s, Indigenous peoples' organizations (IPOs) such as the Inuit Circumpolar Council (ICC) and the Sámi Council have, for instance, been able to influence the United Nations' evolving policy in relation to the development of Indigenous peoples' rights in international law. Among many other activities, this involvement includes their leading role in the negotiation process of the UN Declaration on the Rights of Indigenous peoples and their active engagement in establishing new international bodies to support the development of Indigenous peoples' rights at the UN level. At the regional level, IPOs have also contributed to the development of Arctic policy. When the Arctic Environmental Protection Strategy (AEPS), the predecessor of the Arctic Council, was concluded in 1991, both the ICC and the Sámi Council were among the leading Indigenous organizations involved in the process through their status as observers, a position that evolved to Permanent Participant status when the Arctic Council was established in 1996. Today, the involvement of Arctic Indigenous representatives is also expanding beyond the framework of the Arctic Council through their representation in other fora and decision-making processes which focus on the governance of the Arctic. This involvement is far reaching and has over the years strengthened their unique status and role as non-state actors in the development of Arctic policy.

The purpose of this chapter is to analyse the status and role of Arctic Indigenous peoples and their participation in the governance of the Arctic. Although there are still a number of challenges limiting the efforts of IPOs to influence Arctic policy, this chapter specifically seeks to clarify and underline how Arctic IPOs have built upon their status as Permanent Participants at the Arctic Council to influence the governance of the Arctic, within the Arctic Council and beyond.

For the purpose of this study, the analysis is structured as follows: First, the chapter focuses on two pioneer organizations for Indigenous international influence, namely the Inuit Circumpolar Council and the Sámi Council, as a basis to describe the defining characteristics of IPOs. In the second section, the analysis explores how these IPOs have secured their status of Permanent Participants on the Arctic Council and then addresses the extent of their participation in this capacity. Then, the third section examines the extent to which IPOs have built upon their status as Permanent Participants to consolidate their participation in Arctic international policy processes occurring outside of the Arctic Council. The last section summarizes and concludes the analysis by underlining the pivotal importance of the status of Permanent Participants for the participation of IPOs in the governance of the Arctic, as well as its limitations.

The establishment of Indigenous peoples' organizations in the Arctic

Indigenous peoples have inhabited the Arctic for thousands of years and live today within and across the imposed boundaries of the nation-states that have materialized upon their territories. With the exception of the Inuit in Greenland, they are also now a numerical minority within these states. Consequently, Indigenous peoples have established organizations in order to foster and enhance the voice of their peoples in the governance of their traditional territories. Among these organizations, the Sámi Council and the ICC have gained an important foothold in international policymaking and have also contributed to strengthen the status of Indigenous peoples across the borders that separate them.

The Nordic Sámi Council[1]

Over the course of the last centuries, the establishment of the states of Finland, Norway, Sweden and Russia in the territory of the Sámi people and the impact of international borders have profoundly altered the interaction of the Sámi in their territory and their traditional livelihoods. The state-building process has gradually undermined their cohesion as a group and their ability to preserve and develop their common identity as a people. With the establishment of state borders, the Sámi people became a minority within their states, and their distinct rights and interests as a people have been largely neglected or discounted until recently.

As a means to redress this situation, in 1956, the Sámi created the Nordic Sámi Council, a transnational organization whose purpose has been to strengthen cooperation between the Sámi from Norway, Sweden and Finland. The council was founded "because the Saami in the Nordic countries found it imperative to have a pan-Sámi organization for co-operation in order to promote the Sámi interest and rights as one people" (Henriksen 1999, 27).[2] Since its establishment, the Sámi Council has thus worked to ensure that the Sámi are acknowledged and treated as one people in the Nordic states, with the right to self-determination. In addition, despite the borders separating them and the important military and political antagonisms existing between the Nordic States and the Russian Federation, the Sámi Council has also attempted to involve Russian Sámi in order to strengthen relations between all Sámi people (Berg-Nordlie 2012, 439). In 1989, with the Cold War deescalating, the Soviet Sámi created their first nongovernmental organization (NGO) to represent their interests at the domestic level in Russia and were subsequently included as full members of the Sámi Council in 1992. In 1992, after the integration of Sámi organizations from the Kola Peninsula in Russia, the Nordic Sámi Council was renamed the Sámi Council.

Currently, the Sámi Council is composed of fifteen member organizations (Sámi Council 2019). These organizations nominate candidates for the Sámi Council and the Sámi Conference, which serves as its highest body, with members appointed every four years. An organization can become a member by agreeing to the objectives of the Sámi Council. Today, several Sámi reindeer-herding associations, such as the Reindeer Herders Association of Sweden and Sámi associations such as the Kola Sámi Association or the Norwegian Sámi Association, are among the members of the organization.

At the organizational level, the Sámi from each state are represented in the council proportionally to their number (Sámi Council 2019).This means five representatives appointed from the Norwegian side, four from Sweden and Finland and two from Russia, which makes in total fifteen representatives, including a president, who is elected every four years. The decisions of the council are made by consensus. Since the Sámi do not share a common language, the working language of the Sámi Council varies. In effect, the Sámi Council represents an institutional vehicle for the development of Sámi culture (Cambou 2016, 364). It constitutes a leading actor in the consolidation of pan-Sámi relations and contributes considerably to the maintenance and development of the institutional and cultural relations of the Sámi people across borders between the Nordic states and Russia.

As an NGO, the Sámi Council does not have direct decision-making powers and is not the holder of Sámi rights. Its primary role is to channel the interests and voices of the Sámi through the adoption of resolutions and declarations and to represent the Sámi people through its participation in discussions concerning the Sámi at the national, transnational and international levels. At the national level, the council operates as a spokesperson: it can attend certain governmental meetings and conferences and voice the distinctive interests of the Sámi in discussions concerning them. In this regard, the Council often sits alongside the Sámi Parliaments, which are the elected bodies exercising the right of Sámi self-determination in practice (Kuokkanen, this volume). In accordance with the Sámi Political Programme adopted in 1986,[3] its objective is to protect the rights of the Sámi through legislation at the national level and through agreements between the bodies that formally represent the Sámi and the respective states in which they reside (Henriksen 1999, 26–27). Even though it does not have decision-making power, the Council is therefore an important actor to promote and influence decision-making processes which affect the Sámi people in accordance with their interests and rights. At the transnational level, the Sámi Council also engages in matters affecting the rights of the Sámi communities, such as reindeer crossing, and promotes the recognition of the rights of the Sámi people to self-determination across the borders that separate them (Cambou 2018a; Heinämäki and Cambou 2018). At the international level, it can sit on international fora which address issues concerning the rights of the Sámi people, including the UN and the Arctic Council.

Historically, the Sámi Council has also played an instrumental role in the recognition of the Sámi as an Indigenous people and in developing a legal framework supporting their rights at the international level. In 1973, several Sámi representatives were participants in the first Arctic Peoples' Conference hosted by the International Work Group for Indigenous Affairs, an non-profit organisation whose central goal is to promote the collective rights of the world's Indigenous peoples (Dahl 2009, 37–38). Subsequently, in 1975, the representatives of the council attended the preparatory conference for the foundation of the World Council of Indigenous Peoples (WCIP), which was one of the first international bodies dedicated to the development of the rights of Indigenous peoples in international law (Minde 2003). In 1989, the Sámi Council was granted consultative status by the Economic and Social Council (ECOSOC) of the UN. This status has enabled its participation in multiple political and legal processes affecting the Sámi, such as the drafting process of the UN Declaration. The Sámi Council was also present during the revision process of the International Labour Organization (ILO) Convention No. 169 (Minde 2003, 96) and has been a major proponent of the adoption of a Nordic Sámi Convention to recognize the right of the Sámi people to self-determination within and across the Nordic states (Cambou 2018a; Heinämäki and Cambou 2018; Koivurova 2008).

As an NGO representing the interests of the Sámi as an Indigenous people, the Sámi Council has also been involved in various international and regional processes, such as the 21st Conference of the Parties to the United Nations Framework Convention on Climate Change (UNFCCC), where it has acted alongside Inuit representatives to limit the impact of climate change and find ways to mitigate its negative effects on the livelihoods of their respective communities. In effect, the involvement of the Sámi Council in international processes has helped it to consolidate the relationship between Sámi members and other Indigenous peoples around the world, and this alliance has also enabled it to legitimize the rights of Indigenous peoples at the international level. The Sámi Council, together with the ICC, was, for instance, among the primary advocates for the inclusion of a provision on the rights of Indigenous peoples in the Paris Agreement. It acted in the capacity of an observer during the adoption process of the treaty, which allowed it to favour the inclusion of provisions concerning Indigenous peoples in the treaty.

Hence, the activities of the Sámi Council have been significant to promote and develop the rights of Indigenous peoples and to ensure the recognition of the rights of the Sámi both at the international and domestic levels. As described in the following section, the involvement of the Sámi Council in the activities of the Arctic Council alongside the ICC has also been instrumental in strengthening the voice and interests of the Sámi people in the governance of the Arctic.

The Inuit Circumpolar Council

The ICC constitutes a major international non-governmental organization representing the Inuit of Alaska, Canada, Greenland and Chukotka (Russia). Founded in 1977 by Eben Hopson, the organization aims "to strengthen unity among Inuit of the circumpolar region" despite the borders that separate them and "promote Inuit rights and interests both at the regional and international levels" (Inuit Circumpolar Council 2019). The role of the ICC as a leading pan-Indigenous organization is the fruit of a longstanding engagement of the Inuit in Arctic policy (Dorough 2017, 72–73). Historically, the involvement of the Inuit in Arctic policy can be traced back to the 1960s, a period during which they "turned their attention towards the objective of ensuring that they would become primary actors in the development of Arctic policy and the political discourse concerning this distinct part of the world" (Dorough 2017, 71).

Building upon this engagement, the ICC has grown into a leading actor at the regional and international levels, promoting both Inuit interests and the rights of Indigenous peoples. Even though the ICC, as a non-governmental organization, does not exercise the rights of the Inuit, the organization has shaped its own Arctic policy through various activities and the adoption of declarations which contribute to ascertain the distinctive interests, values and rights of the Inuit to land and resources and their right to self-determination in all fora in which it is involved. However, as underlined in the Circumpolar Inuit Declaration on Sovereignty in the Arctic adopted in 2009, the demands concerning Inuit sovereignty and their right to self-determination do not necessarily challenge state borders but are rather specifically grounded in the human rights of Indigenous peoples. In this regard, the ICC tends to present the Inuit multifariously as a transnational people (Gerhardt 2011) but adopts different national strategies for improving the rights of Inuit at the national level in accordance with their local circumstances. The ICC objectives therefore support the transnational Inuit claim for sovereignty and land rights that precede the Westphalian state conception without contesting the primacy of states as sovereigns (Shadian 2010).

At the organizational level, the ICC is composed of delegates from across their homelands, now the four ICC countries, an Executive Council and the ICC Chair. The General Assembly also includes delegates that represent the concerns of youth and elders to participate for "improving communications and creating synergies" important to all of their communities (Inuit Circumpolar Council 2019). With its inclusive membership, the ICC's work enables the Inuit to consolidate their voice within and across their states. In practice, the engagement of the ICC also covers a various range of issues ranging from language preservation to the protection of the environment (Dorough 2017, 70). In dealing with these challenges, the ICC calls for states to protect and respect the Inuit rights to land and natural resources and to involve them in decision-making processes affecting their territories as a basis to ensure the sustainability and integrity of the Inuit environment, culture and society. This is in accordance with the aims of the ICC to "develop and encourage long-term policies that safeguard the Arctic environment" and to "seek full and active partnership in the political, economic, and social development of circumpolar regions" (Arctic Council 2015).

Besides its engagement in national and regional policymaking, the ICC is also a leading IPO at the international level. With its status as a UN Economic and Social Council Non-Governmental Organization obtained in 1983, the ICC plays an instrumental role in the furtherance of the voices and interests of Indigenous peoples at the UN level. The UN's consultative status allows the organization to be actively involved in the preparation of policy that affects Indigenous peoples generally and the Inuit more specifically. In fact, since its foundation, the ICC has been one of the leading IPOs in the development of the human rights of Indigenous peoples in international law (Dorough, 2017, 74). When the UN Permanent Forum on Indigenous Issues (UNPFII) was established in 2000 as the UN central coordinating body for matters relating to the concerns and rights of the worlds Indigenous peoples, both the ICC and Sámi Council were, for example, given a single seat on the Forum for the Arctic region. The PFII also happens to be the only high-level UN body in which non-state Indigenous-appointed individuals are among the members. In addition to its engagement in Indigenous peoples' fora, the ICC is also actively involved in decision-making process concerning the protection of biodiversity, climate change, Indigenous knowledge and numerous other issues concerning sustainable development (Dorough 2017, 68). The participation of the ICC in international debates concerning these issues has certainly also been an elemental factor contributing to the enhancement of Indigenous perspectives in these domains.

Whereas the consultative status of the ICC at the UN level has enhanced its capacity to promote the rights of Inuit and other Indigenous peoples with varying degrees of success, at the regional level, it is through the status of Permanent Participant at the Arctic Council that the ICC has been able to cement its unique position. With its unique status as Permanent Participant, the ICC, alongside the Sámi Council and other Indigenous organizations, supports the participation of Indigenous peoples in the policymaking of the Arctic Council. Besides the Arctic Council, this status has also allowed Arctic IPOs to increasingly participate in other fora concerned with Arctic policy at the national, regional and international levels. The participation and influence of Arctic IPOs in Arctic policy is further analysed in the following sections.

The status and role of Arctic Indigenous peoples' organizations as Permanent Participants on the Arctic Council

In 1996, the eight Arctic states, namely Canada, the Kingdom of Denmark (including Greenland and the Faroe Islands), Finland, Iceland, Norway, Russia, Sweden and the United States founded the Arctic Council, a regional intergovernmental forum created to enhance cooperation, coordination and interaction among them on common Arctic issues, in particular on issues of sustainable development and environmental protection in the Arctic. In addition to its eight state members, the Arctic Council includes six international Arctic IPOs, which have been designated Permanent Participants. The Permanent Participants include the Aleut International Association (AIA), Arctic Athabaskan Council (AAC), Gwich'in Council International (GCI), Inuit Circumpolar Council, Russian Association of Indigenous peoples of the North (RAIPON) and Sámi Council (SC). In the following sections, the analysis focuses on the significance of Permanent Participant status for Arctic IPOs within the framework of the Arctic Council.

The status of permanent participants

Historically, the Arctic Council is the result of cooperation between the Arctic states with the involvement of Arctic IPOs. In September of 1989, on the initiative of the government of Finland, officials from the eight Arctic countries commenced the negotiation process that was concluded in 1991 in Rovaniemi, Finland, to adopt the Arctic Environmental Protection Strategy. This was an intergovernmental non–legally binding strategy to protect the Arctic environment. Several observers, including the ICC and the Nordic Sámi Council, supported the preparation of the strategy and participated in the earliest ministerial meetings that ultimately outlined the purpose and mandate of the Arctic Council (Dorough 2017, 80). With their observer status in AEPS cooperation, they were also able to push for a status upgrade at the Inuvik ministerial meeting of the AEPS in 1996. As mentioned in the 1996 Inuvik declaration, they became AEPS Permanent Participants, a status they were able to retain in the negotiations over the establishment of the Arctic Council the same year (Scrivener 1999). Under the guidance of the AEPS, the Indigenous peoples Secretariat was then established in 1994 to facilitate and encourage them and other Arctic IPOs to participate actively in the work of the AEPS. After a long process, when the Arctic Council was created in 1996, the contribution of the Indigenous peoples Secretariat was then continued and formalized as a specific entity within the Arctic Council Secretariat with its own board, designated budget and work plan.

Hence, when the Arctic states endorsed the Ottawa Declaration, by which they established the Arctic Council in 1996, governments also confirmed their "recognition of the special relationship and unique contributions to the Arctic of Indigenous people and their communities" at the Arctic Council. To this end, the specific recognition of the interests of Indigenous peoples at the Arctic Council was formalized through recognition of the status of Permanent Participants. Formally, only the ICC; the Sámi Council and the Association of the Indigenous Minorities of the North, Siberia, and the Far East of the Russian Federation were involved in the creation of the Arctic Council. However, in accordance with the Ottawa Declaration, the status of Permanent Participant was equally open to other Arctic IPOs with majority Arctic Indigenous constituency representing either a single Indigenous people resident in more than one Arctic state or more than one Arctic Indigenous people resident in a single Arctic state. This led to the additional recognition of three Arctic IPOs as Permanent Participants, namely the Aleut International Association, Arctic Athabaskan Council and Gwich'in Council International.

The creation of the status of Permanent Participants is connected to the fact that the participation of IPOs contributes to the legitimacy of the Arctic Council, as it can be argued that the Council "represents the values and interest of the original occupants of the region" (Koivurova 2011, 183). Even if the Permanent Participants do not exercise such rights, with the status of Permanent Participants, Arctic IPOs are provided the possibility to promote the rights of Arctic Indigenous peoples who have ownership and use rights over their traditional land and resources and the right to self-determination. From a legal perspective, the status of Permanent Participant is therefore also in accordance with the acceptance of the positionality of Indigenous peoples in the Arctic as distinct peoples with different livelihood systems but with a status that is not equal to sovereign states. Accordingly, Permanent Participants are not granted the right to vote, which is reserved for states, but they can take part in all Arctic Council meetings and voice their views to the eight member states. As stated under the Ottawa Declaration, "the category of Permanent Participation is created to provide for active participation and full consultation with the Arctic Indigenous representatives within the Arctic Council". Thus, as Permanent Participants of the Arctic Council, the ICC and the Sámi Council have full consultation rights in connection with the Council's negotiations and its decisions. They sit at the same table with the Arctic states during the time of negotiations and can table proposals and suggestions for final decisions. In this regard, it is also important to note that Permanent Participants have this status at all levels of decision-making in the Arctic Council, from the political (ministerial and Senior Arctic Officials meetings) to the working level (working groups and other subsidiary bodies). Hence, the status of Permanent Participants is unique, especially as it provides "an avenue to mainstream Indigenous interests in the governance of the Arctic Council without challenging the membership status of sovereign states" (Cambou 2018b, 50).

From a comparative perspective, the peculiarity of the status of Permanent Participants also manifests in several ways. First, the status of Permanent Participants is superior to that of non-Arctic states, which sit as observers in the Council without full consultation rights. As a result, all six Arctic Permanent Participants surpass states such as France, Japan and China with their status at the Arctic Council. Second, the peculiarity of the status of Permanent Participant also manifests in the fact that it is distinct from the representation of Indigenous peoples who participate with state consent in the Arctic Council qua their status as sub-nations and non–self-governing territory. Whereas Permanent Participants own a statutory seat, the participation of self-governing Indigenous peoples can be contested. For example, although it has been a longstanding practice of the delegation of Denmark to include the governments of Greenland and the Faeroe Islands within the Arctic Council where all three bodies have vested interests, both Greenland and the Faeroe Islands representatives were excluded from the executive SAO meetings in 2011–2013, which resulted in a state-centric orientation of the Arctic Council against the interests of Indigenous peoples (Holm Olsen and Shadian 2016, 239). Through another lens, this episode also serves to reaffirm the importance of affording the distinct status of Permanent Participants to Arctic IPOs, which guarantees their statutory participation in the Arctic Council.

Finally, the status of Permanent Participants is also unique insofar as it challenges the classical practice of conflating Indigenous organizations with NGOs with observer status. The status of Indigenous organizations has often been equated to that of NGOs in international institutions. Unlike NGOs, IPOs are, however, sometimes elected decision-making bodies that represent and govern their Indigenous constituencies. In this regard, Koivurova and Heinämäki have argued that "the Arctic Council with its unique model of participation could well serve as a new model enabling Indigenous peoples to find more reasonable status than that of NGOs" (Koivurova and Heinämäki 2006, 105). This is an idea that has recently gained attention at the UN level, as it currently investigates the need to create new means to enable the participation of Indigenous peoples in the UN system beyond their advisory status in the PFII, given that they are not always organized as non-governmental organizations. In this regard, it has also been argued that IPOs should be entitled to a status of Permanent Observer at the UN, which would also allow them to participate and speak at the UN General Assembly, participate in procedural votes and co-sponsor and sign resolutions on the same basis as Palestine or the Holy See.

Hence, the Permanent Participant status of Indigenous peoples at the Arctic Council is unique, and even though they do not have voting rights, it is argued that their contribution in the decision-making process at the Arctic Council is highly influential, making them virtually partners to the member states in the governance of the Arctic Council (Koivurova and Heinämäki 2006). Although the states remain in control of the final decision, Heinämäki and Koivurova have also noted that the status of Indigenous peoples comes, in fact, "close to a de facto power of veto should they all reject a particular proposal" (Koivurova and Heinämäki 2006, 104). This level of power is mainly due to the fact that the Arctic Council uses a consensus approach to decision-making, meaning that individual members can be assigned to block a decision on behalf of Indigenous peoples (Dorough 2017, 82). As explained by Eira, member of the Sámi Council, "although we [the Sámi] do not have a vote, we can block a vote. . . . We can ask Iceland to help or something to say, 'No!'" (Eira in Byers 2013). With the status of Permanent Participants, it can thus be held that the status of Arctic Indigenous organizations has been elevated from the level of simple observer to that of partner in the conduct of Arctic affairs within the Arctic Council.

The domain of influence of the Permanent Participants on the Arctic Council

With the status of Permanent Participants on the Arctic Council, Arctic IPOs have been able to influence to some extent environmental policymaking concerning the Arctic. More specifically, the influence of the Permanent Participants is exercised through their representation in the six working groups of the Arctic Council, which allows them to influence what types of projects are conducted in the Arctic Council, as well as the content of the projects, in a manner that serves their needs, interests and values (Koivurova 2011). In this regard, it is noted that the view of Arctic Indigenous peoples manifests in all scientific documents produced under the auspices of the Arctic Council (Koivurova 2011), as most documents incorporate references concerning them.

For instance, the involvement of Arctic Indigenous peoples has been particularly noticed in relation to the development of policy to address the impacts of contaminants in the Arctic. Since many contaminants end up in the Arctic via the prevailing wind patterns and ocean circulation, it was important for the Arctic Council – and its predecessor, the AEPS – to tackle long-term pollution from outside of the Arctic but settling in Arctic ecosystems. For the Indigenous peoples, this has also been a strong priority, given that many of these contaminants end up in the food chain and therefore threaten their food security. For this reason, Arctic Indigenous peoples, with the lead of the ICC, have been able to influence environmental policy in innovative ways (Dorough 2017). Within the framework of the Arctic Contaminant Action Program (ACAP), which became the sixth permanent working group of the Arctic Council in 2006, the Permanent Participants of the Arctic Council have established the Arctic Council Expert group on Indigenous Contaminant (IPCAP), which is specifically aimed at reducing the exposure and impact of contaminants in Indigenous peoples' communities. Under the auspices of the IPCAP, Arctic Indigenous organizations have been able to launch several projects to tackle issues concerning more specifically the management of waste or the impact of black carbon in Arctic communities.[4] They have also been able to represent the specific views and interests of Indigenous communities through their participation in the ACAP expert group on polychlorinated biphenyl (PCBs).

Interestingly, the work of the Permanent Participants has also contributed to developing international law on the protection of the environment beyond the realm of the Arctic Council mandate. The ICC has, for example, played an instrumental role in the negotiations of the global Stockholm Convention on the Elimination of Persistent Organic Pollutants, as well as the Minamata Mercury Convention (Downie and Fenge 2003; Koivurova et al. 2015). In this context, the assessments of the impacts of organic pollutants on Arctic Indigenous peoples were instrumental to illustrate and demonstrate the danger caused by persistent organic pollutants (POPs) for human health. This situation also resulted in the recognition of the particular risk caused by POPs on Indigenous communities in both the Stockholm Convention on persistent organic pollutants and the Minamata Mercury Convention (Koivurova 2011, 182). Thus, and as will be further evidenced in the following sections, the influence of Arctic IPOs on environmental policymaking concerning the Arctic goes beyond the mere apparatus of the Arctic Council insofar as they also participate in the development of environmental law and policy created outside the auspices of the Arctic Council.

However, the status of Permanent Participants also faces several challenges. The first problem is linked to the lack of funding (Koivurova 2011, 183). Although problems remain, there are nonetheless attempts to meet this challenge with the establishment and functioning of the Álgu fund. In addition, it is also foreseen that the observers of the Council will in the course of time provide funding to support the activities of Permanent Participants, given the specific mention of the Nuuk Criteria, which, to admit observers at the Arctic Council, requires that they "Have demonstrated a political willingness as well as financial ability to contribute to the work of the Permanent Participants and other Arctic Indigenous peoples". The second challenge met by Permanent Participants is connected with their ability to promote their own interests and projects at the Arctic Council. In this regard, a recent study by Chater examined carefully how Permanent Participants have been able to sponsor their own projects (especially speaking to their community interests) and participate in projects run by others (2019). Even if challenges remain and it is still difficult to appraise to what extent Permanent Participants are able to influence Arctic policy in practice, the study concluded nevertheless that, in general, Permanent

Participants have been able to influence to a great extent what has taken place in the Council and more particularly steer its activities towards meeting their local priorities (Chater 2019).

Finally, another more recent challenge met by Permanent Participants is linked to the position of certain states and their capacity to hamper the participation of Indigenous peoples in political decision-making affecting them. In May 2019, a joint ministerial Arctic Council statement was released in place of a ministerial declaration and signed by all eight of the participating foreign ministers without mentioning climate change. This was caused by the objection of the American delegation to include a reference made to fighting climate change in the draft ministerial declaration (Koivurova 2019). As a result, representatives of Indigenous groups at the Arctic Council criticized the meeting's failure to adopt a final statement that did not reflect their view and was adopted without their participation (Erasmus in Quinn 2019). However, it is also important to note that this event did not represent the common practice of the Arctic Council and was ultimately mitigated by the adoption of the joint statement that was supported by the Permanent Participants. Hence, it can nonetheless be concluded that the status and role of Permanent Participants to steer and influence the internal activities of the Arctic Council remains essential and seems unlikely to be further challenged, though their participation would gain from being further consolidated.

The involvement of Arctic Indigenous peoples' organizations outside the Arctic Council

Building upon but beyond their early status as Permanent Participants, Arctic IPOs have also established themselves in other Arctic international governance mechanisms as important actors and contributors. This has taken place through three main avenues. The first avenue involves their representation in the Arctic Council System, which partly consists of treaties negotiated under the Council's auspices and their institutional components (Molenaar 2012, 2016). The second avenue includes their representation in alternative intergovernmental soft-law fora relevant for the governance of the Arctic, such as those of the Barents Euro-Arctic Councils. The third avenue concerns their participation in international negotiations that have led to legally binding agreements concerning the Arctic but which are not connected to the Arctic Council mandate. Each of the following sections examines the basis of this participation and its potential limitations.

The participation of Indigenous peoples' organizations in the Arctic Council system

First, the participation of Arctic IPOs within the expanding system can be noted as a novelty which has contributed to enhancing the influence of IPOs beyond the framework of the Arctic Council. Even if the Arctic Council cannot, as a soft-law intergovernmental forum, adopt its own legally binding decisions, it has proven that it can facilitate the establishment of new organizations and support the adoption of legally binding agreements that work in and apply to the Arctic. On this basis, the Arctic Council has catalyzed the founding of several Arctic organizations, such as the Arctic Economic Council, which has a unique structure reflecting that of the Arctic Council, with Permanent Participants having an equally strong status, but which is formally independent of the Council. In addition, the Arctic Council has also catalyzed several initiatives while providing a forum for the negotiation of three important legally binding agreements among the eight Arctic States, namely the Agreement on Enhancing International

Arctic Scientific Cooperation signed in 2017, the Agreement on Cooperation on Marine Oil Pollution Preparedness and Response in the Arctic signed in 2013 and the Agreement on Cooperation on Aeronautical and Maritime Search and Rescue in the Arctic (SAR) signed in 2011. All three legally binding agreements that have been negotiated under its auspices – via temporary task forces – have included Indigenous participation. However, it is noticeable that among these agreements, two of them only make sporadic references to Indigenous peoples within the operative parts of the treaties or preamble, and the SAR Agreement does not mention Indigenous peoples at all.

While the meaningfulness of the contribution of Arctic IPOs in treaty-making could certainly be further questioned, it can nonetheless be noted that their participation has increasingly expanded beyond the framework of the Arctic Council activities. In fact, the involvement of Arctic IPOs represented as Permanent Participants in meetings that are not, strictly speaking, meetings of the Arctic Council but which are indirectly linked to it has also been observed elsewhere. A good example is the meeting organized by the current chair of the Arctic Council, Finland, convening the meeting of environment ministers in October 2018, where all six Permanent Participants were also invited. Even more interesting was a larger Arctic science ministers' meeting organized in Berlin in November 2018, where some 30 governments and other participants were represented, including the six IPOs that are Permanent Participants in the Arctic Council. As such, the fact that the involvement of Arctic IPOs in the Arctic Council system is usually confirmed indicates that they have managed to gain a specific status beyond the framework of the Arctic Council. Whether this participation can be qualified as meaningful and effectively expand their influence in the governance of the Arctic remains, however, difficult to appraise at this stage.

The representation of Indigenous organizations at the Barents Councils

Second, and beyond the sphere of the Arctic Council System, the unique status of Arctic IPOs as Permanent Participants has also spurred discussions and sparked some inspiration in Barents cooperation. The Barents Euro-Arctic Council (BEAC), together with the Barents Regional Council (BRC), which was created in 1993 to promote cooperation in the Barents region between Sweden, Finland, Norway, Demark, Iceland, the Russian Federation and the EU Commission, also recognizes the specific status of Indigenous peoples in the Barents region (Cambou 2018b, 51). For this purpose, the Working Group on Indigenous Peoples has been established, consisting of representatives of the Sámi, Nenets and Veps peoples, who are the three Indigenous groups living in the Arctic Barents region. The WGIP has a specific position within the Barents cooperation framework. More particularly, the working group is distinguished from other regional working groups because it also has an advisory role to both the BEAC and BRC. With such an advisory function, the WGIP consequently has a political dimension. It can influence the decisions made by the BEAC and the BRC to the extent that the Councils may consider its proposals. In addition, the WGIP participates in all Barents working group sessions. The chair of the WGIP is also a member of the Committee of Senior Officials in the BEAC and the Barents Regional Committee, the executive organ of the BRC, which gives the WGIP a permanent right to attend all meetings organized by the Councils. Since 2011, the right of participation of Indigenous peoples has also been extended, as all three Indigenous peoples of the region can now participate individually in the Committee of Senior Officials without a formal invitation from the Council. Recently, there was also a discussion, at the initiative of the three IPOs, of extending this status to that of Permanent Participants, modelled on the way the Arctic Council functions, but, at least as of yet, this has not taken place (Koivurova 2011).

The participation of Indigenous representatives in the negotiations of the International Agreement to Prevent Unregulated High Seas Fisheries in the Central Arctic Ocean

Finally, the involvement of Arctic Indigenous representatives in decision-making processes concerning the Arctic outside the framework of the Arctic Council has also been observed in the negotiation concerning the adoption of the International Agreement to Prevent Unregulated High Seas Fisheries in the Central Arctic Ocean (CAOFA). With this agreement, the Arctic Ocean coastal states, plus China, Japan, South Korea, Iceland and the European Union, have agreed not to engage in unregulated commercial fishing in the high seas of the Central Arctic Ocean for a specified time frame. Qualitatively, even though the arrangement focuses only on one specific governance issue, this treaty is significant insofar as it involves Arctic and non-Arctic states in the making of an international binding agreement to govern the Arctic region. With regard to Indigenous peoples, the adoption of the agreement is also important both from procedural and substantive perspectives.

First, although none of the IPOs were allowed to participate on the basis of their status as Permanent Participants, three Arctic state delegations included Indigenous representatives in the negotiations as members of their respective government delegations (Moleenar 2016, 163), namely Inuit representatives. While Denmark included representatives of the people of Greenland, alongside the Faeroe Islands, the United States and Canada also included individual Inuit representatives, associated with but not representing the ICC, as members of their state delegations. Russia, on the other hand, did not extend any invitation to Indigenous representatives, which means that no Arctic Indigenous peoples, including Inuit, were part of the Russian delegation during the adoption process of the agreement. In this context, the participation of Indigenous representatives was therefore dependent on their national involvement rather than their differentiated status as Permanent Participants or NGOs. Nonetheless, the participation of the Inuit in the CAOFA process was in direct agreement with the demands of the Inuit to be involved in Arctic policymaking, including more particularly commercial fisheries, a proposal that was expressly underlined in the Kitigaaryuit Declaration adopted by the ICC in 2014.

Furthermore, it is also significant for both the interpretation and spirit of the CAOFA agreement that the treaty makes specific recollection within its preamble of the UN Declaration on the Rights of Indigenous peoples and recognizes "the interests of Arctic residents, including Arctic Indigenous peoples, and the value of Indigenous knowledge in decision-making". In the operative part, the agreement additionally makes particular mention of the interests of "Arctic Indigenous peoples in the long-term conservation and sustainable use of living marine resources" and underlines "the importance of involving them and their communities" in the governance of the Arctic Ocean. Additionally, the document also mentions the state-parties' desire "to promote the use of both scientific knowledge and Indigenous knowledge of the living marine resources of the Arctic Ocean . . . as a basis for fisheries conservation and management" in the Arctic Ocean. Although it has been questioned "whether the incorporation of Indigenous and local knowledge in the CAOFA remains a symbolic victory for Indigenous peoples" (Schatz 2019, 134), the first science meeting programme seems to evince that the use of Inuit traditional knowledge is making valuable inroads into the CAOFA system that goes beyond mere symbolism and taps into effective practices.

Ultimately, one can therefore salute the adoption of the CAOFA and the inclusion of Indigenous peoples' interests. The inclusion of several Inuit representatives in national delegations has been a positive experience for focusing attention on the importance of respecting Indigenous peoples' rights and their traditional knowledge in the drafting of the treaty.

Conclusion

Arctic IPOs, more specifically the ICC and Sámi Council, have been active in Arctic policy-making for many decades and hence were able to establish themselves first as observers and then as Permanent Participants in what became the Arctic Council. With their status as Permanent Participants, Arctic IPOs have acquired a unique status that goes beyond that of mere observers and confers upon them the role of partner in all activities of the Arctic Council. Furthermore, whereas financial and logistical challenges remain, it is established that Permanent Participants can to some extent influence the activities in which they are involved in accordance with their own interests. In this context, the status of Permanent Participants has been praised and has become an example of good state practice for promoting the right of Indigenous peoples to participate in decision-making processes affecting them. In fact, other organizations, such as the Barents-Euro Council, have also opted for a similar approach to engage IPOs in the governance of the Arctic. Although the influence of Permanent Participants certainly remains constrained in effect, this unique status has nonetheless opened an avenue for Indigenous peoples to buttress their interests against state priorities within the Arctic Council.

In addition, while building in some ways upon their status as Permanent Participants, Arctic IPOs have also expanded their participation beyond the framework of the Arctic Council. As this analysis has shown, there is an emerging involvement of Arctic IPOs outside the Arctic Council, which further expands their engagement in policy and law-making processes that affect their rights and interests beyond the Arctic Council framework. Most remarkable of all is the participation of Indigenous peoples in the making of Arctic treaties. Although such engagement does not provide the right for Indigenous peoples to sign these agreements, this experience contributes to cementing their unique status and advancing the development of Indigenous peoples' rights at the Arctic and international levels, at least to some extent.

Yet, whereas the status of IPOs as Permanent Participants has spurred increasing participation of Arctic IPOs beyond the Arctic Council framework, it can be concluded that this status remains unique and will not be easily accommodated elsewhere. The fact that the Arctic Council represents a forum, as opposed to an international organization, certainly contributed to the promotion of such a specific status. Beyond this framework, the state-centric order remains on the other hand firmly in place and allows participation for IPOs only through their status as observers or through their limited invitation by state delegations of Indigenous peoples' representatives who at times happened to be associated with IPOs. Nonetheless, with the recognition of the specific status of IPOs as Permanent Participants at the Arctic Council, it has become increasingly difficult for states to neglect the participation of Indigenous representatives in other fora. In this regard, it can also be held that the status of Permanent Participants has contributed to elevate the status of Arctic IPOs and increased their participation in the governance of the Arctic but within the existing limitations allowed by a system that remains overtly state centred.

Notes

1 This section is based on an extract of the analysis published in Cambou, 2016, 360–366.
2 Its creation happened few years after the establishment of the Nordic Council, an interparliamentary council gathering representatives of Denmark, Finland, Iceland, Norway, Sweden and the autonomous Åland (Finland).
3 Nordic Sámi Council, The Sámi Political Program, Thirteen Nordic Sámi Conference (Finland: Utsjoki, 1986).
4 IPCAP Update for ACAP Working Group Meeting Anchorage, Alaska, November 7–9, 2017.

References

Arctic Council. 2015. "Inuit Circumpolar Council (ICC). Arctic Council." Accessed 29 April 2020. https://arctic-council.org/index.php/en/about-us/permanent-participants/icc.

Berg-Nordlie, M. 2012. "The Iron Curtain through Sápmi. Pan-Sámi Politics, Nordic Cooperation and the Russian Sámi." In *L'image Du Sápmi II* edited by K. Andersson, 436–459. Örebro: Humanistica Oerebroensia.

Byers, M. 2013. *International Law and the Arctic.* Cambridge: Cambridge University Press.

Cambou, D. 2016. "The Legal Significance of the Right of Indigenous Peoples to Self-Determination and Its Implication for the Sámi People." PhD diss. Vrije Univeristeit Brussel.

Cambou, D. 2018a. "The 2005 Draft Nordic Sámi Convention and the Implementation of the Right of the Sámi People to Self-Determination." In *Critical Indigenous Rights Studies New Directions in Indigenous Rights Research.* London: Routledge.

Cambou, D. 2018b. "Enhancing the Participation of Indigenous Peoples at the Intergovernmental Level to Strengthen Self-Determination: Lessons from the Arctic." *Nordic Journal of International Law* (87): 26–55. doi.org/10.1163/15718107–08701002.

Chater, A. 2019. "Change and Continuity Among the Priorities of the Arctic Council's Permanent Participants." In *Leadership for the North: The Influence and Impact of Arctic Council Chairs,* edited by D. C. Nord, 149–166. Cham: Springer.

Dahl, J. 2009. *IWGIA: A History.* Copenhagen, Denmark: IWGIA.

Dorough, D. 2017. "The Rights, Interests and Role of the Arctic Council Permanent Participants." In *Governance of Arctic Shipping Balancing Rights and Interests of States and User States,* edited by R. Beckman et al., 68–103. Leiden: Brill-Nijhoff.

Downie, D., and T. Fenge. 2003. *Northern Lights Against POPs: Combatting Toxic Threats in the Arctic.* Montreal: McGill-Queen's Press.

Gerhardt, H. 2011. "The Inuit and Sovereignty: The Case of the Inuit Circumpolar Conference and Greenland." *Tidsskriftet Politik* 1 (14): 6–14. doi.org/10.7146/politik.v14i1.27469.

Heinämäki, L., and D. Cambou. 2018. *New Proposal for the Nordic Sámi Convention: An Appraisal of the Sámi People's Right to Self-Determination.* Vol. 41, 3–18. Retfærd: nordisk juridisk tidsskrift.

Henriksen, J. B. 1999. *Saami Parliamentary Co-Operation: An Analysis.* Copenhagen: IWGIA.

Holm Olsen, I., and J. Shadian. 2016. "Greenland and the Arctic Council: Sub-National Regions in a Time of Arctic Westphalianisation." In *Arctic Yearbook,* edited by L. Heininen et al., 229–247. Akureyri, Iceland: Northern Research Forum.

Inuit Circumpolar Council, 2019. "About ICC | Inuit Circumpolar Council Canada." Accessed 29 April 2020. www.inuitcircumpolar.com/about-icc/.

Koivurova, T. 2008. "The Draft Nordic Saami Convention: Nations Working Together." *International Community Law Review* 10: 279–293.

Koivurova, T. 2011. "The Status and Role of Indigenous Peoples in Arctic International Governance." In *The Yearbook of Polar Law.* Vol 3, 169–192. Leiden, The Netherlands: Martinus Nijhoff Publishers.

Koivurova T., P. Kankaanpää, and A. Stepien. 2015. "Innovative Environmental Protection: Lessons from the Arctic." *Journal of Environmental Law* 27 (2): 285–311.

Koivurova, T. 2019. "Is This the End of the Arctic Council and Arctic Governance as We Know It?" *The Polar Connection.* Accessed 29 April 2020. http://polarconnection.org/arctic-council-governance-timo-koivurova/.

Koivurova, T., and L. Heinämäki. 2006. "The Participation of Indigenous Peoples in International Norm-Making in the Arctic." *Polar Record* 221 (42): 101–110.

Kuokkanen, R. This volume. "Indigenous Self-Government in the Arctic: Assessing the Scope and Legitimacy in Nunavut, Greenland and Sápmi." In *Handbook of Arctic Indigenous Peoples*, edited by T. Koivurova et al., New York: Routledge.

Minde, H. 2003. "The Challenge of Indigenism: The Struggle for Sámi Land Rights and Self Government in Norway 1960–1990." In *Indigenous Peoples: Resource Management and Global Rights*, edited by S. Jentoft et al., 75–104. Delft: Eburon.

Molenaar, E. 2012. "Current and Prospective Roles of the Arctic Council System within the Context of the Law of the Sea." *The International Journal of Marine and Coastal Law* 27: 553–595. doi. org/10.1163/15718085–12341234.

Molenaar, E. 2016. "Participation in the Central Arctic Ocean Fisheries Agreement." In *Emerging Legal Orders in the Arctic*, edited by A. Shibata et al., Ch. 9. London: Routledge.

Quinn, E. 2009. "U.S. Stonewalling on Climate Language Scuttles Arctic Council Declaration." *The Barents Observer.* Accessed 20 August 2020. https://thebarentsobserver.com/en/arctic/2019/05/us-stone walling-climate-language-scuttles-arctic-council-declaration-0.

Sámi Council. 2019. "About the Sámi Council." www.saamicouncil.net/en/about-saami-council/.

Schatz, V. 2019. "Incorporation of Indigenous and Local Knowledge in Central Arctic Ocean Fisheries Management." *Arctic Review on Law and Politics* (10): 130–134. doi.org/10.23865/arctic.v10.1630.

Scrivener, D. 1999. "Arctic Environmental Cooperation in Transition." *Polar Record* 192 (35): 51–58. doi. org/10.1017/S0032247400026334.

Shadian, J. 2010. "From States to Polities: Reconceptualizing Sovereignty through Inuit Governance." *European Journal of International Relations* 16: 485–510. doi.org/10.1177/1354066109346887.

Other documents

Agreement to Prevent Unregulated High Seas Fisheries in the Central Arctic Ocean (signed in Ilulissat, Greenland 2018) Canada, Denmark (acting on behalf of Greenland and the Faroe Islands), Norway, Russia, and the United States together with China, the European Union (EU), Iceland, Japan, and South Korea (CAOF Agreement).

AEPS, The Inuvik Declaration, Declaration from the Ministerial meeting of the Arctic Environmental Protection Strategy (adopted in Inuvik, Canada1996) (AEPS).

A Circumpolar Inuit Declaration on Sovereignty in the Arctic (signed April 2009, Inuit Circumpolar Council).

Convention Concerning Indigenous and Tribal Peoples in Independent Countries (adopted 27 June 1989, entered into force 5 September 1991) 72 ILO Official Bull. 59 (ILO Convention No. 169).

Draft Nordic Sámi Convention. 2016. Accessed 19 August 2019. www.sametinget.se/111445.

Kitigaaryuit Declaration (signed 24 July 2014, Inuit Circumpolar Council).

Ottawa Declaration, Declaration on the establishment of the Arctic Council (19 September 1996).

United Nations Declaration on the Rights of Indigenous Peoples (13 September 2007) UN Doc A/RES/61/295 (UNDRIP).

20

LEGAL APPRAISAL OF ARCTIC INDIGENOUS PEOPLES' RIGHT TO FREE, PRIOR AND INFORMED CONSENT

Leena Heinämäki

Abstract

Indigenous peoples' right to give or withhold their 'free, prior and informed consent' (FPIC) in decision-making that significantly affects their cultures and lives has rapidly evolved to become one of the key issues in the field of Indigenous peoples' international human rights. In the Arctic, which is home to a number of Indigenous peoples, FPIC has a specific significance due to drastic climatic and environmental change that is currently occurring and already affecting Indigenous peoples' traditional ways of lives in many ways. Since FPIC is still a somewhat new and evolving concept, its normative status and procedural implementation are not totally clear and unambiguous. This chapter studies FPIC as a legal concept, aiming to determine its normative content as well as to define key elements of FPIC as an implementation process. The purpose is to enlighten and alleviate perhaps groundless resistance of states and other actors as well as unfounded expectations of Indigenous peoples regarding the application of FPIC. Although there is no specific Arctic FPIC as a human rights norm, this chapter points out justifications and potential for its wider implementation in the area, encouraging Arctic states to renew their domestic legislation in order to fully endorse FPIC in their policies related to Indigenous peoples.

Introduction

The Arctic has been going through many environmental changes, drastically affecting Indigenous communities, who are in a vulnerable position because of their cultural lifestyles. Early tangible and visible dangers faced by the Arctic environment and Indigenous communities began to appear in the 1960s from northern resource development with hydrocarbon exploitation, production and mining (Stone 2015, 12). In addition to local environmental impacts, Arctic Indigenous peoples have faced global environmental impacts, such as the problems caused by persistent organic pollutants (Hungh et al. 2010, 2854–2873) or the impacts of global climate change (ACIA 2005). As highlighted by Koivurova et al., Indigenous people face climate changes in a social and cultural environment that has already been altered drastically by developments that significantly changed their lifestyles (2008, 7). Hence, climate change is an additional factor affecting already restructured livelihoods, making it more difficult for Indigenous peoples to adjust to the modern world and retain their specific economies and identity (ibid.; Ford et al. 2006). Given the irreversible nature and significant impact of climate change on the human rights of Indigenous peoples in the Arctic, the growing recognition of Indigenous peoples' rights, including their right to meaningful participation in decision-making, is particularly important in this region (Hughes 2018, 15; Heinämäki 2009, 209). Additionally, because of the pressures relating to the development of new industry and government projects, Arctic Indigenous peoples have the feeling that 'they are losing control over their homelands and over their livelihoods' (Nuttall and Wessendorf 2006, 4). For these reasons, it is important that there be proper systems of influence, participation and governance in place for Indigenous peoples in order to adapt their livelihoods to meet the challenge of environmental changes (Nuttall 2017, 19–35).

Although Arctic Indigenous peoples are not all at the same level of development and security in relation to their rights to lands, territories and resources, they have made serious inroads towards the reconceptualization of their relations with the states within which they reside (UNPFII 2012, para 4). Indigenous governance arrangements in the Arctic consist of diverse models from public governments, such as local boroughs in Alaska, municipal-level self-government in the NWT and the governments of Nunavut and Greenland, as well as Indigenous elected assemblies, corporations and resource management regimes (Kuokkanen 2019, 16). Numerous Arctic Indigenous peoples claim ownership and property rights to their territories. Yet states have unilaterally asserted sovereignty over these territories and regard them as state property (ibid.). Although the linguistic and cultural rights of Indigenous peoples are increasingly recognized, land rights remain a subject of debate in many Arctic countries (Koivurova et al. 2008, 3).

Development of Indigenous peoples' engagement has thus far been quite uneven in the Arctic. As described by Kuokkanen, in Canada and Alaska, Indigenous peoples' involvement in resource governance has considerably expanded through negotiating impact and benefit agreements or land claims settlements. In contrast, in many other Arctic regions, little progress has been made (Kuokkanen 2019, 2; Forbes and Kofinas 2014, 255–298; Keeling and Sandlos 2016, 265–268). In Russia, for example, Indigenous peoples' participation in resource governance happens largely on an ad-hoc basis. There are many reasons for this, such as the lack of implementation of laws in place to protect Indigenous peoples' rights and interests, political pressure and the cooptation of local movements by political and industry interests (Wilson 2016, 73–81). Despite the progress in some Arctic regions in recent years regarding Indigenous peoples' greater local autonomy and inclusion in formal procedures and decision-making processes, their participation can nevertheless remain ineffective due to a variety of reasons, including short time frames, lack of financial resources, culturally alien forms of inquiry and disregard for their right to free, prior and informed consent (FPIC) (Kuokkanen 2019, 2; Fjellheim 2006, 8–23).

This chapter assesses Indigenous peoples' right to FPIC as a legal concept in international law, as well as its potential to fill the existing deficiencies in the current consultation procedures and substantive legal protection of Indigenous peoples in the Arctic. Although there is naturally no separate FPIC as a legal standard for the Arctic area, its application and implementation in the Arctic are of particular significance, taking into account the dramatic conditions and future predictions of environmental change in the area.

FPIC is seen as an important means to realize recognition and application of Indigenous peoples' rights and is rapidly becoming one of the most important concepts in contemporary international law concerning Indigenous peoples (Rombouts 2014, 397). However, since FPIC is still in a phase of dynamic development, the full consensus on its application and interpretation is absent, and both its elements and its place in the broader legal framework concerning Indigenous peoples' rights are underexplored (ibid., 20). Nevertheless, this chapter argues that it is not only possible but very important, in fact, to define a basic 'agreed-upon' normative content and procedural implementation of FPIC by relying on the concept's normative basis as well as related interpretations of human rights monitoring bodies, whose specific mandate is to produce substantial content for often abstract and general human rights norms.

Arctic states have embraced their obligation to consult with Indigenous peoples, although each has varying approaches to implementing this requirement. The national legal obligations to consult Indigenous peoples come from a variety of sources. As studied by Hughes (2018, 15–27), for example, in Finland,[1] Russia,[2] Norway[3] and Canada,[4] the requirements have constitutional roots. In these countries, as well as the United States, national and regional legislative[5] and administrative[6] measures specify the consultation process that is required. In Canada, comprehensive land claim agreements and treaties between governments and Indigenous peoples[7] and judicial decisions[8] provide additional important foundations for Indigenous rights (Hughes 2018, 16). The aim of this chapter is not, however, to study the current consultation or land rights regimes of Arctic states or to investigate as such to what degree FPIC has already been implemented in the processes taking place in the Arctic. This has, in fact, already been studied thoroughly, for instance, by Hughes (2018, 15–27), which shows that much will still need to be done by the Arctic states to meet the current international standards.

Independently of the domestic legal systems, all Arctic states are bound by international human rights applicable to Indigenous peoples. The UN Declaration on the Rights of Indigenous Peoples[9] (UNDRIP), adopted by the UN General Assembly in 2007, is considered to codify the existing legally binding human rights and principles related to Indigenous peoples rather than establishing new rights (Anaya 2008, para 86). Despite the original rejection of UNDRIP by some Arctic states (Canada and the United States voting against, Russian abstaining) (Bankes and Koivurova 2014, 235), all Arctic states have since then accepted this international instrument (Davis 2012, 25), which in many ways encourages a paradigm shift in state–Indigenous relations (Heinämäki 2015, 195). The crux of UNDRIP is the recognition of Indigenous peoples' rights to self-determination and free, prior and informed consent (Barelli 2018, 247–269).

FPIC provisions in UNDRIP are meaningful to be taken as a basis for studying and defining current international normative standards related to this legal concept. Despite UNDRIP as such not being formally a legally binding instrument, unlike other UNGA resolutions, this does not mean that many of its provisions would not be reflective or indicative of binding international norms (Åhren 2010, 103). Some authors argue that many essential UNDRIP provisions, including those on self-determination, self-government and autonomy, represent customary international law (Xanthaki 2009). Several institutions,[10] including human rights monitoring bodies,[11] as well as national courts, have already applied UNDRIP as a legal source.[12] UNDRIP has been used as the basis for several guidelines regarding FPIC.[13]

The aim of this chapter is to point out current indisputable minimum legal requirements of FPIC. The first section discusses the justification, purpose and elements of FPIC. In the second section, FPIC's normative background and basis will be analysed. The final section focuses on the implementation of FPIC as a process, aiming at providing practical tools for Arctic actors to implement and run through FPIC as a part of the consultation process.

The justification, purpose and elements of free, prior and informed consent

The right of Indigenous peoples to FPIC in relation to resource extraction and other development projects within the territories traditionally occupied and used by Indigenous peoples is currently a very topical issue internationally, regionally and domestically (Barelli 2012, 1–24). As maintained by the study of the Commission on Human Rights, discussion and standard setting on this topic cover a wide range of bodies and sectors, including the safeguard policies of multilateral development banks and international financial institutions, practices of extractive industries, water and energy development, natural resource management, access to genetic resources and associated traditional knowledge and benefit-sharing arrangements, scientific and medical research and Indigenous peoples' cultural heritage (UNCHR 2005, 3). Additionally, Indigenous peoples' right to FPIC also concerns their participation in law-making that has direct relevance for them (Tobin 2014, 32).

The fundamental justification and purpose of FPIC relate to Indigenous peoples' relationship to their traditional lands that for them are of existential importance (Cobo 1983). In the words of Kreimer: 'territorial rights are a central claim for Indigenous peoples in the world. Those rights are the physical substratum for their ability to survive as peoples, to reproduce their cultures, to maintain and develop their organizations and productive systems' (2003). In the Arctic, territorial rights are important for between 0.4 and 1.5 million Indigenous persons, to whom the Arctic environment is the foundation of their culture and identity, a source of physical and spiritual nourishment and their very existence as Indigenous peoples (Stone 2015, 11).[14]

FPIC is devised as a tool in international law to give Indigenous peoples the power to participate in, and influence the outcome of, decisions that may negatively affect Indigenous peoples' traditional lands, cultures, ways of life and rights (Rombouts 2014, 11). On a basic level, the concept of FPIC is contained within its phrasing: it is the right of Indigenous peoples to make free and informed choices about the development of their culture, lands and resources (Ward 2011, 4). Rombouts specifies the elements of FPIC, supported by UN Indigenous specific mechanisms (UNPFII, EMRIP), as follows: 'free' requires the absence of pressure as well as truthfulness, that is, 'no coercion, intimidation or manipulation' (Rombouts 2014, 187); 'prior' requires that the authorities have interacted with the Indigenous community in question in a timely manner, meaning 'that consent has been sought sufficiently in advance of any authorization or commencement of activities', which means that the FPIC process has to be part of any administrative decision-making process before any permits are issued. In doing so, the state may not use its own experience in decision-making as a guide but has to take into account the 'time requirements of Indigenous consultation/consensus processes' (ibid., 114); 'informed' means that the Indigenous community has to be provided specific types of information, including reasons and effects ('nature, size, pace, reversibility and scope of any proposed project of activity'), including location and duration. Also, assessment of the likely economic, social, cultural and environmental impact, including potential risks and fair and equitable benefit-sharing in a context that respects the precautionary principle and potentially involved persons and groups, as well as potential procedures, should be present (ibid., 115).

From the perspective of Indigenous rights, FPIC processes may be seen as an important means to accomplish more equal and inclusive decision-making (Rombouts 2014, 19). Thus, FPIC allows Indigenous peoples' voices to be heard, even if they constitute a minority, which otherwise cannot compete with the majority in a democratic society dominated by non-Indigenous persons and interests (Heinämäki and Kirchner 2017, 233). Today, Indigenous peoples in many places of the world are trying to renegotiate their relations with states and with new private sector operations seeking access to the resources on their lands (UNCHR 2005, 4). They are asserting their right to FPIC as exercised through their representative institutions in dealing with the many parties interested in their traditional territories. Indigenous peoples are seeking support from international human rights bodies to find new ways of being recognized by international and national laws and systems of decision-making with the recognition of their autonomy and their own values (Colchester and Mackay 2004).

Related to Indigenous peoples' values, their traditional knowledge is another justification for the FPIC and should be incorporated in the process. Arctic communities have lived in harsh environments for millennia and have acquired a great depth of knowledge about the land and waters of their homelands and the species that live in their natural environment which provide food, clothing and meaning to their culture. This knowledge, often referred to as traditional, is extremely valuable in tackling the range of climate, economic and governance issues in the Arctic at present (Arruda and Krutkowski 2017, 515).

Nowadays, FPIC has become also a tool for more effective decision-making. When communities are meaningfully engaged and can affect the outcome, different land-use projects simply go more smoothly. The World Bank Group puts it this way: 'Whether it is entrenched in law and regulations or the result of de facto demands of the affected Indigenous peoples, FPIC is a necessary feature of successful decision making' (Greenspan 2016).

Normative background and basis of free, prior and informed consent

Although FPIC has gained a lot of international and national attention via, and is explicitly articulated in, UNDRIP, it was already referred to in UN bodies as well as inter-American human rights systems well before UNDRIP as a process that would make the most sense in solving conflicts in state–Indigenous relations. In 1999, in his final report, Special Rapporteur Miguel Alfonso Martínez concluded that the process of negotiation and seeking Indigenous peoples' consent is the most appropriate way to approach conflict resolution of Indigenous issues at all levels.[15]

It is important to put FPIC in a larger context related to the overall shift in state–Indigenous relations, which were given a push forward by several developments in the UN during 1970s and 1980s. The studies on Indigenous peoples conducted by UN Special Rapporteur José Martinez Cobo rejected the assimilation approach[16] that was adopted, for instance, in the International Labour Organization's (ILO) Convention No. 107 (1957) Concerning the Protection and Integration of Indigenous and Other Tribal and Semi-Tribal Populations in Independent Countries (ILO 107).[17] In 1982, the Working Group on Indigenous Populations (WGIP) was established by the United Nations Economic and Social Council.[18] Through the process of drafting a declaration on the rights of Indigenous peoples, WGIP engaged states, Indigenous peoples and others in a broad multilateral dialogue on the specific content of norms concerning Indigenous peoples and their rights (Anaya 2004, 63–64).

Indigenous rights discourse focused on allowing Indigenous peoples to continue to exist as distinct polities, with particular attention being directed at their land and natural resource rights (Keal 2008, 317). ILO Convention No. 107 was replaced by the ILO 169 of 1989,[19] reflecting a shift in the general attitude towards Indigenous peoples. ILO No.169 contains not merely substantive protection of the cultural integrity of Indigenous peoples but also recognizes them as political and legal entities that need to be involved in decision-making in matters that concern the group (Barsh 2004, 15, 23).

ILO No.169 refers to the right of FPIC, but only in the context of relocation of Indigenous peoples from their land, in its Article 16. Article 7 of ILO No. 169 recognizes Indigenous peoples' 'right to decide their own priorities for the process of development' and 'to exercise control, to the extent possible, over their own economic, social and cultural development'. In Articles 2, 6 and 15, ILO No. 169 requires that states fully consult with Indigenous peoples and ensure their informed participation in the context of development, national institutions and programs and the management of lands and resources. As a general principle, Article 6 requires that consultation must be undertaken in good faith, in a form appropriate to the circumstances and with the objective of achieving consent. It is thus argued that Articles 6 and 7 of the ILO Convention 'reflect the spirit of prior informed consent and apply to each provision of ILO 169' (Baluarte 2004, 9–10). Anaya states that under Article 6 (2) of ILO 169, 'there is neither a right of *veto*, nor the possibility of carrying out empty consultations, if the Convention is correctly applied. The Convention requires a true dialogue' (Anaya 2009a, 138, 2009b).

The UN Declaration on the Rights of Indigenous Peoples was negotiated over 20 years with the wide participation of Indigenous peoples. By endorsing Indigenous peoples' right to self-determination (Art. 3), it acknowledges that other provisions, including those articulating FPIC, have to be read through the right to self-determination. This alone strengthens the consultation rights of UNDRIP compared to ILO Convention No. 169. One of the defining aspects of self-determination is its recognition of the principle of permanent sovereignty over natural resources and of the associated right to freely dispose of natural wealth and resources (Tobin 2014, 121). UNDRIP also recognizes Indigenous peoples' customary laws, traditional knowledge and own decision-making structures, which are considered important elements of FPIC. Whereas ILO Convention No. 169 talks about consultation (except in the case of relocation), UNDRIP explicitly calls for the FPIC of Indigenous peoples in addition to Article 10 of relocation, as well as in Article 19 when a state is adopting legislative or administrative measures that affect Indigenous peoples and Article 29 regarding the disposal of hazardous waste within their territories. In addition, Article 32 calls for FPIC prior to 'the approval of any project affecting their lands or territories and other resources, particularly in connection with the development, utilisation or exploitation of mineral, water or other resources.' If one compares the language of Articles 19 and 32 on the one hand with Articles 10 and 29 on the other, an argument can be made that the first two articles simply contemplate a good-faith consultative and cooperative process 'in an effort to, but not necessarily' obtain Indigenous peoples' consent. The latter two articles, instead, which don't articulate any such process, provide for an absolute prohibition on certain activities 'unless FPIC has been obtained' (i.e., veto) (Seier 2011, 1–2).

During the lengthy negotiations of UNDRIP, participation rights were some of the most contentious, in large part because of the ambiguity of the definition (Davis 2008, 465). Some Indigenous rights advocates view FPIC as a right to veto projects, while others argue that FPIC is not meant to be a veto right but rather a way of ensuring that Indigenous peoples meaningfully participate in decisions directly impacting their lands, territories and resources (Heinämäki 2016, 224). The former special rapporteur on the rights of Indigenous peoples, James Anaya, states that, except in the previously mentioned Articles 10 and 29, FPIC 'does not provide Indigenous peoples with a "veto power", but rather establishes the need to frame consultation procedures in order to make every effort to build consensus on the part of all concerned' (Anaya 2009a, para 48, 2009b). On the other hand, Anaya admits that 'the strength or importance of the objective of achieving consent varies according to the circumstances and the Indigenous interests involved' and concludes that 'a significant, direct impact on Indigenous peoples' lives or territories establishes a strong presumption that the proposed measure should not go forward without Indigenous peoples' consent' (ibid., para 47).

The UN Expert Mechanism on the Rights of Indigenous Peoples (EMRIP) argues that FPIC has to be obtained, not merely sought in matters of fundamental importance to Indigenous peoples:

> The Declaration on the Rights of Indigenous Peoples requires that the free, prior and informed consent of Indigenous peoples be obtained in matters of fundamental importance to their rights, survival, dignity and well-being. In assessing whether a matter is of importance to the Indigenous peoples concerned, relevant factors include the perspective and priorities of the Indigenous peoples concerned, the nature of the matter or proposed activity and its potential impact on the Indigenous peoples concerned, taking into account, inter alia, the cumulative effects of previous encroachments or activities and historical inequities faced by the Indigenous peoples concerned.
>
> (Human Rights Council, 2011, para 23)

Thus, according to EMRIP, FPIC has to be seen as a process which in itself assesses how important a matter in question is for Indigenous peoples and how significant its impacts, taking into account cumulative impacts, are on Indigenous peoples. The object of the FPIC process is to seek the consent of Indigenous peoples, but it is not a necessity to obtain consent unless the issue is of fundamental importance with significant effects on Indigenous peoples. Doyle and Gilbert share this view by maintaining that FPIC only has to be obtained once 'there is a potential for a profound or major impact on the property rights of an Indigenous people or where their physical or cultural survival may be endangered' (Doyle and Gilbert 2011, 317).

This view is supported by both the Inter-American Court of Human Rights (IACtHR), as well as the UN Human Rights Committee (HRC), both of which have recognized that in the case of significant, large-scale negative impact on the lands and traditional way of life of Indigenous peoples, FPIC has to be obtained (i.e., veto), not only sought.[20] IACtHR particularly uses UNDRIP as the basis to elaborate upon the content and process of FPIC and highlights Indigenous peoples' customary laws and their own decision-making structures to be taken into account in the consultation process.[21] Both IACtHR and the HRC have called for an independent impact assessment in order to define whether the threshold of significant harm may be exceeded.[22]

Anaya points out that FPIC is not primarily about whether the consent of an Indigenous community has been obtained but rather about establishing a due process. According to him, the somewhat different language of UNDRIP compared to ILO Convention No. 169 suggests

> a heightened emphasis on the need for consultations that are in the nature of negotiations towards mutually acceptable arrangements, prior to the decisions on proposed measures, rather than consultations that are more in the nature of mechanisms for providing Indigenous peoples with information about decisions already made or in the-making, without allowing them genuinely to influence the decision-making process.
>
> (Anaya 2009a, para 46)

This is also a crucial point in the Arctic context, where some governments do not require consultation at the earliest stage, such as before tendering concessions or approving exploration for extractive activities (Hughes 2018, 20). Compared to ILO Convention No. 169, UNDRIP, while recognizing Indigenous peoples' right to self-determination, directly connects the right to be consulted in the form of FPIC to territorial rights, tying together procedural and substantial aspects of protection.

Hence, it can be concluded that while ILO Convention No. 169 requires 'the spirit of FPIC', UNDRIP does in fact require the 'body of FPIC', which means that in the FPIC process, Indigenous peoples' consent needs not only to be sought but indeed obtained (i.e., veto) in cases of major importance to Indigenous peoples and where interference may cause a significant negative impact on their traditional lands and way of life. The International Law Association (ILA) has come to similar conclusions (ILA 2010, 24). This view is generally shared by other monitoring bodies of human rights conventions, such as the Committee on Economic, Social and Cultural Rights (CESCR), as well as the Committee on Elimination of Racial Discrimination (CERD) (Heinämäki and Kirchner 2017, 248–253). In their concluding observations on different Arctic states, these committees have expressed their concern about insufficient application of FPIC in relation to issues concerning Indigenous peoples and called for the full implementation of FPIC in their national legislations and policies (ibid.).

After the adoption of UNDRIP, FPIC has started to find its way to international environmental legal regimes and is now referenced in the Biodiversity Convention's (CBD) Nagoya Protocol related to genetic resources, although required only when national legislation grants FPIC.[23] CBD's Conference of Parties (COP) has also adopted voluntary guidelines on FPIC.[24] Apart from CBD, FPIC is currently evolving in the United Nations Framework Convention on Climate Change (UNFCCC) climate change regime related to the programme on 'Reducing Emissions from Deforestation and Forest Degradation in Developing Countries' (REDD and REDD+), which aim at limiting emissions from deforestation in developing countries (Prior et al. 2013, 245). Although there is still no unilateral agreement on how to implement FPIC in these regimes, both the Nagoya Protocol as well as UNFCCC Cancun Decision 1/CP 16[25] refer to UNDRIP, which means that these procedures should follow the guidelines related to implementation of UNDRIP in relation to FPIC. After the adoption of UNDRIP, the World Bank has revised its FPIC provision to require documenting that consent has been obtained. If this cannot be shown, the World Bank will not proceed with the aspects of the project relevant to Indigenous peoples.[26] A number of international financial institutions have recently updated, or are in the process of updating, their standards and safeguard policies to reflect the evolution in the recognition of Indigenous peoples' rights within the international human rights regime, particularly endorsing FPIC.[27]

As indicated, FPIC should not be viewed as a one-directional event but a process (MacKay 2004, 43) in which the state has to act in good faith and with the intention of seeking the consent of the Indigenous community (see Cariño 2005, 19–39). As expressed by Gilbert, FPIC is defined as a process which implies and requires an iterative series of discussions, consultations, meetings and agreements (Gilbert 2014, 203). The absence of consent does not make it impossible to engage in the process, except in cases of significant harm and very survival, but the state has to take the concerns of the Indigenous community into account. There has to be a real, not only theoretical, possibility that a project would be altered or stopped, for example, a permit denied, based on the input of the Indigenous community during the FPIC process (Ward 2011, 58).

During recent years, after the adoption of UNDRIP, it has become a common view to regard FPIC as a right rather than a principle. As Grand Chief John puts it, 'FPIC is a right. It's not simply a concept; it is a right to [a] process. In many cases the natural relation is to lands, territories, and resources' (Grand Chief John 2012, 15–16). It is indeed important to recognize FPIC as a right rather than a principle. As a human rights norm under international law, it poses limitations upon state sovereignty insofar as it tells us that the state's treatment of its citizens is not only an internal matter but also a legitimate matter of international concern (Bankes 2010,

294–295). FPIC could be defined as a right to participate in a qualified process, which embraces the requirements of being free, prior and informed, as specified in the previous section.

Especially since the adoption of UNDRIP, the right to FPIC has been directly related to the right of Indigenous peoples to self-determination. As maintained by Ward, FPIC and other participation rights are not merely administrative processes but are an exercise in and expression of the right to self-determination (Ward 2011, 55). As argued by the report of the UN Commission on Human Rights, the rights to self-determination and FPIC, as collective rights, fundamentally entail the exercise of choices by peoples as rights-bearers and legal persons about their economic, social and cultural development. These cannot be weakened by the consultation of individual constituents about their wishes but rather must enable and guarantee the collective decision-making of the concerned Indigenous peoples and their communities through legitimate customary and agreed-upon processes and through their own institutions (UNCHR 2005, 45). A meaningful process to achieve FPIC must respect the structure of Indigenous communities and their forms of governance; thus, Indigenous peoples should be engaged through their representative institutions.

However, it is not always clear whether groups have formal structures in place for engagement (Hughes 2018, 18). For national governments, the legal requirements for consultation generally identify and define the Indigenous entity governments must engage with by law (ibid.). It is, however, useful to consider not only whether the government recognizes an Indigenous organization as the official representative of the community but also whether the community itself shares this perception (ibid., 19). Hence, communities should be allowed to define FPIC representatives among themselves. According to the ILO Handbook, where there is a diversity of competing institutions, the identification of a single representative institution may not be possible, and it may be best to take an inclusive approach, allowing for participation of the diversity of organizational expressions (ILO 2013, 14). As noted by Hughes, other international organizations, such as the UN, have also created guidance for identifying Indigenous peoples, and by using this guidance to identify with whom to engage, entities conducting Arctic activities can ensure that their engagement comports with international standards (Hughes 2018, 19).

Implementation of free, prior and informed consent as a process

Rombouts asserts that the conflicting interpretations and lack of clarity as to its scope and content hamper effective implementation of FPIC (2014, 11). However, as Franco has commented, as a right, 'FPIC is neither self-interpreting nor self-implementing' (2014, 3), which means that actors that engage in an FPIC process must give specific content to FPIC. Related to implementation of FPIC processes, there are a number of studies and guidelines on how to run through the process properly. Although practice reveals shortcomings in implementation of FPIC processes, it also provides good examples and indicates the potential of FPIC to generate mutually beneficial agreements (Rombouts 2014, 23).

Based on UNDRIP and other UN mechanisms on the rights of Indigenous peoples, many international bodies and organizations have created guidelines on how to implement FPIC (ibid.). The Forest Stewardship Council (FSC) has adopted a comprehensive 'step by step' approach in implementing FPIC, which is based on seven steps: 1) identification of rights holders and their rights through engagement, 2) preparation for further engagement and agreement on the scope of the FPIC process, 3) undertaking participatory mapping and impact assessment, 4) informing affected rights holders, 5) negotiating and allowing rights holders to decide on an

FPIC proposal, 6) verifying and formalizing the FPIC agreement, 7) implementing and monitoring the FPIC agreement (Forest Stewardship Council 2018, 11).

Anaya and Puig summarize some key procedural and material elements of a comprehensive consultation process that should guide the implementation of FPIC. One of the major issues in appropriate implementation is the impact assessment. The authors point out that consultation processes shall actively contribute to the prior assessment of all potential impacts of the proposed measure, including the extent to which Indigenous peoples' substantive rights and interests may be affected. In addition, consultation procedures are key to the search for less harmful alternatives, obtaining tangible benefits and advancing the enjoyment of Indigenous peoples' human rights (Anaya and Puig 2017, 457–458). The authors raise a crucial point by stating that in order for consultation procedures to adequately fulfil their safeguard role, the power imbalances (technical, political or economic) among the actors involved must be mitigated. Practical measures include, inter alia, employing independent facilitators for consultations or negotiations, establishing funding mechanisms that allow Indigenous peoples to have access to independent technical assistance and advice or developing standardized procedures for the flow of information (ibid., 458). Lack of resources is also one of the major general challenges for Arctic Indigenous peoples to adequately participate in processes that impact them. The Inuit Circumpolar Conference recommends that the participation of Indigenous peoples 'should be facilitated through adequate training and funding'.[28] Anaya and Pruig continue that Indigenous peoples should enjoy full access to the information gathered in impact assessments that are done by state agencies or business enterprises. Indigenous peoples should have the opportunity to participate in impact assessments in the course of consultations or otherwise. States should ensure the impartiality of the information and the objectivity of impact assessments either by subjecting them to independent review or by requiring that the assessments be performed free from potential conflict of interest (ibid.). Substantive requirements, according to the previous authors, include, for example, mitigation of impacts, compensation and benefit sharing, joint management agreements and adequate grievance procedures (ibid., 459–460).

Indigenous peoples' customs, customary laws and values plays an important role in any meaningful FPIC process. Indigenous peoples' worldview and distinct epistemologies underlie their systems of laws, customs and tradition, which are rooted in land, spirituality and culture (Tobin 2014, 29). The focus of Indigenous peoples' legal regimes tends to be towards the restoration of community well-being rather than retribution (ibid.). Tsosie identifies key elements related to customary law regimes, including communal and collective aspects of ownership, rights and responsibilities grounded in a spiritual value base and the ethic 'that resources must be used in a way that is productive and beneficial to all members, including future generations' (Tsosie 1997, 5). This is in stark contrast to the current western focus on maximum exploitation of resources and the enforcement of property rights (Tobin 2014, 32). Hence, viewed from this perspective, FPIC has the potential not only to ensure the rights of Indigenous peoples but to positively contribute to the sustainability of the environment. A genuine FPIC process must take into account Indigenous peoples' traditional knowledge, which can significantly contribute to the assessment of the impacts of proposed actions taking place or having effects on Indigenous peoples.

As described by the EMRIP, FPIC should be documented, capturing the steps for accomplishing such consent and the essence of the agreement reached by the concerned parties, in accordance with Indigenous peoples' customary norms and traditional methods of decision-making, including diverging opinions and conditional views. Guidelines or models for seeking free, prior and informed consent that are developed by either states or private actors should not prevail over Indigenous peoples' own community protocols or traditional practices of capturing or recording agreements (EMRIP 2018, para 42). Forms of expressing consent may include, for example, treaties, agreements and contracts. Often terms are commemorated in a memorandum of agreement or understanding or other document that is satisfactory to the Indigenous peoples (ibid., para 43). The process of creating a community protocol can make communities aware of international human rights standards and help develop their negotiation skills in asserting their rights (Cultural Survival Quarterly 2012, 17).

EMRIP notes that, as a dynamic process, the implementation of FPIC should also be monitored and evaluated regularly. Such agreements should include mechanisms for participatory monitoring. The implementation of FPIC should also include accessible recourse mechanisms for disputes and grievances, devised with the effective participation of Indigenous peoples, including judicial review (EMRIP 2018, para 45). Also, the ILO's Committee of Experts on the Application of Conventions and Recommendations (CEACR) stresses the need for periodic evaluation of the operation of the consultation mechanisms, with the participation of the peoples concerned to continue to improve their effectiveness (ILO 2010, 7).

FPIC, when implemented in the Arctic between relevant Indigenous communities and by states as well as other actors operating in traditional lands of Indigenous peoples, has great potential for several reasons. First of all, when implemented according to international human rights norms, the concept of FPIC interlinks substantive and procedural rights of Indigenous peoples. Indigenous peoples' right to FPIC cannot, according to the principles built into FPIC, be viewed in isolation from their substantive land or cultural rights. Therefore, when fully implemented by Arctic states, FPIC not only requires improvements in consultation procedures but calls for the adequate securing of the land and cultural rights for successful subsistence and other traditional land use purposes. Second, and related, since FPIC is viewed as an integral part of actualization of Indigenous peoples' right to self-determination, it has the potential to expand the current consultation and self-governing regimes of Arctic states to include decision-making and governance powers regarding traditional lands and natural resources embedded in the areas that Indigenous peoples use for carrying on their traditional livelihoods. The third basis for FPIC's great potential in the Arctic relates to the vast experience of Arctic Indigenous peoples to have a dialogue with Arctic states and other actors, significant examples being their permanent participation in the Arctic Council or in the Barents Euro-Arctic Council. In these regimes, Arctic Indigenous peoples' traditional knowledge, an important element for a successful FPIC process, has already been embraced and implemented in many levels of policymaking. The workshop statement of an ongoing Arctic Council Sustainable Development Working Group Initiative called 'Good Practice Recommendations for Environmental Impact Assessment and Public Participation in the Arctic' elaborates on meaningful participation of Arctic Indigenous peoples in EIA procedures. It emphasizes how meaningful engagement should happen early, before project scoping has occurred, throughout the entire process, and Indigenous knowledge holders should have equitable and meaningful roles in the utilization of their knowledge. Indigenous peoples should be helped to determine a decision which reflects their values.[29]

Conclusion

FPIC should be understood as Indigenous peoples' right to a qualified consultation and assessment process, guided by and honouring their right to self-determination, where Indigenous peoples, via their representative institutions, have a genuine possibility to share their knowledge and views related to the process. In the process, there should be a true possibility for Indigenous peoples to influence the outcome. FPIC is not considered a right to veto by the mainstream interpretation of the current human rights law, except in cases where Indigenous peoples' survival and well-being are at stake and where there might be large-scale impacts on their lands, cultures and traditional livelihoods. Indigenous peoples' decision on giving or withholding consent should be based on assessment of cumulative impacts on Indigenous peoples' societies, culture, environment and rights, which must be understood as an inherent element of an appropriate FPIC process.

The Arctic can be viewed as a specific context for the application of FPIC as an international human rights norm due to its vulnerability to environmental and sociocultural changes accelerated by global warming that are drastically changing the Arctic environment and lives of people, particularly of Indigenous communities. On the other hand, Indigenous peoples have shown great resilience and adaptive capacity, as well as valuable knowledge related to the environment and its changes. Arctic Indigenous peoples do not want to be viewed as victims of these environmental changes but have expressed willingness to share their knowledge and views when allowed to meaningfully participate in decision-making at all relevant levels.

FPIC requires Indigenous peoples' values, customs and customary laws, as well as decision-making structures, to be taken into account when establishing FPIC procedures. Therefore, FPIC challenges the current normative framework to understand and implement Indigenous peoples' own ontologies, which often rely on different premises than the 'Western' worldview (Heinämäki et al. 2020, 111–118). FPIC as a legal concept and as a process, co-designed and agreed upon between two parties – for example, state/some other actor and an Indigenous community – has a capacity to embrace Indigenous peoples' own views on their participation and rights, contributing to and enriching Arctic democracy and policies on the environment and sustainable development. Arctic states should renew their existing legal and administrative consultation procedures and sufficiently protect the substantial basis of Indigenous peoples' traditional, land-based livelihoods aligned with the requirements set by the current human rights law regarding the application of FPIC.

Notes

1 Act on Sámi Parliament, Art. 9.
2 Konstitutsiia Rossiiskoi Federatsii [Russian Constitution], Art 69.
3 Kongeriget Norges Grundlov [Norwegian Constitution] art 108. Consultation duty is also based on two agreements: Prosedyrer for konsultasjoner mellom statlige myndigheter og Sametinget the basic Consultation Agreement – [Procedures for Consultations between State Authorities and the Sámi Parliament]; Avtale mellom Sametinget og Miljøverndepartementet om retningslinjer for verneplan arbeid etter naturvernloven i samiska områder [Consultation agreement concerning nature conservation matters in traditional Sámi areas].
4 Constitution Act, 1982, Schedule B to the Canada Act 1982 (U.K.) c 11[Canada Constitution], sec.35.
5 See, for example, National Historic Preservation Act, 16 U.S.C.§ 470 et seq. (US); Finnmark Act, LOV-2005–06–17–85 § 3 (17 June 2005) (Norway).
6 See, for example, Department of Aboriginal Affairs and Northern Development Canada, 'Aboriginal Consultation and Accommodation Updated Guidelines' (2011) (INAC Guidelines); Exec. Order No. 13,175, 65 Fed. Reg. 67,249 (9 November 2000) (US).
7 See, for example, Indian and Northern Affairs Canada, Gwich'in Comprehensive Land Claim Agreement (22 April 1992) para 21.1.3.
8 See, for example, Beckman v Little Salmon/Carmacks First Nation [2010] 3 S.C.R. 103 (Canada).
9 UN Declaration Convention on the Rights of Indigenous Peoples (2007) UN Doc A/61/L.67.
10 See UN Development Group, Guidelines on Indigenous Peoples' Issues (1 February 2008). Similarly, the 31 UN specialized agencies that make up the Inter-Agency Support Group to the UN Permanent Forum on Indigenous Issues have announced to advance the spirit and letter of the UNDRIP. See Statement on the UN Declaration on the Rights of Indigenous Peoples, adopted at the Annual

Meeting of the Inter-Agency Support Group in September 2007; FAO, Policy on Indigenous and Tribal Peoples (2010) 2 and Chapter IV, 'Objectives for Engagement with Indigenous Peoples'; UN Global Compact, 'UN Declaration on the Rights of Indigenous Peoples; A Business Reference Guide' (December 2013); OEA/SerK/XVI, GT/DADIN/doc321/08, 3.

11 See, Inter-American Court of Human Rights, *Saramaka People v. Suriname* (Judgement of 28 November 2007) IACHR Series C, No. 172; Inter-American Court of Human Rights, *Kichwa Indigenous People of Sarayaku v. Ecuador* (Judgment of 27 June 2012) IACHR Series C, No. 245.

12 The Supreme Court of Belize also applied the principles of the Declaration as a framework for determining land rights. Shortly after the adoption of the Declaration by the UN General Assembly, the Supreme Court of Belize made a decision relating to the rights of the Maya community to their lands and resources, applying the Declaration: Supreme Court of Belize, *Aurelio Cal v. Attorney-General of Belize* Claim 121/2007 (18 October 2007). The Constitutional Court of Colombia applied UNDRIP in its decision T-129 (Constitutional Court of Colombia, Case T-129, Decision of 3 March 2011). See also examples in Canadian Supreme Court referring to UNDRIP. See Allard, C. chapter 12: Canada, in L. Heinämäki et al. (2017, 391), referring to information gathered from a PowerPoint presentation (October 2016) by Nigel Bankes and a synopsis of case law prepared by the law student Amy Matychuk. Note that in Manitoba, there is a new legislation that refers to the principles of UNDRIP; see the Path to Reconciliation Act 2016.

13 FAO (2018); Oxfam (2010); Rainforest Alliance (2017).

14 United Nations Permanent Forum on Indigenous Issues estimates the number as 0.4. million, while the Arctic Council Indigenous Peoples Secretariat estimates 1.5. million. The range reflects different definitions of the circumpolar Arctic and of the term Indigenous.

15 UN Doc E/CN.4/Sub.2/1999/20, para. 263.

16 The UN Sub-Commission on Prevention of Discrimination and Protection of Minorities (1986) UN Doc E/CN.4/RES/1986/38.

17 ILO Convention No 107 Concerning the Protection and Integration of Indigenous and Other Tribal and Semi-Tribal Populations in Independent Countries (adopted 26 June 1957, entered into force 2 June 1959) 328 UNTS 247.

18 Economic and Social Council Resolution 1982/34, Study of the Problem of Discrimination against Indigenous Populations (7 May 1982). The WGIP is an organ of the Sub-Commission on the Promotion and Protection of Human Rights.

19 ILO Convention No. 169 Concerning Indigenous and Tribal People in Independent Countries (adopted 27 June 1989, entered into force 5 September 1991) 1650 UNTS 383.

20 *Saramaka People v. Suriname*, Inter-American Court of Human Rights (Judgment of 28 November 2007) Series C, No 172; *Poma Poma v. Peru*, Human Rights Committee, Communication No. 1457/2006 (27 March 2009) UN Doc CCPR/C/95/D/1457/2006.

21 *Saramaka People v. Suriname*, Inter-American Court of Human Rights (Judgment of 28 November 2007) Series C, No 172; *Poma Poma v. Peru*, Human Rights Committee, Communication No. 1457/2006 (27 March 2009) UN Doc CCPR/C/95/D/1457/2006.

22 *Saramaka People v. Suriname*, Inter-American Court of Human Rights (Judgment of 28 November 2007) Series C, No 172; *Poma Poma v. Peru*, Human Rights Committee, Communication No. 1457/2006 (27 March 2009) UN Doc CCPR/C/95/D/1457/2006.

23 Nagoya Protocol on Access to Genetic Resources and the Fair and Equitable Sharing of Benefits Arising from Their Utilization to the Convention on Biological Diversity (29 October 2010) UNEP/CBD/COP/DEC/X/1 of 29, Art. 6.2.

24 Mo'otz kuxtal1 voluntary guidelines (17 December 2016) CBD/COP/DEC/XIII/18.

25 Framework Convention on Climate Change, Conference of the Parties, Report of the Conference of the Parties on its sixteenth session, held in Cancun from 29 November to 10 December 2010, Addendum, Part Two: Action taken by the Conference of the Parties at its sixteenth session (15 March 2011) FCCC/ CP/2010/7/Add.1 Annex I, 2. (c).

26 The World Bank, Review and Update of the World Bank Safeguard Policies (2016).

27 See, for example, IFC Performance Standard No 7, World Bank Operational Policy on Indigenous Peoples No. 4.1; Office of the Compliance Advisor/Ombudsman (CAO) IFC and MIGA, World Bank Group Advisory Note 'IFC's Policy and Performance Standards on Social and Environmental Sustainability and Disclosure Policy, Commentary on IFC's Progress Report on the First 18 Months of Application', 17 December 2007; Asian Development Bank, The Safeguard Policy Statement (Second Draft) October 2008, 11, 19.

28 Arctic Council, 'Kiruna Declaration' (15 May 2013) ('recognizing the rights of the Indigenous peoples and interests of all Arctic inhabitants, and emphasizing that a fundamental strength of the Council is the unique role played by Arctic Indigenous peoples'); ICC, 'Inuit Arctic Policy', www.inuitcircumpolar.com/uploads/3/0/5/4/30542564/inuit_arctic_policy.pdf
29 Arctic EIA project 2018, www.sdwg.org/activities/sdwg-projects-2017-2019/arctic-eia/arctic-eia-new/

References

ACIA. 2005. *Arctic Climate Impact Assessment. ACIA Overview Report*. Cambridge: Cambridge University Press.

Agnes Portalewska. 2012. "Free, Prior and Informed Consent, Protection of Indigenous Peoples to Self-Determination, Participation and Decision-Making." *Cultural Survival Quarterly* 36 (4). Accessed 25 August 2020. https://www.culturalsurvival.org/publications/cultural-survival-quarterly/free-prior-and-informed-consent-protecting-Indigenous.

Åhren, M. 2010. "The Saami Traditional Dress and Beauty Pageants: Indigenous Peoples' Rights of Ownership and Self-Determination Over Their Cultures." Avhandling leverert for graden Philosophiae Doctor i rettsvitenskap [Thesis supplied for the degree of Philosophiae Doctor of Law] unpublished.

Anaya, J. 2004. *Indigenous Peoples in International Law*. Oxford: Oxford University Press.

Anaya, J. 2008. Report to the Human Rights Council, A/HRC/9/9, 11 Aug. 2008 (the Declaration on the Rights of Indigenous Peoples in context; implementation measures required).

Anaya, J. 2009a. *International Human Rights and Indigenous Peoples*. New York: Aspen.

Anaya, J. 2009b. UN Special Rapporteur James Anaya on the Rights of Indigenous Peoples, Promotion and Protection of All Human Rights, Civil, Political, Economic, Social and Cultural Rights, Including the Right to Development: Report of the Special Rapporteur on the Situation of Human Rights and Fundamental Freedoms of Indigenous People, UN Doc. A/HRC/12/34, 15 July 2009.

Anaya, J., and S. Puig. 2017. "Mitigating State Sovereignty: The Duty to Consult with Indigenous Peoples." *University of Toronto Law Journal* 67 (4): 435–464.

Arruda, G. M., and S. Krutkowski. 2017. "Arctic Governance, Indigenous Knowledge, Science and Technology in Times of Climate Change, Self-Realization, Recognition, Representativeness." *Journal of Enterprising Communities: People and Places in the Global Economy* 11 (4): 514–528.

Baluarte, D. C. 2004. "Balancing Indigenous Rights and a State's Right to Develop in Latin America: The Inter-American Rights Regime and ILO Convention 169." *Sustainable Development Law and Policy* (Summer): 9–15.

Bankes, N. 2010. "International Human Rights Law and Natural Resources Project within the Traditional Territories of Indigenous Peoples." *Alberta Law Review* 47 (2): 494–495.

Bankes, N., and T. Koivurova. 2014. "Legal Systems." In *Arctic Human Development Report II*, edited by J. N. Larsen and G. Fondahl, 223–254. Copenhagen: Norden.

Barelli, M. 2012. "Free, Prior and Informed Consent in the Aftermath of the UN Declaration on the Rights of Indigenous Peoples: Development and Challenges Ahead." *International Journal of Human Rights* 16 (1): 1–24.

Barelli, M. 2018. "Free, Prior and Informed Consent in the UNDRIP: Articles 10, 19, (29 (2) and 32 (2)." In *The UN Declaration on the Rights of Indigenous Peoples, A Commentary*, edited by J. Hohmann and M. Weller, 247–269. Oxford: Oxford University Press.

Barsh, R. L. 2004. "Indigenous Peoples in the 1990's: From Object to Subject of International Law?" *Harvard Human Rights Journal* 7: 33–86 (1994), reprinted in *Indigenous Peoples, the Environment and Law*, edited by L. Watters, 15, Durham: Carolina Academic Press.

Cariño, J. 2005. "Indigenous Peoples' Right to Free, Prior and Informed Consent: Reflections on Concepts and Practice." *Arizona Journal of International and Comparative Law* 22 (1): 19–39.

Cobo, M. J. 1983. Study of the Problem of Discrimination against Indigenous Populations, UN Doc. E/CN/4 Sub.2/1986/7/Add. 4, 28 June 1983.

Colchester, M., and F. Mackay. 2004. "In Search of Middle Ground: Indigenous Peoples, Collective Representation and the Right to Free, Prior and Informed Consent." *Forest Peoples Programme*. Accessed 1 April 2019. www.forestpeoples.org/sites/fpp/files/publication/2010/08/fpicipsaug04eng.pdf.

Commission on Human Rights, Sub-Commission on Prevention of Discrimination and Protection of Minorities. 1999. Study on Treaties, Agreements and Other Constructive Arrangements between States and Indigenous Populations. Final report by Miguel Alfonso Martínez, Special Rapporteur E/CN.4/Sub.2/1999/20, 22 June 1999.

Davis, M. 2008. "Indigenous Struggles in Standard-Setting: The United Nations Declaration on the Rights of Indigenous Peoples." *Melbourne Journal of International Law* 9 (2): 439–471.

Davis, M. 2012. "To Bind or Not to Bind: The United Nations Declaration on the Rights of Indigenous Peoples Five Years on." *Australian International Law Journal* 19: 17–48.

Doyle, C. and Gilbert, J. 2011. "A New dawn over the land: Shedding light on collective ownership and consent." *Reflections on the UN Declaration on the Rights of Indigenous Peoples*, edited by S. Allen & A. Xanthaki, 289–328. Oxford: Hart Publishing.

FAO. 2018. Respecting Free, Prior and Informed Consent, Practical Guidance for Governments, Companies, NGOs, Indigenous Peoples and Local Communities in Relation to Land Acquisition, Governance and Tenure, Technical Guide No. 3, 2018. Accessed 1 April 2019. www.fao.org/3/a-i3496e.pdf.

Fjellheim, R. S. 2006. "Arctic Oil and Gas – Corporate Social Responsibility." *Journal of Indigenous Peoples Rights* 4: 8–23.

Forbes, B., and G. Kofinas. 2014. "Resource Governance." In *Arctic Human Development Report II*, edited by J. N. Larsen and G. Fondahl, 255–298. Copenhagen: Norden.

Ford, J. D., B. Smith, and J. Wandel, J. 2006. "Vulnerability to Climate Change in the Arctic: A Case Study from Arctic Bay, Canada." *Global Environmental Change* 16 (2006): 145–160.

Forest Stewardship Council. 2018. "Implementing Free, Prior and Informed Consent (FPIC), a Forest Stewardship Council Discussion Paper." FSC-DIS-003 V1 EN, March.

Franco, J. 2014. "Reclaiming Free Prior and Informed Consent (FPIC) in the Context of Global Land Grabs." Transnational Institute for Hands off the Land Alliance. Accessed 5 February 2019. www.tni.org/files/download/reclaiming_fpic_0.pdf.

Gilbert, J. 2014. *Nomadic Peoples and Human Rights*. London and New York: Routledge.

Grand Chief John. 2012. "Free, Prior and Informed Consent, Protection of Indigenous Peoples to Self-Determination, Participation and Decision-Making." *Cultural Survival Quarterly* 36 (4).

Greenspan, E. 2016. "Why Should the World Bank Support Free, Prior and Informed Consent? Their Report Explains." *Oxfam*. Accessed 25 August 2020. https://politicsofpoverty.oxfamamerica.org/why-should-the-world-bank-support-free-prior-and-informed-consent-their-report-explains/.

Heinämäki, L. 2009. "Rethinking the Status of Indigenous Peoples in International Environmental Decision-Making: Pondering the Role of Arctic Indigenous Peoples and the Challenge of Climate Change." In *Climate Governance in the Arctic*, edited by T. Koivurova, E. C. H. Keskitalo, and N. Bankes, 207–262. Environment & Policy, Vol. 50. Dordrecht: Springer.

Heinämäki, L. 2015. "The Rapidly Evolving International Status of Indigenous Peoples: The Example of the Sámi People in Finland." In *Autonomous Sámi Law: Indigenous Rights in Scandinavia*, edited by C. Allard et al., 189–206. Ashgate: Juris Divsitas.

Heinämäki, L. 2016. "Global Context – Arctic Importance: Free, Prior and Informed Consent – A New Paradigm in International Law Related to Indigenous Peoples." In *Indigenous Peoples' Governance and Protected Areas in the Circumpolar Arctic*, edited by T. M. Herrmann and T. Martin, 209–242. Dordrecht: Springer.

Heinämäki, L., C. Allad, S. Kirchner, A. Xanthaki, S. Valkonen, and U. Mörkenstam. 2017. *Saamelaisten oikeuksien toteutuminen: kansainvälinen vertaileva tutkimus [Actualizing Sámi People's Rights in Finland: International Comparative Research]*. Prime Minister's Office, Publications of the Government's analysis, assessment and research activities 4/2017, Finnish Government Publication Series, Helsinki.

Heinämäki, L., and S. Kirchner. 2017. "Assessment on Recent Developments Regarding Indigenous Peoples' Legal Status and Rights in International Law with Special Focus on Free, Prior and Informed Consent." In *Saamelaisten oikeuksien toteutuminen: kansainvälinen vertaileva tutkimus [Actualizing Sámi People's Rights in Finland: International Comparative Research]*, edited by L. Heinämäki et al., 224–284. Prime Minister's Office, Publications of the Government's analysis, assessment and research activities 4/2017, Finnish Government Publication Series, Helsinki.

Heinämäki, L., S. Valkonen, and J. Valkonen. 2020. "Legal (Non)Recognition of Sámi Customary Relationship with the Land in Finland: Challenges So Far and Prospects in the Modern Human Rights Era." In *Philosophies of Polar Law*, edited by D. Bunikowski and A. D. Hemmings, 101–118. London and New York: Routledge, Series Routledge Research in Polar Law.

Hughes, L. 2018. "Relationships with Arctic Indigenous Peoples: To What Extent Has Prior Informed Consent Become a Norm?" *Review of European, Comparative and International Environmental Law* 27 (1): 15–27.

Human Rights Council. 2011. Final Report of the Study on Indigenous Peoples and the Right to Participate in Decision-Making. Report of the Expert Mechanism on the Rights of Indigenous Peoples, 17 August 2001, UN Doc A/HRC/18/42

Hung, H., R. Kallenborn, K. Breivik, Y. Su, E. Brorström-Lundén, K. Olafsdottir, J. M. Thorlacius, S. Leppänen, R. Bossi, H. Skov, S. Manø, G. W. Patton, G. Stern, E. Sverko, and P. Fellin. 2010. "Atmospheric Monitoring of Organic Pollutants in the Arctic Under the Arctic Monitoring and Assessment Programme (AMAP): 1993–2006." *Science of the Total Environment* 408: 2854–2873.

ILA. 2010. *Interim Report: The Rights of Indigenous peoples at the Hague Conference.* Interim Report to the 74th ILA Conference in The Hague, 15–20 August 2010. Accessed 23 May 2019. www.ila-hq.org/en/committees/index.cfm/cid/1024.

ILO. 2010. *ILO Observation 2010/81.* Accessed 27 August 2020. https://www.ilo.org/wcmsp5/groups/public/---ed_norm/---normes/documents/meetingdocument/wcms_305958.pdf.

ILO. 2013. *Understanding the Indigenous and Tribal Peoples Convention. Handbook for ILO Tripartite Constituents.* Geneva: International Labour Office.

Keal, P. 2008. "Indigenous Sovereignty." In *Re-Envisioning Sovereignty – The End of Westphalia?* edited by T. Jacobsen, C. Champford, and R. Thakur, 315–330. Ashgate: Farnham.

Keeling, A., J. Sandlos. 2016. "Introduction: Critical Perspectives on Extractive Industries in Northern Canada." *The Extractive Industries and Society* 3 (2): 265–268.

Koivurova, T., H. Tervo, and A. Stepien. 2008. "Indigenous Peoples in the Arctic, Background Paper." *Arctic Transform.* Accessed 1 April 2019. https://arctic-transform.eu/download/IndigPeoBP.pdf.

Kreimer, O. 2003. Report of the Rapporteur, Meeting of the Working Group on the Fifth Section of the Draft Declaration with Special Emphasis on "Traditional Forms of Ownership and Cultural Survival, Right to Land and Territories", OAS Doc. No. GT/DADIN/doc.113/03 rev. 1, 20 February 2003, 1.

Kuokkanen, R. J. 2019. "At the Intersection of Arctic Indigenous Governance and Extractive Industries: A Survey of Three Cases." *The Extractive Industries and Society* 6 (1): 15–21.

MacKay, F. 2004. "Indigenous Peoples' Right to Free, Prior and Informed Consent and the World Bank's Extractive Industries Review." *Sustainable Development Law and Policy Journal* 4 (2): 43–65.

Nuttall, M. 2017. "An Environment at Risk: Arctic Indigenous Peoples, Local Livelihoods and Climate Change." In *Arctic Alpine Ecosystems and People in a Changing Environment*, edited by J. B. Ørbæk, R. Kallenborn, I. Tombre, E. N. Hegseth, S. Falk-Petersen, and A. H. Hoel, 19–35. Dordrecht: Springer.

Nuttall, M., and K. 2006. "Editorial." In *Indigenous Affairs: Arctic Oil and Gas Development.* Copenhagen: IWGIA.

Oxfam. 2010. "Guide to Free, Prior and Informed Consent." Accessed 1 April 2019. www.culturalsurvival.org/sites/default/files/guidetofreepriorinformedconsent_0.pdf.

Prior, T., S. Duyck, L. Heinämäki, T. Koivurova, and A. Stepien. 2013. "Addressing Climate Vulnerability: Promoting the Participatory Rights of Indigenous Peoples and Women through Finnish Foreign Policy." *Juridica Lapponica* 38.

Rainforest Alliance. 2017. "Guide for Free, Prior and Informed Consent (FPIC) Processes." July. Accessed 1 April 2019. www.rainforest-alliance.org/business/wp-content/uploads/2017/11/07_fpic-guide_en.pdf.

Rombouts, S. J. 2014. *Having a Say, Indigenous Peoples, International Law and Free, Prior and Informed Consent.* Oisterwijk: Wolf Legal Publishers (WLP)

Seier, F. 2011. "'Free, Prior and Informed Consent" under UNDRIP: What Does it Really Mean?', 1 Right 2 Respect, Business and Human Rights Advisors. Accessed 20 June 2017. http://www.right2respect.com/2011/06/%E2%80%98free-prior-and-informed-consent%E2%80%99-under-the-un-declaration-onthe-rights-of-indigenous-peoples-what-does-it-really-mean/.

Stone, D. P. 2015. *The Changing Arctic Environment: The Arctic Messenger.* Cambridge: Cambridge University Press

Tobin, B. 2014. *Indigenous Peoples, Customary Law and Human Rights – Why Living Law Matters.* London and New York: Routledge.

Tsosie, R. 1997. "Indigenous Peoples' Claims to Cultural Property: A Legal Perspective." *Museum Anthropology* 21 (3): 5–11.

UN Commission on Human Rights, Sub-Comm. on the Promotion and Protection of Human Rights Working Group on Indigenous Populations. "Working Paper: Standard-Setting: Legal Commentary on the Concept of Free, Prior and Informed Consent." 14 July 2005 UN Doc. E/CN.4/Sub.2/AC.4/2005/WP.1.

UN Expert Mechanism on the Rights of Indigenous Peoples (EMRIP). 2018. "Free, Prior and Informed Consent: A Human Rights-Based Approach." UN Doc. A/HRC/39/62, 10 August.

UN Permanent Forum on Indigenous Issues (UNPFII) "Indigenous Participatory Mechanisms in the Arctic Council, the Circumpolar Inuit Declaration on Resource Development Principles in Inuit Nunaat and the Laponia Management System." prepared by Dalee Sambo Doroug, 3 March 2012 UN Doc. E/C.19/2012/10.

UNPFII. "Report on the 10th Session UNPFII." UN Docs E/2011/43, E/C.19/2011/14 (2011).

UN Sub-Commission on Prevention of Discrimination and Protection of Minorities. "Study of the Problem of Discrimination against Indigenous Populations." (1986) UN Doc E/CN.4/Sub.2/1986/7, Adds 1–4.

Ward, T. 2011. "The Right to Free, Prior and Informed Consent: Indigenous Peoples' Participation Rights within International Law." *Northwestern Journal of International Human Rights* 10 (2): 54–84.

Wilson, E. 2016. "What Is the Social Licence to Operate? Local Perceptions of Oil and Gas Projects in Russia's Komi Republic and Sakhalin Island." *The Extractive Industries and Society* 3 (1): 73–81.

Xanthaki, A. 2009. "Indigenous Rights in International Law Over the Last 10 Years and Future Developments" *Melbourne Journal of International Law* 27: 35–36.

EPILOGUE

Alaska natives and climate change

Paul C. Ongtooguk

As I think about Alaska Native cultures within the context of climate change and the opening of the Arctic, I am reminded of Captain Cook's 1778 exploration of Alaskan waters (Beaglehole 1999; Price 1971; Rickman 1966).[1] Captain Cook dreamed of finding a northwest passage and ventured into dangerous Alaskan coastal areas in his quest for open water. Captain Cook's dream of an Arctic passage is now a contemporary reality that is reverberating throughout the world. The warming of the Arctic has opened the area to economic and political opportunities of enormous consequence. Northern nations are vying for position. The Arctic Council, formed in 1991 and composed of eight Arctic nations, is front-page news. Cruise ships are advertising Arctic adventure travel. Russia is increasing the size of its icebreaker fleet. Oil companies are gearing up for exploration. Environmentalists are fighting for polar bear survival and also to be at the table of managing Arctic lands and waters.

Ironically, however, while the climate has changed dramatically over the last 200 years, the perception of Alaska's Indigenous peoples in the Arctic remains frozen in the past. In 1878 (almost 140 years ago), Captain Cook introduced Alaska's Arctic residents to the world as uncivilized and passive societies. While Alaska Native political and social influence is rising from being completely ignored on matters concerning the policies of the Arctic, Alaska's Indigenous residents continue to be marginalized, ignored and excluded as Western powers jockey for maximum exploitation of resources. Indigenous people being invited to comment on already foreordained policies is asking for birds to sing; it is not significant participation.

In the canon of Western history, Europe is cast as the 'actor' in the role of the discoverer. The 19th-century Western sense of superiority was buoyed by the development of technologies that enabled the exploration of the world and the conquering of other societies. Indigenous peoples were inevitably viewed as inferior, uncivilized and passive. Donald Keene (1969), however, in his text called the *Japanese Discovery of Europe, 1720–1830*, challenges this characterization. It was Keene's title, with its twist on the roles, that first caught my attention. In fact, I was struck from the first reading, and my many re-readings, that what Keene did was remarkable. Absent in his book was the focus on the tall ships, the intrepid explorers, the European heroes, the myopic vision of the civilized and uncivilized and the all-too-familiar accounts told from the perspective of those who were on board the ships. Significantly, in the accounts collected by Keene, it is the Japanese who are writing among themselves about these strange creatures who came on sailing ships into their lands, into their country. The Japanese accounts present a fascinating and multilayered look at the interactions that occurred and provide evidence that they were not confused, or unintelligent, observers. Rather, it is clear that they were thoughtful and salient people capable of understanding and connecting the powerful technologies that the Europeans brought into their harbours to their society. Keene weaves together one theme of the Japanese reaction to the question, "What do we do with the Europeans and the gun?" and further develops this thread through accounts that demonstrate the astute realization by the Japanese that consequences of these warfare technologies could turn their societies upside down, as in the case that a peasant armed with a gun who could kill a powerful Japanese samurai (1969). In short, the *Japanese Discovery of Europe, 1720–1830* changes the narrative vis-à-vis Japanese society and the Europeans by breaking stereotypic notions of passive non-European societies whose peoples are perplexed by the overwhelming superiority of the conquering European nations.

Such stereotypical notions of passive Indigenous peoples are particularly insidious, as they are the point of reference from which Westerners interpret Indigenous histories, achievements and evolutions into modern societies. Clearly, generalizations of Indigenous societies as backward and passive persist today despite many examples to the contrary. In Alaska, Captain Cook in 1778 played a significant role in the trajectory of European exploration and 'discovery' and in the establishment of the narrative of Alaska Native peoples as hopelessly uncivilized and incompetent. During this time, Alaska Natives, as far as we know, were not using a written language, and thus one-half of our understanding of this pivotal point in Alaska Native history is missing. However, in considering Cook's voyage, Western accounts can be used to reverse the mirror so that events can be interpreted from an Indigenous point of view. Doing so reveals that from this early period of contact, Alaska Native cultures demonstrated remarkable abilities to interpret, to adapt and to maintain their cultural identities. This same theme characterizes the responses of Alaska Native societies in the modern era through various waves of exploitation, colonization and natural resource extractions. In fact, Alaska Native cultures from the period of 'discovery' to contemporary times have demonstrated a powerful sense of imagination and the ability to affect that imagination within our cultural frameworks.

One legacy of Captain Cook's third voyage is the body of written artefacts that chronicle the encounters and contacts that members of his expedition had with various Alaska Native societies. An examination of the several diaries, journals and pictorial representations that were kept by Cook and others on this voyage puts to rest the theory that Alaska Natives were passive and intellectually overwhelmed. To the contrary, the records reveal descriptions of Alaska Natives interacting intelligently and confidently with these strangers who appeared on their shores. For example, an event occurred on 19 June as recounted in the Kippis narrative (1883, chap. VI). On this date, Captain Clerke, on the *Discovery*, noted, in the vicinity of the Schumagin Islands, that several Natives were following his ship in canoes and making signs that they wanted to communicate with those on board.

> On Captain Clerke's coming on board the Resolution, he related that several of the natives had followed his ship; that one of them had made many signs, taking off his cap, and bowing after the manner of Europeans; and that, at length, he had fastened to a rope, which was handed down to him, a small thin wooden case or box. Having delivered his parcel safe, and spoken something, accompanied with more signs, the canoes dropped astern, and left the Discovery.
>
> (Kippis 1883)

Following the departure of the Natives, the box was opened and found to contain a note written in what was assumed to be the Russian language, although there was no member of the crew who could "decipher the alphabet of the writer" (Kippis 1883). The date 1778 was prefixed on the note, and there was reference inside the note to 1776. Captain Clerke signalled to Captain Cook, aboard the *Resolution*, that he wished to speak. Captain Cook pulled alongside, but after examining the note and speculating on its content decided that the matter was not of any significance and pushed on.

The Kippis account reports the event from the European point of view and at first reading seems to provide little detail regarding Alaska Natives. However, a closer reading allows an extrapolation of some reasonable suppositions regarding Alaska Natives. For example, there is a certain confidence, and lack of fear, expressed by people who would follow the European ships in canoes and initiate contact. There is also clear purpose expressed in this sequence of events related both to the steps that were taken by the Natives to initiate contact and to the delivery of the written document. Greeting the Europeans with a "bow" demonstrates that the manners and customs of an alien group had been analysed, understood and applied appropriately in a new context. The fact that the note was carefully delivered and encased in a wooden box probably indicates that the Native peoples believed it to be of value and at least indicates that it was not treated casually. Since the note, theoretically, could have been read by the Europeans, it is clear that it had been cared for, that is, it was not overly crumpled, spotted, torn or otherwise abused. It is also reasonable to presume that the Native people had identified Captain Cook's group as being similar to other groups (or group) that they had encountered and that had, perhaps, delivered the note originally. Thus, the Kippis entry is actually quite rich descriptively and allows the construction of a picture of an intentional society that sought encounters with other societies.

In fact, multiple notes and references from journals and diaries provide evidence of Native societies who were actively observing, exploring and intentionally interacting with others by comparing and contrasting, analysing and making thoughtful decisions about trade. What was it about the British ships that caught their attention? How were the implications of the longer British knife perceived? Why did Natives zealously protect their spears and bows? What were the cultural patterns of interaction that had evolved, and how had they evolved? While frustratingly incomplete, it is clear that Native peoples were curious and purposeful in their encounters with the Cook expedition. Thus, when we reverse the mirror of the Western accounts and view them from the Native perspective, we are able to obtain a glimpse into the functioning of some Alaska Native societies in 1778. The extrapolation of information about relationships, encounters and intent counters the backward and passive labels that continue to be indiscriminately applied to Alaska Native peoples.

Captain Cook's third voyage pulled Alaska and Alaska Natives into the information of the age, into the known world of the Europeans. For Cook, this was surely of little consequence. His journals are mostly filled with rich descriptions of geography, navigational challenges, ship repair, weather conditions and explorations of promising passages. And while Native cultures were not his primary interest, the written and pictorial accounts emerging from his voyage about Alaska's Native peoples significantly impacted the Western world's understanding of the earth's inhabitants, the Eskimo, for example.

There is another family of the earth, concerning which new information has been derived from the voyages of our British navigators. That the Esquimaux, who had hitherto only been found seated on the coasts of Labrador and Hudson's Bay, agreed with the Greenlanders in every circumstance of customs, manners and language, which could demonstrate an original identity of nation, had already been ascertained. But that the same tribe now actually inhabited the islands and coasts on the west side of North America, opposite Kamchatka, was a discovery, the completion of which was reserved for Captain Cook. From his account, it appears that these people had extended their migrations to Norton Sound, Oonalashka and Prince William's Sound, that is, to nearly the distance of fifteen hundred leagues from their stations in Greenland and the coast of Labrador. Nor does this curious fact rest merely on the evidence arising from similitude of manners: for it stands confirmed by a table of words, exhibiting such an affinity of language as will remove every doubt from the mind of the most scrupulous inquirer (Kippis 1883, chap. VII).

It can be noted from the language of the text that it appears the migration direction is the reverse of the actual migration.

However, the worlds of Captain Cook and Alaska Natives intersected briefly in 1778. For Captain Cook, the intersection resulted in taking a moment of time to sail past and to name a geographic location or to make a diary entry. (The coastline of Alaska is now littered with names placed on maps however the officers of the expedition saw fit. Sledge Island, Bligh Reef, Turnagain arm, Cape Prince of Wales, Cook Inlet, Point Possession, Prince William Sound, Norton Sound and many more.) For Alaska Natives, the intersection was likewise short in time, with the most probable understanding that strangers had momentarily sailed through their world. Captain Cook's expedition, particularly this third voyage, is a part of the culminating adolescent period of scientific knowledge in European history. The Age of Exploration was marked by searches to find and use technological knowledge in the pursuit of power. Unfortunately, Western societies lacked a sense of imagination regarding its limits. Hence, Cook and his officers were confident that as educated, Christian Englishmen they could record accurate and significant information about Indigenous peoples and share it with others. They did not view themselves as amateurs, although this naïve confidence, combined with their sense of Christian superiority, caused them not to see or to be curious about the internal dynamics of Native cultures. This same perspective would later cost the lives of the entire crews of both ships of the Franklin Expedition of 1846. To today's reader, the journal accounts from this period project an edge of simple confidence that is in contrast to the lack of any attempt to explore Native peoples' understanding of the world, their relationship to the lands and waters or their relations with each other. Societies that differed in technological capacity of Europe were judged as backward, inferior and ignorant. And so, despite the fact that Alaska Native peoples occupied and lived on this Northern land for thousands of years, the legacy of the Cook maps, the place names that he coined to signal European ownership and his perceptions of 'backward' Indigenous peoples shaped the world's understanding. Captain Cook, a short-term visitor to Alaska's waters, substantially contributed to the shift of Alaska Native communities from being the centre of their world to marginalized members of other people's worlds.

In fact, Alaska Native societies have been remarkably able to adapt to the waves of change that have swept through the north during the last three hundred years. Native peoples have weathered Russian occupation, American colonialism and the stripping of much of our lands and waters. Capitalism, corporate greed and the imposition of foreign political systems were powerful tides that could have, but did not, overwhelm Native societies. Rather, Native peoples have able to unlock the dynamics of cultural change and respond to Western dominance by strategically identifying and adapting certain elements of Western structures within our various cultural frameworks. This ability is an essential theme in understanding the cultural survival of the Indigenous peoples of the North and an important thread in understanding the changing face of the Arctic.

One way to understand the development of societies is to study people who made significant contributions to the advancement of those societies. Captain Cook is an icon in Western history. He was an intrepid explorer who showcased the scientific knowledge of the period and who represented the confidence of the age. In Alaska, he sailed into waters unknown to Europeans along the jagged coast and found refuge in bays previously unknown to Europeans. He navigated some of the world's most powerful tidal zones and rode out fierce Arctic storms. The accuracy and completeness of the maps resulting from Cook's voyage are related to his goal of locating a northwest passage. Any promising body of water was approached with the hope and determination that this would be the long-awaited passage, and mapping was completed with a doggedness of detail of inlets, bays and rivers. While denied his dream of finding a northwest passage, his third voyage pulled Alaska Natives into the information of the age and thus into the world of the Europeans.

While Captain Cook sailed the world buoyed by European technology, resources, hope and dreams, Alaska Native leaders, following the advent of the Europeans, emerged within a very different context. The most predictable outcome following contact would have been the assimilation of Alaska Natives by the dominant Western cultures that imposed their economic, political, educational and religious systems in communities throughout the territory. Alaska Natives faced this alien culture with a lack of equivalent resources; a lack of European formal education preparing for leadership and advocacy and active efforts to undermine traditional ways of cultures, knowledge and languages. Yet despite these handicaps, powerful leaders emerged, and Native cultures were transformed and reimagined. Eben Hopson was a giant in Alaska Native history, and his life was at least as significant for Native peoples as Cook's was for Western peoples.

Hobson was an Inupiat Eskimo from Barrow born in 1922 and schooled at the Bureau of Indian Affairs' Barrow Day School. At the age of fifteen, he wrote to the Commissioner of Indian Affairs in Washington, D.C., to complain that the principal in Barrow was using students to work on public work projects without compensation. The BIA notified the principal about the complaint, and Hopson, now labelled a troublemaker, was denied access to further education. He became a labourer and was drafted into the U.S. Army by an all-white Barrow draft board during the Second World War. Eben Hobson's remarkable political career began following his return to Barrow in 1946 and his service as a member of the Barrow City Council (www.ebenhopson.com/paper/1980/FinalBio.html).

In 1956, Hobson was elected to the Alaska Territorial Legislature, and following statehood, he was elected to the State Senate. In contrast, however, to a figure like Cook who rode at the top of the dominant Western cultural ladder, Hobson was an outsider. Tribal political systems had been sidelined and replaced by a complex overlay of territorial, federal and later state structures. As a state legislator representing Alaska Native communities, Hobson understood the limitations of his role as an outnumbered minority in a majority-driven political process. He was, however, an incredibly astute observer and saw that the real opportunity was to adapt Western political structures into tools that would promote Alaska Native communities. In contrast to many who viewed Alaska as a short-term exploitation opportunity, Hobson belonged to Alaska. His history, family and culture were integrally related to the land and the waters, and his perspective was long term.

In 1965, Hobson became the first executive director of the Arctic Slope Native Association, which was one of the organizations that launched the Alaska Native Land Claims Movement in 1965. In 1968, he became the executive director of the Alaska Federation of Natives (AFN), and during his tenure, the Native land claims lobby successfully brokered the Alaska Native Claims Settlement Act of 1971, the terms of which included a cash settlement of nearly one billion dollars and forty million acres of land title. Eben Hobson left his position at the AFN in

1970 to become special assistant for Native affairs to Governor William Egan. He was now in a place where he could promote the welfare of Alaska Native communities through the development of local government, and he began that initiative despite the intense objections of the oil companies and the state and federal governments. "We have found that local government for our people is not an inevitable result of Arctic oil and gas development, however. Our people in Canada, Greenland and elsewhere in Alaska will have to struggle hard against great resistance to govern themselves" (Eben Hopson 1976). Nevertheless, from his seat in the governor's office, Hobson was able to secure the state's promise to work in support of the Arctic Slope Native Association's plan to organize a borough for northern Alaska – an area of some eighty-eight thousand square miles and four thousand people spread among eight northern villages. Significantly, funding for the borough would come from taxes assessed on the oil companies developing the fields on Prudhoe Bay. 'Oil' was furious and eventually convinced the state legislature to cap the Borough's taxing authority. Nevertheless, the North Slope Borough was established. The vision and imagination of Eben Hobson proved not only able to explore and understand Western political institutions but also to sort through the complex intersections of government, capitalism and principles of self-government. He imagined a template, a map of sorts, and he fought to attach that vision to a huge area of the north so that some Alaska Natives would be accorded political power and an economic base.

Hobson left the governor's office in 1972 to campaign for approval of the North Slope Borough and then to run for the office of borough mayor. In 1974, Hobson entered the race for the Democratic Party's gubernatorial nomination. He pulled out of the race, however, after he negotiated a nine-point political agreement for rural Alaska that included the development of modern communications. Hobson continued in his role as mayor and continued his fight against dominance by big oil and even national ambitions – this time going international.

> There is only one Beaufort Sea. It is a single Arctic ecological system shared by the North Slope Borough and the Northwest Territories. We Inupiat are a single Beaufort Sea community living under two national flags. We must contend with two different political systems, and two sets of rules governing oil and gas development, to protect our environmental values within our larger Beaufort coastal community.
>
> (1976)

Hobson again successfully applied political acuity to lobby for the creation of a formal structure, in this case an international consortium that would establish rules for the industrial development of the Arctic. The Inuit Circumpolar Conference (ICC) met in 1977 in Barrow, also attended by Inuit leaders from Canada and Greenland, with a place set aside for Inuit from Russia, who were not allowed to participate. The delegates drafted resolutions on land claims, Arctic environmental protection, health and Inuit culture and education. The success of the ICC won Eben Hobson international acclaim and marks the culmination of a triumphal career. Ironically, almost two hundred years after Captain Cook's third voyage, Eben Hobson, a marginalized, self-educated Alaska Native from a small Northern village, created the first international Indigenous organization to address issues related to the Northwest Passage.

Ironically, the lens that framed the English understanding of Native societies may have resulted in Cook missing an opportunity to learn more about the existence of the Northwest Passage that he so relentlessly pursued. In October, while the Cook expedition lay for several weeks at Oonalashka repairing the ships and preparing for the final journey south, Cook (Beaglehole 1999, 468) made extensive observations about the local inhabitants. Among those observations were notes on the language used by the Natives.

> To this specimen I have added such corresponding words as I have been able to collect out of the language of the Greenlanders, and Esquimaux; from all of which there is great reason to believe that all these nations are of the same extraction, and if so there can be little doubt but there is a northern communication of some sort by sea between this ocean and Baffins Bay, but it may be effectually shut up against Shipping, by ice and other impediments, such at least was my opinion at this time.

Captain Cook came to Alaska in search of the Northwest Passage, and in trying to accomplish that task, he visited Norton Sound, Little Diomede, Wales and places further north. Then, in October, in Oonalashka, he concludes, really on the scantest of evidence, that the languages on this side of the Arctic Ocean and those on the other Baffin Bay and Greenland sides are in fact the same. The cultural thread that he recognized actually points to his one great insight about the Northwest Passage; that is, that it is the same peoples who are the residents along its thousands of miles and that in this way it transcends Captain Cook and Alaska Natives and recognizes that it is the Inuit people who cross boundaries and lines of demarcation among the Americans, the British, the Danes and the Russians. The Esquimaux language left him little doubt about northern communication of some sort, and so while he did not bring his ship successfully through the Northwest Passage, he at least perceived a way in which he could surmise its existence.

Now, more than two hundred years since Cook's third voyage, Inuit people of Alaska retain their connections to the Inuit of Canada, Greenland and other European nations. Political boundaries continue to separate the Inuit, but the Northwest Passage that Captain Cook sought is actually the natural trace line, the water backbone of our people who are today preparing for the next chapter in the history of a potential international route as climate change diminishes the sea ice and as open water flows.

As the Western world and its institutions struggle to understand the impact of climate change in the North, the Alaska Native Inupiat people and other Inuit will heed our histories and traditional cultures as one way to respond to these dramatic changes. The Inuit have lived across the North for thousands of years. The land and the water are homelands of ancient, contemporary and proud cultures. We do not aspire to relocate; our identity is integrally tied to these northern places. William Oquilluk, in *The People of Kauwerak*, describes some aspects of the traditional history, including periods of climate change, among the Inuit of the Mary's Igloo region of northwest Alaska. The history, in Oquilluk's account, is divided into a series of disasters and describes people's responses to those disasters. The first disaster is the reverse of today, as it was a time when the Arctic began to freeze.

One morning Ekeuhnick went out into the countryside with many problems to think about heavy in his mind. This was because the world itself was changing. He thought about how his people must face this new climate without ever having known any way of living except the time when the country was warm. He pondered deep in his head about it. The people were not prepared to protect themselves from the cold or to find the food they needed now that the plants and animals were scarce. He could not follow Aungayoukuksuk's instructions about the journey to the distant Mountain until his people were able to survive in the changed times (Oquilluk and Bland 1981, 30).

Inuit did survive the first disaster, and Oquilluk's rich account through three disasters is testimony to the intelligence and resilience of us as peoples.

After this time, the people followed everything that Ekeuhnick said to them. They had seen how he used his mind. Everything that came to his mind, he thought about and looked at in his head through the power of imagination. Whenever Ekeuhnick thought of a plan, he tried it out. When it worked, he went and told the people about it. The people looked around. They could see that those who survived the first disaster had warm clothing and a warm home. In those long-ago, ancient times, people's knowledge and thinking were slow to come. Everything was new and different. The people were learning to survive and to live with the changes that had come to their land. Their minds begin to work one way or another, and they begin to live by their thinking. It was hard for them to start thinking to do things so differently at first. Now, each one's own mind could do that which they could never do before. The living had really changed for both the earth and the people (Oquilluk and Bland 1981, 31).

In contrast, climate change today has resulted in a race to the Arctic today for land, military presence and resource extraction that seems characterized by thoughtless reactions from cultures propelled by a colonial appetite for financial gain and political power. The destructive results of this heedless rush seem not to be considered, even though the consequences of industrialism offer stark and irrefutable evidence of the altering of the planet and the unsustainability of current practices. In contrast, *People of Kauwerak* tells a history that may well serve as a guide not only to Alaska Natives but to the world, as the struggle to adapt to the changing face of the lands and waters becomes ever more imperative. The wisdom of Kauwerak lies not in the particular adaptations that were made in response to the freezing or warming of the earth but rather in the understanding that responding to change demanded certain dispositions. The people of Kauwerak demonstrated the ability to observe, to accept, to adapt and to sustain. The Arctic Council would do well to adapt the foundational wisdom of the People of Kauwerak to guide its deliberations.

- They begin to live by their thinking.
- Now, each one's own mind could do that which they could never do before.
- The people were learning to survive and to live with the changes that had come to their land.
- . . . the Power of Imagination.

(Oquilluk and Bland 1981)

In addition to providing us with dispositional wisdom, *People of Kauwerak* teaches us that inventiveness and action are an integral part of responding to change. Inupiaq did not wait to see where things landed during periods of calamity. Rather we have a history of planned, prepared and engaged-in activities that would ensure our futures. Climate change today demands action. How can the northern Indigenous peoples of Alaska meet these challenges?

In May of 2015, Secretary of Defence Ashton Carter testified to Congress regarding the need for a better Arctic defence policy. The impressive size of the Russian icebreaker fleet is well known, as are the Russian large-scale military training exercises being conducted in the Arctic. Russian Deputy Prime Minister Dmitry Rogozin has referred to the Arctic as "a Russian Mecca", and in 2012, he warned that the Russian Arctic interests needed to be reinforced.[2] At the Congressional hearing in May, Secretary Carter offered to work with Senator Murkowski to review current U.S. policy.[3]

The Alaska Territorial Guard is a model for one arm of this military effort. The Guard was formed during 1942 despite fears from some about arming Alaska Natives given the context of racial segregation that existed in many public places, including schools and housing. The concerns were unfounded. Alaska Natives overwhelmingly volunteered to serve in a dedicated, patriotic fashion on a foundation of generations of knowledge about living in these lands and water. The Alaska Territorial Guard rested on this foundation. This included quite a few Native women. Laura Beltz Wright is noted as one of the volunteers who scored well in her rifle qualifications, as in 49 out 50 bull's-eyes. Guard members were proud to have those powerful 30–06 bolt action rifles for training and to learn the elements of formation marching. They eagerly learned how to take and pass orders and wore the Alaska Territorial Guard patch sewn onto the shoulder of the white cloth–covered fur hunting parka as professionals. From a military standpoint, it is clear that it would have been impossible to place regular soldiers up and down the western coast of Alaska, as the conditions, local food, knowledge of the areas and safe movement could not have been taught at military speed. Alaska Natives brought personal winter equipment, clothing and survival skills informed by generations of insight and development. In support of Inuit and U.S. interests in the Arctic, perhaps it is time to reimagine and reconstitute a new Alaska Territorial Guard.

Victoria Hermann (2015) has written about how the Lower 48 view the Arctic from an extractive narrative.[4] She points out that a closer look, however, at northern nature across the Arctic reveals the development of sustainable resources, such as hydropower in Norway, wind in Canada, geothermal in Iceland and energy efficiency in Finland. Hermann calls on U.S. Arctic Council leaders to reimagine the Arctic as an area with enormous renewable energy resources that can be developed to support remote Arctic communities. "In short, it provides the potential to transform Alaska's economy from an extractive, carbon-intensive industry into sustainable, human-centric development" (Hermann 2015).

Alaska Natives are eager to support sustainable energy. Current prices for heating fuel in the north are among the highest in the country. The extraction of oil in the Arctic poses an existential threat to the very survival of Arctic communities and cultures. Coastal erosion from climate change is necessitating the relocation of several communities. Renewable energy sources, such as wind-diesel hybrid systems, have been installed in some communities with Department of Energy support. These opportunities need to be expanded. Opportunities associated with relocation should be seized and structural designs and building materials carefully considered. We Alaska Natives are keen observers of our lands. We have tradition and history to know where and how to build. We have direct observation and knowledge of the wind, the water and changes in the environment that should be incorporated into sustainable energy planning.

Canada has recognized that the warming of the Arctic provides an opportunity to support their northern Indigenous communities with policies and assistance that support their expansion and wellbeing. With similar support, northern communities in Alaska would prosper from new economic opportunities and become vibrant and dynamic beacons marking the Alaskan coastline. With the eroding of coastline, there is some sentiment in Alaska that this is a time to dismantle and diminish village communities, but the opposite is true, for without claim to the lands and waters of the North – the footprint of time argument – there is little reason for other states or nations to regard these lands as anything but vacant. The scattered urban spots in Alaska do not, by themselves, represent a future for the state. It is only in combination with the places that Alaska Natives call home that the economic, social and cultural futures of the state will be assured. Traditional wisdom has demonstrated that the challenges of change can be reversed by the power of imagination. As future generations examine our response to the warming of the Arctic, the highest accolade that could be applied would be, "they began to live by their thinking".

It is not natural resources nor a shorter transportation route between Asia and Europe that may be the most important contribution of the Arctic for the world in the long term. Our greatest contribution as Inuit may be a view of the natural world and the place of humans in this.

A careful understanding of Inuit concepts involves learning from the natural world as much as benefiting from the gifts the natural world provides humans who respect and are grateful.

Unlike the Genesis accounts, we are not handed dominion by God over the world to name it according to our needs and desires. In our traditional belief, humans are marginal to the natural world. Inuit understand we are not as fast or as strong; we do not have even our own clothing to protect us from the weather. We cannot hear or see as well as the other tribes who happen not to be human (animals). We are also not separate from animals, as we live in relation to and not above them.

Every nation brings strengths and blind spots unique to each. One of our strengths has been that the Arctic has taught, without mercy, how we can well disappear and the world will continue without us. William Oquilluk represents a small window on this disposition, sense, and context about the Arctic and our place in it. William Oquilluk wrote about a fifth disaster we might face in the Arctic.

"The Fifth Disaster is maybe now. . . . The rules and stories of our ancestors are being forgotten. The people do not know who their relations are" (1981, 225). These relations are about human families, but they are also about our relations to tribes not human and our relations with the lands and waters of the Arctic.

The world at large seems to be in the process of strip-mining nature in unsustainable ways. The great discovery of the Arctic may not be in the records of Captain Cook or other Arctic visitors. The great discovery may be what we Inuit learned over generations and ages.

The best way to move forward is not simply continuing the actions of this modern system without thought but more use of imagination and forward thinking about humans adapting ourselves and bearing in mind our relations in ways not yet imagined.

Notes

1 5 Aug 1778. Online website of the Captain Cook Society, accessed 16 May 2020. www.captaincook society.com/home/detail/articleid/52/cook-was-searching-the-coasts-of-asia-and-america-for-a-pas sage-to-the-atlantic

2 Alaska Daily News, 21 April 2015, "The Arctic Is Russia's Mecca, Says Top Moscow Official," accessed 16 May 2020. www.adn.com/nation-world/article/arctic-russias-mecca-says-top-moscow- official/2015/04/21/

3 Alaska Daily News, 7 May 2015. "Carter: US Arctic Defence Policy Falls Short," accessed 16 May 2020. www.adn.com/military/article/carter-arctic-defense-policy-falls-short/2015/05/06/

4 Alaska Dispatch News, 13 June 2015, "Arctic Energy Debate Can Be More Than Shell Rigs and Greenpeace Protests," accessed 16 May 2020. www.adn.com/commentary/article/arctic-energy-debate-can-be- more-shell-rigs-and-greenpeace-protests/2015/06/13/

References

Beaglehole, J. C. (ed.). 1999. *The Journals of Captain James Cook on His Voyages of Discovery: Volume III, Part I: The Voyage of the Resolution and Discovery 1776–1780*. Hakluyt Society, Reprinted by The Boydell Press, Cambridge: Woodbridge.

Eben Hopson, E. 1976. *The Berger Speech: Mayor Eben Hopson's Testimony*. Eben Hopson Memorial Archives Celebrate the Life and Leadership of the Late Eben Hopson. Accessed 20 May 2020. http://ebenhopson.com/the-berger-speech/.

Hermann, V. 2015. *You Can't Have Your Baked Alaska and Eat It Too*. The Arctic Institute. Accessed 18 April 2020. www.thearcticinstitute.org/alaska-arctic-energy/.

Keene, D. 1969. *The Japanese Discovery of Europe, 1720–1830*. 1st ed. Stanford: Stanford University Press.

Kippis, A. 1883. *A Narrative of the Voyages round the World Performed by Captain James Cook: With an Account of His Life, During the Previous and Intervening Periods*. London: Bickers & Son.

Oquilluk, W. A., and L. L. Bland. 1981. *People of Kauwerak: Legends of the Northern Eskimo*. Anchorage, Alaska: Alaska Pacific University Press.

Price, A. G. (ed.). 1971. *The Explorations of Captain James Cook in the Pacific As Told by Selections of His Own Journals 1768–1779*. New York: Dover Publications.

Rickman, J. 1966. *Journal of Captain Cook's Last Voyage to the Pacific Ocean*. Ann Arbor, MI: University Microfilms.

INDEX

Page numbers in *italic* indicate a figure and page numbers in **bold** indicate a table on the corresponding page.

Index

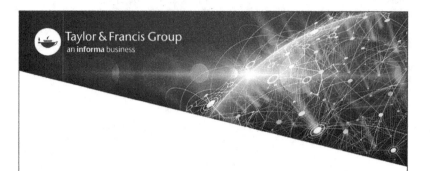